Calculus of Several Variables

CONTEMPORARY UNDERGRADUATE MATHEMATICS SERIES
Robert J. Wisner, Editor

MATHEMATICS FOR THE LIBERAL ARTS STUDENT
Fred Richman, Carol Walker, and Robert J. Wisner, ***New Mexico State University***

MODERN MATHEMATICS: AN ELEMENTARY APPROACH
Ruric E. Wheeler, ***Samford University***

ANALYTIC GEOMETRY
James E. Hall, ***The University of Wisconsin***

INTRODUCTORY GEOMETRY: AN INFORMAL APPROACH
James R. Smart, ***San Jose State College***

A PRIMER OF COMPLEX VARIABLES WITH AN INTRODUCTION TO ADVANCED TECHNIQUES
Hugh J. Hamilton, ***Pomona College***

THEORY AND EXAMPLES OF POINT-SET TOPOLOGY
John Greever, ***Harvey Mudd College***

INTRODUCTION TO ANALYSIS
Edward Gaughan, ***New Mexico State University***

CALCULUS OF SEVERAL VARIABLES
E. K. McLachlan, ***Oklahoma State University***

Calculus of Several Variables

E. K. McLACHLAN
Oklahoma State University

BROOKS/COLE PUBLISHING COMPANY
Belmont, California
A Division of Wadsworth Publishing Company, Inc.

The Lissajous harmonogram shown on the cover was made at the Oklahoma State University Engineering Exposition of 1967 by James Muckleroy, Jr., using a double elliptic pendulum.

Brooks/Cole Publishing Company, Belmont, California. A division of Wadsworth Publishing Company, Inc.

L.C. Cat. Card No.: 68-11088
Printed in the United States of America

to my parents

Preface

Several years ago the Committee on the Undergraduate Program in Mathematics, CUPM, recommended a mathematics curriculum for engineers and physicists. This program, which seems sound mathematically and reasonable pedagogically, called for a new course, Functions of Several Variables. While the new course does not contain new mathematics—the content is what might best be described as the "core" of the traditional advanced calculus courses—it does represent a new packaging of mathematics subject matter. None of the texts available when the course was outlined could fit its requirements. Therefore, this book has been written to fill an obvious need.

Great effort has been made to provide a nice blend of the simply manipulative aspects of the calculus of several variables and the definition-theorem-proof aspects. Some of the theorems are proved completely, others are proved for restricted circumstances; at other times the proof is left as an exercise with a suggested procedure, and occasionally the proof of a theorem is referenced. Besides numerous detailed examples and illustrations, the book includes more than the usual amount of expository discussion. There are many problems of both the mechanical and the theoretical type and answers are given for almost all of the manipulative problems.

The organization of the book, which follows very closely the CUPM recommendation, except for the chapter on topology, deals with the basic topics of the differential and the integral portions of the calculus of several variables. The chapter on topology is included because serious consideration of the calculus is not possible without some attention to topology.

Almost every section is organized so that the theoretical discussion comes near the end. Sections usually begin with some discussion, examples, and illustrations designed to give the student a feeling for the motivation for the theorems. This order lends itself to the omission of proofs when time is limited.

The problems at the end of each section are similarly arranged, so that problems of the mechanical type come first and theoretical problems appear later. The problems tend to be graduated in difficulty, and manipulative problems usually have four or more parts, so that assignments can be varied from one semester to the next.

Although the book would require six semester hours for a comprehensive treatment, it can be used profitably for a more abbreviated treatment. For example, in a three-semester-hour course it is possible to de-emphasize greatly the theory and cover the following sections: 1–1, 1–2, 1–3, and 1–5; 2–1, 2–2, 2–3, and 2–7; 3–1, 3–2, 3–4, 3–5, and 3–6; 4–1, 4–2, 4–4, and maybe 4–3; 5–1, 5–2, 5–3, 5–4, and perhaps 5–5 or 5–6; 6–2, 6–4, and 6–5. The selection could be varied to include more topics of Chapters 5 and 6 by omitting more sections of Chapters 1 and 2. A careful treatment of the topology in Chapter 2 is not necessary if the desire is to de-emphasize theory.

Although CUPM suggests that a course in linear algebra precede the course in Functions of Severable Variables, it is not absolutely essential for the use of this book. The need is most acutely felt in Chapter 1. After this, a basic knowledge of linear algebra is seldom called upon.

The label "Theorem" is reserved for the more basic, or pivotal, results and "Proposition" is used for the less significant results. The symbol ▌ indicates the end of a proof; log means logarithm to the base e, and $\{(x, y, z): x \geq 0,\ y \geq 0, \cdots\}$ means the set of (x, y, z) such that $x \geq 0$ *and* $y \geq 0$, etc. Thus, in the last notation, when the qualifiers are separated by a comma, it is understood that *and* is the conjunction between them.

Of the many people who have given valuable assistance in the writing of this book, I first wish to acknowledge the help of my former colleague of Baylor University, the late R. L. Clayton, with whom I spent many pleasant hours as we taught ourselves the calculus of several variables. Thanks is due to Dr. Walter J. Williams, also of Baylor University, under whose leadership there were many opportunities for teaching calculus. The author is grateful to Professor L. Wayne Johnson of Oklahoma State University, who, as department head, helped make possible the classroom use of the preliminary notes of this manuscript. A special thanks goes also to my colleagues R. L. Caskey, J. E. Hoffman, J. M. Jobe, R. D. McDole, and J. V. Robison and to the students who made many valuable suggestions while using the preliminary notes. I am also very much indebted to Professors Karl Stromberg of the University of Oregon, James Scroggs of the University of Arkansas, Donald Benson of the University of California at Davis, Robert J. Wisner of New Mexico State University, and others who made very helpful suggestions for the improvement of the manuscript. Finally, it is most appropriate to acknowledge the valuable typing and editing assistance given by my wife, Polly, and to recognize the assistance in proofreading given by my daughter, Kay.

Contents

3 Differential Calculus of Functions of Several Variables

4 Applications of Differential Calculus

5 Integral Calculus of Functions of Several Variables

6 Differential and Integral Calculus of Vector Fields

1

Preliminaries

1–1. Vector Spaces, R^n

The vector spaces used here are commonly referred to as R^n spaces. That is, the vectors $\mathbf{x}$ of R^n consist of the real n-tuples, $(x_1, x_2, \cdots, x_n)$, where x_i, the ith coordinate of $\mathbf{x}$ in R^n, is a real number for each i, $1 \leq i \leq n$. If $\mathbf{x} = (x_1, x_2, \cdots, x_n)$ and $\mathbf{y} = (y_1, y_2, \cdots, y_n)$ are two vectors in R^n, then the addition of $\mathbf{x}$ and $\mathbf{y}$ in R^n is defined by the following equation:

$$\mathbf{x} + \mathbf{y} = (x_1 + y_1, x_2 + y_2, \cdots, x_n + y_n). \tag{1–1.1}$$

Just as the sum of two real numbers is a real number, $\mathbf{x} + \mathbf{y}$ is a vector in R_n.

Since the addition of two vectors is accomplished by coordinatewise addition (that is, the ith coordinate of $\mathbf{x}$ is added to the ith coordinate of $\mathbf{y}$) and since reals are commutative and associative, it follows that

$$\mathbf{x} + \mathbf{y} = \mathbf{y} + \mathbf{x} \tag{1–1.2}$$

and

$$\mathbf{x} + (\mathbf{y} + \mathbf{z}) = (\mathbf{x} + \mathbf{y}) + \mathbf{z}. \tag{1–1.3}$$

Let $\mathbf{0} = (0, 0, \cdots, 0)$ be called the *zero vector*, and if $\mathbf{x} = (x_1, \cdots, x_n)$, then let $-\mathbf{x} = (-x_1, \cdots, -x_n)$ be the *inverse vector* of $\mathbf{x}$. It follows from (1–1.1) that for any $\mathbf{x}$,

$$\mathbf{x} + \mathbf{0} = \mathbf{x} \tag{1–1.4}$$

and

$$\mathbf{x} + (-\mathbf{x}) = \mathbf{0}. \tag{1–1.5}$$

Clearly, $\mathbf{0}$ and $-\mathbf{x}$ are the only vectors that have these two properties; hence, it is correct to refer to them as *the* zero vector and *the* inverse vector of $\mathbf{x}$, respectively.

The scalar field over which R^n is taken is the real field; that is, the set of real numbers. If α is any real number, then the *scalar multiplication* of $\mathbf{x}$ by α is defined by

$$\alpha\mathbf{x} = (\alpha x_1, \alpha x_2, \cdots, \alpha x_n). \tag{1–1.6}$$

Again this new vector $\alpha\mathbf{x}$ is a real n-tuple and thus belongs to R^n. This multiplication can be described by saying that each coordinate of $\mathbf{x}$ is multiplied by α.

Since the distributive law holds for real numbers and since scalar multiplication and addition are accomplished by multiplying and adding in each coordinate, it is easy to see that the following two distributive laws hold:

$$\alpha(\mathbf{x} + \mathbf{y}) = \alpha\mathbf{x} + \alpha\mathbf{y} \tag{1–1.7}$$

and

$$(\alpha + \beta)\mathbf{x} = \alpha\mathbf{x} + \beta\mathbf{x}. \tag{1–1.8}$$

It is apparent also from (1–1.6) and from the associativity of multiplication of the reals that

$$(\alpha\beta)\mathbf{x} = \alpha(\beta\mathbf{x}). \tag{1–1.9}$$

Finally, we can notice that

$$1(\mathbf{x}) = \mathbf{x}. \tag{1–1.10}$$

Any set of vectors V with scalars R, the reals, is a *vector space over the reals*, written $V(R)$, if we have the properties (1–1.1) through (1–1.10). This vector space of real n-tuples with this structure is denoted simply R^n instead of $V(R)$.

Notice that if $\alpha = -1$, then $(-1)(\mathbf{x}) = -\mathbf{x}$, the inverse of $\mathbf{x}$. Therefore, it is convenient to write $\mathbf{x} - \mathbf{y}$ for $\mathbf{x} + (-1)\mathbf{y} = \mathbf{x} + (-\mathbf{y})$; that is,

$$\mathbf{x} - \mathbf{y} = (x_1 - y_1, x_2 - y_2, \cdots, x_n - y_n). \tag{1–1.11}$$

A *linear combination* of a set of vectors S in R^n is a sum of the form

$$\textstyle\sum_{i=1}^{m} \alpha_i \mathbf{x}_i$$

where each α_i is a real number and each $\mathbf{x}_i \in S$ for $1 \leq i \leq m$. For a linear combination the sum must be a finite sum. As a result of the finiteness, the sum defines a vector in R^n by virtue of (1–1.1) and (1–1.6). Some of the scalars α_i may be zero since no restriction to the contrary is placed on a linear combination.

A set S in R^n is a *linearly dependent* set if there exists a linear combination of a set of vectors of S,

$$\textstyle\sum_{i=1}^{m} \alpha_i \mathbf{x}_i = \mathbf{0},$$

such that not all α_i, $1 \leq i \leq m$, are zero. Any set S in R^n that is not linearly dependent is *linearly independent.*

For example, the set $\{(1, 2, 3), (2, -1, 2), (-4, 7, 0)\}$ in R^3 is linearly dependent since

$$2(1, 2, 3) - 3(2, -1, 2) - (-4, 7, 0) = \mathbf{0}.$$

The set $\{(1, 0, 0), (0, 1, 0), (0, 0, 1)\}$ in R^3 is linearly independent, since

$$\alpha_1(1, 0, 0) + \alpha_2(0, 1, 0) + \alpha_3(0, 0, 1) = (\alpha_1, \alpha_2, \alpha_3) = \mathbf{0}$$

implies α_1, α_2, and $\alpha_3 = 0$.

A set V_1 is a *subspace* of R^n if the set V_1 is closed under addition and scalar multiplication; that is, if $\mathbf{x}$ and $\mathbf{y}$ belong to V_1, then $\mathbf{x} + \mathbf{y}$ belongs to V_1, and if α is real, $\alpha\mathbf{x}$ belongs to V_1. It is not difficult to show that if V_1 is a subspace, then the set of vectors in V_1 satisfies each of the equations (1–1.1) through (1–1.10).

Given a set S in R^n, then the set of vectors *spanned* by S is the set S_1 of vectors that are a linear combination of the set of vectors S. Notice that S_1 is a subspace, since a sum of two linear combinations of S and a scalar multiple of a linear combination of vectors of S is again a linear combination of vectors of S. For this reason it is appropriate to speak of the *subspace spanned by S.*

A set of vectors S of R^n is a *basis* of R^n if S is a linearly independent set and if S spans R^n. Such a set is the set $\mathbf{u}_1 = (1, 0, \cdots, 0)$, $\mathbf{u}_2 = (0, 1, \cdots, 0)$, $\cdots$, $\mathbf{u}_n = (0, 0, \cdots, 1)$. This set is linearly independent, as was illustrated earlier in the case of R^3. Also it spans R^n since

$$\mathbf{x} = (x_i, \cdots, x_n) = \textstyle\sum_{i=1}^{n} x_i \mathbf{u}_i. \tag{1–1.12}$$

This basis of R^n, which is a very particular basis, shall be referred to as the *usual basis* of R^n. That R^n can have other bases is illustrated by the following set of n vectors: $\{(1, 0, 0, \cdots, 0), (1, 1, 0, \cdots, 0), (1, 1, 1, \cdots, 0), \cdots, (1, 1, 1, \cdots, 1)\}$. The following proposition summarizes the main points about bases of R^n.

Proposition 1–1.1. *If S is a linearly independent subset of R^n that is not a basis of R^n, then there is a set S_1 that contains S and is a basis of R^n. The number of elements of a basis of a given subspace of R^n is unique. In particular, any basis of R^n itself contains exactly n vectors.*

The first part of this proposition indicates that any linearly independent set can be enlarged into a basis for R^n. The second statement indicates the uniqueness of the number of elements of a basis of a subspace of R^n. This number is called the *dimension* of the subspace. Thus R^n is of dimension n. We define the dimension of the subspace $\{\mathbf{0}\}$ to be zero.

All of these various definitions—linear combination, $\cdots$, dimension—could be stated to apply to any vector space over the reals. However, since our attention will be exclusively with R^n, the definitions are given in that setting.

In R^n, with the usual basis $\mathbf{u}_1, \mathbf{u}_2, \cdots, \mathbf{u}_n$, it follows from (1–1.12) that the ith coordinate of $\mathbf{x}$ is the correct real scalar by which $\mathbf{u}_i$ is multiplied in order to have $\mathbf{x}$ expressed as a linear combination of this basis. If $\mathbf{v}_1, \mathbf{v}_2, \cdots, \mathbf{v}_n$ is another basis of R^n, then

$$\mathbf{x} = (x_1, x_2, \cdots, x_n) = \textstyle\sum_{i=1}^n \alpha_i \mathbf{v}_i$$

where we now have a new set of real scalars, $(\alpha_1, \cdots, \alpha_n)$. This n-tuple will be called the *coordinates of* $\mathbf{x}$ *relative to the basis* $\mathbf{v}_1, \mathbf{v}_2, \cdots, \mathbf{v}_n$. These coordinates of $\mathbf{x}$ are unique for the basis $\mathbf{v}_1, \mathbf{v}_2, \cdots, \mathbf{v}_n$; for if $(\alpha_1', \alpha_2', \cdots, \alpha_n')$ is such that

$$\mathbf{x} = \textstyle\sum_{i=1}^n \alpha_i \mathbf{v}_i = \sum_{i=1}^n \alpha_i' \mathbf{v}_i,$$

then

$$\textstyle\sum_{i=1}^n (\alpha_i - \alpha_i')\mathbf{v}_i = \mathbf{0}$$

implies $\alpha_i - \alpha_i' = 0$ or $\alpha_i = \alpha_i'$ for $1 \le i \le n$. Unless some emphasis is made to the contrary, the phrase "the coordinates of x" will mean the coordinates of $\mathbf{x}$ relative to the usual basis, $\mathbf{u}_1, \mathbf{u}_2, \cdots, \mathbf{u}_n$.

Problems

1. Let $\mathbf{x} = (1, -2, 4)$ and let $\mathbf{y} = (-2, 3, -1)$. Find the following:
   (a) $\mathbf{x} + \mathbf{y}$, (b) $\mathbf{x} - \mathbf{y}$, (c) $4\mathbf{x} + \mathbf{y}$,
   (d) $x + (\frac{1}{2})(\mathbf{y} - \mathbf{x})$, (e) $2\mathbf{x} - 3\mathbf{y}$, (f) $\mathbf{x} + 2(\mathbf{y} - 3\mathbf{x})$.
2. Assuming all of the properties of the reals, prove the following:
   (a) the uniqueness of the zero vector,
   (b) the uniqueness of the inverse of x, and
   (c) the two distributive laws, (1–1.7) and (1–1.8).

3. If $\mathbf{x}_1 = (x_{11}, \cdots, x_{1n}), \cdots, \mathbf{x}_m = (x_{m1}, \cdots, x_{mn})$, then $\{\mathbf{x}_1, \cdots, \mathbf{x}_m\}$ is a linearly independent set if the $m \times n$ matrix

$$\begin{pmatrix} x_{11} & \cdots & x_{1n} \\ \cdot & & \cdot \\ \cdot & & \cdot \\ \cdot & & \cdot \\ x_{m1} & \cdots & x_{mn} \end{pmatrix}$$

has rank m. Use this fact to determine whether the following sets are linearly dependent or not:

(a) $\{(1, 0, -1), (2, 2, 2), (-3, 1, 1)\}$;
(b) $\{(1, 0, 0, 0), (0, 0, 1, 0), (0, 0, 0, 1)\}$;
(c) $\{(1, 2, 2, 3), (1, -1, -1, 1), (0, 0, 1, 1), (-1, 7, 8, 4)\}$;
(d) $\{(1, -2, -2, 2, 2), (2, 0, 1, 0, 3), (0, -4, -5, 4, 1), (2, 8, 11, -8, 1)\}$.

4. Show that $\{\mathbf{x}_1, \cdots, \mathbf{x}_m\}$ is a linearly dependent set of nonzero vectors if and only if for some k, $2 \leq k \leq m$, $\mathbf{x}_k$ is a linear combination of the set of vectors $\{\mathbf{x}_1, \cdots, \mathbf{x}_{k-1}\}$.

5. By elementary row transformations on the matrix (x_{ij}) of Problem 3, we get a similar matrix. Prove that the vectors whose coordinates are entries on the row of a matrix similar to (x_{ij}) are in the space spanned by $\{\mathbf{x}_1, \cdots, \mathbf{x}_m\}$. One may use elementary row transformations to find a matrix similar to (x_{ij}) that has all zeros in as many of the rows as possible. The remaining nonzero rows will be vectors that form a linearly independent set, a basis for the subspace spanned by $\{\mathbf{x}_1, \cdots, \mathbf{x}_m\}$. Use these facts to find a basis for the subspaces spanned by the following sets of vectors:

(a) $\{(1, 1, -1), (2, 0, -3), (2, 2, 2)\}$;
(b) $\{(1, 3, 2), (2, -1, 1), (6, 4, 2), (-5, 13, 10)\}$;
(c) $\{(-1, 5, -1, 2), (1, 1, -1, -3), (2, 0, 3, 2), (-4, 4, -3, 3)\}$;
(d) $\{(3, 0, 6, 4, 10), (1, 1, 0, -1, -2), (1, 2, 3, 4, 5), (-3, 1, -3, 1, -3)\}$.

6. Show that if S is a nonempty set in R^n, then the set S_1 of linear combinations of the vectors of S is a subspace of R^n that contains S.

7. Find the coordinates of $(3, -2, 4)$ relative to the basis $\{(1, 0, 0), (1, 1, 0), (1, 1, 1)\}$ of R^3. Do the same for $(4, -4, 1)$.

8. Enlarge the set $\{(1, 0, 1, 0), (0, 1, 0, 1)\}$ into a basis for R^4.

9. Let S_1 and S_2 be subspaces of R^n. Show that $S_1 \cap S_2$ is a subspace of R^n. If $S_1 + S_2$ is the set of all vectors $\mathbf{x} + \mathbf{y}$ where $\mathbf{x} \in S_1$ and $\mathbf{y} \in S_2$, show $S_1 + S_2$ is a subspace of R^n. If dim ( ) means "dimension of," show that

$$\dim (S_1) + \dim (S_2) = \dim (S_1 \cap S_2) + \dim (S_1 + S_2).$$

10. Given the linearly independent set of vectors $\{\mathbf{y}_1, \cdots, \mathbf{y}_m\}$ in R^n where $m < n$, explain why the set $\{\mathbf{y}_1, \cdots, \mathbf{y}_m, \mathbf{u}_1, \cdots, \mathbf{u}_n\}$ is a linearly dependent set of vectors. Use the results of Problem 4 to show that there is a smallest index i such that $\mathbf{u}_i$ can be expressed as a linear combination of the set of vectors that precede $\mathbf{u}_i$. Explain why $\{\mathbf{y}_1, \cdots, \mathbf{y}_m, \mathbf{u}_1, \cdots, \mathbf{u}_{i-1}, \mathbf{u}_{i+1}, \cdots, \mathbf{u}_n\}$ spans R^n. If this new set of vectors with $\mathbf{u}_i$ removed is a linearly independent set, we have a basis for R^n. Otherwise explain how, by repeating the argument, one can get a basis for R^n that contains $\{\mathbf{y}_1, \cdots, \mathbf{y}_m\}$. This result proves the first statement of Proposition 1–1.1.

Answers

1. (a) $(-1, 1, 3)$; (b) $(3, -5, 5)$; (c) $(2, -5, 15)$; (d) $(-\tfrac{1}{2}, \tfrac{1}{2}, \tfrac{3}{2})$; (e) $(8, -13, 11)$; (f) $(-9, 16, -22)$.
3. (a) independent; (b) independent; (c) dependent; (d) dependent.
5. There are many possible sets of correct answers. These sets are obtained by successively making zero the first row, second row, etc., below the diagonal and then clearing fractions.
   (a) $\{(1, 1, -1), (0, -2, -1), (0, 0, 4)\}$;
   (b) $\{(1, 3, 2), (0, -7, -3), (0, 0, -4)\}$;
   (c) $\{(-1, 5, -1, 2), (0, 6, -2, -1), (0, 0, 13, 23)\}$;
   (d) $\{(3, 0, 6, 4, 10)\ (0, 3, -6, -7, -16), (0, 0, 15, 22, 37)\}$.
7. $(5, -6, 4), (8, -5, 1)$.
8. There are many possible answers. Here are two vectors that can be added to the given set to form a basis: $\{(0, 0, 1, 0), (0, 0, 0, 1)\}$.

1–2. Inner Products and Norms

Let ϕ be a function that maps pairs of vectors $(\mathbf{x}, \mathbf{y})$ from $R^n \times R^n$ into the reals; that is, $\phi: (\mathbf{x}, \mathbf{y}) \to \phi(\mathbf{x}, \mathbf{y})$ where $\mathbf{x}$ and $\mathbf{y}$ are vectors in R^n and $\phi(\mathbf{x}, \mathbf{y}) \in R$, the set of real numbers. If ϕ is such that for all $\mathbf{x}$, $\mathbf{y}$, $\mathbf{x}_1$, and $\mathbf{x}_2$ in R^n

$$\phi(\mathbf{x}, \mathbf{y}) = \phi(\mathbf{y}, \mathbf{x}), \tag{1–2.1}$$

$$\phi(\alpha_1 \mathbf{x}_1 + \alpha_2 \mathbf{x}_2, \mathbf{y}) = \alpha_1 \phi(\mathbf{x}_1, \mathbf{y}) + \alpha_2 \phi(\mathbf{x}_2, \mathbf{y}), \tag{1–2.2}$$

and

$$\phi(\mathbf{x}, \mathbf{x}) \geq 0 \text{ for all } \mathbf{x} \in R^n \text{ and } \phi(\mathbf{x}, \mathbf{x}) = 0 \quad \text{if and only if } \mathbf{x} = \mathbf{0}, \tag{1–2.3}$$

then the vector space R^n with a mapping ϕ as above is an *inner product space*. The image of $(\mathbf{x}, \mathbf{y})$, $\phi(\mathbf{x}, \mathbf{y})$, is the ϕ *inner product of* $\mathbf{x}$ *and* $\mathbf{y}$.

Let the function ϕ be defined as follows:

$$\phi(\mathbf{x}, \mathbf{y}) = \textstyle\sum_{i=1}^{n} x_i y_i .$$

This function ϕ maps $R^n \times R^n$ into R in such a way as to satisfy (1–2.1), (1–2.2), and (1–2.3); hence ϕ defines an inner product on R^n. This particular inner product is called the *dot product* of $\mathbf{x}$ and $\mathbf{y}$, and we write

$$\mathbf{x} \cdot \mathbf{y} = \textstyle\sum_{i=1}^{n} x_i y_i . \tag{1–2.4}$$

Writing the above three properties of the dot product in the "dot" notation, we have

$$\mathbf{x} \cdot \mathbf{y} = \mathbf{y} \cdot \mathbf{x}, \tag{1–2.5}$$

$$(\alpha_1 \mathbf{x}_1 + \alpha_2 \mathbf{x}_2) \cdot \mathbf{y} = \alpha_1(\mathbf{x}_1 \cdot \mathbf{y}) + \alpha_2(\mathbf{x}_2 \cdot \mathbf{y}), \tag{1–2.6}$$

and

$$\mathbf{x} \cdot \mathbf{x} \geq 0 \text{ for all } \mathbf{x} \in R^n \text{ and } \mathbf{x} \cdot \mathbf{x} = 0 \tag{1–2.7}$$

if and only if $\mathbf{x} = \mathbf{0}$.

Let ψ be a function that maps R^n into the reals; that is, $\psi\colon \mathbf{x} \to \psi(\mathbf{x})$ where $\mathbf{x} \in R^n$ and $\psi(\mathbf{x}) \in R$. If ψ is such that

$$\psi(\mathbf{x}) \geq 0 \text{ for all } \mathbf{x} \in R^n \text{ and } \psi(\mathbf{x}) = 0 \quad \text{if and only if } \mathbf{x} = \mathbf{0}, \tag{1–2.8}$$

$$\psi(\alpha\mathbf{x}) = |\alpha|\, \psi(\mathbf{x}) \text{ for all real } \alpha \text{ and all } \mathbf{x} \in R^n, \tag{1–2.9}$$

and

$$\psi(\mathbf{x} + \mathbf{y}) \leq \psi(\mathbf{x}) + \psi(\mathbf{y}) \text{ for all } \mathbf{x} \text{ and } \mathbf{y} \text{ in } R^n, \tag{1–2.10}$$

then ψ is a *norm* on R^n. The vector space $V(R)$ with a mapping ψ as above is a *normed vector space* over the reals.

If ϕ is a mapping from $R^n \times R^n$ into R such that $\phi(\mathbf{x}, \mathbf{y})$ is an inner product on R^n, then ψ such that

$$\psi(\mathbf{x}) = \sqrt{\phi(\mathbf{x}, \mathbf{x})}$$

can be easily seen to satisfy (1–2.8) and (1–2.9). Later (Problem 6), it will be shown that ψ satisfies (1–2.10) also. If, in particular, the inner product of $\mathbf{x}$ and $\mathbf{y}$ is given by (1–2.4), the dot product, we write

$$\|\mathbf{x}\| = \sqrt{\mathbf{x} \cdot \mathbf{x}}. \tag{1–2.11}$$

Thus,

$$\|\mathbf{x}\| = \sqrt{\textstyle\sum_{i=1}^{n} x_i^2}. \tag{1–2.12}$$

This is the usual norm that is taken on R^n.

The vector space R^n with the dot product of (1–2.4) and the norm of (1–2.11) is *Euclidean n-space*, written E^n.

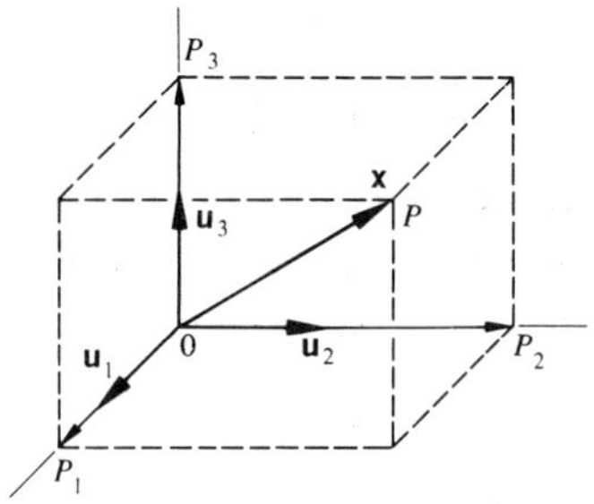

Figure 1–2.1

In Figure 1–2.1, the vectors $\mathbf{x} = (x_1, x_2, x_3)$, $\mathbf{u}_1 = (1, 0, 0)$, $\mathbf{u}_2 = (0, 1, 0)$, and $\mathbf{u}_3 = (0, 0, 1)$ are pictured. The vectors are shown as directed line segments—that is, line segments with an arrow on one end to indicate a direction.

In mathematics literature, vectors in R^n are treated in three different ways: as an n-tuple of real numbers, as a point in a space, and as an equivalence class of directed line segments.

In our presentation, a vector was taken as an n-tuple of real numbers, but in much of the physical science literature there is the desire to think of a vector $\mathbf{x}$ as a *point*, particularly if $n \leq 3$. For example, there are statements like "the position of a particle is at $\mathbf{x}$," meaning that a "particle" has coordinates (x_1, x_2, x_3) in a picture of E^3 like Figure 1–2.1. To be precise one could say something like "the point given by the vector $\mathbf{x}$," but this is seldom done in practice. Writers usually just expect the reader to apply correctly this interpretation from the context.

In E^3, for example, the interpretation of a vector as an equivalence class of directed line segments, one takes as given the three mutually perpendicular lines, the axes. Then notions like points, lines, line segments, and parallel are accepted on an intuitional basis. The idea of a directed line segment is indicated by a picture like Figure 1–2.2(a). The line segment, AB, has end points A and B. It is directed from A toward B as indicated by the arrow. To indicate this directed line segment we write $\overrightarrow{AB}$. The end points A and B are the *initial* and *terminal* end points respectively. Two directed line segments are equivalent, written $=$, if they have the same direction and the same norm, as for example $\overrightarrow{AB} = \overrightarrow{CD}$ in Figure 1–2.2(b). In fact, all the directed line segments pictured in Figure 1–2.2(b) are equivalent. Since this relation "=" between directed line segments is: *reflexive* ($\overrightarrow{AB} = \overrightarrow{AB}$), *symmetric* ($\overrightarrow{AB} = \overrightarrow{CD}$ implies $\overrightarrow{CD} = \overrightarrow{AB}$), and *transitive* (if $\overrightarrow{AB} = \overrightarrow{CD}$ and $\overrightarrow{CD} = \overrightarrow{EF}$, then

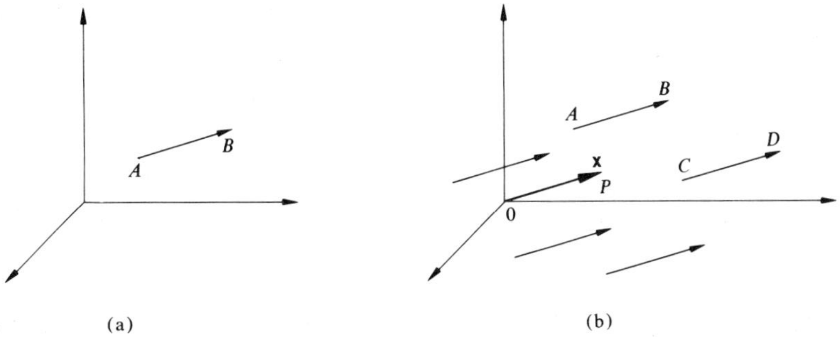

Figure 1–2.2

$\overrightarrow{AB} = \overrightarrow{EF}$), it follows that "=" is an *equivalence relation* on the set of directed line segments. In Figure 1–2.2(b) all of those directed line segments pictured belong to the same *equivalence class*. In particular, $\overrightarrow{OP}$ belongs to this equivalence class. It is the only directed line segment in this class that has its initial end point at the origin—that is, in *standard position*. Any one of the vectors pictured can be a *representative* of the equivalence class. A *vector* in this setting is an equivalence class of these directed line segments. The vector is pictured by any directed line segment of the equivalence class. Addition of vectors, which in reality is addition of equivalence classes, is accomplished by taking a representative from each of these classes that have common initial end points and adding by the "parallelogram law." Figure 1–2.3 shows the addition of $\overrightarrow{AB}$ to $\overrightarrow{AC}$ obtaining $\overrightarrow{AD}$ by the parallelogram law.

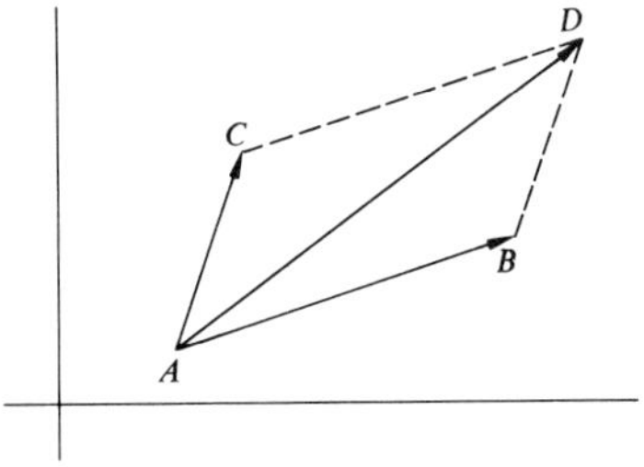

Figure 1–2.3

Scalar multiplication is handled in a somewhat geometric manner also. Notice that in Figure 1–2.2(b),

$$\|\mathbf{x}\| = \sqrt{x_1^2 + x_2^2 + x_3^2}$$

(the norm of **x**), assigns a "length or magnitude" to the vector **x**. This is taken as the *magnitude* of the vector represented by $\overrightarrow{OP}$, written $\overline{OP}$. If

$\alpha \geq 0$, the new vector represented by $\alpha(\overline{OP})$ is a vector in the same direction as $\overrightarrow{OP}$ of magnitude $\alpha(\overline{OP})$. If $\alpha < 0$, the new vector $\alpha(\overrightarrow{OP})$ is in the direction directly opposite to $\overrightarrow{OP}$ and is of magnitude $|\alpha|$ $(\overline{OP})$.

This approach to vectors is particularly useful in picturing physical quantities such as force, velocity, acceleration, etc. In this text we shall have occasion to use vectors with each of these three meanings. We trust that this brief warning will be sufficient to alert the reader to these three shades of meanings of vectors.

The following proposition states the useful Cauchy-Schwarz inequality.

Proposition 1–2.1. *If $x_1, x_2, \cdots, x_n$ and $y_1, y_2, \cdots, y_n$ are arbitrary real numbers, then*

$$(\sum_{i=1}^{n} x_i y_i)^2 \leq (\sum_{i=1}^{n} x_i^2)(\sum_{i=1}^{n} y_i^2). \tag{1–2.13}$$

The equality holds if and only if there is a real constant λ such that either $\lambda x_i = y_i$ for all i or $x_i = \lambda y_i$ for all i.

Proof. If all $x_i = 0$ then both sides of the inequality are zero, and the inequality (1–2.13) holds.

Consider now the case of not all $x_i = 0$; that is,

$$\sum_{i=1}^{n} x_i^2 > 0.$$

For any real number λ, $(x_i \lambda + y_i)^2 \geq 0$ for all i. Thus

$$\sum_{i=1}^{n} (x_i \lambda + y_i)^2 \geq 0. \tag{1–2.14}$$

Squaring and collecting terms, the inequality of (1–2.14) becomes $a\lambda^2 + 2b\lambda + c \geq 0$ where

$$a = \sum_{i=1}^{n} x_i^2,$$

$$b = \sum_{i=1}^{n} x_i y_i,$$

and

$$c = \sum_{i=1}^{n} y_i^2.$$

Now $a\lambda^2 + 2b\lambda + c \geq 0$ with $a > 0$ implies that $(2b)^2 - 4ac$ (the discriminant of the quadratic) <0 or $b^2 \leq ac$, which is the required inequality. The part of the proof showing when the inequality of (1–2.13) is an equality is done in Problems 7 and 8 at the end of this section. ∎

If $\mathbf{x} = (x_1, x_2, \cdots, x_n)$ and $\mathbf{y} = (y_1, y_2, \cdots, y_n)$, then the Cauchy-Schwarz inequality, (1–2.13), becomes

$$|\mathbf{x} \cdot \mathbf{y}| \leq \|\mathbf{x}\| \, \|\mathbf{y}\| \tag{1–2.15}$$

where equality holds if and only if there is a real number λ such that $\mathbf{x} = \lambda\mathbf{y}$ or $\mathbf{y} = \lambda\mathbf{x}$.

If $\mathbf{x}$ and $\mathbf{y}$ are nonzero vectors such that $\{\mathbf{x}, \mathbf{y}\}$ is a linearly independent subset, then the linear combinations of $\{\mathbf{x}, \mathbf{y}\}$ give a two dimensional subspace of E^n, denoted $[\{\mathbf{x}, \mathbf{y}\}]$. The norm, $\| \quad \|$, on E^n induces a norm on $[\{\mathbf{x}, \mathbf{y}\}]$ which is identical with the usual norm for a plane, this particular two dimensional subspace. Thus the Law of Cosines holds in this plane for the triangle whose vertices are $\mathbf{0}$, $\mathbf{x}$, and $\mathbf{y}$. That is, if θ is the interior angle of this triangle at $\mathbf{0}$, then

$$\|\mathbf{x} - \mathbf{y}\|^2 = \|\mathbf{x}\|^2 + \|\mathbf{y}\|^2 - 2\,\|\mathbf{x}\|\,\|\mathbf{y}\|\cos\theta \tag{1–2.16}$$

where $\|\mathbf{x} - \mathbf{y}\|$ is the length of the side opposite the angle θ. However, by expanding $\|\mathbf{x} - \mathbf{y}\|^2$ using (1–2.11), (1–2.5), and (1–2.6), we have

$$\begin{aligned}\|\mathbf{x} - \mathbf{y}\|^2 &= (\mathbf{x} - \mathbf{y}) \cdot (\mathbf{x} - \mathbf{y}),\\ &= \mathbf{x} \cdot \mathbf{x} - 2(\mathbf{x} \cdot \mathbf{y}) + \mathbf{y} \cdot \mathbf{y},\\ &= \|\mathbf{x}\|^2 - 2(\mathbf{x} \cdot \mathbf{y}) + \|\mathbf{y}\|^2.\end{aligned} \tag{1–2.17}$$

Equating (1–2.16) and (1–2.17) and canceling give

$$\mathbf{x} \cdot \mathbf{y} = \|\mathbf{x}\|\,\|\mathbf{y}\|\cos\theta. \tag{1–2.18}$$

This approach to obtaining (1–2.18) makes some demands on one's geometrical intuition. If the above derivation of (1–2.18) is satisfactory to the reader, then the definition below is a natural consequence. If the reader dislikes the reliance on geometrical intuition, he may simply regard the foregoing as motivation for the following definition.

Definition 1–2.1. *If* $\mathbf{x}$ *and* $\mathbf{y}$ *are two nonzero vectors in* E^n, *the angle* θ *between* $\mathbf{x}$ *and* $\mathbf{y}$ *is the solution* θ, $0 \le \theta \le \pi$, *of the equation*

$$\cos\theta = \frac{\mathbf{x} \cdot \mathbf{y}}{\|\mathbf{x}\|\,\|\mathbf{y}\|}. \tag{1–2.19}$$

The Cauchy-Schwarz inequality assures us that

$$-1 \le \frac{\mathbf{x} \cdot \mathbf{y}}{\|\mathbf{x}\|\,\|\mathbf{y}\|} \le 1.$$

From Proposition 1–2.1, it follows that $|\cos\theta| = 1$ if and only if there is a real number λ such that $\mathbf{x} = \lambda\mathbf{y}$ (or $\mathbf{y} = \lambda\mathbf{x}$); that is, $\mathbf{x}$ and $\mathbf{y}$ have the same direction if $\cos\theta = 1$ or the opposite direction if $\cos\theta = -1$.

For nonzero vectors $\mathbf{x}$ and $\mathbf{y}$, if $\mathbf{x} \cdot \mathbf{y} = 0$, then $\cos\theta = 0$ and $\theta = \pi/2$.

This suggests that we say that $\mathbf{x}$ and $\mathbf{y}$ are *orthogonal* (perpendicular) if $\mathbf{x} \cdot \mathbf{y} = 0$. Note that this definition allows $\mathbf{0}$ to be orthogonal to any vector.

A set X of vectors in E^n is said to be an *orthogonal set* if for any two vectors $\mathbf{x}$ and $\mathbf{y}$ in X we have $\mathbf{x} \cdot \mathbf{y} = 0$. If X is an orthogonal set in E^n and if, in addition, $\|\mathbf{x}\| = 1$ for each $\mathbf{x}$ in X, then X is *orthonormal.* The set of vectors $\mathbf{u}_1 = (1, 0, \cdots, 0)$, $\mathbf{u}_2 = (0, 1, \cdots, 0), \cdots, \mathbf{u}_n = (0, 0, \cdots, 1)$ is an orthonormal set in E^n. Since $\mathbf{u}_1, \mathbf{u}_2, \cdots, \mathbf{u}_n$ is a basis of E^n, then we say it is an *orthonormal basis* of E^n.

With the Cauchy-Schwarz inequality it is now possible to show in the following proposition that the norm, $\|\quad\|$, satisfies (1–2.10).

Proposition 1–2.2. *For any two vectors* $\mathbf{x}$ *and* $\mathbf{y}$ *in* E^n,

$$\|\mathbf{x} + \mathbf{y}\| \leq \|\mathbf{x}\| + \|\mathbf{y}\|.$$

Proof. From (1–2.11), (1–2.5), and (1–2.6) it follows that

$$\begin{aligned} \|\mathbf{x} + \mathbf{y}\|^2 &= (\mathbf{x} + \mathbf{y}) \cdot (\mathbf{x} + \mathbf{y}), \\ &= \mathbf{x} \cdot \mathbf{x} + 2(\mathbf{x} \cdot \mathbf{y}) + \mathbf{y} \cdot \mathbf{y}, \\ &= \|\mathbf{x}\|^2 + 2(\mathbf{x} \cdot \mathbf{y}) + \|\mathbf{y}\|^2. \end{aligned} \tag{1–2.20}$$

Applying the Cauchy-Schwarz inequality, (1–2.15), to equation (1–2.20) it follows that

$$\|\mathbf{x} + \mathbf{y}\|^2 \leq \|\mathbf{x}\|^2 + 2\|\mathbf{x}\|\,\|\mathbf{y}\| + \|\mathbf{y}\|^2 = (\|\mathbf{x}\| + \|\mathbf{y}\|)^2,$$

or

$$\|\mathbf{x} + \mathbf{y}\| \leq \|\mathbf{x}\| + \|\mathbf{y}\|. \quad \blacksquare$$

Notice that from the definition of subtraction in (1–1.11) and from the definition of $\|\quad\|$ in (1–2.12),

$$\|\mathbf{x} - \mathbf{y}\| = \|\mathbf{y} - \mathbf{x}\|. \tag{1–2.21}$$

The set of vectors *between* the vectors $\mathbf{x}$ and $\mathbf{y}$ is the set

$$\{(1 - \lambda)\mathbf{x} + \lambda\mathbf{y} : 0 < \lambda < 1\}.$$

The motivation of this definition is apparent if we write

$$(1 - \lambda)\mathbf{x} + \lambda\mathbf{y} = \mathbf{x} + \lambda(\mathbf{y} - \mathbf{x})$$

and observe Figure 1–2.4. A set S in E^n is *convex* if for any two points $\mathbf{x}$ and $\mathbf{y}$ the points between $\mathbf{x}$ and $\mathbf{y}$ belong to S. The set of points $\{\mathbf{x}: \|\mathbf{x}\| \leq 1\}$, the *unit sphere* centered at the origin, is a convex set.

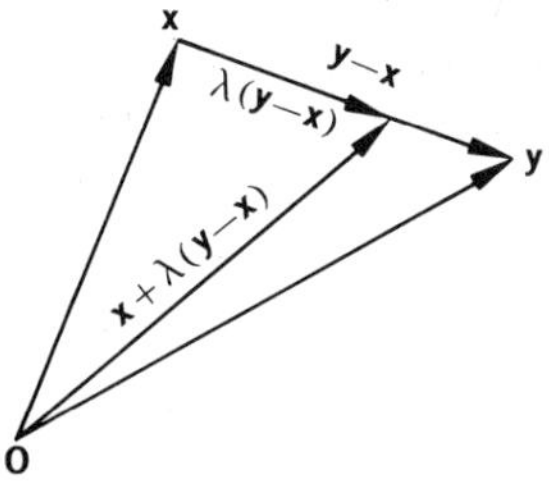

Figure 1–2.4

The spaces R^2 and R^3 consist, respectively, of ordered pairs and ordered triples. Oftentimes the pairs and triples are designated (x, y) and (x, y, z) instead of (x_1, x_2) and (x_1, x_2, x_3). Occasionally there will be a reference to (x, y)-space or maybe (x, y)-plane (or (x, y, z)-space). In such cases we shall mean that the variables are designated x and y (or x, y, and z) instead of x with subscripts; but, moreover, *it is understood that the spaces have the dot product and norm,* $\|\quad\|$, *making them* E^2 *(and* E^3*) spaces.*

Problems

1. For the following pairs of vectors, find the dot product, the norm of each vector, and the cosine of the angle between the two vectors. Sketch (a) and (b).

   (a) $(0, 1), (3, 4)$. (b) $(1, -1, 2), (-2, 3, -1)$.
   (c) $(1, 5, -2, -2), (2, 0, 4, -1)$.
   (d) $(1, 1, -1, 1, 1), (0, -3, 2, 1, 2)$.

2. Prove that the dot product defined by (1–2.4) satisfies (1–2.5), (1–2.6), and (1–2.7).

3. Prove that if

$$\psi(\mathbf{x}) = \sqrt{\phi(\mathbf{x}, \mathbf{x})}$$

   where ϕ is an inner product on R^n, then ψ satisfies (1–2.8) and (1–2.9).

4. Let ψ be defined as in Problem 3. Show that for any real λ

$$[\psi(\lambda\mathbf{x} + \mathbf{y})]^2 = [\psi(\mathbf{x})]^2\lambda^2 + 2\phi(\mathbf{x}, \mathbf{y})\lambda + [\psi(\mathbf{y})]^2.$$

5. From the equality of Problem 4 show that

$$|\phi(\mathbf{x}, \mathbf{y})| \leq \psi(\mathbf{x})\psi(\mathbf{y}),$$

   which is a generalization on the Cauchy-Schwarz inequality, (1–2.15).

6. Show that if ψ is defined as in Problem 3, then from Problem 5 we can get (1–2.10).

7. Assume that $\lambda x_i = y_i$, $1 \leq i \leq n$, where $\mathbf{x} = (x_1, \cdots, x_n)$ and $\mathbf{y} = (y_1, \cdots, y_n)$; that is, $\lambda\mathbf{x} = \mathbf{y}$. Show by substituting for $\mathbf{y}$ on both sides of (1–2.15) that (1–2.15) becomes an equality.

8. Suppose $|\mathbf{x} \cdot \mathbf{y}| = \|\mathbf{x}\| \, \|\mathbf{y}\|$; that is, (1–2.15), or equivalently (1–2.13), is an equality. If $\mathbf{x} = \mathbf{0}$ then $\mathbf{x} = 0\mathbf{y}$. Suppose $\mathbf{x} \neq \mathbf{0}$, and show $\|\lambda\mathbf{x} - \mathbf{y}\|^2 = 0$ where $\lambda = (\mathbf{x} \cdot \mathbf{y})/\|\mathbf{x}\|^2$ and thus $\mathbf{y} = \lambda\mathbf{x}$.

9. In R^n show that $\{\mathbf{x}, \mathbf{y}\}$ is a linearly dependent set if and only if $|\mathbf{x} \cdot \mathbf{y}| = \|\mathbf{x}\| \, \|\mathbf{y}\|$.

10. Let $\rho(\mathbf{x}, \mathbf{y}) = \|\mathbf{x} - \mathbf{y}\|$. Show that E^n with ρ given as indicated is a *metric space*. That is, show that for all $\mathbf{x}$, $\mathbf{y}$, and $\mathbf{z}$ in E^n

    (a) $\rho(\mathbf{x}, \mathbf{y}) \geq 0$, and $\rho(\mathbf{x}, \mathbf{y}) = 0$ if and only if $\mathbf{x} = \mathbf{y}$;
    (b) $\rho(\mathbf{x}, \mathbf{y}) = \rho(\mathbf{y}, \mathbf{x})$;
    (c) $\rho(\mathbf{x}, \mathbf{y}) \leq \rho(\mathbf{x}, \mathbf{z}) + \rho(\mathbf{z}, \mathbf{y})$.

11. Show that the following sets in E^n are convex. Make a sketch of the set by shading for the case $n = 2$.

    (a) $\{\mathbf{x}: \|\mathbf{x}\| \leq 1\}$, (b) $\{\mathbf{x}: \|\mathbf{x} - \mathbf{x}_0\| < \delta, \delta > 0\}$,
    (c) $\{\mathbf{x}: \sum_{i=1}^{n} |x_i| \leq 1\}$, (d) $\{\mathbf{x}: \max\{|x_i|: 1 \leq i \leq n\} \leq 1\}$.

12. Let $\mathbf{x}_1, \mathbf{x}_2, \cdots, \mathbf{x}_n$ be a basis for E^n. Then $\mathbf{x}_i \neq \mathbf{0}$ for all i, $1 \leq i \leq n$. Let

$$\mathbf{y}_1 = \mathbf{x}_1/\|\mathbf{x}_1\|$$

and

$$\mathbf{z}_2 = \mathbf{x}_2 - (\mathbf{x}_2 \cdot \mathbf{y}_1)\mathbf{y}_1.$$

Show $\{\mathbf{y}_1, \mathbf{z}_2\}$ is an orthogonal set in E^n. Let $\mathbf{y}_2 = \mathbf{z}_2/\|\mathbf{z}_2\|$ and thus make $\{\mathbf{y}_1, \mathbf{y}_2\}$ an orthonormal set. Let

$$\mathbf{z}_3 = \mathbf{x}_3 - (\mathbf{x}_3 \cdot \mathbf{y}_1)\mathbf{y}_1 - (\mathbf{x}_3 \cdot \mathbf{y}_2)\mathbf{y}_2$$

and show $\{\mathbf{y}_1, \mathbf{y}_2, \mathbf{z}_3\}$ is an orthogonal set; then show how to change this set to an orthonormal set. Make a sketch in E^3 showing each of the above vectors. Show how to obtain an orthonormal basis for E^n.

Answers

1. (a) $4; 1, 5; 4/5$. (b) $-7; \sqrt{6}, \sqrt{14}; -(1/2)\sqrt{7/3}$.
   (c) $-4; \sqrt{34}, \sqrt{21}; -4/\sqrt{714}$. (d) $-2; \sqrt{5}, \sqrt{18}; -\sqrt{10}/15$.

1–3. Vector Products in E^3

For the particular E^n space E^3 and only E^3, there is another product of vectors that is in use: the *vector product* (or *cross product*) of $\mathbf{x}$ and $\mathbf{y}$, written $\mathbf{x} \times \mathbf{y}$. Below, this product is defined precisely.

Definition 1–3.1. *Given* $\mathbf{x} = (x_1, x_2, x_3)$ *and* $\mathbf{y} = (y_1, y_2, y_3)$, *then*

$$\mathbf{x} \times \mathbf{y} = (x_2 y_3 - x_3 y_2, x_3 y_1 - x_1 y_3, x_1 y_2 - x_2 y_1) \qquad (1\text{–}3.1)$$

is the vector product of the vectors $\mathbf{x}$ *and* $\mathbf{y}$.

This product of two vectors, which is slightly more complicated than the scalar product, has one other difference that is more basic. The product $\phi(\mathbf{x}, \mathbf{y})$ or, in particular, $\mathbf{x} \cdot \mathbf{y}$, is a real number, whereas the product $\mathbf{x} \times \mathbf{y}$ is a vector in E^3.

The following proposition gives some of the main properties of the vector product.

Proposition 1–3.1. *Let* $\mathbf{x}$, $\mathbf{y}$, *and* $\mathbf{z}$ *be vectors in* E^3, *and let* α *and* β *be real numbers, then*

$$\mathbf{x} \times \mathbf{y} = -(\mathbf{y} \times \mathbf{x}), \qquad (1\text{–}3.2)$$

$$\mathbf{x} \times (\alpha\mathbf{y} + \beta\mathbf{z}) = \alpha(\mathbf{x} \times \mathbf{y}) + \beta(\mathbf{x} \times \mathbf{z}), \qquad (1\text{–}3.3)$$

and

$$\mathbf{x} \cdot (\mathbf{x} \times \mathbf{y}) = 0. \qquad (1\text{–}3.4)$$

The properties (1–3.2) and (1–3.3) are referred to by saying the vector product is *skew-symmetric* and *linear in the second argument*, respectively.

Proof. Let $\mathbf{x} = (x_1, x_2, x_3)$ and $\mathbf{y} = (y_1, y_2, y_3)$. The first coordinate of $-(\mathbf{y} \times \mathbf{x})$ is

$$-(y_2 x_3 - y_3 x_2) = x_2 y_3 - x_3 y_2,$$

which is the first coordinate of $\mathbf{x} \times \mathbf{y}$ (cf. (1–3.1)). Similarly, the second and third coordinates of $\mathbf{x} \times \mathbf{y}$ and $-(\mathbf{x} \times \mathbf{y})$ can be compared and found to be equal. Hence $\mathbf{x} \times \mathbf{y} = -(\mathbf{y} \times \mathbf{x})$.

If $\mathbf{z} = (z_1, z_2, z_3)$, the ith coordinate of $\alpha\mathbf{y} + \beta\mathbf{z}$ is $\alpha y_i + \beta z_i$. Thus the first coordinate of $\mathbf{x} \times (\alpha\mathbf{y} + \beta\mathbf{z})$ is

$$x_2(\alpha y_3 + \beta z_3) - x_3(\alpha y_2 + \beta z_2) = \alpha(x_2 y_3 - x_3 y_2) + \beta(x_2 z_2 - x_3 z_2).$$

The expression on the right is clearly the first coordinate of $\alpha(\mathbf{x} \times \mathbf{y}) + \beta(\mathbf{x} \times \mathbf{z})$. The other coordinates can be checked in a similar way, and hence (1–3.3) holds.

Lastly, by applying the definition of the dot product and the scalar product, we have

$$\begin{aligned} \mathbf{x} \cdot (\mathbf{x} \times \mathbf{y}) &= x_1(x_2 y_3 - x_3 y_2) + x_2(x_3 y_1 - x_1 y_3) + x_3(x_1 y_2 - x_2 y_1) \\ &= 0. \quad \blacksquare \end{aligned} \qquad (1\text{–}3.5)$$

From these properties several additional properties can be obtained. For example, if $\mathbf{y} = \mathbf{x}$ in (1–3.2), then $\mathbf{x} \times \mathbf{x} = -(\mathbf{x} \times \mathbf{x})$ and, hence,

$$\mathbf{x} \times \mathbf{x} = \mathbf{0} \tag{1–3.6}$$

for all $\mathbf{x}$ in E^3. By using (1–3.2) and (1–3.3), it follows that the vector product is also linear in the first argument; that is,

$$(\alpha\mathbf{y} + \beta\mathbf{z}) \times \mathbf{x} = \alpha(\mathbf{y} \times \mathbf{x}) + \beta(\mathbf{z} \times \mathbf{x}) \tag{1–3.7}$$

for all $\mathbf{x}$, $\mathbf{y}$, and $\mathbf{z}$ in E^3.

With three vectors, $\mathbf{x}$, $\mathbf{y}$, and $\mathbf{z}$, the only meaningful way to have a product of these three vectors in this order ($\mathbf{x}$, $\mathbf{y}$, and then $\mathbf{z}$) that involves both the dot product and the vector product is: $\mathbf{x} \cdot (\mathbf{y} \times \mathbf{z})$, the *mixed triple product*. Expanding according to the definitions gives

$$\mathbf{x} \cdot (\mathbf{y} \times \mathbf{z}) = x_1(y_2 z_3 - y_3 z_2) + x_2(y_3 z_1 - y_1 z_3) + x_3(y_1 z_2 - y_2 z_1). \tag{1–3.8}$$

Hence,

$$\mathbf{x} \cdot (\mathbf{y} \times \mathbf{z}) = \begin{vmatrix} x_1 & x_2 & x_3 \\ y_1 & y_2 & y_3 \\ z_1 & z_2 & z_3 \end{vmatrix}. \tag{1–3.9}$$

It is convenient to write $(\mathbf{x}, \mathbf{y}, \mathbf{z}) = \mathbf{x} \cdot (\mathbf{y} \times \mathbf{z})$. If $(\mathbf{x}, \mathbf{y}, \mathbf{z}) > 0$, then $\{\mathbf{x}, \mathbf{y}, \mathbf{z}\}$ is called a *positive triple*.

Thus, (1–3.4) can be interpreted as a determinant with two identical rows. Using this same reasoning, it follows that $\mathbf{x} \cdot (\mathbf{x} \times \mathbf{y}) = 0$, and that $\mathbf{y} \cdot (\mathbf{x} \times \mathbf{y}) = 0$, and hence $\mathbf{x} \times \mathbf{y}$ *is orthogonal to both* $\mathbf{x}$ *and* $\mathbf{y}$.

Since an interchange of rows in a determinant changes the sign of the determinant, it follows that the six arrangements of the three rows of (1–3.9) group themselves into two groups of three each. The following equalities indicate the equality between the members of each group:

$$\mathbf{x} \cdot (\mathbf{y} \times \mathbf{z}) = \mathbf{z} \cdot (\mathbf{x} \times \mathbf{y}) = \mathbf{y} \cdot (\mathbf{z} \times \mathbf{x})$$

and

$$\mathbf{x} \cdot (\mathbf{z} \times \mathbf{y}) = \mathbf{z} \cdot (\mathbf{y} \times \mathbf{x}) = \mathbf{y} \cdot (\mathbf{x} \times \mathbf{z}).$$

The following identity, of which one member is a mixed product involving four vectors, is known as the *Lagrange identity*,

$$(\mathbf{w} \times \mathbf{x}) \cdot (\mathbf{y} \times \mathbf{z}) = (\mathbf{w} \cdot \mathbf{y})(\mathbf{x} \cdot \mathbf{z}) - (\mathbf{w} \cdot \mathbf{z})(\mathbf{x} \cdot \mathbf{y}). \tag{1–3.10}$$

This identity can be established by first writing out each vector in component form and by then expanding both sides of the equation until the identity is apparent. This identity enables us to obtain the norm, $\| \quad \|$, of the vector product of two vectors.

Proposition 1–3.2. *If* $\mathbf{x}$ *and* $\mathbf{y}$ *are two nonzero vectors in* E^3, *then*

$$\|\mathbf{x} \times \mathbf{y}\| = \|\mathbf{x}\| \, \|\mathbf{y}\| \sin \theta, \tag{1–3.11}$$

where θ *is the angle between* $\mathbf{x}$ *and* $\mathbf{y}$.

Proof. Using in order (1–2.11), the Lagrange identity, (1–2.11), and Definition 1–2.1, it follows that:

$$\begin{aligned}\|\mathbf{x} \times \mathbf{y}\|^2 &= (\mathbf{x} \times \mathbf{y}) \cdot (\mathbf{x} \times \mathbf{y}) \\ &= (\mathbf{x} \cdot \mathbf{x})(\mathbf{y} \cdot \mathbf{y}) - (\mathbf{x} \cdot \mathbf{y})^2 \\ &= \|\mathbf{x}\|^2 \, \|\mathbf{y}\|^2 - \|\mathbf{x}\|^2 \, \|\mathbf{y}\|^2 \cos^2 \theta \\ &= \|\mathbf{x}\|^2 \, \|\mathbf{y}\|^2 \, (1. - \cos^2 \theta) \\ &= \|\mathbf{x}\|^2 \, \|\mathbf{y}\|^2 \sin^2 \theta.\end{aligned}$$

Taking the square root of both sides gives (1–3.11), since $|\sin \theta| = \sin \theta$ for $0 \leq \theta \leq \pi$. ▌

If $\|\mathbf{x} \times \mathbf{y}\| = 0$ for nonzero vectors $\mathbf{x}$ and $\mathbf{y}$, then $\sin \theta = 0$, and hence $|\cos \theta| = 1$. Thus there exists λ such that $\mathbf{x} = \lambda \mathbf{y}$.

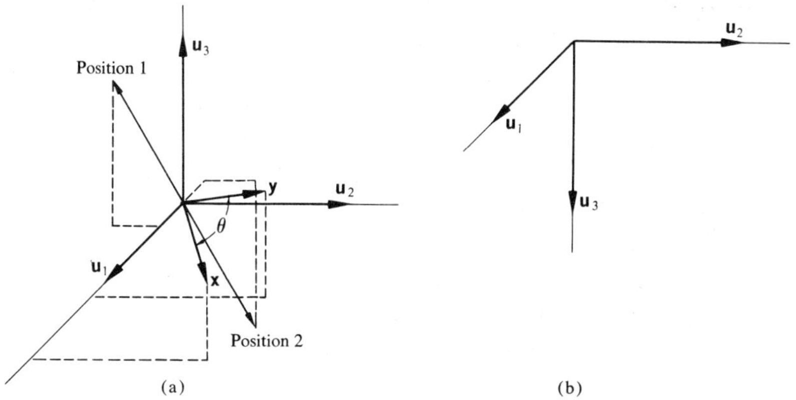

Figure 1–3.1

Definition 1–3.1 gives a way of calculating $\mathbf{x} \times \mathbf{y}$ for two known vectors $\mathbf{x}$ and $\mathbf{y}$ in E^3. It is now known that $\mathbf{x} \times \mathbf{y}$ is a vector that is orthogonal to the vectors $\mathbf{x}$ and $\mathbf{y}$ and has magnitude given by (1–3.11). Thus if $\mathbf{x} \times \mathbf{y} \neq \mathbf{0}$, then $\mathbf{x}$ and $\mathbf{y} \neq \mathbf{0}$ and each of the vectors $\mathbf{x}$, $\mathbf{y}$, and $\mathbf{x} \times \mathbf{y}$ can be sketched in

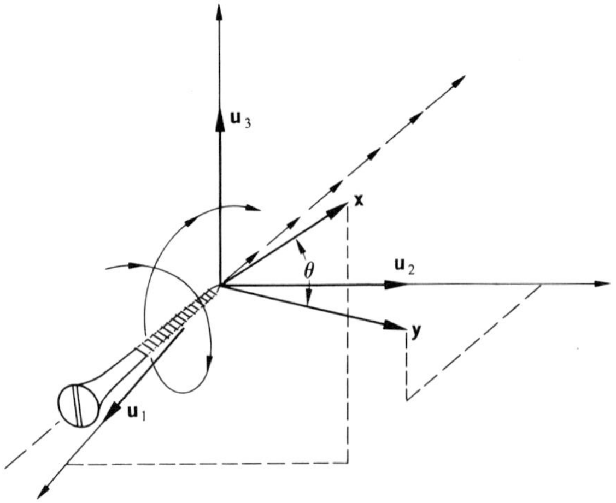

Figure 1–3.2

standard position (cf. Figure 1–3.1(a)). Thinking of **x** and **y** as being given, we could sketch these two vectors first. Then knowing the magnitude of $\mathbf{x} \times \mathbf{y}$ and knowing that $\mathbf{x} \times \mathbf{y}$ is perpendicular to **x** and **y**, then $\mathbf{x} \times \mathbf{y}$ in standard position must be in Position 1 or Position 2 in Figure 1–3.1(a). Given $\mathbf{u}_1$, $\mathbf{u}_2$, and $\mathbf{u}_3$ as shown in Figure 1–3.1(a), the ordered triple $(\mathbf{u}_1, \mathbf{u}_2, \mathbf{u}_3)$ forms a *right-hand system.* Then $\mathbf{x} \times \mathbf{y}$ as defined in Definition 1–3.1 will be in Position 1, which is in the direction given by the right-hand rule.

The *right-hand rule* is simply the following: Given vectors **x** and **y** as shown in Figure 1–3.2, the direction of $\mathbf{x} \times \mathbf{y}$ is the direction that a right-hand screw would move if the screw is placed such that its axis is perpendicular to **x** and **y** and the screw is rotated such that the line segment **x** is rotated toward **y** through the shorter angle (θ) between them (cf. Figure 1–3.2).

Figure 1–3.1(b) shows a *left-hand system.* The ordered triple of vectors $(\mathbf{u}_1, \mathbf{u}_2, \mathbf{u}_3)$ gives a right-hand (left-hand) system if by the right-hand rule $\mathbf{u}_2 \times \mathbf{u}_2 = \mathbf{u}_3$, $(-\mathbf{u}_3)$.

Definition 1–3.1 could be given such that each of the components has the signs exactly opposite to those of (1–3.1). In such a case each of the properties up to this point, except the right-hand rule, would hold.

There is one other meaningful product of three vectors **x**, **y**, and **z**, namely, $\mathbf{x} \times (\mathbf{y} \times \mathbf{z})$, the *vector triple product.* By writing out each of the vectors in component form and then expanding both sides, it is easy to see that

$$\mathbf{x} \times (\mathbf{y} \times \mathbf{z}) = (\mathbf{x} \cdot \mathbf{z})\mathbf{y} - (\mathbf{x} \cdot \mathbf{y})\mathbf{z}. \qquad (1\text{–}3.12)$$

If, for example, $\{\mathbf{x}, \mathbf{y}, \mathbf{z}\}$ is a linearly independent set in E^3, then $\mathbf{x} \times (\mathbf{y} \times \mathbf{z})$ is perpendicular to $\mathbf{y} \times \mathbf{z}$, but **y** and **z** are perpendicular to $\mathbf{y} \times \mathbf{z}$. Hence

$\mathbf{x} \times (\mathbf{y} \times \mathbf{z})$ is in the plane determined by $\mathbf{y}$ and $\mathbf{z}$. This is exactly what (1–3.12) shows, since the right side is a linear combination of $\{\mathbf{y}, \mathbf{z}\}$.

Problems

1. Calculate the following vector products. Make a sketch each time showing each of the three vectors.

   (a) $(1, 2, -1) \times (-2, 3, 4)$, (b) $(-2, 3, 4) \times (1, 2, -1)$,
   (c) $(-3, -2, 0) \times (1, -1, -1)$, (d) $(-4, 2, 3) \times (1, -1, 5)$.

2. Calculate all of the pairs $\mathbf{u}_i \cdot \mathbf{u}_j$ and $\mathbf{u}_i \times \mathbf{u}_j$ for $i = 1, 2, 3$ and $j = 1, 2, 3$.

3. Show that $\|\mathbf{x} \times \mathbf{y}\|$ is equal to the area of a parallelogram that has the length of two of its adjacent sides $\|\mathbf{x}\|$ and $\|\mathbf{y}\|$.

4. Show that $|\mathbf{x} \cdot (\mathbf{y} \times \mathbf{z})|$ is equal to the volume of a parallelepiped (a six-sided solid having three pairs of opposite faces parallel to one another) such that the length of the three edges that meet at a vertex are $\|\mathbf{x}\|$, $\|\mathbf{y}\|$, and $\|\mathbf{z}\|$.

5. Show
   $$(\alpha\mathbf{x}_1 + \beta\mathbf{x}_2, \mathbf{y}, \mathbf{z}) = \alpha(\mathbf{x}_1, \mathbf{y}, \mathbf{z}) + \beta(\mathbf{x}_2, \mathbf{y}, \mathbf{z}).$$

6. Prove the Lagrange identity, which is (1–3.10).

7. Prove the identity (1–3.12).

8. Is multiplication according to the vector product associative? In other words, prove or disprove that
   $$\mathbf{x} \times (\mathbf{y} \times \mathbf{z}) = (\mathbf{x} \times \mathbf{y}) \times \mathbf{z}$$
   for all $\mathbf{x}$, $\mathbf{y}$, and $\mathbf{z}$ in E^3.

9. Given that $\mathbf{x} = (2, -1, 0)$, $\mathbf{y} = (3, 1, -4)$, $\mathbf{z} = (-1, 3, 3)$, and $\mathbf{w} = (-2, -2, 5)$, calculate the following:

   (a) $\mathbf{x} \times (\mathbf{y} \times \mathbf{z})$, (b) $\mathbf{y} \cdot (\mathbf{w} \times \mathbf{x})$, (c) $(\mathbf{w} \times \mathbf{y}) \times \mathbf{z}$,
   (d) $(\mathbf{w} \times \mathbf{z}) \cdot \mathbf{y}$, (e) $\mathbf{y} \times (\mathbf{y} \times \mathbf{x})$, (f) $\mathbf{z} \cdot (\mathbf{x} \times \mathbf{z})$.

10. Demonstrate the reasonableness of the following statements:

    (a) $\{\mathbf{x} + \lambda\mathbf{y} : \lambda \text{ real}\}$ is a line through $\mathbf{x}$ parallel to $\mathbf{y}$.
    (b) $\{\mathbf{x} : (\mathbf{x} - \mathbf{a}) \cdot \mathbf{b} = 0\}$ is a plane through $\mathbf{a}$ perpendicular to the direction of $\mathbf{b}$.

11. Find a vector formula that will give the distance from the point $\mathbf{z}$ to

    (a) the line of Problem 9(a),
    (b) the plane of Problem 9(b).

12. The *moment of the vector* $\mathbf{x}$ *applied at* $\mathbf{b}$ *about the point* $\mathbf{a}$ is
    $$\mathbf{M}(\mathbf{x}, \mathbf{b}, \mathbf{a}) = (\mathbf{b} - \mathbf{a}) \times \mathbf{x}.$$

Show that $\mathbf{M}(\mathbf{x}, \mathbf{c}, \mathbf{a}) = \mathbf{M}(\mathbf{x}, \mathbf{b}, \mathbf{a})$ if $\mathbf{c}$ is any point on the line through $\mathbf{b}$ in the direction of $\mathbf{x}$.

13. Let l be a directed line in E^3 whose direction is parallel to the unit vector $\mathbf{u}$. The *moment of the vector* $\mathbf{x}$ *applied at* $\mathbf{b}$ *about the line* l is

$$M_{\mathbf{u}}(\mathbf{x}, \mathbf{b}) = \mathbf{u} \cdot (\mathbf{b} \times \mathbf{x}).$$

Show that if $\mathbf{c}$ is any point on the line through $\mathbf{b}$ in the direction $\mathbf{u}$, then $M_{\mathbf{u}}(\mathbf{x}, \mathbf{c}) = M_{\mathbf{u}}(\mathbf{x}, \mathbf{b})$.

Answers

1. (a) $(11, -2, 7)$, (b) $(-11, 2, -7)$,
   (c) $(2, -3, 5)$, (d) $(13, 23, 2)$.
2. $\mathbf{u}_i \cdot \mathbf{u}_j = 1$ if $i = j$, and 0 otherwise.
   $\mathbf{u}_1 \times \mathbf{u}_1 = \mathbf{0}$, $\mathbf{u}_1 \times \mathbf{u}_2 = \mathbf{u}_3$, $\mathbf{u}_1 \times \mathbf{u}_3 = -\mathbf{u}_2$,
   $\mathbf{u}_2 \times \mathbf{u}_1 = -\mathbf{u}_3$, $\mathbf{u}_2 \times \mathbf{u}_2 = \mathbf{0}$, $\mathbf{u}_2 \times \mathbf{u}_3 = \mathbf{u}_1$,
   $\mathbf{u}_3 \times \mathbf{u}_1 = \mathbf{u}_2$, $\mathbf{u}_3 \times \mathbf{u}_2 = -\mathbf{u}_1$, $\mathbf{u}_3 \times \mathbf{u}_3 = \mathbf{0}$.
9. (a) $(-10, -20, 5)$, (b) 1, (c) $(9, -13, 16)$,
   (d) -30, (e) $(-37, 31, -20)$, (f) 0.

1–4. Indicial Notation

There is a certain type of notation that is frequently used in mathematics books and articles called indicial notation. When authors speak of indicial notation, they customarily have in mind (a) the summation convention, (b) the Kronecker delta, and (c) the epsilon notation. A great amount of economy in writing can be effected once a person becomes familiar with this notation.

An example of a place where the summation convention is useful is in the case of a system of n linear equations in the n variables, $(x_1, \cdots, x_n)$. Written out in detail, the system looks as follows:

$$\begin{aligned} a_{11} x_1 + a_{12} x_2 + \cdots + a_{1n} x_n &= b_1, \\ a_{21} x_1 + a_{22} x_2 + \cdots + a_{2n} x_n &= b_2, \\ \cdots\cdots\cdots\cdots\cdots\cdots\cdots \\ a_{n1} x_1 + a_{n2} x_2 + \cdots + a_{nn} x_n &= b_n. \end{aligned}$$

If we write the coefficients with one subscript and one superscript, instead of with two subscripts, and if we use the summation symbol, we could write our system as

$$\begin{aligned} \textstyle\sum_{j=1}^n a_1{}^j x_j &= b_1, \\ \cdots\cdots\cdots \\ \textstyle\sum_{j=1}^n a_n{}^j x_j &= b_n; \end{aligned}$$

or we could simply write

$$\textstyle\sum_{j=1}^{n} a_i^{\,j} x_j = b_i, \quad i = 1, 2, \cdots, n.$$

With the *summation convention* we would write only

$$a_i^{\,j} x_j = b_i, \quad i = 1, 2, \cdots, n.$$

The idea is that when an index is repeated in a product, one upper and one lower, as is j in our case, a summation is understood over all the indices. In our case the summation is from 1 to n. The index j is referred to as a *dummy index*, since the choice of j or any other letter does not affect the answer.

The same system could be written with the coefficients having two superscripts, but in that case it is customary to write the b terms with superscripts; that is,

$$a^{ij} x_j = b^i, \quad i = 1, 2, \cdots, n.$$

If we write the variables with superscripts, then we could write our system as

$$a_{ij}\, x^j = b_i$$

or

$$a_j^{\,i} x^j = b^i$$

for $i = 1, 2, \cdots, n$.

The *Kronecker delta* is the lower-case delta with two subscripts, two superscripts, or one of each—that is, δ_{ij}, δ^{ij}, or $\delta_j^{\,i}$. In any case,

$$\delta_{ij} = \delta^{ij} = \delta_j^{\,i} = \begin{cases} 1, & \text{if } i = j, \\ 0, & \text{if } i \neq j, \end{cases}$$

where i and j range over the integers $1, 2, \cdots, n$. For example, if we write a matrix $A = (a_{ij})$ where a_{ij} is the entry on the ith row and in the jth column, then the matrix (δ_{ij}) is the identity matrix.

The third type of indicial notation is the *epsilon symbol*, which is like ε^{ijk} where i, j, and k are 1, 2, or 3. If at least two of the superscripts agree, then the symbol has the value zero. The symbol has the value $+1$ if all superscripts are distinct and if an *even* number of interchanges of pairs is necessary to bring the superscripts to the order 1-2-3. For example, 312 becomes 132, becomes 123 on interchanging successively the indicated pairs. If an *odd* number of interchanges of pairs is necessary, the symbol has the value -1. Therefore,

$$\varepsilon^{123} = \varepsilon^{312} = \varepsilon^{231} = +1$$

and

$$\varepsilon^{213} = \varepsilon^{321} = \varepsilon^{132} = -1.$$

It can be shown that oddness or evenness does not depend on a particular sequence of interchanges.

Using the epsilon symbol and the summation convention, we can write the 3 × 3 determinant $|a_j{}^i|$ as

$$\begin{vmatrix} a_1{}^1 & a_2{}^1 & a_3{}^1 \\ a_1{}^2 & a_2{}^2 & a_3{}^2 \\ a_1{}^3 & a_2{}^3 & a_3{}^3 \end{vmatrix} = \varepsilon^{ijk} a_i{}^1 a_j{}^2 a_k{}^3,$$

since

$$\varepsilon^{ijk} a_i{}^1 a_j{}^2 a_k{}^3 = a_1{}^1 a_2{}^2 a_3{}^3 + a_1{}^2 a_2{}^3 a_3{}^1 = a_1{}^3 a_2{}^1 a_3{}^2 - (a_1{}^3 a_2{}^2 a_3{}^1 + a_1{}^2 a_2{}^1 a_3{}^3 + a_1{}^1 a_2{}^3 a_3{}^2).$$

With the same understanding about ε_{ijk} as ε^{ijk}, we could write the same 3 × 3 determinant as

$$\varepsilon_{ijk}\, a_1{}^i a_2{}^j a_3{}^k.$$

Similarly, the $n \times n$ determinant could be written

$$\begin{aligned} |a_j{}^i| &= \varepsilon^{i_1 i_2 \cdots i_n} a_{i_1}{}^1 a_{i_2}{}^2 \cdots a_{i_n}{}^n, \\ &= \varepsilon_{i_1 i_2 \cdots i_n} a_1{}^{i_1} a_2{}^{i_2} \cdots a_n{}^{i_n}. \end{aligned}$$

The following examples further illustrate how these notations are used.

Example 1–4.1. Consider the general 3 × 3 determinant $|a_j{}^i|$. Then

$$\begin{vmatrix} a_1{}^1 & a_2{}^1 & a_3{}^1 \\ a_1{}^2 & a_2{}^2 & a_3{}^2 \\ a_1{}^3 & a_2{}^3 & a_3{}^3 \end{vmatrix} = \varepsilon_{ijk}\, a_1{}^i a_2{}^j a_3{}^k = \varepsilon^{ijk} a_i{}^1 a_j{}^2 a_k{}^3.$$

If we interchange rows or columns, we get instead

$$\begin{vmatrix} a_1{}^1 & a_3{}^1 & a_2{}^1 \\ a_1{}^2 & a_3{}^2 & a_2{}^2 \\ a_1{}^3 & a_3{}^3 & a_2{}^3 \end{vmatrix} = \varepsilon_{ijk}\, a_1{}^i a_3{}^j a_2{}^k = \varepsilon_{ikj}\, a_1{}^i a_2{}^k a_3{}^j$$

or

$$\begin{vmatrix} a_1{}^3 & a_2{}^3 & a_3{}^3 \\ a_1{}^2 & a_2{}^2 & a_3{}^2 \\ a_1{}^1 & a_2{}^1 & a_3{}^1 \end{vmatrix} = \varepsilon^{ijk} a_i{}^3 a_j{}^2 a_k{}^1 = \varepsilon^{kji} a_k{}^1 a_j{}^2 a_i{}^3.$$

Since $\varepsilon_{ijk} = -\varepsilon_{ikj}$ and $\varepsilon^{ijk} = -\varepsilon^{kji}$, we see that interchanging of columns or rows interchanges the sign of the determinant.

Example 1–4.2. Consider the general 4×4 determinant $|a_k{}^i|$. Then

$$\begin{vmatrix} a_1{}^1 & a_2{}^1 & a_3{}^1 & a_4{}^1 \\ a_1{}^2 & a_2{}^2 & a_3{}^2 & a_4{}^2 \\ a_1{}^3 & a_2{}^3 & a_3{}^3 & a_4{}^3 \\ a_1{}^4 & a_2{}^4 & a_3{}^4 & a_4{}^4 \end{vmatrix} = \varepsilon^{ijkl} a_i{}^1 a_j{}^2 a_k{}^3 a_l{}^4.$$

Let $A_i{}^j$ be the cofactor of $a_j{}^i$, then recall the following identity which holds for all i and j:

$$a_\alpha{}^i A_j{}^\alpha = |a_q{}^p|\, \delta_j{}^i.$$

The identity is written here with the summation convention. If we let $i = j = 4$, we get

$$a_1{}^4 A_4{}^1 + a_2{}^4 A_4{}^2 + a_3{}^4 A_4{}^3 + a_4{}^4 A_4{}^4 = |a_q{}^p|. \tag{1–4.1}$$

From above, we get

$$\begin{aligned} |a_q{}^p| &= \varepsilon^{ijkl} a_i{}^1 a_j{}^2 a_k{}^3 a_l{}^4 \\ &= a_l{}^4 (\varepsilon^{ijkl} a_i{}^1 a_j{}^2 a_k{}^3), \\ &= a_1{}^4 (\varepsilon^{ijk1} a_i{}^1 a_j{}^2 a_k{}^3) + a_2{}^4 (\varepsilon^{ijk2} a_i{}^1 a_j{}^2 a_k{}^3) \\ &\quad + a_3{}^4 (\varepsilon^{ijk3} a_i{}^1 a_j{}^2 a_k{}^3) + a_4{}^4 (\varepsilon^{ijk4} a_i{}^1 a_j{}^2 a_k{}^3), \\ &= a_1{}^4 (-\varepsilon_{ijk}\, a_2{}^i a_3{}^j a_4{}^k) + a_2{}^4 (\varepsilon_{ijk}\, a_1{}^i a_3{}^j a_4{}^k) \\ &\quad + a_3{}^4 (-\varepsilon_{ijk}\, a_1{}^i a_2{}^j a_4{}^k) + a_4{}^4 (\varepsilon_{ijk}\, a_1{}^i a_2{}^j a_3{}^k). \end{aligned}$$

The quantities in the parentheses are in order, respectively, the cofactors $A_4{}^1$, $A_4{}^2$, $A_4{}^3$, and $A_4{}^4$, and hence we have re-established the identity of (1–4.1).

Problems

1. If the matrix $A = (a^{ij})$ is skew-symmetric—that is, $a^{ij} = -a^{ji}$, then show that the quadratic form $a^{ij} x_i x_j$ is identically zero.
2. Use a summation symbol to explain the meaning of the following:

   (a) $dy_i = \dfrac{\partial y_i}{\partial x_j} dx_j$, (b) $\dfrac{\partial z_i}{\partial x_j} = \dfrac{\partial z_i}{\partial y_k} \dfrac{\partial y_k}{\partial x_j}$.
3. Show that if

   $$z_i = b_i{}^1 y_1 + \cdots + b_i{}^n y_n \quad \text{and} \quad y_j = a_j{}^1 x_1 + \cdots + b_j{}^n x_n,$$

   then $z_i = b_i{}^k a_k{}^j x_j$.

4. Show the following:

   (a) $\delta_i{}^i = n$, (b) $\varepsilon^{ijk} = -\varepsilon^{kji}$,

   (c) $|\delta_j{}^i| = 1$, (d) $\varepsilon^{ijk} = \varepsilon^{ijk4}$,

   (e) $\delta^{ij}\delta_{ij} = n$, (f) $\varepsilon^{1ijk}a_{1i}\, a_{2j}\, a_{3k} = \varepsilon^{ijk}a_{i2}\, a_{j3}\, a_{k4}$.

5. Show

$$\begin{vmatrix} a_{11} & a_{12} \\ a_{21} & a_{22} \end{vmatrix} = \varepsilon^{ij}a_{1i}\, a_{2j}.$$

6. Use indicial notation to establish the equality $a_3{}^\alpha A_\alpha{}^3 = |a_j{}^i|$ for the determinant considered in Example 1–4.2.

7. Prove the following identities with determinants using the ε symbol. For simplicity let $n = 3$.

   (a) If two rows (columns) are identical, the determinant is zero.

   (b) If a row (or a column) is multiplied by a constant, then the determinant is multiplied by the same constant.

   (c) The scalar multiple of any row (or column) can be added to any other row (or column) without changing the value of the determinant.

8. Let $(a_j{}^i(x))$ be a 3×3 determinant of differentiable functions $a_j{}^i$ on some open interval of the real line. Use indicial notation to find

$$\frac{d}{dx}\,|a_j{}^i(x)|.$$

9. Let A and B be 3×3 matrices. Use indicial notation to prove $|AB| = |A|\,|B|$.

Answers

2. (a) $dy_i = \sum_{j=1}^n \frac{\partial y_i}{\partial x_j}\, dx_j$, (b) $\frac{\partial z_i}{\partial x_j} = \sum_{k=1}^n \frac{\partial z_i}{\partial y_k}\,\frac{\partial y_k}{\partial x_j}$.

1–5. Functions of a Single Variable

Let f be a function whose domain is the closed interval $[a, b] = \{x\colon a \le x \le b\}$ where a and b are real numbers, $a < b$, and whose range is in $R = E^1$. That is, we write $f\colon [a, b] \to R$, and we say f maps $[a, b]$ into R. Then f is *continuous at* $x_0 \in [a, b]$ if, given $\varepsilon > 0$, there exists $\delta > 0$ such that $|f(x) - f(x_0)| < \varepsilon$ for $|x - x_0| < \delta$ and $x \in [a, b]$. The function f is *continuous on* $[a, b]$ if f is continuous at each point of $[a, b]$, and we write $f \in C([a, b])$.

If x_0 belongs to the open interval $(a, b) = \{x: a < x < b\}$ and if

$$\lim_{x\to x_0} \frac{f(x) - f(x_0)}{x - x_0}$$

exists (meaning exists and is finite), then f is said to be differentiable at x_0. The value of this limit at x_0 is denoted by $f'(x_0)$ and is called the *derivative* of f at x_0. The function obtained from f in this manner, $\{(x, f'(x)): a < x < b\}$, is the *derivative of* f, written f'. The derivative of f', or the *second derivative* of f, is denoted f''. Similarly, the third derivative is f''' and the nth derivative is $f^{(n)}$.

If f has a derivative at $x_0 \in (a, b)$, it is possible to show that f *must be* continuous at x_0. Thus if $f'(x)$ exists for all $x \in (a, b)$, then f is necessarily continuous on (a, b), but there is no assurance about what happens at the end points, as for example in the function f defined as follows on $[-1, 1]$:

$$f(x) = \begin{cases} 1 + (1 - x^2)^{1/2}, & x \in (-1, 1), \\ 0, & \text{elsewhere.} \end{cases}$$

If we restrict our attention to an open interval (a, b) and $f^{(n)}$, $n > 1$, is continuous on (a, b), so that we write $f \in C^n((a, b))$, then $f^{(n-1)}$ exists and is continuous on (a, b), and we write $f \in C^{n-1}((a, b))$. Continuing, we see that $f \in C^n((a, b))$ implies that $f \in C^k((a, b))$ for $k = 0, 1, 2, \cdots, n$ where $C^0((a, b)) = C((a, b))$.

A function f defined on (a, b) such that f is continuous on (a, b) is *extendable to a continuous function on* $[a, b]$, if it is possible to define f at a and at b such that f is continuous on $[a, b]$. The function f defined by $f(x) = (1 - x^2)^{1/2}$ on $(-1, 1)$ is extendable to a continuous function on $[-1, 1]$, but the derivative of this function is not.

If $f \in C([a, b])$, $f' \in C((a, b))$, and f' is extendable to a continuous function on $[a, b]$, then f is *continuously differentiable on* $[a, b]$.

There is another type of continuity on an interval that is less demanding than continuity. Let $a_0, a_1, a_2, \cdots, a_n$ be a set of real numbers such that $a = a_0 < a_1 < a_2 < \cdots < a_n = b$; that is, $\{a_0, a_1, \cdots, a_n\}$ is a *partition* of the interval $[a, b]$. Suppose that f is a function whose domain, Do $(f) \subset [a, b]$ and range, Ra $(f) \subset R$ such that for some partition $\{a_0, a_1, \cdots, a_n\}$ of $[a, b]$ $f \in C((a_i, a_{i+1}))$ for $i = 0, 1, \cdots, n - 1$ and f is extendable into a continuous function on $[a_i, a_{i+1}]$ for $i = 0, 1, \cdots, n - 1$. In such a case, f is said to be *piecewise continuous on* $[a, b]$. Note that, by the definition, the domain of f must contain each open interval (a_i, a_{i+1}), $i = 0, 1, 2, \cdots, n - 1$, but need not contain the end points of these intervals.

Similarly, if f is a function whose domain, Do $(f) = [a, b]$, and whose range, Ra $(f) \subset R$ such that for some partition $\{a_0, a_1, \cdots, a_n\}$ of $[a, b]$, f is continuously differentiable on each $[a_i, a_{i+1}]$, $i = 0, 1, 2, \cdots, n-1$, then f is *piecewise differentiable* on $[a, b]$. For a piecewise differentiable function, f must be continuous on each subinterval $[a_i, a_{i+1}]$, and hence f is continuous on $[a, b]$. Notice also that if f is piecewise differentiable on $[a, b]$, then f' is piecewise continuous on $[a, b]$.

Example 1–5.1. If $[x]$ means the largest integer less than or equal to x, then the function f defined by $f(x) = x - [x]$ on $[0, 3]$ is piecewise continuous on $[0, 3]$, whereas the function g defined by $g(x) = 2\,|f(x) - \tfrac{1}{2}|$ on $[0, 3]$ is piecewise differentiable on $[0, 3]$. Figure 1–5.1(a) and (b) represent f and g respectively.

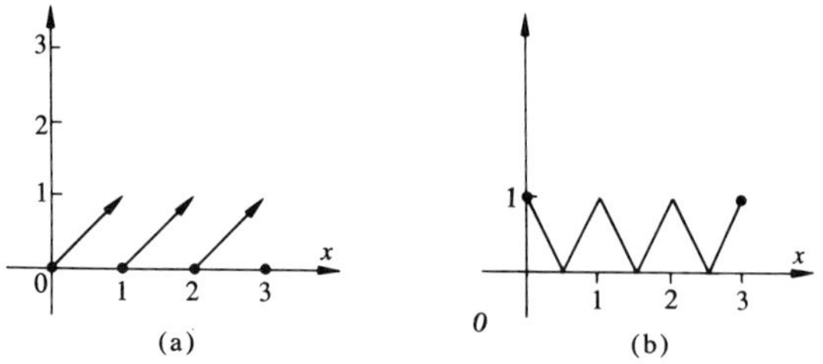

Figure 1–5.1

Now let us consider a vector-valued function—that is, a function $\mathbf{f} = (f_1, \cdots, f_n)$ such that $\mathbf{f}\colon [a, b] \to E^n$. The components of the function f_i are such that $f_i\colon [a, b] \to R$. The function $\mathbf{f}$ is *continuous on* $[a, b]$ if each f_i is continuous on $[a, b]$, $1 \le i \le n$, and we write $\mathbf{f} \in C([a, b])$. The function $\mathbf{f}$ has a derivative at $x_0 \in (a, b)$ if each f_i has a *derivative at* x_0. The value of this derivative at x_0 is denoted by

$$\mathbf{f}'(x_0) = (f_1'(x_0), \cdots, f_n'(x_0)).$$

The function obtained from $\mathbf{f}$ in this manner, $\{(x, \mathbf{f}'(x))\colon a < x < b\}$, is the derivative of $\mathbf{f}$, written $\mathbf{f}'$. Similarly, we get the higher derivatives of $\mathbf{f}$; that is, $\mathbf{f}''$, $\mathbf{f}'''$, etc. As in the simpler circumstance, we write $\mathbf{f} \in C^k((a, b))$, if each $f_i \in C^k((a, b))$, $1 \le i \le n$. Furthermore, $\mathbf{f}$ is *continuously differentiable*, *piecewise continuous*, or *piecewise differentiable* on $[a, b]$, if each f_i, $1 \le i \le n$, is continuously differentiable, piecewise continuous, or piecewise differentiable, respectively, on $[a, b]$.

The function whose components are the functions f and g of Example 1–5.1 is a piecewise continuous function. The function $\mathbf{f} = (f_1, f_2)$, illustrated

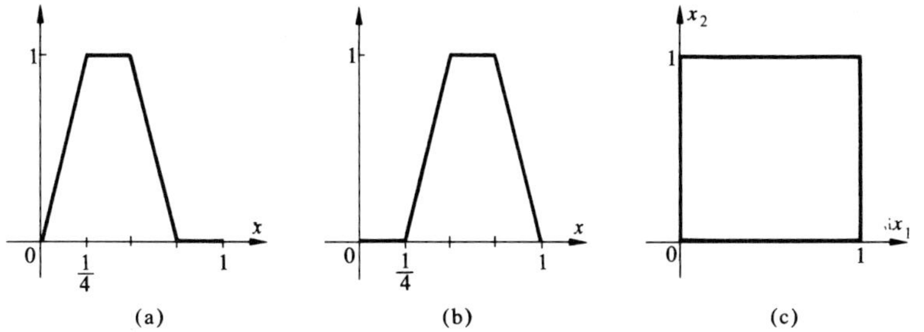

Figure 1–5.2

in Figure 1–5.2, is a piecewise differentiable function. In (a) we have the graph of f_1, in (*b*) the graph of f_2, and in (*c*) the image of f.

Let $\mathbf{f}$: $[a, b] \to E^n$. The space $R = E^1$ that contains $[a, b]$ is our *parameter space*, and the variable of that space (let us take t) is the *parameter* of the function $\mathbf{f}$. If the parameter t is time and $n = 3$, then

$$\mathbf{f}(t) = (f_1(t), f_2(t), f_3(t))$$

could be giving the coordinates of some "particle" in E^3 at time t. Thus the vector $\mathbf{f}(t)$ would be the *position vector* of a particle at time t. The vectors $\mathbf{f}'(t_0)$ and $\mathbf{f}''(t_0)$, $a < t_0 < b$, if they exist, are known, respectively, as the *velocity* and *acceleration vectors* of $\mathbf{f}$ at t_0. These vectors are customarily denoted $\mathbf{v}(t_0)$ and $\mathbf{a}(t_0)$. Figure 1–5.3 illustrates at $t = 1$ these two vectors for

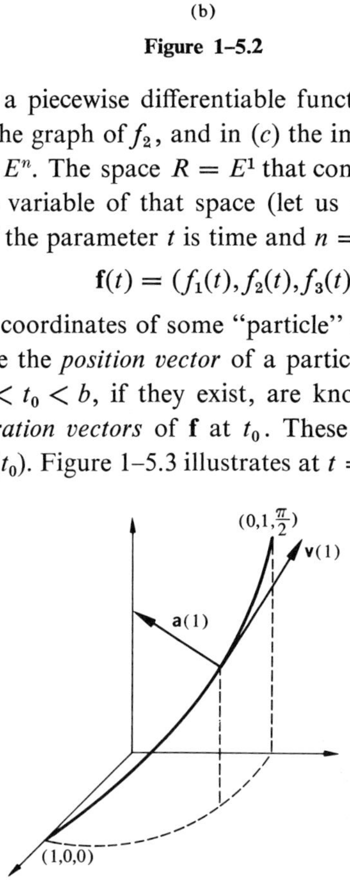

Figure 1–5.3

the function $\mathbf{f}(t) = (\cos t, \sin t, t)$ on $[0, \pi/2]$. In this case, $\mathbf{f}'(1) = \mathbf{v}(1) = (-\sin 1, \cos 1, 1)$ and $\mathbf{f}''(1) = \mathbf{a}(1) = (-\cos 1, -\sin 1, 0)$.

Regardless of whether t is time or not, if at t_0, $\mathbf{f}'(t_0)$ exists and is not zero, the vector

$$\mathbf{t}(t_0) = \frac{\mathbf{f}'(t_0)}{\|\mathbf{f}'(t_0)\|}$$

is the *unit tangent* of $\mathbf{f}$ at t_0. The reasonableness of choosing this name becomes apparent as we recall that $\mathbf{f}'(t_0) = (f_1'(t_0), \cdots, f_n'(t_0))$ and that

$$f_i'(t_0) = \lim_{t \to t_0} \frac{f_i(t) - f_i(t_0)}{t - t_0}.$$

Then for $t \neq t_0$, consider

$$\frac{\mathbf{f}(t) - \mathbf{f}(t_0)}{t - t_0}$$

in Figure 1–5.4. Let $\overrightarrow{OA} = \mathbf{f}(t_0)$ and $\overrightarrow{OB} = \mathbf{f}(t)$; then $\mathbf{f}(t) - \mathbf{f}(t_0) = \overrightarrow{AB}$. Dividing by $t - t_0$ gives a vector like $\overrightarrow{AC}$, that is, parallel to $\overrightarrow{AB}$. As $t \to t_0$, each $f_i(t) \to f_i(t_0)$, since each f_i is continuous at t_0 if $f_i'(t_0)$ exists. Thus, B

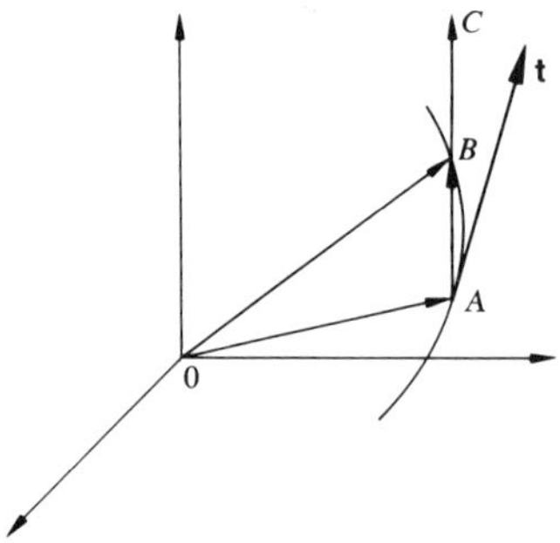

Figure 1–5.4

will move back toward A, and conceivably the limiting vector $\mathbf{f}'(t_0)$ would be a vector in the same direction as the vector $\mathbf{t}$ shown in the figure. Dividing by $\|\mathbf{f}'(t_0)\|$ only makes a vector of length one, in the same direction as $\mathbf{f}'(t_0)$.

The two products of vectors, the dot product and the vector product, can appear with functions $\mathbf{f}$ and $\mathbf{g}$ like those that we have been discussing. If $\mathbf{f}\colon [a, b] \to E^n$ and $\mathbf{g}\colon [a, b] \to E^n$, then

$$\mathbf{f}(t) \cdot \mathbf{g}(t) = \textstyle\sum_{i=1}^{n} f_i(t) g_i(t), \tag{1–5.1}$$

and for $n = 3$

$$\mathbf{f}(t) \times \mathbf{g}(t) = (f_2(t)g_3(t) - f_3(t)g_2(t), f_3(t)g_1(t) - f_1(t)g_3(t), \\ f_1(t)g_2(t) - f_2(t)g_1(t)). \tag{1–5.2}$$

Since $\mathbf{f}(t) \cdot \mathbf{g}(t) \in R$ and $\mathbf{f}(t) \times \mathbf{g}(t) \in E^3$, we have two new functions defined. Call them $\mathbf{f} \cdot \mathbf{g}$ and $\mathbf{f} \times \mathbf{g}$, respectively. Then $\mathbf{f} \cdot \mathbf{g}$ and $\mathbf{f} \times \mathbf{g}$ map $[a, b]$ into R and E^3, respectively. In both cases it is appropriate to consider the derivatives of these functions. According to the definitions already given, it is easy to see that if $\mathbf{f}'(t)$ and $\mathbf{g}'(t)$ exist for $a < t < b$, then

$$\frac{d}{dt}\mathbf{f}(t) \cdot \mathbf{g}(t) = \mathbf{f}(t) \cdot g'(t) + \mathbf{f}'(t) \cdot \mathbf{g}(t) \qquad (1\text{–}5.3)$$

and

$$\frac{d}{dt}(\mathbf{f}(t) \times \mathbf{g}(t)) = \mathbf{f}(t) \times \mathbf{g}'(t) + \mathbf{f}'(t) \times \mathbf{g}(t). \qquad (1\text{–}5.4)$$

Recall again that a vector product is not commutative, and so the order of writing a vector product is important.

The idea of a "curve" is very much related to the above discussion, but until now the words "curve" and "arc" have been carefully avoided. Intuitively we view a curve as a set of points. The approach here is to define "curve" so that our intuition and definition agree. The words "curve" and "arc" are sometimes given different meanings, but here we shall have them be one and the same.

A set of points Γ in E^n is a *curve* (or *arc*) if there exists a continuous function $\mathbf{f}$ on an interval $[a, b]$ such that Ra $(\mathbf{f}) = \Gamma$. A set of points Γ in E^n is a *smooth curve* (or *smooth arc*) if there exists a continuously differentiable function $\mathbf{f}$ on an interval $[a, b]$ such that Ra $(\mathbf{f}) = \Gamma$ and $\mathbf{f}'(t) \neq 0$ for $a < t < b$ and $\lim_{t\to a^+} \mathbf{f}'(t)$ and $\lim_{t\to b^-} \mathbf{f}'(t) \neq \mathbf{0}$. A set of points Γ in E^n is a *piecewise smooth curve* (or *piecewise smooth arc*) if there is a piecewise differentiable function $\mathbf{f}$ on an interval $[a, b]$ such that Ra $(\mathbf{f}) = \Gamma$ and such that $\lim_{t\to t_0^+} \mathbf{f}'(t)$ and $\lim_{t\to t_0^-} \mathbf{f}'(t) \neq \mathbf{0}$ for any $t_0 \in [a, b]$. Notice that this implies that $\mathbf{f}'(t) \neq \mathbf{0}$ for any $t \in [a, b]$. Thus any curve Γ must have associated with it a nonempty set of functions of a certain type. The curve is always the image of a certain closed interval of the parameter space by one of these functions of the required type. Each such function that satisfies the appropriate restrictions is a *parametric representation* of Γ. If any parametric representation $\mathbf{f}$ of Γ is such that $\mathbf{f}(a) = \mathbf{f}(b)$ where $[a, b] =$ Do $(\mathbf{f})$, then Γ is *closed.* If Γ is closed, there are functions $\mathbf{f}$ such that the image of $[a, b]$ by $\mathbf{f}$ is Γ, but $\mathbf{f}(a) \neq \mathbf{f}(b)$. We shall not accept those $\mathbf{f}$ as parametric representations of Γ.

If Γ is not a closed curve and there is a parametric representation, $\mathbf{f}$ of Γ such that $\mathbf{f}$ is 1–1 on $[a, b] =$ Do $(\mathbf{f})$, then Γ is a *simple curve.* We assume that if Γ is a simple curve, then $\mathbf{f}$ is not a parametric representation unless $\mathbf{f}$ is 1–1 on $[a, b]$. Lastly, if Γ is a closed curve and there is a parametric representation of Γ, $\mathbf{f}$ such that $\mathbf{f}$ is 1–1 on $[a, b)$ where $[a, b] =$ Do $(\mathbf{f})$, then Γ is a *simple closed curve.* Again, we shall assume that if Γ is a simple closed curve and $\mathbf{f}$ is not a parametric representation unless $\mathbf{f}$ is 1–1 on $[a, b)$ where $[a, b] =$ Do $(\mathbf{f})$.

If Γ is a curve with parametric representation $\mathbf{f}$ on $[a, b]$, then *with respect to* $\mathbf{f}$, $\mathbf{f}(a)$ and $\mathbf{f}(b)$ are the *initial* and *terminal* points of Γ, and Γ is said to *join* $\mathbf{f}(a)$ to $\mathbf{f}(b)$.

From the given definition of a curve very unusual things can happen. It is possible to define a continuous function $\mathbf{f}$ on $[0, 1]$ into E^2 such that Ra $(\mathbf{f})$ is a unit square. Such a curve is called a *space-filling curve*. An example of such was first given by Peano in 1890 (cf. Problem 10).

The following example illustrates the importance of the derivative not being zero for a smooth curve.

Example 1–5.2. Let $\mathbf{f}(t) = (t^3, t^2)$ for $-1 \leq t \leq 1$. Then at $t = 0$, $\mathbf{f}'(0) = \mathbf{0}$. In Figure 1–5.5 the curve Γ that has $\mathbf{f}$ as a parametric representation is shown. At (0, 0) the curve violates what our intuition would have us think of as a "smooth" curve.

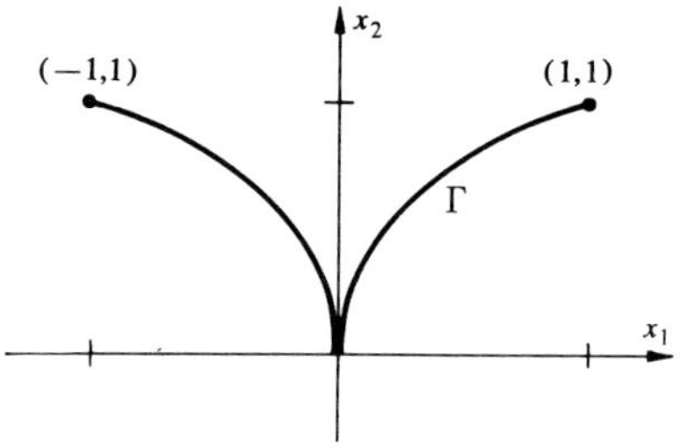

Figure 1–5.5

Figure 1–5.2(c) illustrates a piecewise smooth curve. In fact, the curve of Example 1–5.2 is piecewise smooth, for we could take

$$\mathbf{f}(t) = \begin{cases} (-(-t)^{3/2}, t), & -1 \leq t < 0, \\ (t^{3/2}, t), & 0 \leq t \leq 1. \end{cases}$$

If in the definition of a piecewise smooth curve we had allowed $\{t: \mathbf{f}'(t) = \mathbf{0}\}$ to be an infinite set, then several things could happen. The following two examples demonstrate this.

Example 1–5.3. Let $\mathbf{f}$ be given by $\mathbf{f}(t) = (f_1(t), f_2(t))$ where

$$f_i(t) = \begin{cases} 16t^2\left(\left(\frac{3}{4}\right) - t\right), & 0 \leq t \leq \frac{1}{2}, \\ 1, \quad \frac{1}{2} < t \leq 1, \end{cases}$$

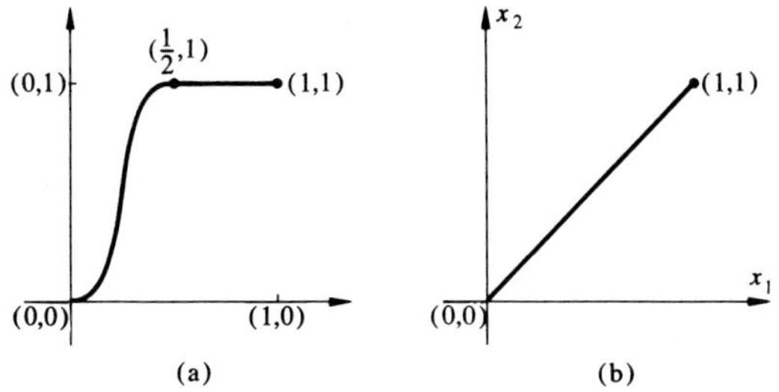

Figure 1–5.6

for $i = 1, 2$. For $t \in [\frac{1}{2}, 1]$, $\mathbf{f}(t) = (1, 1)$; that is, $\mathbf{f}(t)$ remains fixed on this half of the interval. In Figure 1–5.6(a) f_i is pictured, and in Figure 1–5.6(b) the image of $\mathbf{f}$ is pictured.

Example 1–5.4. Let $\mathbf{f}$ be given by $\mathbf{f}(t) = (f_1(t), f_2(t))$ where

$$f_1(t) = t^3, \qquad f_2(t) = t^2, \qquad 0 \le t \le \sqrt[3]{\frac{1}{4}},$$

$$f_1(t) = \left(t - \frac{1}{2}\right)^3 + \frac{1}{2}, \quad f_2(t) = \left(t - \frac{1}{2}\right)^2, \quad \frac{1}{2} - \sqrt[3]{\frac{1}{8}} \le t \le \frac{1}{2} + \sqrt[3]{\frac{1}{8}},$$

$$f_1(t) = \left(t - \frac{3}{4}\right)^3 + \frac{3}{4}, \quad f_2(t) = \left(t - \frac{3}{4}\right)^2, \quad \frac{3}{4} - \sqrt[3]{\frac{1}{16}} \le t \le \frac{3}{4} + \sqrt[3]{\frac{1}{16}},$$

and in general let

$$f_1(t) = \left(t - \frac{2^n - 1}{2^n}\right)^3 + \frac{2^n - 1}{2^n},$$

$$f_2(t) = \left(t - \frac{2^n - 1}{2^n}\right)^2,$$

for

$$\frac{2^n - 1}{2^n} - \sqrt[3]{\frac{1}{2^{n+2}}} \le t \le \frac{2^n - 1}{2^n} + \sqrt[3]{\frac{1}{2^{n+2}}}.$$

For the interval $(\sqrt[3]{(1/4)}, \frac{1}{2} - \sqrt[3]{(1/8)})$ let

$$f_1(t) = a_1 + b_1 t + c_1 t^2 + d_1 t^3$$

$$f_2(t) = a_2 + b_2 t + c_2 t^2 + d_2 t^3$$

where a_i, b_i, c_i, and d_i are constants chosen such that f_i is continuous and has a derivative at the two points $t = \sqrt[3]{(1/4)}$ and $t = (1/2) - \sqrt[3]{(1/8)}$, $i = 1, 2$. Similarly, define f_1 and f_2 on the intervals $((1/2) + \sqrt[3]{(1/8)},\ (3/4) - \sqrt[3]{(1/16)})$, $((3/4) + \sqrt[3]{(1/16)},\ (7/8) - \sqrt[3]{(1/32)})$, etc. Let $f_1(1) = f_2(1) = 0$. Then **f** is continuously differentiable on $[0, 1]$, but $\mathbf{f}'(t) = \mathbf{0}$ at $t = 1/2, 3/4, 7/8$, etc. Then, Γ, the image of **f**, is not smooth. Since any partition of $[0, 1]$ is a finite set, it is not possible to absorb these points where $\mathbf{f}(t) = \mathbf{0}$ as points of a partition, as is possible in the case of Example 1–5.2, in order to render Γ as a piecewise smooth curve. Figure 1–5.7 shows the image of **f**.

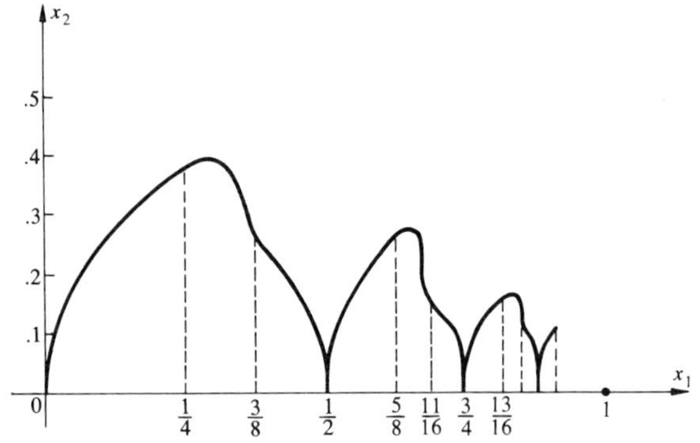

Figure 1–5.7

The following example shows several parametric representations of a particular simple curve.

Example 1–5.5. The simple curve shown in Figure 1–5.8(a) may have any one of the following three parametric representations **f**, **g**, or **h** given by

$$\mathbf{f}(t) = (\cos t, \sin t), \quad \left[0, \frac{\pi}{2}\right];$$

$$\mathbf{g}(t) = (\cos 2t, \sin 2t), \quad \left[0, \frac{\pi}{4}\right];$$

$$\mathbf{h}(t) = (\sin t, \cos t), \quad \left[0, \frac{\pi}{2}\right].$$

The parametric representations **f** and **g** have the same initial point and terminal point—namely, (1, 0) and (0, 1) respectively. There are clearly more parametric representations possible.

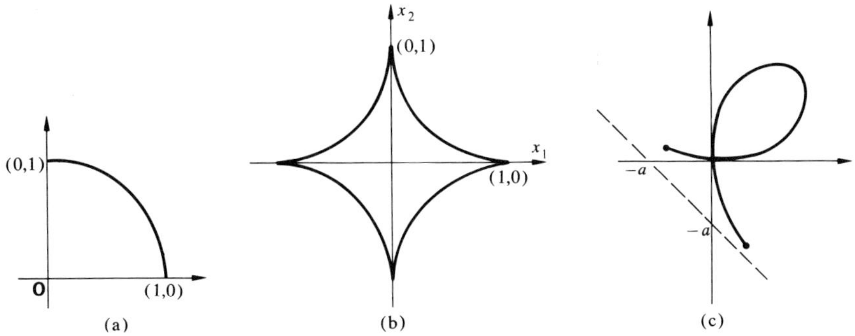

Figure 1–5.8

Below is an example of a simple closed curve.

Example 1–5.6. Let $\mathbf{f}(t) = (\cos^3 t, \sin^3 t)$ for $t \in [0, 2\pi]$. The curve for which this $\mathbf{f}$ is a parametric representation is shown in Figure 1–5.8(b). This is not a smooth curve for there is trouble at the image of the points where $t = 0, \pi/2, \pi, 3\pi/2$, and 2π, but the curve is piecewise smooth.

The following example taken from the Folium of Descartes is an example of a curve that is not simple.

Example 1–5.7. Let the function $\mathbf{f}$ be given as follows:

$$\mathbf{f}(t) = \left(\frac{a \sin t \cos^2 t}{\cos^3 t + \sin^3 t}, \frac{a \sin^2 t \cos t}{\cos^3 t + \sin^3 t}\right)$$

for $t \in [-\pi/6, 2\pi/3]$. This function $\mathbf{f}$ is not 1–1 on $[-\pi/6, 2\pi/3]$ since $\mathbf{f}(0) = \mathbf{f}(\pi/2) = \mathbf{0}$. The curve for which $\mathbf{f}$ is a parametric representation is shown in Figure 1–5.8(c).

It is sometimes desirable to give a parametric representation of a smooth curve such that the parameter of the curve is arc length, written s. The meaning is that $\mathbf{f}$ is such that $\Gamma = \{\mathbf{f}(s): a \le s \le b\}$ and $\mathbf{f}$ is such that for $a \le s_1 < s_2 \le b$ the distance along Γ from $\mathbf{f}(s_1)$ to $\mathbf{f}(s_2)$,

$$\int_{s_1}^{s_2} \|\mathbf{f}'(s)\| \, ds,$$

is the distance from s_1 to s_2, $s_2 - s_1$. For example, let

$$\Gamma = \{\mathbf{g}(t): g_1(t) = t^3, \qquad g_2(t) = t^2, 0 \le t \le 1\}$$

(cf. Example 1–5.2). Then, to find a parametric representation such that the parameter is arc length, let

$$\begin{aligned} s &= \int_0^t \sqrt{(g_1'(\tau))^2 + (g_2(\tau))^2}\, d\tau \\ &= \int_0^t \sqrt{9\tau^4 + 4\tau^2}\, d\tau \\ &= 3\int_0^t \tau\sqrt{\tau^2 + (2/3)^2}\, d\tau \\ &= \frac{1}{3}\left(t^2 + \frac{4}{9}\right)^{(3/2)} - \frac{8}{81}. \end{aligned} \tag{1–5.5}$$

Then

$$t = H(s) = \sqrt{(3s + (81/27))^{2/3} - (4/9)}.$$

Substituting for t gives

$$\Gamma = \{\mathbf{g}(H(s)) : 0 \le s \le s_2\}$$

where $s_2 = (13\sqrt{13} - 8)/81$.

Notice from (1–5.5) that

$$\frac{ds}{dt} = \|\mathbf{g}'(t)\|. \tag{1–5.6}$$

Hence, if $t = s$, it follows that $\mathbf{g}'(s)$ is always a unit vector.

Problems

1. Let $f(x) = |x - \frac{1}{2}|$ for $0 \le x \le 1$. Show $f \in C([0, 1)]$ but $f \notin C^1((0, 1))$. Let $f_1(x) = \int_0^x f(t)\, dt$. Find f_1 by performing the indicated integration. Show $f_1 \in C^1((0, 1))$, but $f_1 \notin C^2((0, 1))$. Find a function f_2 that belongs to $C^2((0, 1))$, but not to $C^3((0, 1))$.
2. Notice that for a given interval $[a, b]$, {cont. diff.} $\subset$ {piece. diff.} $\subset$ {cont.} $\subset$ {piece. cont.}; that is, the class of continuously differentiable functions is contained in the class of piecewise differentiable functions, etc. For each of the following functions on the indicated interval, find the most restrictive of the above classes to which it belongs. Explain your answer and sketch each function.

   (a) $f(x) = x^2$, $[0, 1]$. (b) $f(x) = x^{1/3}$, $[0, 1]$.
   (c) $f(x) = [x]$, $[0, 2]$. (d) $f(x) = x(1 - x)^{1/2}$, $[0, 1]$.

   (e) $f(x) = \left\{\begin{array}{rl} 2x + 1, & -1 \le x < 0, \\ 1, & 0 \le x < 1, \\ 2 - x, & 1 \le x \le 2, \end{array}\right\}$, $[-1, 2]$.

   (f) $f(x) = \left\{\begin{array}{rl} -1 + x^2, & -1 \le x < 1, \\ x - 1, & 1 \le x \le 2, \end{array}\right\}$, $[-1, 2]$.

3. Find a function $\mathbf{f} = (f_1, f_2)$ on [0, 1] similar to the function illustrated in Figure 1–5.2 such that the image of [0, 1] by $\mathbf{f}$ is:
   (a) the triangle whose vertices are (0 0), (1, 0), and (0, 1),
   (b) the two line segments joining $(-1, 1)$ and $(1, 1)$ to $(0, 0)$ and the shortest arc of a circle of radius $\sqrt{2}$ and center at (0, 0) joining (1, 1) to $(-1, 1)$.
   (c) Find a parametric representation of (a) such that the parameter is arc length.

4. Find the velocity and acceleration and the magnitudes of each of these vectors at the time t indicated for the several cases, of the position vector $\mathbf{f}(t)$ defined below. Also find the unit tangent at the same point. Plot the curves and each of the vectors. For (a) and for $t \in [0, 1]$, find a parametric representation such that the parameter is arc length.
   (a) $\mathbf{f}(t) = (t, t^2, t^3)$, $t = 1$.
   (b) $\mathbf{f}(t) = (2 \cos t, 3 \sin t, t)$, $t = \pi/2$.
   (c) $\mathbf{f}(t) = (v_0(\cos \theta)t, v_0(\sin \theta)t - 16t^2)$, at time t where

   $f_2(t) = v_0(\sin \theta)\, t - 16t^2$ is a maximum and where $v_0 > 0$, $0 < \theta < \pi/2$.
   (d) $\mathbf{f}(t) = (2 \sec t, \tan t)$, $t = \pi/4$.

5. Find

$$\frac{d}{dt}\mathbf{f}(t) \cdot \mathbf{g}(t) \quad \text{and} \quad \frac{d}{dt}\mathbf{f}(t) \times \mathbf{g}(t)$$

   at $t = 1$ by equations (1–5.3) and (1–5.4) for the cases $\mathbf{f}$ and $\mathbf{g}$ given below:

   (a) $\mathbf{f}(t) = (t^2, 2t - 1, t - t^3)$, $\quad \mathbf{g}(t) = \left(\frac{1}{t}, 3t, t + 5\right)$;

   (b) $\mathbf{f}(t) = \left(\sin\left(\frac{\pi}{2}\right)t, \cos\left(\frac{\pi}{2}\right)t, t\right)$,

$$\mathbf{g}(t) = \left(\cot\left(\frac{\pi}{4}\right)t, 2\csc\left(\frac{\pi}{4}\right)t, t^2\right).$$

   Check your answer by first taking the dot or cross products and then differentiating.

6. Find three parametric representations of each of the following curves:
   (a) $\Gamma = \{(x_1, x_2): x_2 = x_1^2, |x_1| < 1\}$,
   (b) $\Gamma = \{(x_1, x_2): 4x_1^2 + 9x_2^2 = 36\}$,
   (c) $\Gamma = \{(x_1, x_2, x_3): x_1^2 + x_2^2 = 4, x_1 = x_3\}$,
   (d) $\Gamma = \{(x_1, x_2): (x_1^2 + x_2^2)^2 = 4a^2(x_1^2 - x_2^2)\}$, $a > 0$,
   (e) $\Gamma = \{(x_1, x_2): |x_1| + |x_2| = 1\}$,
   (f) $\Gamma = \{(x_1, x_2): x_1^4 + x_2^4 = 1\}$.

7. For each of the parametric representations given in Problem 6 give the initial and terminal points. Classify each of the curves of Problem 6 according to whether it is closed or not closed, whether it is simple or

not simple, and whether it is a smooth curve, a piecewise smooth curve, or simply a curve.

8. In Example 1–5.4 there is an assertion of essentially this sort: If $t_1 < t_2$ with $f(t) = a + bt + ct^2 + dt^3$, it is possible to choose a, b, c, and d such that $f(t_1)$, $f'(t_1)$, $f(t_2)$, and $f'(t_2)$ take on four preassigned real finite values. Demonstrate that this is indeed true.

9. If $\Gamma = \{\mathbf{f}(t): a \leq t \leq b\}$ is a smooth curve in E^n, then let $\{\mathbf{f}(t_0) + \lambda\mathbf{f}'(t_0): \lambda, \text{real}\}$ be the *tangent line* to Γ at $\mathbf{f}(t_0)$, $a < t_0 < b$. Extend this definition to the case of $t_0 = a$ or b. At $t_0 = a$ (b) we could restrict $\lambda \geq 0$ $(\lambda \leq 0)$ and have a "half-line," a *tangent ray*. Extend this idea to the case of a piecewise smooth curve. If $\Gamma = \{\mathbf{f}(t): -1 \leq t \leq 1\}$ is a curve in E^2 where $\mathbf{f}$ is continuously differentiable on $[-1, 1]$ and $\mathbf{f}(0) = \mathbf{0}$, but $\mathbf{f}'(t) \neq \mathbf{0}$ for $t \neq 0$, then show that it is possible that

$$\lim_{t\to t_0^+} \mathbf{f}'(t)/\|f'(t)\|$$

does not exist and thus the "limiting tangent line to Γ at $\mathbf{f}(0)$" need not exist.

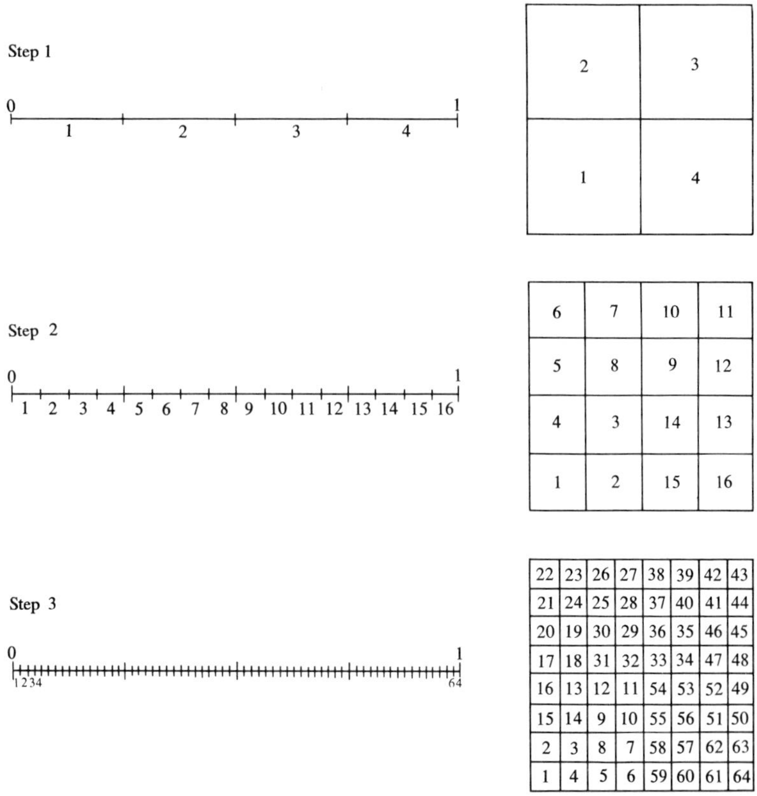

Figure 1–5.9

10. The following is an example of a "space-filling curve." Figure 1–5.9 illustrates the first three steps in an infinite sequence of steps. On Step 1, identify each of the closed intervals numbered 1, 2, 3, and 4 with the subsquares of the same number—that is, in the square beside the interval in Figure 1–5.9. Similarly, in Step 2 and Step 3, identify corresponding subintervals and subsquares. Notice that on Step 2 the four subintervals of the original interval numbered 1 in Step 1—that is, intervals 1, 2, 3, and 4 of Step 2—are identified with four subsquares of the original square number 1 in Step 1; similarly for the subintervals numbered 5, 6, 7, and 8 of the original square numbered 2. This same pattern continues as we proceed from Step 2 to Step 3. Also note that if two intervals abut, then the squares with which they are identified also abut. That is, the squares are numbered in such a way as always to move smoothly from one square to an adjacent square. Show Step 4 in this sequence.

    Now the object is to define a continuous mapping **f** of the interval [0, 1] into a given square that might very well be the square in E^2 having (0, 0) and (1, 1) as opposite vertices. Each point $x \in [0, 1]$ is in a closed interval in Step 1, Step 2, Step 3, etc. It may be in two closed intervals as for example $x = 1/4$ in Step 1. In such a case, take the one to the left for definiteness. In this way each $x \in [0, 1]$ determines a unique sequence of closed intervals such that x is the only point in all of these closed intervals. Let $\mathbf{f}(x)$ be the point of the square that is the only point in the corresponding sequence of closed squares. Show that f defined in this way is continuous but not simple.

Answers

1. $f_1(x) = \begin{cases} \frac{1}{2}(x - x^2), & x < \frac{1}{2}, \\ \frac{1}{2}(x^2 - x) + (\frac{1}{4}), & x \geq \frac{1}{2}, \end{cases} \qquad f_2(x) = \int_0^x f_1(t)\, dt.$

2. (a) Continuously differentiable. (b) Continuous.
   (c) Piecewise continuous. (d) Continuous.
   (e) Piecewise differentiable. (f) Piecewise differentiable.

4. (a) $\mathbf{f}(1) = (1, 2, 3)$, $\mathbf{f}''(1) = (0, 2, 6)$, $\|\mathbf{f}'(1)\| = \sqrt{14}$,
   $\|\mathbf{f}''(1)\| = 2\sqrt{10}$, $\mathbf{t}(1) = (1, 2, 3)\sqrt{14}$.
   (b) $(-2, 0, 1)$, $(0, -3, 0)$, $\sqrt{5}$, 3, $(1/\sqrt{5})(-2, 3, 1)$.
   (c) $(v_0 \cos\theta, 0)$, $(0, -32)$, $v_0 \cos\theta$, 32, (1, 0).
   (Note that $f_2(t)$ has a maximum value at $t = (\frac{1}{32})\, v_0 \sin\theta$.)
   (d) $(2\sqrt{2}, 2)$, $(6\sqrt{2}, 4)$, $2\sqrt{3}$, $2\sqrt{22}$, $(\frac{1}{3})(\sqrt{6}, \sqrt{3})$.

5. (a) $\frac{d}{dt}\mathbf{f}(1) \cdot \mathbf{g}(1) = -2, \qquad \frac{d}{dt}\mathbf{f}(1) \times \mathbf{g}(1) = (19, -15, 8);$

   (b) $3 - \left(\sqrt{2} + \left(\frac{1}{8}\right)\right)\pi, \left(\pi\frac{(1 - \sqrt{2})}{2\sqrt{2}}, -1 - \left(\frac{\pi}{8}\right), \pi\frac{(\sqrt{2} - 1)}{2\sqrt{2}}\right).$

7. (a) Not closed, simple, smooth curve.
   (b) Closed, simple, smooth curve.
   (c) Closed, simple, smooth curve.
   (d) Closed, not simple, smooth curve.
   (e) Closed, simple, piecewise smooth.
   (f) Closed, simple, smooth.

2

Topology on Euclidean n-Space

2–1. Neighborhoods, Intervals, and Bounded Sets

The idea of a neighborhood of a point in E^n has a meaning similar to the meaning of a neighborhood in general—that is, the points of E^n that are "near" to a given point. The following definition gives the precise meaning.

Definition 2–1.1. *Let* $\mathbf{a}$ *be a point of* E^n *and let* $r > 0$. *The open neighborhood (simply neighborhood) of* $\mathbf{a}$ *of radius* r, *written* $N(\mathbf{a}; r)$ *is the set of* $\mathbf{x}$ *such that* $\|\mathbf{x} - \mathbf{a}\| < r$; *that is,* $N(\mathbf{a}; r) = \{\mathbf{x}: \|\mathbf{x} - \mathbf{a}\| < r\}$. *Similarly,* $\bar{N}(\mathbf{a}; r) = \{\mathbf{x}: \|\mathbf{x} - \mathbf{a}\| \leq r\}$ *and* $N'(\mathbf{a}; r) = \{\mathbf{x}: 0 < \|\mathbf{x} - \mathbf{a}\| < r\}$ *are closed and deleted neighborhoods of radius* r *respectively.*

Thus for $n = 3$, a neighborhood of $\mathbf{a}$ is the set of points *within* a spherical surface of radius r centered at $\mathbf{a}$. In the case of a closed neighborhood, the spherical surface is included in the set. For the deleted neighborhood, the set is the same as the open neighborhood, but the center point $\mathbf{a}$ has been removed. The concepts in E^3 are very excellent guides to the intuition for $n > 3$.

For $n = 2$, a neighborhood of $\mathbf{a}$ is the set of points within a circle centered at $\mathbf{a}$; that is, the points form a "disk with the points on the edge removed." The closed neighborhood includes the points on the edge, while in a deleted neighborhood the point at the center is removed. Figure 2–1.1 illustrates these sets; (a) is a neighborhood of $\mathbf{a}$, (b) is a closed neighborhood, and (c) is a deleted neighborhood. For $n = 1$, $N(\mathbf{a}; r)$ is a line segment of length $2r$ centered at $\mathbf{a}$ that has its end points removed. Again, for the closed neighborhood the end points are included, and for the deleted neighborhood the midpoint is removed.

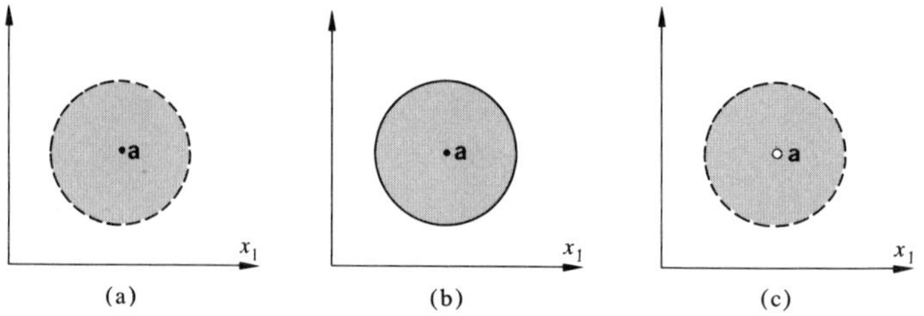

Figure 2–1.1

Very much related to the idea of a neighborhood in E^n is that of an interval. By way of review, given two real numbers a and b, $a < b$, then $(a, b) = \{x: a < x < b\}$ is an *open interval*, $[a, b] = \{x: a \leq x \leq b\}$ is a *closed interval*, and $(a, b] = \{x: a < x < b\}$ and $[a, b) = \{x: a \leq x < b\}$ are *half-open intervals*. Each of these is a *finite interval* in E^1 with end points a and b. The possible *semi-infinite intervals* are of the following types: the open intervals $(a, \infty) = \{x: x > a\}$ and $(-\infty, a) = \{x: x < a\}$ and the closed intervals $[a, \infty) = \{x: x \geq a\}$ and $(-\infty, x] = \{x: x \leq a\}$. There is one *infinite interval*, namely $(-\infty, \infty)$. The degenerate closed interval $[a, a]$ consists of the single point a.

For $n > 1$, a great variety of intervals could be defined, but only two types will be defined here.

Definition 2–1.2. *Let* $\mathbf{a} = (a_1, a_2, \cdots, a_n)$ *and* $\mathbf{b} = (b_1, b_2, \cdots, b_n)$ *be two points of* E^n *such that* $a_i < b_i$ *for* $i = 1, 2, \cdots, n$. *Then*

$$(\mathbf{a}, \mathbf{b}) = \{\mathbf{x}: a_i < x_i < b_i, 1 \leq i \leq n\}$$

and

$$[\mathbf{a}, \mathbf{b}] = \{x: a_i \leq x_i \leq b_i, 1 \leq i \leq n\}$$

are the open and closed intervals, respectively, in E^n *determined by* $\mathbf{a}$ *and* $\mathbf{b}$.

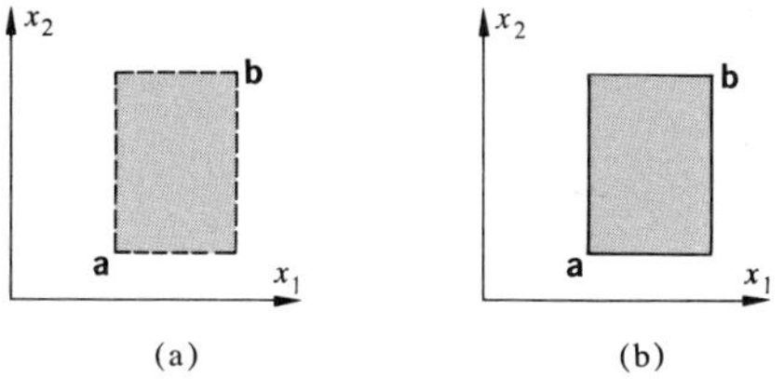

Figure 2–1.2

For $n = 2$, Figure 2–1.2(a) illustrates $(\mathbf{a}, \mathbf{b})$, and Figure 2–1.2(b) illustrates $[\mathbf{a}, \mathbf{b}]$.

In E^1 a set S is said to be *bounded above* if there exists a real number b such that $s \leq b$ for all $s \in S$. The number b is an *upper bound* for S. Similarly, S is *bounded below* if there exists a real number a such that $a \leq s$ for all $s \in S$. The number a is a *lower bound* for S. A set S in E^1 is *bounded* if S is bounded above and below. That is, $S \subset [a, b]$. Let $m = \max(|a|, |b|)$. Then $-m \leq -|a| \leq a$ and $b \leq |b| \leq m$, and hence $S \subset [-m, m]$. Conversely, if there exists a positive number m such that $S \subset [-m, m]$, then S is bounded. Hence, an equivalent alternate definition is that S is bounded if there exists $m > 0$ such that $|s| \leq m$ for all $s \in S$.

For $n > 1$, the idea of a bounded set is extended by the following definition. When $n = 1$ this definition, by the above remarks, is equivalent to the definition that has already been given.

Definition 2–1.3. *A set S in E^n is bounded if there exists an $r > 0$ such that $S \subset \{\mathbf{x}: \|\mathbf{x}\| \leq r\}$.*

In Problem 1 it will be shown that, as in E^1, an alternate definition can be given. That is, a set S in E^n is bounded if there exists a closed interval $[\mathbf{a}, \mathbf{b}]$ that contains S.

If $\bar{b}$ is an upper bound of a set S and if $\bar{b}$ is such that for any upper bound b of S, it follows that $b \geq \bar{b}$, then $\bar{b}$ is a *least upper bound* or *supremum* of S. If b' is also a least upper bound of S, then $b' \leq \bar{b}$, $\bar{b} \leq b'$, and hence $\bar{b} = b'$. Thus, a set can have only one least upper bound. This least upper bound of S, $\bar{b}$, is designated as l.u.b. (S) or sup (S). *The existence of a least upper bound for every bounded-above set of real numbers is a postulated property of the real numbers.* Either the existence is postulated directly, or it follows as a theorem from the postulates that are taken.

If $\underline{a}$ is a lower bound of a set S and if $\underline{a}$ is such that for any lower bound a of S it follows that $a \leq \underline{a}$, then a is a *greatest lower bound* or *infimum* of S. As in the case of the least upper bound, a set can have at most one greatest lower bound. The number $\underline{a}$ is written, $\underline{a}$ = g.l.b. (S) or inf (S). There is no need to postulate also the existence of a greatest lower bound for a bounded-below set (cf. Problem 4).

Problems

1. Show that for any given neighborhood $N(\mathbf{a}; r)$ in E^n there is an open (closed) interval that contains $N(\mathbf{a}; r)$. Show that the remark is still true if the roles of neighborhood and interval are interchanged.
2. An interval $(\mathbf{a}, \mathbf{b})$ (or $[\mathbf{a}, \mathbf{b}]$) has $\frac{1}{2}(\mathbf{a} + \mathbf{b})$ as its *center*. Show that, given any neighborhood $N(\mathbf{c}; r)$ and any point $\mathbf{x} \in N(\mathbf{c}; r)$, there is an open interval $(\mathbf{a}, \mathbf{b})$ with $\mathbf{x}$ as its center such that $(\mathbf{a}, \mathbf{b}) \subset N(\mathbf{c}; r)$. Show that the statement still holds when the roles of neighborhood and interval are interchanged.
3. Find sup (S) and inf (S) if they exist for the following sets:
   (a) $\{1/n: n = 1, 2, 3, \cdots\}$; (b) $\{x: x^2 > 3\}$;
   (c) $(-2, 3)$; (d) $\{4 - x^2: x, \text{real}\}$;
   (e) $\{n^2: n = 1, 2, 3, \cdots\}$; (f) $\{5\}$.
   If the sup (S) *and* inf (S) *fail to exist, that is S is not bounded above or below, respectively, then it is customary to take* sup $(S) = +\infty$ *and* inf $(S) = -\infty$ *respectively.*
4. If the set S is bounded below by a, then show that the set $S' = \{x: -x \in S\}$ is bounded above by $-a$.
5. Assume the postulate: Every bounded-above set of real numbers has a least upper bound. Show that every bounded-below set of real numbers has a greatest lower bound.
6. Show that the set of rationals x such that $x^2 < 2$ has no least upper bound among the rational numbers. Thus, the postulate of Problem 4 does not hold in the set of rational numbers.

Answers

3. (a) sup $(S) = 1$, inf $(S) = 0$; (b) $\infty, -\infty$; (c) $3, -2$;
   (d) $4, -\infty$; (e) $\infty, 1$; (f) $5, 5$.

2–2. Open Sets

The sets $N(\mathbf{a}; r)$ are sometimes called open spheres. Thus, besides the open intervals, we now have two types of sets in which the adjective open is used. The following definition of an open set will be such that an open interval and an open sphere will be open sets, but many other sets will be open sets as well.

Definition 2–2.1. *A point* **a** *belonging to a set S in E^n is an interior point of S if there exists $N(\mathbf{a}; r)$ for some $r > 0$ such that $N(\mathbf{a}; r) \subset S$. The set of interior points of S is called the interior of S, written S°. If every point of S is an interior point of S, that is, $S^{\circ} = S$, then S is an open set in E^n.*

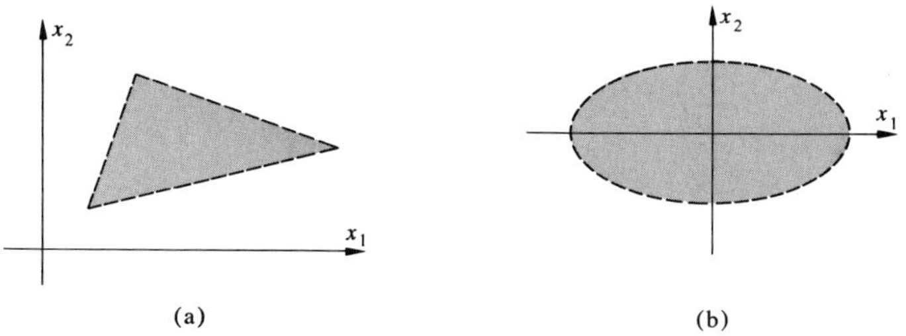

Figure 2–2.1

It is at least intuitively clear that in E^2 the set of points on the "inside" of a triangle or on the "inside" of an ellipse are examples of open sets in E^2 (cf. Figure 2–2.1). Similarly, in E^3 the points "inside" a tetrahedron or an ellipsoid are open sets in E^3. However, note that in E^3 the set of points inside a triangle is not an open set in E^3. These sets are illustrated in Figure 2–2.2.

Certain operations of the set algebra on collections of open sets give us again open sets. The following proposition summarizes the situation.

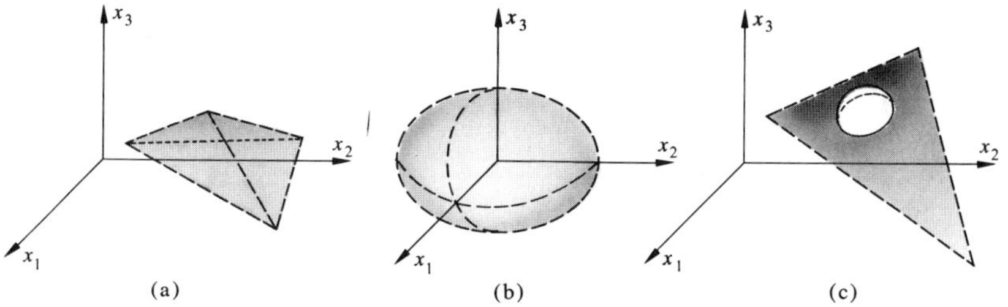

Figure 2–2.2

Proposition 2–2.1. *In E^n,* (i) *any union of open sets is an open set, and* (ii) *any finite intersection of open sets is an open set.*

Proof. Let G_α be an open set in E^n for any $\alpha \in A$, an index set. Thus A might be $\{1, 2, \cdots, n\}$, a finite set of indices; A might be $\{1, 2, \cdots\}$ a countably infinite set of indices; or A might be something like the set $[0, 1]$, an uncountable set. Let

$$\mathbf{x} \in \bigcup_{\alpha \in A} G_\alpha ,$$

then $\mathbf{x} \in G_\alpha$ for some index α. Since G_α is open, there is a neighborhood of $\mathbf{x}$, $N(\mathbf{x}; r)$ such that $N(\mathbf{x}; r) \subset G_\alpha$. Hence

$$N(\mathbf{x}; r) \subset \bigcup_{\alpha \in A} G_\alpha ,$$

and thus $\mathbf{x}$ is an interior point of $\bigcup_{\alpha \in A} G_\alpha$.

For the proof of (ii), let $G_1, G_2, \cdots, G_m$ be a collection of m open sets in E^n. Let

$$\mathbf{x} \in \bigcap_{i=1}^m G_i .$$

Then $\mathbf{x} \in G_i$, for each i, and since G_i is open there exists $N(\mathbf{x}; r_i)$ such that $N(\mathbf{x}; r_i) \subset G_i$. If $r = \min(r_1, r_2, \cdots, r_m)$, then $r > 0$, $N(\mathbf{x}; r) \subset N(\mathbf{x}; r_i)$ for each i, $1 \leq i \leq m$ and $N(\mathbf{x}; r) \subset \bigcap_{i=1}^m G_i$. ∎

Since $\emptyset$, the empty set, has no points, every point of $\emptyset$ is an interior point of $\emptyset$. Thus the empty set is an open set in E^n. Also the set E^n itself is an open set in E^n since for any $\mathbf{x} \in E^n$, $N(\mathbf{x}; 1) \subset E^n$.

This collection of sets in E^n, the sets that are designated as open sets, are said to give a topology on E^n. The precise meaning of a topology on an arbitrary set is given by the following definition.

Definition 2–2.2. *Given an arbitrary set M and a nonempty collection of subsets $\mathscr{H} = \{H_\alpha : \alpha \in A\}$: If*

(i) any arbitrary union of sets of $\mathscr{H}$ belongs to $\mathscr{H}$,

(ii) any finite intersection of sets of $\mathscr{H}$ belongs to $\mathscr{H}$, and

(iii) $\emptyset$ and $M \in \mathscr{H}$,

then the collection of subsets $\mathscr{H}$ is a topology (or topological structure) on M, and M with the collection of sets $\mathscr{H}$ having (i), (ii), and (iii) is a topological space. The sets of $\mathscr{H}$ are called the open sets of the topology.

Clearly, E^n with the collection of open sets above is a topological space. This topology on E^n is the topology *generated* by these spherical neighborhoods, in that an open set has been defined to be any set S such that if $\mathbf{x} \in \delta$, then there is a neighborhood $N(\mathbf{a}; \delta) \subset S$. A topology can be generated on E^n by other kinds of neighborhoods—for example, rectangular neighborhoods and square neighborhoods. A rectangular and a square neighborhood of $\mathbf{a}$ in E^n are the respective sets $N_R(\mathbf{a}; \mathbf{b}) = (\mathbf{a} - \mathbf{b}, \mathbf{a} - \mathbf{b})$, where $b_i > 0$

for all i, $1 \leq i \leq n$; and

$$N_S(\mathbf{a}; \delta) = \{\mathbf{x}: |x_i - a_i| < \delta, 1 \leq i \leq n\}$$

where $\delta > 0$. In Problems 5 and 6 it will be shown that in E^n the same topology is generated by each of these three types of neighborhoods. In other words, we say that the topologies are *equivalent.*

Notice that the spherical neighborhoods are defined using the norm. Thus, to use the spherical neighborhood, $N(\mathbf{a}; \delta)$, we must have E^n space or at least an R^n space with a norm. The definition of a rectangular neighborhood makes use of vector addition, and hence the space must be R^n, or at least the addition operation must be defined. *The square neighborhoods make no use of any of the vector operations, nor do they require that the space have a norm. The space can consist of simply the ordered real n-tuples. Yet open sets are defined, and a topology is given.*

The topology of E^n generated by spherical (or rectangular or square) neighborhoods is used quite often, so often in fact that it is referred to as the *usual topology* of E^n. Unless there is some statement to the contrary, the topology on E^n is assumed to be the usual topology.

Problems

1. Show that in E^n open spheres and open intervals are open sets.
2. Find the interior of the following sets, and tell whether the set is open or not open. Make a sketch of each set.
   (a) $[0, 1]$, (b) $\{(x_1, 0): 0 \leq x_1 \leq 1\}$,
   (c) $\{1/n: n = 1, 2, 3, \cdots\}$, (d) $\{(x_1, 0): x_1 \neq 0\}$,
   (e) $\{(x_1, x_2): x_1^2 + x_2^2 < 1\}$, (f) $\{(x_1, x_2, x_3): x_1^2 + x_2^2 \leq 1\}$,
   (g) $\{(x_1, x_2): x_1 + x_2 \leq 1\}$, (h) $\{(x_1, x_2): x_1 x_2 < 1\}$.
3. Show that S^o is always an open set, that is, $(S^o)^o = S^o$.
4. Let S_1 and S_2 be subsets of E^n. Show $S_1{}^o \cap S_2{}^o = (S_1 \cap S_2)^o$.
5. Show that $S_1{}^o \cup S_2{}^o \subset (S_1 \cup S_2)^o$. Give examples to show that the inclusion may be proper or equality may hold.
6. Suppose that S is an open set in the topology generated by spherical neighborhoods; if $\mathbf{x} \in S$, then there exists a $\delta > 0$ such that $N(\mathbf{x}; \delta) \subset S$. Show that there exists $\mathbf{b}$ such that $b_i > 0$, $1 \leq i \leq n$, such that

$$N_R(\mathbf{x}: \mathbf{b}) = (\mathbf{x} - \mathbf{b}, \mathbf{x} + \mathbf{b}) \subset N(\mathbf{x}; \delta) \subset S;$$

and hence S is an open set in the topology generated by rectangular neighborhoods; and thus every open set in the topology generated by spherical neighborhoods is an open set in the topology generated by rectangular neighborhoods. Show the converse and thus show that these two topologies are equivalent.

7. Show in R^n that the topologies generated by rectangular neighborhoods and the topologies generated by square neighborhoods are equivalent. What is a simple criterion for the equivalence of topologies generated by neighborhoods?

8. In E^2, let $N_V(\mathbf{a}; r) = \{(x_1, x_2): |x_1 - a_1| < r\}$ where $r > 0$; and let $\mathscr{H}_V$ be the collection of set S in E^2 such that if $\mathbf{a} \in S$, there exists $N_V(\mathbf{a}; r) \subset S$. Thus $\mathscr{H}_V$ is the topology generated by the neighborhoods $N_V(\mathbf{a}; r)$. Show that $\mathscr{H}_V$ is a proper subcollection of the open sets of E^2 in the usual topology on E^2.

9. On any nonempty set M let $\mathscr{H}_I$ be the collection that consists of the two sets $\{\emptyset, M\}$ and let $\mathscr{H}_D$ be the collection that consists of all the subsets of M. Show that both collections of sets give a topology on M. Show that if $\mathscr{H}$ is any collection of sets on M that gives a topology on M, then $\mathscr{H}_I \subset \mathscr{H} \subset \mathscr{H}_D$. The topologies on M given by $\mathscr{H}_I$ and $\mathscr{H}_D$ are termed the *indiscrete* and *discrete* topologies, respectively.

10. Let M_1 be a subset of M, a topological space with the topology on M given by the collection of sets $\mathscr{H}$. Let $\mathscr{H}_1 = \{M_1 \cap H: H \in \mathscr{H}\}$. Show that $\mathscr{H}_1$ is a collection of subsets of M that has properties (i), (ii), and (iii) of Definition 2–2.2, and hence $\mathscr{H}_1$ gives a topology on M_1. This topology, $\mathscr{H}_1$, is the *topology induced* on M_1 by $\mathscr{H}$—that is, the *induced topology* (or *relative topology*).

11. Show that the usual topology on E^1 is the topology induced on E^1 by the usual topology on E^2. Show that the similar relationship holds between E^n and E^{n+1}, $n > 1$.

12. In the topology induced on [0, 1] by the usual topology in E^1, show that the open sets in the topology need not be open sets of E^1 in the usual topology of E^1.

13. In E^1 let $\mathscr{H}_r$ be the collection of sets that are the arbitrary unions of sets of the form $[x, \infty)$ where x is real. By convention it is understood that the meaning of "arbitrary union" includes the union of no sets, and hence $\emptyset \in \mathscr{H}_r$. Show that $\mathscr{H}_r$ is a topology for E^1.

Answers

2. (a) (0, 1), not open; (b) $\emptyset$, not open;
(c) $\emptyset$, not open; (d) $\{(x_1, 0): x_1 \neq 0\}$, open;
(e) $\{(x_1, x_2): x_1^2 + x_2^2 < 1\}$, open;
(f) $\{(x_1, x_2, x_3): x_1^2 + x_2^2 < 1\}$, not open;
(g) $\{(x_1, x_2): x_1 + x_2 < 1\}$, not open;
(h) $\{(x_1, x_2): x_1 x_2 < 1\}$, open.

2–3. Closed Sets

The word "accumulate" means collect, amass, or pile up; hence, a point of accumulation would mean a place where certain things collect. Following this thinking gives exactly the correct intuitive meaning of an accumulation point of a set.

Definition 2–3.1. *Let S be a set in E^n. A point* $\mathbf{a}$ *is an accumulation point of S if every $N'(\mathbf{a}; r)$ contains a point of S.*

If in E^1, $S = [0, 1]$, then each point of S is an accumulation point of S. If in E^1, S is the set of rational points, then every real point is an accumulation point of S. A set may have a finite number of accumulation points, as in the case of the set $\{(-1)^n(1 + (1/n)) : n = 1, 2, 3, \cdots\}$. There are only two accumulation points of this set, namely -1 and $+1$, neither of which belongs to the set. Any set having only a finite number of points will have no accumulation points.

Thus the set of accumulation points of a given set may take a variety of forms. However, those sets like $[0, 1]$, which contain all of their accumulation points, get special attention.

Definition 2–3.2. *A set S in E^1 is a closed set if S contains all of its accumulation points.*

Thus $[0, 1]$ is a closed set in E^1, $\{(x_1, x_2) : x_1^2 + x_2^2 \leq 1\}$ is a closed set in E^2, and $\{(1/n, 0, 0) : n = 1, 2, 3, \cdots\} \cup \{\mathbf{0}\}$ is a closed set in E^3. These sets are illustrated in Figure 2–3.1. Any finite set in E^n is a closed set.

Some care should be given to the understanding of the definitions of an

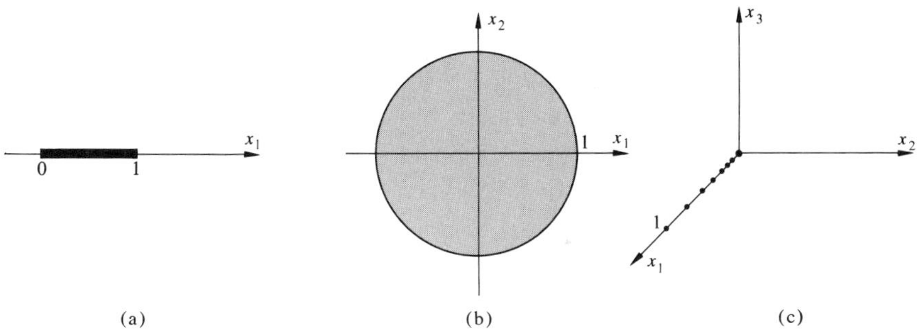

Figure 2–3.1

open set and a closed set. The common error is to think that a set is either an open set or a closed set. This is not the case at all. In E^1 the set $(0, 1]$ is *neither* an open set *nor* a closed set; the set E^1 itself is *both* an open *and* a closed set; the set $(0, 1)$ is an open set that is not a closed set; and the set $[0, 1]$ is a closed set that is not an open set. The following proposition provides a relationship between the open and closed sets.

Proposition 2–3.1. *The complement of any open set in E^n is a closed set, and the complement of any closed set in E^n is an open set.*

Proof. Let S be any open set in E^n, and let S^c be the set of points in E^n that do not belong to S; that is, the complement of S in E^n. Let $\mathbf{a}$ be an accumulation point of S^c. If $\mathbf{a} \in S$, then $\mathbf{a}$ is an interior point of S, and there exists $N(\mathbf{a}; r)$ such that $N(\mathbf{a}; r) \subset S$. Thus $N'(\mathbf{a}; r)$ would not contain a point of S^c which would contradict the assumption that $\mathbf{a}$ is an accumulation point of S^c. Therefore $\mathbf{a} \notin S$ or $\mathbf{a} \in S^c$. This shows that S^c contains all of its accumulation points, and hence S^c is a closed set.

If now S is a closed set in E^n, then any $\mathbf{a} \in S^c$ is not an accumulation point of S. Thus, there exists $N'(\mathbf{a}; r)$ such that $N'(\mathbf{a}; r)$ does not contain a point of S. With $a \notin S$, it follows that $N(\mathbf{a}; r)$ does not contain a point of S. That is, $N(\mathbf{a}; r) \subset S^c$. Hence any point of S^c is an interior point of S^c, and S^c is an open set. ∎

This proposition and one of the fundamental laws of set theory, De Morgan's law, allow us to consider unions and intersections of collections of closed sets.

Proposition 2–3.2. *In E^n,* (i) *any intersection of closed sets is a closed set, and* (ii) *any finite union of closed sets is a closed set.*

Proof. Let F_α be a closed set in E^n for any $\alpha \in A$, an index set. Then $(\bigcap_{\alpha \in A} F_\alpha)^c = \bigcup_{\alpha \in A} F_\alpha{}^c$ by De Morgan's law. By Proposition 2–3.1 $F_\alpha{}^c$ is an open set for each α. Then from Proposition 2–2.1(i) we find $\bigcup_{\alpha \in A} F^c$, a union of open sets, is open. Thus $(\bigcap_{\alpha \in A} F_\alpha)^c$ is open. Again using Proposition 2–3.1, we have $\bigcap_{\alpha \in A} F_\alpha$, a closed set.

Similarly, let $F_1, F_2, \cdots, F_m$ be closed sets in E^n. Then by the other set identity of De Morgan's law, it follows that

$$(\textstyle\bigcup_{i=1}^m F_i)^c = \bigcap_{i=1}^m F_i{}^c.$$

Again notice that $F_i{}^c$ is an open set for each i. The set on the right side of the equation is a finite intersection of open sets. Hence $(\bigcup_{i=1}^m F_i)^c$ is open, and $\bigcup_{1=i}^m F_i$ is closed. ∎

Example 2–3.1. The collection of open intervals $(-1/n, 1/n)$ for $n = 1, 2, 3, \cdots$ is such that $\bigcap_{n=1}^{\infty} (-1/n, 1/n) = \{0\}$. Thus, there are collections of *infinitely many* open sets such that the intersection of the sets of the collection is not open. Therefore, Proposition 2–2.1(ii) gives the best possible result. Similarly, Proposition 2–3.2(ii) gives the best possible results. The complements of these open intervals, $(-1/n, 1/n)^c$, are closed sets. The union of these closed sets, $\bigcup_{n=1}^{\infty} (-1/n, 1/n)^c = \{0\}^c$, is an open set.

The set E^n itself is a closed set in E^n in the usual topology. Thus, given any set S in E^n, there is always a closed set that contains S. If S is nonempty, the intersection of all the closed sets of E^n that contain S is a closed set that contains S. This particular closed set determined by S is frequently used, and hence it is given a name.

Definition 2–3.3. *Let S be a subset of E^n. The intersection of all closed sets containing S, $\bigcap_{S \subset F} F$ is the closure of S in E^n, written $\bar{S}$.*

If F_1 is any closed set in E^n containing S, then clearly $\bar{S} \subset \bigcap_{S \subset F} F \subset F_1$. Thus $\bar{S}$ is frequently referred to as the "smallest closed set containing S." If S is a closed set, then clearly $\bar{S} = S$. Thus $\bar{\bar{S}} = \bar{S}$.

In everyday references to maps of plots of land, one often speaks of the "boundary" (of the land). In the definition below, the boundary of an arbitrary set S in E^n is defined. For the sets S of E^2 that might represent a plot of land, the boundary of S will agree with one's intuition.

Definition 2–3.4. *Let S be a set in E^n. The boundary of S in E^n, written* bdy (S), *is the set $\bar{S} \cap \overline{(S^c)}$. A point of* bdy (S) *is a boundary point of S.*

Notice that the boundary of a set S may be the empty set as in the case $S = E^1$ in E^1. If $S = E^1$, and we are considering E^1 as a particular subset of E^2, then each point of S is a boundary point of S. If $S = [0, 1]$ or $(0, 1)$ in E^1, the bdy $(S) = \{0, 1\}$. If S in E^1 is the set of rationals, then the bdy $(S) = E^1$. Thus, the boundary of a set sometimes can be a set that we intuitively would want to call the "boundary" of a set, and at other times our intuition can fail us. The definition however, is explicit in all cases.

Problems

1. Find the set of accumulation points of the following sets:
   (a) $\{(n - 1)/n : n = 1, 2, 3, \cdots\}$ in E^1,

(b) $\{(-1)^n(1/n) + \sin(n\pi/2)\colon n = 1, 2, 3, \cdots\}$ in E^1,
(c) $\{m/2^n\colon m$, an integer; $n = 1, 2, 3, \cdots\}$ in E^1,
(d) $\{\mathbf{x}\colon \|\mathbf{x}\| = m + (1/n)$, m and $n = 1, 2, 3, \cdots\}$ in E^2,
(e) $\{(m, n)\colon m$ and n integers$\}$ in E^2.

Any point of a set S that is not an accumulation point of S is an *isolated point* of S. Find the isolated points of the above sets.

2. For each of the sets in Problem 1, find the closure and the boundary.
3. Tell whether the sets of Problem 2, page 45, are closed or not closed.
4. Show that if $S \subset E^n$, $\bar{S}$ equals the union of the set S and the set of accumulation points of S.
5. Show that a closed interval and a closed sphere in E^n are both closed sets in E^n.
6. Let S_1 and S_2 be subsets of E^n, then show $\overline{S_1 \cup S_2} = \bar{S}_1 \cup \bar{S}_2$. Does the equality still hold for an infinite number of sets?
7. Show that $\overline{S_1 \cap S_2} \subset \bar{S}_1 \subset \bar{S}_2$. Give examples to show that the inclusion may be proper or equality may hold.
8. Show that the bdy $(S) =$ bdy (S^c).
9. Show the boundary of a set is always a closed set.
10. Show $(S^o)^c = \overline{(S^c)}$, and $(\bar{S})^c = (S^c)^o$.
11. Show $S -$ bdy $(S) = S^o$.
12. Show that if $S = \{\mathbf{x}\colon \|\mathbf{x} - \mathbf{x}_0\| \le r, r > 0\}$, then

$$\text{bdy}\,(S) = \{\mathbf{x}\colon \|\mathbf{x} - \mathbf{x}_0\| = r\},$$

which is the surface of the n-sphere in E^n.

Answers

1. Accumulation points of the set:
(a) $\{1\}$, (b) $\{-1, 0, 1\}$, (c) E^1,
(d) $\{\mathbf{x}\colon \|\mathbf{x}\| = m + 1/n$, m and $n = 1, 2, 3, \cdots\}$, (e) $\emptyset$.
Each point of the given set is an isolated point of the set for (a), (b), and (e); for examples (c) and (d) there are no isolated points.
2. Let S be the given set, then for
(a) $\bar{S} = S \cup \{1\}$, (b) $\bar{S} = S \cup \{-1, 0, 1\}$, (c) $\bar{S} = E^1$,
(d) $\bar{S} = S$, and (e) $\bar{S} = S$.
For
(a) bdy $(S) = \bar{S}$, (b) bdy $(S) = \bar{S}$, (c) bdy $(S) = E^1$,
(d) bdy $(S) = \bar{S}$, (e) bdy $(S) = S$.
3. The sets of (a), (b), (f), and (g) are closed, and the others are not closed.

2–4. Bolzano-Weierstrass Theorem

The following theorem, which is credited to the two mathematicians Bernard Bolzano and Karl Weierstrass, is very essential because it is used in proving some of the basic theorems of analysis.

Theorem 2–4.1. *Every infinite bounded set in E^n has an accumulation point in E^n.*

Proof for E^1. Let S be an infinite set in E^1 that is contained in the interval $[a, b]$. At least one of the two intervals $[a, \frac{1}{2}(a + b)]$, $[\frac{1}{2}(a + b), b]$ contains infinitely many points of the set S since S is an infinite set. Label one of these two intervals that contains infinitely many points of S, $[a_1, b_1]$. Then $a \leq a_1 < b_1 \leq b$, and the length of the interval $[a_1, b_1]$ is $\frac{1}{2}(b - a)$.

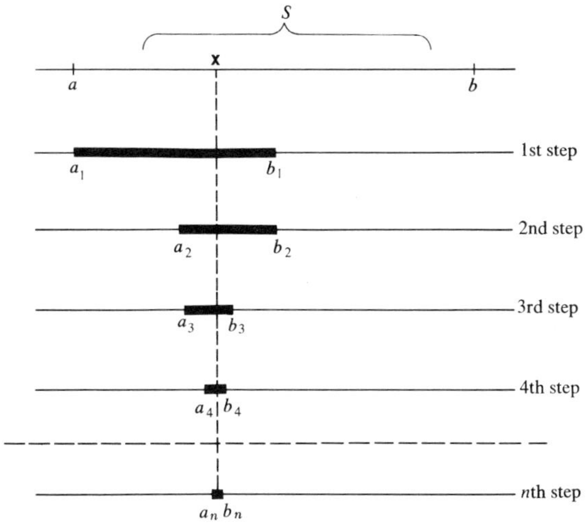

Figure 2–4.1

Suppose this same process has been repeated n times until an interval $[a_n, b_n]$ has been obtained such that $[a_n, b_n]$ contains infinitely many points of S, $a_{n-1} \leq a_n < b_n \leq b_{n-1}$, and the length of

$$[a_n, b_n] = \tfrac{1}{2}(b_n - a_n) = \left(\frac{1}{2^n}\right)(b - a).$$

Then at least one of the two intervals $[a_n, \frac{1}{2}(a_n + b_n)]$, $[\frac{1}{2}(a_n + b_n), b_n]$ contains infinitely many points of S. Label one of these two intervals that contains infinitely many points of S, $[a_{n+1}, b_{n+1}]$. Then $a_n \leq a_{n+1} < b_{n+1} \leq b_n$, and the length of

$$[a_{n+1}, b_{n+1}] = \tfrac{1}{2}(b_n - a_n) = (\tfrac{1}{2}^{n+1})(b - a).$$

Thus there exists an infinite sequence of intervals $\{[a_n, b_n]\}$ such that each interval $[a_n, b_n]$ is of length $(\tfrac{1}{2}^n)(b - a)$, each interval contains infinitely many points of S, and each interval is contained in the previous interval (cf. Figure 2–4.1).

Notice that

$$a \leq a_1 \leq \cdots \leq a_n \leq a_{n+1} \leq \cdots \leq a_{n+k} < b_{n+k} \leq \cdots$$
$$\leq b_{n+1} \leq b_n \leq \cdots \leq b_1 \leq b$$

for $k = 1, 2, 3, \cdots$. Thus b_n is an upper bound for $A = \{a, a_1, a_2, \cdots\}$, and a_n is a lower bound for $B = \{b, b_1, b_2, \cdots\}$. Let $\bar{x} = \sup (A)$ and let $\bar{x}' = \inf (B)$. Suppose $\bar{x}' < \bar{x}$. If there exists any $a' \in A$ such that $\bar{x}' < a'$, then a', which is a lower bound of B, would be greater than $\bar{x}'$, the greatest lower bound of B. However, if there is no element of A greater than $\bar{x}'$, then $\bar{x}$ would not be the least upper bound of A. Therefore, $\bar{x} \leq \bar{x}'$.

Now $a_n \leq \bar{x} \leq \bar{x}' \leq b_n$ for any positive integer n. Thus

$$0 \leq \bar{x}' - \bar{x} \leq b_n - a_n = \left(\frac{1}{2^n}\right)(b - a).$$

Since $(\tfrac{1}{2}^n)(b - a)$ can be made arbitrarily small, then $\bar{x}' = \bar{x}$. The point $\bar{x}$ belongs to $[a_n, b_n]$ for all n since $a_n \leq \bar{x} \leq b_n$. This point $\bar{x}$ will be shown to be an accumulation point of S.

Let $N'(\bar{x}; \varepsilon)$ be any deleted neighborhood of $\bar{x}$. Take n such that

$$b_n - a_n = \left(\frac{1}{2^n}\right)(b - a) < \varepsilon.$$

Then since $\bar{x} \in [a_n, b_n]$, $[a_n, b_n] \subset N(\bar{x}; \varepsilon)$. There are infinitely many points of S contained in $[a_n, b_n]$, and therefore there exists a point of S in $N'(\bar{x}; \varepsilon)$. Consequently, $\bar{x}$ is an accumulation point of S. ∎

The proof for E^n can be given similarly. For example, if $n = 2$, the infinite set S is bounded, and hence S is contained in a closed interval $[\mathbf{a}, \mathbf{b}]$, a rectangle. At least one of the four intervals (rectangles)

$$I_{11} = [\mathbf{a}, \tfrac{1}{2}(\mathbf{a} + \mathbf{b})],$$

$$I_{12} = [\tfrac{1}{2}(a_1 + b_1), a_2), (b_1, \tfrac{1}{2}(a_2 + b_2)],$$

$$I_{13} = [(a_1, \tfrac{1}{2}(a_2 + b_2)), (\tfrac{1}{2}(a_1 + b_1), b_2)],$$

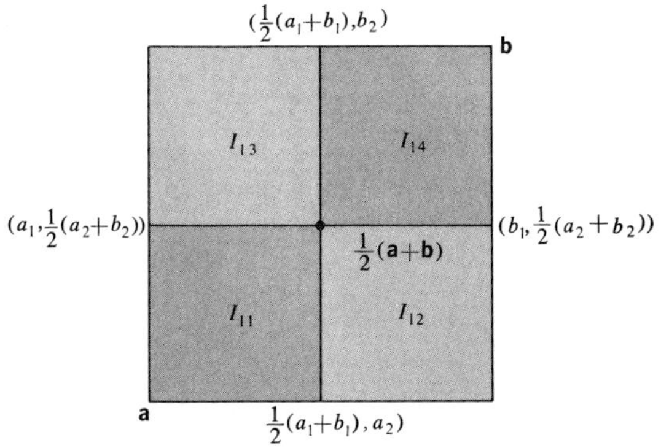

Figure 2–4.2

or

$$I_{14} = [\tfrac{1}{2}(\mathbf{a} + \mathbf{b}), \mathbf{b}]$$

contains infinitely many points of S (cf. Figure 2–4.2). Call one of these intervals that contains infinitely many points of S, $[\mathbf{a}^1, \mathbf{b}^1]$. Then proceed to obtain a sequence of these closed intervals as in the case of $n = 1$.

Problems

1. For the following subsets of the interval [0, 1] give a sequence of subintervals of [0, 1] as constructed in the proof of Theorem 2–4.1 whose intersection is an accumulation point of the given set.
   (a) $S = \{1/k: k = 1, 2, 3, \cdots\}$,
   (b) $S = \{k/(k + 1): k = 1, 2, 3, \cdots\}$,
   (c) $S = (\frac{1}{2}, \frac{1}{2} - \frac{1}{4}, \frac{1}{2} - \frac{1}{4} + \frac{1}{8}, \cdots\}$,
   (d) $S = \{$the set of rationals between 0 and 1$\}$.
2. Give an example to show each of the following:
   (a) an infinite set in E^n that has no accumulation points in E^n,
   (b) an infinite unbounded set in E^n that has an accumulation point in E^n,
   (c) an infinite bounded set in E^n that has exactly three accumulation points in E^n,
   (d) an infinite bounded set in Q^n, where Q^n are those points of E^n whose coordinates are all rational, such that the set has an accumulation point that is not in Q^n.
3. Prove the Bolzano-Weierstrass theorem for $n = 2$.

4. The following is an alternate method of proving the Bolzano-Weierstrass theorem. Let S_1 be a set such that $x \in S_1$ if there are infinitely many points s of S such that $x < s$. Show S_1 is nonempty and is bounded above by b. Let $\bar{x} = \sup(S_1)$. Show that $\bar{x}$ is an accumulation point of S.

5. For the sets S in Problem 1 and the given sequence of subintervals, what will be the point $\bar{x}$ of Problem 4?

Answers

1. (a) $[0, 1/2], [0, 1/4], [0, 1/8], \cdots, [0, 1/2^n], \cdots;$
   (b) $[1/2, 1], [3/4, 1], [7/8, 1], \cdots, [(2^n - 1)/2^n, 1], \cdots;$
   (c) $$[0, 1/2], [1/4, 1/2], [1/4, 3/8], \cdots, \left[\frac{2n-3}{2^{2n-2}}, \frac{2n-1}{2^{2n-1}}\right],$$
   $$\left[\frac{2n+1}{2^{2n}}, \frac{2n-1}{2^{2n-1}}\right], \cdots;$$
   (d) Any sequence of the above subintervals would be suitable.
5. (a) $\bar{x} = 0$, (b) 1, (c) $1/3$.

2–5. Cauchy Sequences

One of the applications of the Bolzano-Weierstrass theorem is in connection with Cauchy sequences. First let us review the meaning of a convergent sequence.

A sequence $\{\mathbf{x}_k\}$ in E^n is a *convergent* sequence if there exists $\mathbf{a} \in E^n$ such that $\lim_{k\to\infty} \mathbf{x}_k = \mathbf{a}$, that is, if given $\varepsilon > 0$ there is an integer $k_0 > 0$, dependent upon ε, such that $\|\mathbf{x}_k - \mathbf{a}\| < \varepsilon$ for all $k > k_0$. Such a sequence is said to *converge* to $\mathbf{a}$. When a sequence is not convergent, it is *divergent*. The symbol, $\to$, is read: converges to. Thus as an alternative to $\lim_{k\to\infty} \mathbf{x}_k = a$, we may write $\mathbf{x}_k \to \mathbf{a}$ as $k \to \infty$. The point $\mathbf{a}$ is a *limit* of the sequence.

Given a sequence $\{\mathbf{x}_k\}$ and a point $\mathbf{a}$, then it is possible to check to see if $\|\mathbf{x}_k - \mathbf{a}\| \to 0$ as $k \to \infty$. If not, then we know that $\mathbf{a}$ is not a limit of the sequence, but we do not know if $\mathbf{a}$ was just a poor choice and perhaps some other point is a limit of the sequence. It would be desirable to know whether or not a given sequence has any point as a limit. The Cauchy sequences which are defined in the following definition are precisely the convergent sequences, as we shall see in the subsequent theorem. Notice that the definition of a Cauchy sequence omits any mention of a limit of the sequence.

Definition 2–5.1. *A sequence* $\{\mathbf{x}_k\}$ *in* E^n *is a Cauchy sequence if given* $\varepsilon > 0$ *there exists* k_0 *which depends upon* ε *such that* $\|\mathbf{x}_p - \mathbf{x}_q\| < \varepsilon$ *when* p *and* $q \geq k_0$.

The conditions of a Cauchy sequence are sometimes referred to as *Cauchy's criterion* or *Cauchy's condition.*

The sequence $\{x_k\}$ where $x_k = 1/2^k$ is such that

$$|x_p - x_q| = \left|\frac{1}{2^p} - \frac{1}{2^q}\right| < \frac{1}{2^{p_1}}$$

where $p_1 = \min(p, q)$. Then for $\varepsilon > 0$, $1/2^{p_1} < \varepsilon$ for $p_1 \geq k_0 > \log_2(1/\varepsilon)$. Hence $\{x_k\}$ is a Cauchy sequence. The sequence $\{x_k\}$ where

$$x_k = \frac{(-1)^k k}{k+1}$$

is such that $|x_{2p+1} - x_{2p}| \geq 1$ for $p \geq 1$. Hence this sequence is not a Cauchy sequence.

Theorem 2–5.1. *A necessary and sufficient condition that a sequence in* E^n *be a convergent sequence is that the sequence be a Cauchy sequence.*

Proof. Let $\{\mathbf{x}_k\}$ be a sequence in E^n such that $\mathbf{x}_k \to \mathbf{a} \in E^n$ as $k \to \infty$. Then given $\varepsilon > 0$, there exists k_0 such that when $k \geq k_0$ we have $\|\mathbf{x}_k - \mathbf{a}\| < \varepsilon/2$. Let p and q be $\geq k_0$, then

$$\|\mathbf{x}_p - \mathbf{x}_q\| \leq \|\mathbf{x}_p - \mathbf{a}\| + \|\mathbf{a} - \mathbf{x}_q\| < \frac{\varepsilon}{2} + \frac{\varepsilon}{2} = \varepsilon.$$

Thus, each convergent sequence is necessarily a Cauchy sequence.

Let $\{\mathbf{x}_k\}$ be a Cauchy sequence in E^n. Then, given $1 > 0$, there exists a positive integer k' such that $\|\mathbf{x}_p - \mathbf{x}_{k'}\| < 1$ for $p \geq k'$. The set of vectors $\{\mathbf{x}_k : k \geq k'\}$ is contained in the sphere of radius one with center at $\mathbf{x}_{k'}$.

Suppose $\mathbf{x}_k$ for infinitely many indices $k \geq k'$ is a particular vector $\mathbf{a}$. In this case we shall show that $\mathbf{x}_k \to \mathbf{a}$ as $k \to \infty$. Given $\varepsilon > 0$, there exists k_0 dependent upon ε such that when p and q are $\geq k_0$, then $\|\mathbf{x}_p - \mathbf{x}_q\| < \varepsilon$. Since $\mathbf{x}_k = \mathbf{a}$ for infinitely many k, there exists $k'' \geq k_0$ such that $\mathbf{x}_{k''} = \mathbf{a}$. Then for $k \geq k_0$,

$$\|\mathbf{x}_k - \mathbf{a}\| \leq \|\mathbf{x}_k - \mathbf{x}_{k''}\| + \|\mathbf{x}_{k''} - \mathbf{a}\| < \varepsilon + 0 = \varepsilon.$$

Thus $\lim_{k\to\infty} \mathbf{x}_k = \mathbf{a}$.

Suppose now the other case: that there is no vector in E^n such that $\mathbf{x}_k$ is repeatedly equal to this vector for infinitely many indices $k \geq k'$. Thus

$\{\mathbf{x}_k : k \geq k'\}$ is an infinite set of vectors contained in the sphere $N(\mathbf{x}_{k'}; 1)$; that is, $\{\mathbf{x}_k : k \geq k'\}$ is an infinite bounded set in E^n. By the Bolzano-Weierstrass theorem, this set has at least one accumulation point. Call it $\mathbf{a}$. We shall show also in this case that $\lim_{k\to\infty} \mathbf{x}_k = \mathbf{a}$.

First we shall get a subsequence of $\{\mathbf{x}_k\}$ that converges to $\mathbf{a}$. Since $\mathbf{a}$ is an accumulation point of $\{\mathbf{x}_k : k \geq k'\}$, there exists $\mathbf{x}_{k_1}$ such that

$$\mathbf{x}_{k_1} \in N'(\mathbf{a}; \tfrac{1}{2}).$$

Then there exists $\mathbf{x}_{k_2}$, $k_2 > k_1$ such that

$$\mathbf{x}_{k_2} \in N(\mathbf{a}, \delta_2)$$

where $\delta_2 = \min(\tfrac{1}{2^2}, \|\mathbf{x}_{k_1} - \mathbf{a}\|)$. Note $\mathbf{x}_{k_2} \neq \mathbf{x}_{k_1}$. Suppose $\mathbf{x}_{k_1}, \cdots, \mathbf{x}_{k_{j-1}}$ have been chosen such that $k_1 < k_2 < \cdots < k_{j-1}$ and

$$\|\mathbf{x}_{k_i} - \mathbf{a}\| < \min\left(\|\mathbf{x}_{k_{i-1}} - \mathbf{a}\|, \frac{1}{2^i}\right), \quad 1 < i \leq j-1.$$

Then again since $\mathbf{a}$ is an accumulation point of $\{\mathbf{x}_k : k \geq k'\}$ it is possible to choose $\mathbf{x}_{k_j}$ such that $k_j > k_{j-1}$ and $\|\mathbf{x}_{k_j} - \mathbf{a}\| < \min(\|\mathbf{x}_{k_{j-1}} - \mathbf{a}\|, \tfrac{1}{2^j})$. Thus we have the subsequence $\{\mathbf{x}_{k_j}\}$ such that if we are given $\varepsilon > 0$, there exists $j_0 \geq \log_2(2/\varepsilon)$ such that if $j \geq j_0$, then

$$\|\mathbf{x}_{k_j} - \mathbf{a}\| < \frac{1}{2^j} < \frac{\varepsilon}{2}.$$

Therefore, $\lim_{j\to\infty} \mathbf{x}_{k_j} = \mathbf{a}$.

Now we shall show that the sequence $\{\mathbf{x}_k\}$ has the same limit as this subsequence. Using the fact that $\{\mathbf{x}_k\}$ is a Cauchy sequence, we have with the above, given $\varepsilon > 0$, that there exists k_0' such that when p and $q \geq k_0'$, then $\|\mathbf{x}_p - \mathbf{x}_q\| < \varepsilon/2$. Take $k \geq k_0'$ and take $k_j \geq \max(k_0', k_{j_0})$. Then

$$\|\mathbf{x}_k - \mathbf{a}\| \leq \|\mathbf{x}_k - \mathbf{x}_{k_j}\| + \|\mathbf{x}_{k_j} - \mathbf{a}\| < \frac{\varepsilon}{2} + \frac{\varepsilon}{2} = \varepsilon.$$

Therefore, $\lim_{k\to\infty} \mathbf{x}_k = \mathbf{a}$, and $\{\mathbf{x}_k\}$ is a convergent sequence. ∎

Problems

1. If $\{\mathbf{x}_k\}$ is a sequence in E^n, then $\lim_{k\to\infty} \mathbf{x}_k = \mathbf{a}$ if and only if $\lim_{k\to\infty} x_{ik} = a_i$, $1 \leq i \leq n$, where $\mathbf{x}_k = (x_{1k}, \cdots, x_{nk})$ and $\mathbf{a} = (a_1, \cdots, a_n)$.
2. Use the results of Problem 1 to find $\lim_{k\to\infty} \mathbf{x}_k$ in E^3 if $\mathbf{x}_k = (x_{1k}, x_{2k}, x_{3k})$ where

$$x_{1k} = \sin(\pi/k),$$
$$x_{2k} = \textstyle\sum_{i=1}^{k} (-1)^{i+1}(\tfrac{1}{2^i}),$$
$$x_{3k} = \textstyle\sum_{i=0}^{k} (1/i!).$$

3. Prove by using directly Definition 2–5.1 that the following sequences are Cauchy sequences:
   (a) $\{x_k\}$ if $x_k = \sum_{i=1}^{n} (-1)^{i+1}(1/i)$,
   (b) $\{x_k\}$ if $x_k = \sum_{i=1}^{n} (-1)^{i+1}(\frac{1}{3^i})$,
   (c) $\{x_k\}$ if $x_k = a_1 - a_2 + \cdots + (-1)^{k+1} a_k$ where $a_{i+1} \geq a_i > 0$ and $\lim_{i\to\infty} a_i = 0$.
4. Prove in E^3 the sequence $\{\mathbf{x}_k\}$ whose components are given below is a Cauchy sequence by first using Problem 1 to show that it is a convergent sequence:
$$x_{1k} = k^2/e^k,$$
$$x_{2k} = (1 + 3k)^{(1/2k)},$$
$$x_{3k} = k^3 \sin (1/k) - k^2.$$
5. Give an example of a convergent sequence $\{\mathbf{x}_k\}$ in E^n such that the set $\{\mathbf{x}_1, \mathbf{x}_2, \mathbf{x}_3, \cdots\}$ does not have an accumulation point.
6. In E^n if $\{\mathbf{x}_k\}$ is a convergent sequence, prove that the set $\{\mathbf{x}_1, \mathbf{x}_2, \mathbf{x}_3, \cdots\}$ is a bounded set. Give an example to disprove the converse.
7. In E^n if $\{\mathbf{x}_k\}$ is a convergent sequence, show that the sequence converges to only one point.
8. Let $\{\mathbf{x}_k\}$ be a sequence in E^n. Then $\mathbf{y}_k = (1/k) \sum_{i=1}^{k} \mathbf{x}_i$ is the *center of mass* of k unit masses placed at each of the points $\mathbf{x}_1, \cdots, \mathbf{x}_k$. Show that if $\mathbf{x}_k \to \mathbf{a}$ as $k \to \infty$, then $\mathbf{y}_k \to \mathbf{a}$. Find an example to show that the converse is not true.

Answers

2. $\lim_{k\to\infty} \mathbf{x}_k = (1, \frac{1}{3}, e)$.

2–6. Heine-Borel Theorem

The Heine-Borel theorem is the second of two main topological theorems necessary for proving some of the basic theorems of analysis. The first was the Bolanzo-Weierstrass theorem.

First, it is necessary to define the idea of an open covering.

Definition 2–6.1. *A collection $\mathscr{G}$ of sets G_α written $\mathscr{G} = \{G_\alpha : \alpha \in A\}$ is an open covering of a set S in E^n if G_α is an open set for each $\alpha \in A$ and if $S \subset \bigcup_{\alpha\in A} G_\alpha$.*

For example, if $S = (0, 3)$ in E^1 and $\mathscr{G} = \{N(x; 1): x \in S\}$, then $\mathscr{G}$ is an open covering of S in E^1. Notice that the finite subcollection of $\mathscr{G}$ consisting of $\{N(1; 1), N(2; 1)\}$ is also an open covering of S. There are other open coverings of the same set $S = (0, 3)$ for which there are no finite subcollections that cover S. For instance, let $\mathscr{G} = \{(x/2, 3) : x \in S\}$. Since $(x/2, 3)$ is an open set for each $x \in S$ and since for each $x \in S$ we have $x \in (x/2, 3)$, it follows that $\mathscr{G}$ is an open covering of S. Any finite subcollection of $\mathscr{G}$ is of the form $\{(x_i/2, 3) : i = 1, 2, \cdots, m\}$. Take $\delta = \frac{1}{2} \min (x_1, \cdots, x_m)$. Now $\delta > 0$, and thus the nonvoid subset of S, $(0, \delta]$ is not covered by the above finite subcollection. Hence there is no finite subcollection of $\mathscr{G}$ that covers $S = (0, 3)$.

Special attention is given to those subsets S of E^n such that from any collection of open sets providing an open covering of S there is always a finite subcollection that covers S. The following definition names these sets.

Definition 2–6.2. *A set S in E^n is a compact set if for any collection $\mathscr{G}$ of open sets that provide an open covering of S there exists a finite subcollection of $\mathscr{G}$ that also is an open covering of S.*

Next, the Heine-Borel theorem assures us that there are sets that are compact.

Theorem 2–6.1. *Every closed and bounded set in E^n is a compact set.*

Proof for $n = 2$. Suppose S is a closed and bounded subset of E^2 and that $\mathscr{G} = \{G_\alpha : \alpha \in A\}$ is an open covering of S that contains no finite subcollection that is an open covering of S. Since S is bounded, there exists a closed interval $[\mathbf{a}, \mathbf{b}]$ in E^2 that contains S. Due to the original assumptions about S and $\mathscr{G}$, at least one of the four closed intervals I_{11}, I_{12}, I_{13}, or I_{14} (cf. Figure 2–4.2) is such that $S \cap I_{1i}$, $1 \leq i \leq 4$, requires infinitely many sets G_α of $\mathscr{G}$ to be an open covering of it. Call one of the four intervals having this property $[\mathbf{a}^1, \mathbf{b}^1]$. Repeat the argument again on $[\mathbf{a}^1, \mathbf{b}^1]$ obtaining an interval $[\mathbf{a}^2, \mathbf{b}^2]$ (cf. Figure 2–6.1) such that $S \cap [\mathbf{a}^2, \mathbf{b}^2]$ requires infinitely many sets G_α of $\mathscr{G}$ to be an open covering of it.

Suppose that the intervals $[\mathbf{a}, \mathbf{b}]$, $[\mathbf{a}^1, \mathbf{b}^1], \cdots, [\mathbf{a}^n, \mathbf{b}^n]$ have been obtained such that each interval is a subset of the preceding interval and such that $S \cap [\mathbf{a}^j, \mathbf{b}^j]$, $1 \leq j \leq n$, each require infinitely many subsets G_α of $\mathscr{G}$ for an open covering. Then again, since $S \cap [\mathbf{a}^n, \mathbf{b}^n]$ requires infinitely many sets G_α of $\mathscr{G}$ for an open covering, at least one of the intervals I_{n1}, I_{n2}, I_{n3}, or I_{n4} (cf. Figure 2–6.2) is such that $S \cap I_{ni}$, $1 \leq i \leq 4$, requires infinitely many sets G_α of $\mathscr{G}$ for an open covering. Thus by induction we have an infinite sequence of closed intervals, $[\mathbf{a}, \mathbf{b}]$, $[\mathbf{a}^1, \mathbf{b}^1], \cdots, [\mathbf{a}^n, \mathbf{b}^n], \cdots$, such that

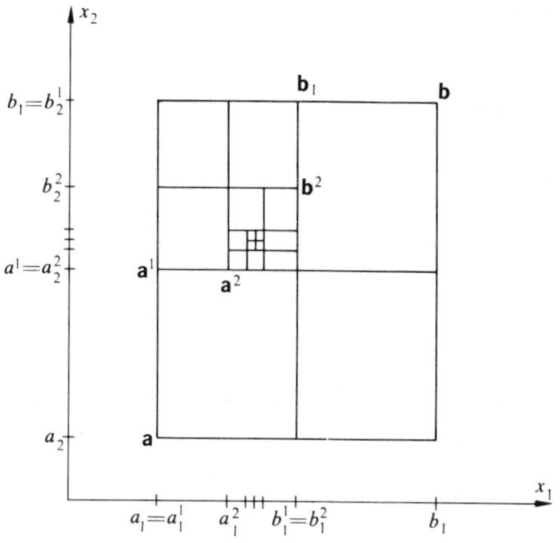

Figure 2–6.1

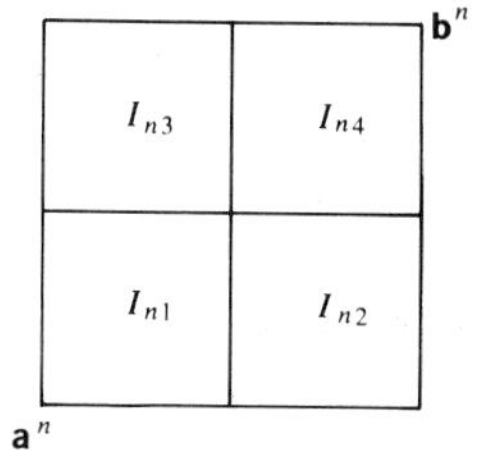

Figure 2–6.2

each is contained in the preceding one and such that $S \cap [\mathbf{a}^n, \mathbf{b}^n]$ requires infinitely many subsets G_α of $\mathscr{G}$ for an open covering.

Notice that on the x^i-axis, $i = 1, 2$, we will have a sequence of closed intervals $[a_i, b_i]$, $[a_i^1, b_i^1], \cdots, [a_i^n, b_i^n], \cdots$, such that each is contained in the preceding one, and such that the length of the $(n + 1)$st interval of the sequence is $b_i^n - a_i^n = (1/2^n)(b_i - a_i)$ (cf. Figure 2–6.1). This is precisely the circumstance of the similar construction in the proof of the Bolzano-Weierstrass theorem. Thus on the x_i-axis there exists a unique point $\bar{x}^i$ such that $\bar{x}^i \in [a_i^n, b_i^n]$ for all $n = 1, 2, 3, \cdots$, and $i = 1, 2$. Hence

$$\bar{\mathbf{x}} = (\bar{x}^1, \bar{x}^2) \in [\mathbf{a}^n, \mathbf{b}^n]$$

for $n = 1, 2, 3, \cdots$. We shall use this point $\bar{\mathbf{x}}$ to obtain the contradiction.

First we shall show that $\bar{\mathbf{x}}$ is an accumulation point of S and, hence, belongs to S, since S is a closed set. Let $N(\bar{\mathbf{x}}; r_1)$ be any neighborhood of $\bar{\mathbf{x}}$. Notice that the diagonal of $[\mathbf{a}^n, \mathbf{b}^n]$ is of length $\|\mathbf{b}^n - \mathbf{a}^n\| = (1/2^n)\,\|\mathbf{b} - \mathbf{a}\|$. Thus it is possible to take $n = n_1$, a sufficiently large index n_1, such that $(1/2^{n_1})\,\|\mathbf{b} - \mathbf{a}\| < r_1$. Thus, since $\bar{\mathbf{x}} \in [\mathbf{a}^{n_1}, \mathbf{b}^{n_1}]$, it follows that

$$[\mathbf{a}^{n_1}, \mathbf{b}^{n_1}] \subset N(\bar{\mathbf{x}}; r_1).$$

Now $[\mathbf{a}^{n_1}, \mathbf{b}^{n_1}]$ contains infinitely many points of S, or otherwise $S \cap [\mathbf{a}^{n_1}, \mathbf{b}^{n_1}]$ could not require infinitely many sets G_α of $\mathscr{G}$ for an open covering. Hence there must be a point of S different from $\bar{\mathbf{x}}$ belonging to $N(\bar{\mathbf{x}}; r_1)$. Therefore, $\bar{\mathbf{x}}$ is an accumulation point of S and thus belongs to S.

Since $\bar{\mathbf{x}} \in S$, then $\bar{\mathbf{x}} \in \bigcup_{\alpha \in A} G_\alpha$ and there exists at least one of the sets G_α, G_{α_0} such that $\bar{\mathbf{x}} \in G_{\alpha_0}$. The set G_α is open, and hence there exists a neighborhood $N(\bar{\mathbf{x}}; r_2) \subset G_{\alpha_0}$. Again it is possible to take $n = n_2$, a sufficiently large index such that

$$[\mathbf{a}^{n_2}, \mathbf{b}^{n_2}] \subset N(\bar{\mathbf{x}}; r_2) \subset G_{\alpha_0}.$$

Thus we have our contradiction; that is,

$$S \cap [\mathbf{a}^{n_2}, \mathbf{b}^{n_2}] \subset G_{\alpha_0} \quad \text{and} \quad S \cap [\mathbf{a}^{n_2}, \mathbf{b}^{n_2}]$$

do not require infinitely many sets G_α of $\mathscr{G}$ for an open covering. ∎

Thus the sets, like closed intervals, closed spheres, and finite sets, are all closed and bounded sets in E^n and hence are compact sets. Conceivably, there might be sets in E^n other than closed and bounded sets that are compact. However, the following theorem, the converse of the Heine-Borel theorem, assures us that this is not the case.

Theorem 2–6.2. *Every compact set in E^n is closed and bounded.*

Proof. Suppose S is a compact set of E^n. We shall show first that S is bounded. The collection of open sets, $\{N(\mathbf{x}; 1): \mathbf{x} \in S\}$ is an open covering of S. Thus there exists a finite subcollection, $\{N(\mathbf{x}_i; 1): 1 \leq i \leq k\}$, that is an open covering of S. Let

$$r = \max \{\|\mathbf{x}_i\| : 1 \leq i \leq k\}.$$

Then $S \subset N(\mathbf{0}; 1 + r)$, since $\mathbf{x} \in S$ implies $\|\mathbf{x} - \mathbf{x}_i\|$ for some i, $1 \leq i \leq k$, and

$$\|\mathbf{x}\| \leq \|\mathbf{x} - \mathbf{x}_i\| + \|\mathbf{x}_i\| \leq 1 + r.$$

We shall now show that S is closed. Suppose that $\mathbf{x}$ is an accumulation point of S that does not belong to S; that is, we deny that S is closed. Then since $\bar{N}(\bar{\mathbf{x}}; 1/k)$ is closed for $k = 1, 2, 3, \cdots$, the complement, $(\bar{N}(\bar{\mathbf{x}}; 1/k))^c$, is open. Furthermore, since

$$\bigcap_{k=1}^{\infty} N\left(\mathbf{x}; \frac{1}{k}\right) = \{\bar{\mathbf{x}}\},$$

it follows by De Morgan's law that

$$\{\bar{\mathbf{x}}\}^c = \bigcup_{k=1}^{\infty} \left(N\left(\mathbf{x}; \frac{1}{k}\right)\right)^c.$$

Since $\bar{\mathbf{x}} \notin S$, $S \subset \{\bar{\mathbf{x}}\}^c$ and $\{(\bar{N}(\mathbf{x}; 1/k))^c; k = 1, 2, 3, \cdots\}$ is an open covering of S. Thus there is a finite subcollection $(\bar{N}(\mathbf{x}; 1/k_1))^c, \cdots, (\bar{N}(\mathbf{x}; 1/k_m))^c$

that gives an open covering of S. Let

$$\delta = \min\left\{\frac{1}{k_i}\,;\,1 \le i \le m\right\}.$$

Then $S \subset (\bar{N}(\mathbf{x};\,\delta))^c$ and $N(\mathbf{x};\,\delta) \subset \bar{N}(\mathbf{x};\,\delta) \subset S^c$. Thus $N'(\mathbf{x};\,\delta)$ contains no point of S. This contradicts the assumption that $\bar{\mathbf{x}}$ was an accumulation point of S. Therefore, all accumulation points of S belong to S and S is closed. ▎

Problems

1. Determine in the following examples in E^1 whether or not the given collection of sets is an open covering of the indicated set.
   (a) $\{(x - (1/10),\, x + (1/10)) : x \in (0, 1)\}$, $S = (0, 1)$;
   (b) $\{(-n/(n + 1),\, n/(n+1)) : n = 1, 2, \cdots\}$, $S = (-1, 1)$;
   (c) $\{(-1, n) : n = 1, 2, 3, \cdots\}$, $S = [0, \infty)$;
   (d) $\{((1/n) - (1/n^2),\, (1/n) + (1/n^2)) : n = 2, 3, 4, \cdots\}$,
   $S = \{1/n : n = 1, 2, 3, \cdots\}$;
   (e) $\{[0, 1/2^n) : n = 1, 2, 3, \cdots\}$, $S = \{1/n : n = 3, 4, 5, \cdots\}$;
   (f) $\{(\log(1/x), \log x) : x > 1\}$, $S = \{-5, 2, 0, 6, 10\}$.
2. Which of the collections of sets of Problem 1 contain a finite subcollection of sets that cover the indicated set? Demonstrate the reasons for your answer.
3. Use Problem 1(c) and (d) to show that both parts of the hypothesis of Theorem 2–6.1 are essential.
4. Which of the following sets in E^2 are compact? Give reasons for your answer. Make a sketch of each of the sets.
   (a) $\{(x_1, x_2) : x_1^2 - x_2^2 \le 1\}$;
   (b) $\{(x_1, x_2) : x_2 = \sin(1/x_1),\, |x_1| \le \pi\}$;
   (c) $\{(r, \theta) : r = e^{-\theta},\, \theta \ge 0\}$, where r and θ are polar coordinates;
   (d) $\{(x_1, x_2) : x_2 = \log x_1,\, 0 < x_1 \le 1\}$;
   (e) $\{(x_1, x_2) : x_1^6 + x_2^6 \le 1\}$;
   (f) $\{(x_1, x_2) : x_1 \text{ is rational and } 0 \le x_1 \le 1,\, |x_2| \le 1\}$.
5. State and prove the Heine-Borel theorem for E^1.
6. Let S be a nonempty closed and bounded subset of E^1. Suppose $S \subset [a, b]$. Let $a_1 = \inf(S)$ and $b_1 = \sup(S)$. Show that a_1 and $b_1 \in S$ and that $a \le a_1 \le b_1 \le b$.
7. Let $\mathscr{G} = \{G_\alpha\}$ be a collection of open sets covering the set S of Problem 6. Let S_1 be the set of x such that $x \in S$ and $S \cap [a, x]$ can be covered by a finite number of sets from the collection $\mathscr{G}$. Show S_1 is not empty. Let $x_1 = \sup(S_1)$. Show that $a_1 < x_1 \le b_1$ and that $x_1 \in S$. Since $x_1 \in S$, then there exists an open set $G_{\alpha 0}$ such that $x_1 \in G_{\alpha 0}$. Since $G_{\alpha 0}$ is open, there

exists a $\delta > 0$ such that $a_1 \leq x_1 - \delta$ and $(x_1 - \delta, x_1 + \delta) \subset G_{\alpha 0}$. For any x' such that $x_1 - \delta < x' < x_1$, the set $[a_1, x'] \cap S$ can be covered by a finite number of open sets, $G_{\alpha 1}, G_{\alpha 2}, \cdots, G_{\alpha m}$ from the collection $\mathscr{G}$. If $x_1 + \delta > b_1$ then $S = S \cap [a_1, x_1 + \delta) \subset \bigcup_{i=0}^{m} G_{\alpha i}$, and thus S is compact. Show $x_1 + \delta \leq b_1$ leads to a contradiction.

Answers

1. Yes, for each case except (e).
2. The collections of (a), (e), and (f) contain a finite subcollection of sets that cover the indicated set. The other collections do not.
4. Only the set of part (e) is compact.

2–7. Connectedness and Regions

There are several different types of connectedness that have bearing on the problems we shall consider. Below is a formal definition of three types of connectedness: *connected*, *arcwise connected*, and *polygonally connected.*

Definition 2–7.1. *A set S in E^n is:*

(a) *connected, if there do not exist nonempty sets S_1 and S_2 in E^n such that $S = S_1 \cup S_2$, $S_1 \cap \bar{S}_2$ is empty and $\bar{S}_1 \cap S_2$ is empty:*

(b) *arcwise connected, if any two points of S can be joined by an arc (curve) that lies entirely in S; or*

(c) *polygonally connected, if any two points of S can be joined by a curve that lies entirely in S where the curve is the union of a finite number of line segments.*

To illustrate the meaning of these definitions we shall consider some examples.

Example 2–7.1. The following three sets in E^2 are not arcwise connected or connected (cf. Figure 2–7.1).

(a) $\{(x_1, x_2): x_1 x_2 = 1\}$,

(b) $\{(x_1, x_2): |x_1 + 1| + |x_2| < 1\} \cup \{(x_1, x_2): |x_1 - 1| + |x_2| < 1\}$,

(c) $\{(x_1, x_2): x_2 = \sin(1/x_1), 0 < x_1 \leq 1/\pi\} \cup \{(x_1, 0): x_1 < 0\}$

The following three sets, which are only slight modifications of the first

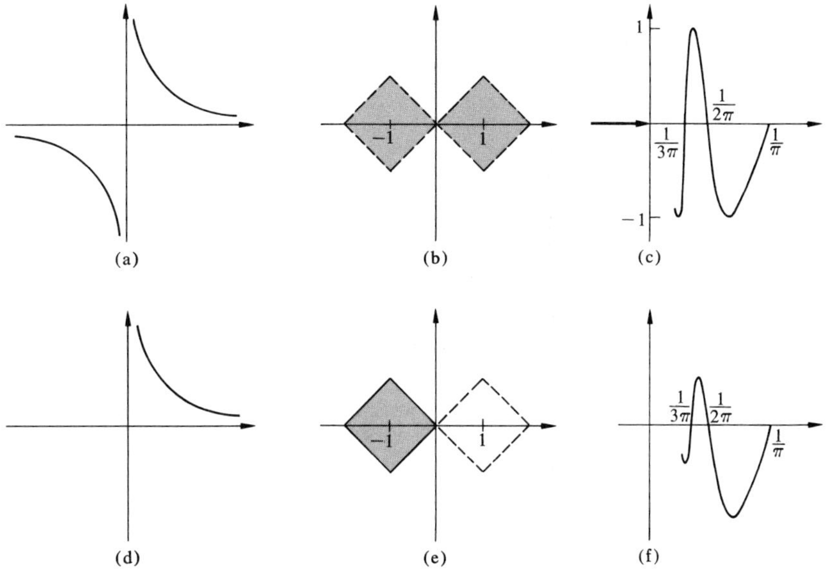

Figure 2–7.1

three, are arcwise connected and connected:

(d) $\{(x_1, x_2): x_1 x_2 = 1, x_1 > 0\}$,

(e) $\{(x_1, x_2): |x_1 + 1| + |x_2| \leq 1\} \cup \{(x_1, x_2): |x_1 - 1| + |x_2| < 1\}$,

(f) $\{(x_1, x_2): x_2 = x_1 \sin (1/x_1), 0 < x_1 < 1/\pi\} \cup \{(x_1, 0): x_1 \leq 0\}$.

Only (e) is polygonally connected.

It is clear from the definition that any polygonally connected set is arcwise connected. The following example shows that there are connected sets that are not arcwise connected.

Example 2–7.2. In E^2 let $S = S_1 \cup S_2$ where

$$S_1 = \{(x_1, x_2): x_2 = \sin (1/x_1), 0 < x_1 \leq 1/\pi\}, \qquad S_2 = \{(0, x_2): |x_2| \leq 1\}.$$

This set S, which is illustrated in Figure 2–7.2, is not arcwise connected since we cannot, for example, join (0, 0) to $(1/\pi, 0)$ by an arc within S. However, if we try to get two nonempty sets T_1 and T_2 such that $S = T_1 \cup T_2$ and $T_1 \cap \bar{T}_2$ and $\bar{T}_1 \cap T_2$ are empty, then it is clearly impossible if both T_1 and T_2 meet S_1 (or S_2). The only hope is that we might take $T_1 = S_1$ and $T_2 = S_2$, but this is also impossible, since $\bar{S}_1$ meets S_2. Therefore, S is connected.

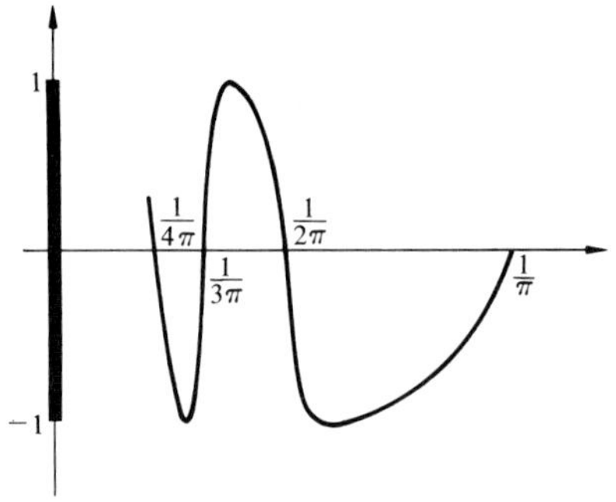

Figure 2–7.2

The following proposition summarizes the relationship between these three types of connectedness. The above two examples demonstrate part of the proof of the proposition, and the remainder of the proof is included in the problems.

Proposition 2–7.1. *In E^n, a polygonally connected set is an arcwise connected set; an arcwise connected set is a connected set. For $n > 1$, the converse statements of both of the above statements are not true. However, if S is an open set in E^n, then the statements "S is polygonally connected," "S is arcwise connected," and "S is connected" are equivalent statements.*

Since the three types of connectedness are equivalent for open sets, it is desirable to have a unique name for a set of this kind rather than at one time calling it "an arcwise connected open set," another time "a polygonally connected open set," and another time "a connected open set."

The terminology we shall use is *open connected set.* Thus an open connected set in E is an open set that is connected (arcwise connected or polygonally connected). An open connected set is sometimes called a "domain" or an "open region" by other writers. The reasoning behind this name "open region" becomes clear in light of the following definition.

Definition 2–7.2. *A set S in E^n is a region if S is an open connected set plus some, none, or all of its boundary points.*

The two extreme cases—that is, a region with no boundary points and a region with all of its boundary points—are called an *open region* and a *closed region*, respectively. Thus an open region is a connected set and an open set. Since the boundary of an open set contains all of the accumulation points of the set that the set does not already have, a closed region is a closed set.

In Figure 2–7.1(e), the set illustrated is not a region, since there is no open connected set such that some boundary points could be added to it to obtain this set. Thus Definition 2–7.2 implies that if all boundary points of a region S are removed, that is, $S - \text{bdy}(S)$, which is S°, it is an open connected set (cf. Problem 10 of Section 2–3).

Next are some examples of sets that will help to explain better the meaning of a region.

Example 2–7.3. The following sets are all regions in E^2.

(a) $\{(x_1, x_2): x_1^2 + 4x_2^2 \leq 4\}$,

(b) $\{(x_1, x_2): x_1 x_2 < 1\}$,

(c) $\{(x_1, x_2): 0 \leq x_1 \leq 1, |x_2| \leq 1\} - \bigcup_{n=2}^{\infty} \{(1/n, 0)\}$,

(d) $\{(x_1, x_2): 0 \leq x_1 \leq 1, 0 \leq x_2 \leq 1\}$
$$- \bigcup_{n=2}^{\infty} \{(1/n, x_2): 0 \leq x_2 \leq \tfrac{1}{2}\},$$

(e) $\{(r, \theta): r_1(\theta) \leq r \leq r_2(\theta), 0 \leq \theta\}$ where r and θ are the polar coordinates and where

$$r_1(\theta) = 2\pi\left[\frac{1}{\theta + 2\pi} - \frac{1}{2}\left|\frac{1}{\theta + 4\pi} - \frac{1}{\theta + 2\pi}\right|\right]$$

and

$$r_2(\theta) = 2\pi\left[\frac{1}{\theta + 2\pi}\right],$$

(f) $\{(x_1, x_2): 0 \leq x_1 \leq 1, 0 \leq x_2 \leq 1\} \cup (\bigcup_{n=1}^{\infty} T_n)$ where T_n is the closed triangular region with vertices

$$\left(\frac{1}{2^n} - \frac{1}{2^{n+2}}, 1\right), \quad \left(\frac{1}{2^n} + \frac{1}{2^{n+2}}, 1\right), \quad \text{and} \quad \left(\frac{1}{2^n}, 2\right).$$

Each of these sets is illustrated in Figure 2–7.3. The region in (a) is a closed region, while the set in (b) is an open region. All of the other regions are neither open nor closed.

A *bounded region* is a region that is a bounded set. A *compact region* is a region that is a compact set—that is, a closed and bounded region.

Now we wish to define a special kind of region in E^2.

Definition 2–7.3. *Any compact region of the plane whose boundary is a simple piecewise smooth closed curve is a plane regular region.*

Of the regions in Example 2–7.3, only (a) is a plane regular region. The region in (b) is not closed, but even if we take the closure of this region we

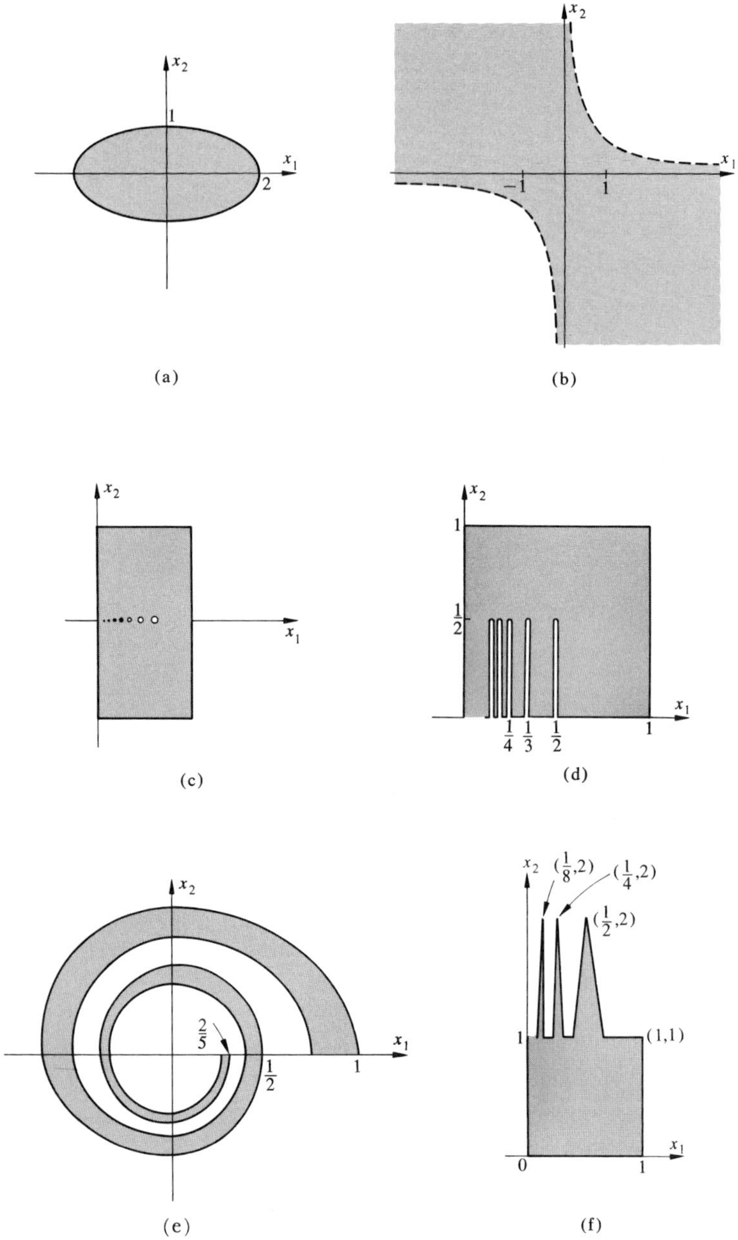
x_2
1
x_1
2
(a)
x_2
x_1
−1
1
(b)
x_2
x_1
(c)
x_2
1
$\frac{1}{2}$
x_1
$\frac{1}{4}$ $\frac{1}{3}$ $\frac{1}{2}$
1
(d)
x_2
$\frac{2}{5}$
$\frac{1}{2}$
1
x_1
(e)
x_2
$(\frac{1}{8},2)$
$(\frac{1}{4},2)$
$(\frac{1}{2},2)$
1
(1,1)
x_1
0
1
(f)

Figure 2–7.3

would not have a compact region. However, the closure of the region in (c) and (d) would be plane regular regions. The region in (e) would be closed if we added the point at the origin, and the region thus closed would be a plane regular region. The closure of the region in (f) would yield a compact set, but its boundary would not be of the required type; it would not be piecewise smooth, for one thing.

Lastly, we wish to establish one more concept about plane regular regions: the idea of a plane regular region being the sum of plane regular regions.

Definition 2–7.4. *A region S in a plane is a sum of the plane regular regions $S_1, S_2, \cdots, S_m$, written $S = S_1 + \cdots + S_m$, if $S = \bigcup_{i=1}^{m} S_i$ and if each set $S_i \cap S_j$, $1 \leq i < j \leq m$, is a simple smooth curve that belongs to the* bdy (S_i) *and the* bdy (S_j); *is one or both end points of a simple smooth curve that belong to the* bdy (S_i) *and the* bdy (S_j); *or is empty. Such a set S is a finite plane region.*

In Figure 2–7.4 there are some examples that illustrate the meaning of this definition. In each of the figures (a), (b), (c), and (d), the shaded set is S, a finite plane region. In (a) and (b), S is a sum of the plane regular regions S_1, S_2, S_3, and S_4. Notice in (a) that S_1 and S_2 meet in a simple

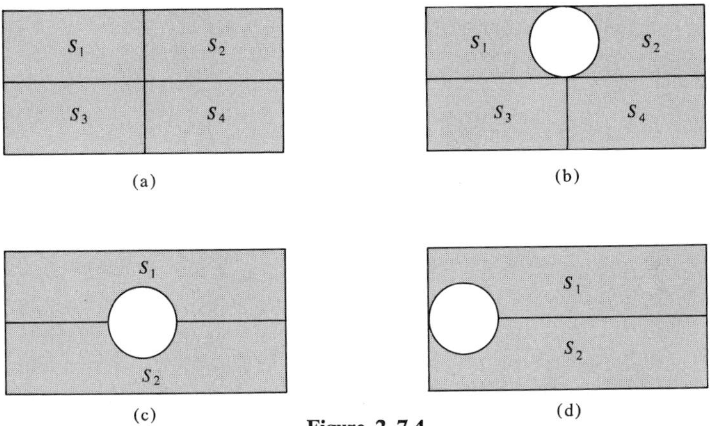

Figure 2–7.4

smooth curve that is in the boundary of S_1 and S_2, that S_1 and S_3 meet in a simple smooth curve, and that S_1 and S_4 meet in the end point of a simple smooth curve that is in the boundary of both S_1 and S_4. Similarly, as we check the other pairs, we find that the conditions of Definition 2–7.4 are satisfied. In (b) we have an example to show how two regions, S_1 and S_2, can have in common two end points of a simple smooth curve that lies in the bdy (S_1) and the bdy (S_2). In (c) and (d), S is not the sum of regions S_1 and S_2, for in (c), $S_1 \cap S_2$ consists of two simple smooth curves in the boundary of each region and in (d), $S_1 \cap S_2$ consists of a simple smooth curve and an end point of another simple smooth curve that are in the boundary of S_1 and S_2.

The idea of a finite plane region being a sum of plane regular regions is in a sense a decomposition of a plane regular region into smaller plane regular regions. The whole idea is akin to taking a partition of an interval $[a, b]$. The partition gives a decomposition of the interval into a "sum" of smaller subintervals. Use will be made of this idea of a sum of plane regular regions in subsequent chapters.

Problems

1. Show that in E^2 the following sets have the properties indicated.
   (a) $\{(x_1, x_2): x_2 \neq 0\}$, not connected;
   (b) $\{(x_1, x_2): x_1 x_2 > 0\} \cup \{(0, 0)\}$, polygonally connected;
   (c) $\{(x_1, x_2): x_1^2 + x_2^2 = 1\}$, arcwise connected.
2. Classify the sets shaded in Figure 2–7.5 as to connectedness. In (c) consider the following two possibilities: that

$$\{(0, x_2): (-1/9) \leq x_2 \leq (10/9)\}$$

   belongs to the shaded set; does not belong to the shaded set.
3. The regions of Example 2–7.3 are all arcwise connected. Consider the closure of each of these regions. Which of them are arcwise connected? What is the boundary of each of the given regions?
4. In Figure 2–7.4 the plane regular regions S in (c) and (d) are not sums of S_1 and S_2. Decompose S into a sum of plane regular regions. Try to do this with the minimum number of subregions. Is it possible to have only two subregions? Suppose that S in (c) is of length two units and of width one unit with the hole of diameter $1/2$ in the center. Decompose S into a finite sum of plane regular regions such that the boundary of each subregion consists of exactly three simple smooth curves and such that each subregion is small enough to be contained in a disk of radius $1/4$.
5. In E^1 prove that the three types of connectedness are equivalent and that a connected set in E^1 is an interval.
6. Prove that an open arcwise connected set is connected.
7. Prove that an open arcwise connected set is polygonally connected.

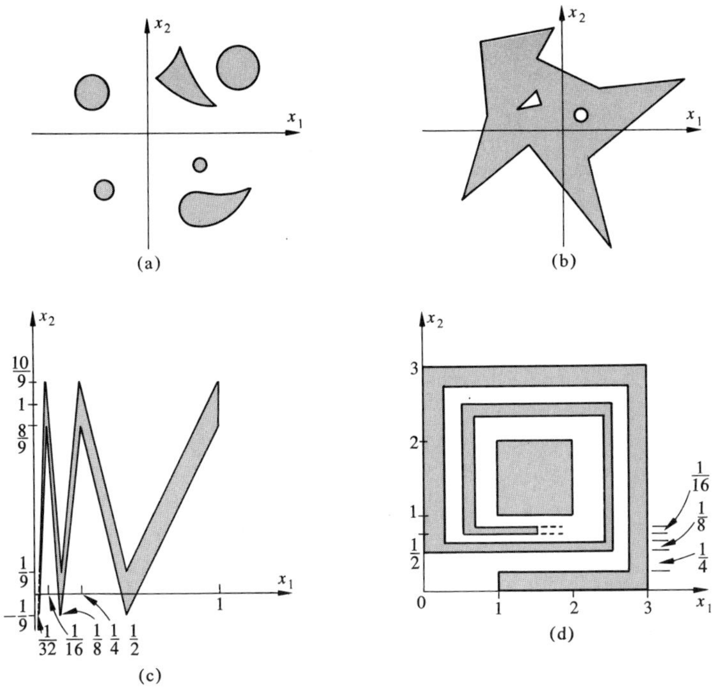

Figure 2–7.5

8. Suppose S is an open connected set in E^n. Let $\mathbf{x}_0$ belong to S, let S_1 be all of those points of S that can be joined to $\mathbf{x}_0$ by an arc in S and let S_2 be the points of S not in S_1. Show that the sets S_1 and S_2 are open sets. Show that S_2 is empty and hence $S_1 = S$, and thus S is arcwise connected.
9. Prove that a compact region is arcwise connected.
10. Is the intersection of two plane regular regions a plane regular region?

Answers

2. (a) Not connected, arcwise connected, nor polygonally connected.
   (b) Connected, arcwise connected, and polygonally connected.
   (c) If the set on the x_2-axis belongs to the shaded set, then the set is connected but not arcwise nor polygonally connected. If it does not belong, then the shaded set is connected, arcwise connected, and polygonally connected.
   (d) Connected, but not arcwise nor polygonally connected.

3

Differential Calculus of Functions of Several Variables

3–1. Limits and Continuity

Let f be a real-valued function whose domain, Do $(f) \subset E^n$; that is, f is a function of n real variables. Thus, the range of f, Ra $(f) \subset R$, the set of real numbers. If $S =$ Do (f), then it is said that f maps S into R, which is denoted by $f\colon S \to R$. The set of ordered pairs $(\mathbf{x}, f(\mathbf{x}))$ in E^{n+1} where $\mathbf{x} \in S$ is the *graph* of f on S.

For functions of several variables, as in the case of the calculus of a real-valued function of a single variable, the concept of a limit is fundamental.

Definition 3–1.1. *Let $f\colon S \to R$ where S is a subset* of E^n. *If $\mathbf{x}_0$ is an accumulation point of S, the domain of f, then* $\lim_{\mathbf{x}\to\mathbf{x}_0} f(\mathbf{x}) = b$; *that is, the limit of f equals b as $\mathbf{x}$ approaches $\mathbf{x}_0$ if, given an arbitrary positive number* $\varepsilon > 0$, *there exists a positive number $\delta > 0$ such that $|f(\mathbf{x}) - b| < \varepsilon$ when* $\mathbf{x} \in N'(\mathbf{x}_0\,; \delta) \cap S$.

Note that the above definition could be rephrased slightly as follows: ... $\lim_{\mathbf{x}\to\mathbf{x}_0} f(\mathbf{x}) = b$ if, given a neighborhood $N(b; \varepsilon)$, there exists a neighborhood $N(\mathbf{x}_0; \delta)$ such that $f(\mathbf{x}) \in N(b; \varepsilon)$ when $\mathbf{x} \in N'(\mathbf{x}_0; \delta) \cap S$. Stating it crudely, it means that the function f has the limit b if $f(\mathbf{x})$ is "close to b" when $\mathbf{x}$ is "close but not equal to $\mathbf{x}_0$."

With the idea of a limit, it is now possible to define the continuity property of a function.

Definition 3–1.2. *Let $S \subset E^n$ and let $f: S \to R$. Then $\mathbf{x}_0$, a point of S, the domain of f, is a point of continuity of f if*

(i) $\mathbf{x}_0$ *is an accumulation point of S and* $\lim_{\mathbf{x}\to\mathbf{x}_0} f(\mathbf{x}) = f(\mathbf{x}_0)$, *or if*

(ii) $\mathbf{x}_0$ *is not an accumulation point of S, that is, $\mathbf{x}_0$ is an isolated point of S. If $\mathbf{x}_0$ is a point of continuity of f, then f is said to be* continuous at $\mathbf{x}_0$. *The function f is* continuous on S_1, *a subset of S, if f is continuous at each point of S_1. The function f is discontinuous at all points where it is not continuous.*

The following examples will serve to illustrate the above definitions.

Example 3–1.1. Consider the function f defined by

$$f(x_1, x_2) = \sqrt{4 - x_1^2 - x_2^2}.$$

The graph of the function is the upper half of a hemispherical surface of

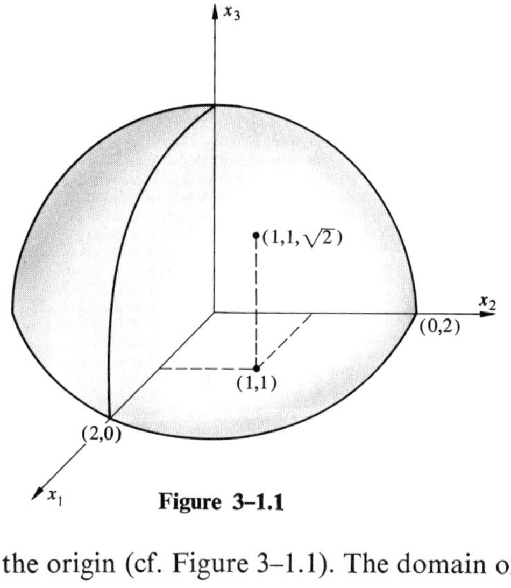

Figure 3–1.1

radius 2 centered at the origin (cf. Figure 3–1.1). The domain of the function is $\{(x_1, x_2): x_1^2 + x_2^2 \leq 4\}$, since otherwise $f(x_1, x_2)$ would be imaginary.

The function is continuous at each point of its domain. In particular, at the point $(1, 1), f(1, 1) = \sqrt{2}$. Taking ε to be an arbitrary small positive number, the set of points in E^3 such that $|x_3 - \sqrt{2}| < \varepsilon$ is a horizontal slab of thickness 2ε with its faces missing. The faces are missing because of the strict inequality. The set $N((1, 1); \delta)$ in E^3 is an infinite open cylinder of radius δ parallel to the x_3-axis and centered at (1, 1). The intersection of the horizontal slab and the infinite cylinder is a finite cylinder resembling a tin can (cf. Figure 3–1.2). The can is of height 2ε and of radius δ centered at $(1, 1, \sqrt{2})$.

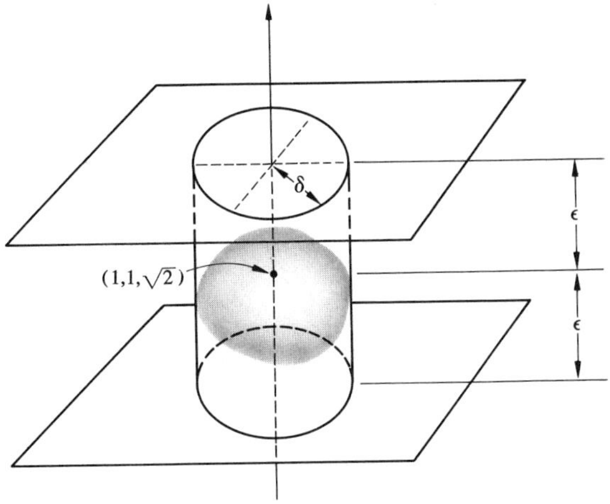

Figure 3–1.2

Thus the definition of continuity at (1, 1) says geometrically that given $\varepsilon > 0$ it is possible to pick the radius δ of the infinite cylinder small enough that the intersection of this cylinder and the hemisphere lies below $\sqrt{2} + \varepsilon$ and above $\sqrt{2} - \varepsilon$; in other words, it lies inside this finite can of height 2ε and radius δ centered at $(1, 1, \sqrt{2})$.

Notice in the definitions: . . . $\mathbf{x} \in N'(\mathbf{x}_0; \delta) \cap S$. The function f is only defined on the points S. Hence $f(\mathbf{x})$ in $|f(\mathbf{x}) - b| < \varepsilon$ would be undefined and thus meaningless if $\mathbf{x} \notin S$. For instance, in the above example, if the point to consider were (2, 0) instead of (1, 1), then $N((2, 0); \delta)$ would always contain points not in the domain of f—that is, outside of the circle given by $x_1^2 + x_2^2 = 4$. The function f has no meaning at these points, and hence they are of no significance in the consideration of the continuity of f.

The following two examples illustrate functions that show various things that can happen relative to limits and continuity.

Example 3–1.2. Let f be defined by the following equation:

$$f(x_1, x_2) = \frac{(x_1^2 + x_2^2)\sqrt{4 - x_1^2 - x_2^2}}{x_1^2 + x_2^2}.$$

The domain of definition is now $\{(x_1, x_2): 0 < x_1^2 + x_2^2 \leq 4\}$ since at (0, 0) the denominator is zero, and thus the value of the fraction is undefined. Since at all points other than (0, 0) the factors $x_1^2 + x_2^2$ can be canceled, the graph of f is a hemispherical surface with a point missing at (0, 0, 2). Since the point (0, 0) is not in the domain of f, then (0, 0) is not a point of continuity of f. However,

$$\lim_{\mathbf{x} \to (0,0)} f(\mathbf{x}) = 2.$$

Thus it is possible for a function to have a limit at a point and still not be continuous at the point. This particular discontinuity could be remedied if f were defined at (0, 0) to be 2. For this reason, such a discontinuity is called a *removable discontinuity*.

Example 3–1.3. Let the function f be defined as follows

$$f(x_1, x_2) = \begin{cases} \dfrac{x_1^2 - 4x_2^2}{x_1^2 + 4x_2^2}, & (x_1, x_2) \neq (0, 0), \\ 0, & (0, 0). \end{cases}$$

The domain of the function is all of E^2. In particular, (0, 0) belongs to the domain and is, of course, an accumulation point of the domain. Yet the function is discontinuous at (0, 0), since it has no limit at (0, 0). This can best be observed by considering the value of the function along lines $x_2 = mx_1$ through the origin. Hence, except at the origin,

$$f(x_1, mx_1) = \frac{x_1^2 - 4m^2x_1^2}{x_1^2 + 4m^2x_1^2}$$

$$= \begin{cases} \dfrac{1 - 4m^2}{1 + 4m^2}, & x_2 = mx_1, \\ -1, & x_1 = 0. \end{cases}$$

The family of lines in the surface is indicated by Figure 3–1.3. Thus in any deleted neighborhood of (0, 0), $N'((0, 0), \delta)$ there would be points such that $f(\mathbf{x})$ evaluated at those points has the value 1—namely, on the line $x_2 = 0$. Also on the line $x_1 = 0$ there would be points where $f(\mathbf{x})$ has the

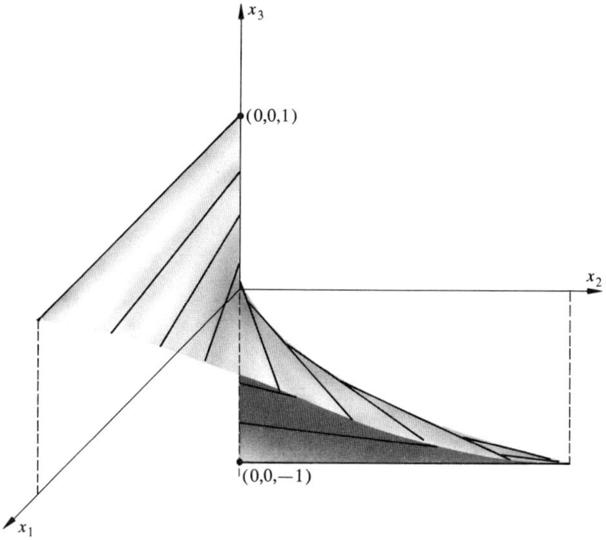

Figure 3–1.3

value −1. Thus take $\varepsilon = \frac{1}{2}$. Then

$$\begin{aligned} |f(\mathbf{x}) - f(\mathbf{x}_0)| &= |f(\mathbf{x}) - 0| \\ &= |f(\mathbf{x})| \\ &= 1 \nless \varepsilon \end{aligned}$$

for some $\mathbf{x} \in N'((0, 0); \delta)$ and for any given $\delta > 0$. Furthermore, the function cannot be redefined at (0, 0) and make the function continuous at (0, 0). Regardless of how f is defined at (0, 0), there are points of $N'((0, 0); \delta)$ for any $\delta > 0$ such that $|f(\mathbf{x}) - f((0, 0))| \geq 1$.

Example 3–1.4. Consider the function defined by the equation

$$f(x_1, x_2) = \sqrt{-((x_1 - 2)^2 + (x_2 - 1)^2)}.$$

The domain of this function consists of a single point, namely, (2, 1). Since (2, 1) is a point of the domain but not an accumulation point of the domain, f is continuous at (2, 1) (cf. (ii) of Definition 3–1.2). This part of the definition is something of a convention that deals only with these exceptional points of the domain. However, the reader should bear in mind that this convention has been made.

The following theorem, which will not be proved, is the fundamental theorem of limits for real-valued functions of several variables. It is analogous to the earlier theorems on limits for real-valued functions of a single variable. Proofs could be given in an analogous manner.

Theorem 3–1.1. *Let f and g map S, a subset of E^n, into R, and let $\mathbf{x}_0$ be an accumulation point of S. If*

$$\lim_{\mathbf{x}\to\mathbf{x}_0} f(\mathbf{x}) = a$$

and

$$\lim_{\mathbf{x}\to\mathbf{x}_0} g(\mathbf{x}) = b,$$

then

$$\text{(i)}\quad \lim_{\mathbf{x}\to\mathbf{x}_0} [f(\mathbf{x}) + g(\mathbf{x})] = a + b,$$

$$\text{(ii)}\quad \lim_{\mathbf{x}\to\mathbf{x}_0} [f(\mathbf{x})g(\mathbf{x})] = ab,$$

$$\text{(iii)}\quad \lim_{\mathbf{x}\to\mathbf{x}_0} [f(\mathbf{x})/g(\mathbf{x})] = \frac{a}{b} \quad \text{if } b \neq 0.$$

From this theorem on limits, it is very easy to get the following fundamental theorem about continuity.

Theorem 3–1.2. *Let f and g be functions with domain in E^n and range in R. If $\mathbf{x}_0$ is a point of continuity of f and g, then $f + g$ and fg are continuous at $\mathbf{x}_0$. If $g(\mathbf{x}_0) \neq 0$ then f/g is continuous at $\mathbf{x}_0$.*

In short, the last theorem says that the sums and products of functions that are continuous at a point are also continuous at the same point. The same is true for quotients except at points where the denominator is zero.

With the foregoing theorem about sums, products, and quotients and the following theorem about composites of continuous functions, it is possible to affirm that many of the functions that the reader encounters are, indeed, functions whose points of discontinuity are somewhat rare.

Theorem 3–1.3. *Let $f_i : S \to R$ where S is a subset of E^n for $i = 1, \cdots, m$. Let $g: T \to R$ where T is a subset of E^m. Let*

$$\{\mathbf{f}(\mathbf{x}) = (f_1(\mathbf{x}), \cdots, f_m(\mathbf{x})): \mathbf{x} \in S\} \subset T.$$

Furthermore, let $\mathbf{x}_0$ be an accumulation point of S such that each f_i is continuous at $\mathbf{x}_0$ and such that

$$\mathbf{f}(\mathbf{x}_0) = (f_1(\mathbf{x}_0), \cdots, f_m(\mathbf{x}_0))$$

is an accumulation point of T, and g is continuous at $\mathbf{f}(\mathbf{x}_0)$. Then the composite function defined by

$$g(\mathbf{f}(\mathbf{x})) = g(f_1(\mathbf{x}), \cdots, f_m(\mathbf{x}))$$

is continuous at $\mathbf{x}_0$.

The last theorem can be used in the following manner. Is the function h defined by

$$h(x_1, x_2) = \sin(x_1^2 + 2x_1 x_2 + x_2^3)$$

continuous for all $\mathbf{x} \in E^2$? The answer is yes. Let $g(u) = \sin u$ and let $f(x_1, x_2) = x_1^2 + 2x_1 x_2 + x_2^3$. For the moment, grant that polynomials of any finite number of variables and any degree are continuous everywhere. Recall that $g = \sin$ is continuous everywhere. Thus the hypothesis of the above theorem is satisfied. In this case, $n = 2$ and $m = 1$.

In Definition 3–1.3, which follows, we see that a function f such that $f: S \to R$, where $S \subset E^n$ is continuous at a point $\mathbf{x}_0 \in S$ if, given $\varepsilon > 0$, there exists a $\delta > 0$ such that $|f(\mathbf{x}) - f(\mathbf{x}_0)| \leq \varepsilon$ for all $\mathbf{x} \in S \cap N(\mathbf{x}_0; \delta)$. The number δ will depend upon ε and the choice of the point $\mathbf{x}_0$. In the proofs of certain basic theorems in integration theory, it is useful to know that if f is continuous on a set S, then, given $\varepsilon > 0$, there exists a sufficiently small $\delta > 0$ such that for any $\mathbf{x}_0 \in S$,

$$|f(\mathbf{x}) - f(\mathbf{x}_0)| \leq \varepsilon$$

for all $\mathbf{x} \in S \cap N(\mathbf{x}_0; \delta)$. In such cases the function f is said to be *uniformly continuous* on S.

Definition 3–1.3. *Let f be a function such that $f: S \to R$ where $S \subset E^n$. If, given $\varepsilon > 0$, there exists $\delta > 0$ such that*

$$|f(\mathbf{x}_1) - f(\mathbf{x}_2)| < \varepsilon$$

for any $\mathbf{x}_1$ and $\mathbf{x}_2$ in S such that $\|\mathbf{x}_1 - \mathbf{x}_2\| < \delta$, then f is uniformly continuous on S.

The functions f and g defined by

$$f(x) = e^x, \quad g(x) = \frac{1}{x}$$

on the sets R and $(0, 1]$, respectively, are not uniformly continuous on their domain sets, but the function h defined by

$$h(x) = x^2$$

is uniformly continuous on $[0, 1]$ (cf. Figure 3–1.4). In the case of the function h, it is intuitively clear that the point $x = 1$, where $f'(x) = 2x$ approaches 2 (its largest value on $[0, 1]$), is the point where the smallest δ

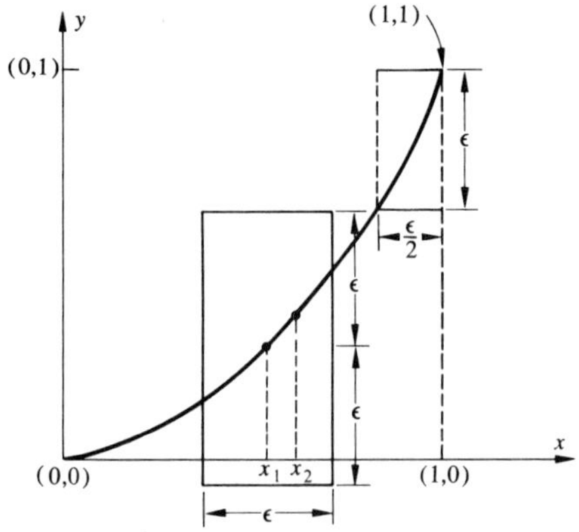

Figure 3–1.4

would be required. That is, if $\delta = \varepsilon/2$, then we can show that $|f(x_1) - f(x_2)| \le \varepsilon$ when x_1 and $x_2 \in [0, 1]$ and $|x_1 - x_2| < \delta$.

The following theorem gives a sufficient condition on the set S—that is, compactness—such that a continuous function on S is uniformly continuous.

Theorem 3–1.4. *Let f be a function such that $f\colon S \to R$ where $S \subset E^n$. If f is continuous on S and if S is compact, then f is uniformly continuous on S.*

Proof. Since f is continuous on S it follows that for any given $\varepsilon > 0$ and for any point $\mathbf{a} \in S$, there exists $\delta(\mathbf{a}) > 0$, ($\delta(\mathbf{a})$ depends upon $\mathbf{a}$), such that $|f(\mathbf{x}) - f(\mathbf{a})| < \varepsilon/2$ when $\|\mathbf{x} - \mathbf{a}\| < \delta(\mathbf{a})$ and $\mathbf{x} \in S$. The collection of sets, $\{N(\mathbf{a}; \delta(\mathbf{a})/2)\colon \mathbf{a} \in S\}$, give an open covering of S. Since S is compact, there exists a finite subcollection,

$$\left\{N\left(\mathbf{a}_i ; \delta\left(\frac{\mathbf{a}_i}{2}\right)\right) : \mathbf{a}_i \in S,\ 1 \le i \le m\right\},$$

that gives an open covering of S. Let $\delta = \min\{\delta(\mathbf{a}_i)/2\colon 1 \le i \le m\}$. If $\mathbf{x}_1$ and $\mathbf{x}_2 \in S$ and $\|\mathbf{x}_1 - \mathbf{x}_2\| < \delta$, then $\mathbf{x}_1 \in N(\mathbf{a}_i ; \delta(\mathbf{a}_i)/2)$ for some $1 \le i \le m$ and

$$\|\mathbf{x}_2 - \mathbf{a}_i\| \le \|\mathbf{x}_2 - \mathbf{x}_1\| + \|\mathbf{x}_1 - \mathbf{a}\| < \frac{\delta(\mathbf{a}_i)}{2} + \frac{\delta \mathbf{a}_i}{2} < \delta(\mathbf{a}_i).$$

Thus from the way that $N(\mathbf{a}_i ; \delta(\mathbf{a}_i))$ was chosen, it follows that

$$|f(\mathbf{x}_1) - f(\mathbf{x}_2)| \le |f(\mathbf{x}_1) - f(\mathbf{a}_i)| + |f(\mathbf{a}_i) - f(\mathbf{x}_2)| < \frac{\varepsilon}{2} + \frac{\varepsilon}{2} = \varepsilon.$$

Hence, f is uniformly continuous on S. ∎

Problems

1. Let f be defined by the following equations:
   (a) $f(x_1, \cdots, x_n) = b$, a real constant,
   (b) $f(x_1, \cdots, x_n) = x_i$, $1 \le i \le n$.
   In each case, use the definition directly to show that $\lim_{\mathbf{x} \to \mathbf{x}_0} f(\mathbf{x}) = f(\mathbf{x}_0)$, and thus show that each of the functions is continuous in E^n.

2. Let f be defined by $f(x_1, \cdots, x_n) = x_i^k$ where k is a positive integer and $0 < i \le n$. Use Problem 1, Theorem 3–1.2, and induction to show that f is continuous everywhere in E^n for all k.

3. Let f be defined by $f(x_1, \cdots, x_n) = c x_1^{k_1} \cdots x_n^{k_n}$. Use Problem 2, induction, and Theorem 3–1.2 to show f is continuous everywhere in E^n for all positive integers k_i and all constants c.

4. Use induction and Theorem 3–1.2 to show that all polynomials in any number of variables are continuous everywhere.

5. Use Problem 4 and Theorem 3–1.2 to prove that all rational functions (in any number of variables) are continuous at all points except those where the denominator is zero. Recall that rational functions are quotients of polynomials.

6. Find the domains of definition and the sets of points upon which the functions defined below are continuous.

   (a) $f(\mathbf{x}) = \dfrac{x_1 + x_2 + 1}{x_1^2 + x_2^2}$, (b) $f(\mathbf{x}) = \dfrac{(x_2 - 1)(x_2 - 3)}{(x_1 - 4)(x_1 + 2)}$,

   (c) $f(\mathbf{x}) = \dfrac{x_1^2 - 2x_1 x_2 + x_2^2}{x_1^2 - x_2^2}$, (d) $f(\mathbf{x}) = \dfrac{x_1 + x_2}{x_1^2 + x_2^2 - 1}$.

   Make a sketch of Do (f) for each function. For the function f of part (c), does $\lim_{\mathbf{x} \to (1,1)} f(\mathbf{x})$ exist?

7. Assume that it has been shown that g defined by $g(u) = u^{(1/m)}$ is continuous for all u if m is an odd positive integer and that g is continuous for all $u \ge 0$ if m is an even positive integer. Follow the same instructions as for Problem 6 for the functions defined below.

   (a) $f(\mathbf{x}) = \sqrt{36 - 4x_1^2 - 9x_2^2}$,

   (b) $f(\mathbf{x}) = \dfrac{(x_1 \sqrt{36 - 4x_1^2 - 9x_2^2})}{x_1}$,

   (c) $f(\mathbf{x}) = \sqrt{1 + x_1^2 - x_2^2}$,
   (d) $f(\mathbf{x}) = (1 - x_1^2 - x_2^2 - x_3^2)^{1/3}$.
   Explain how Theorem 3–1.3 is being used in each case.

8. Assume that sine and cosine are continuous for all values of the argument. Continue the instructions of Problem 6 for the functions defined below.

(a) $f(\mathbf{x}) = \dfrac{\sin (x_1^2 + x_2^2)^{1/2}}{(x_1^2 + x_2^2)^{1/2}}$,

(b) $f(\mathbf{x}) = \begin{cases} \dfrac{\sin (x_1^2 + x_2^2)^{1/2}}{(x_1^2 + x_2^2)^{1/2}}, & (x_1, x_2) \neq (0, 0), \\ 1, \ (x_1, x_2) = (0, 0), & \end{cases}$

(c) $f(\mathbf{x}) = \tan (x_1 - \pi), \quad x \in E^2$,

(d) $f(\mathbf{x}) = \csc \left(\dfrac{\pi}{x_1^2 + x_2^2 + 1} \right)$.

9. Assume that an exponential function is continuous for all values of the argument and that the logarithmic function is defined and continuous for all positive values of the argument. Continue the instructions of Problem 6 for the functions defined below.

(a) $f(\mathbf{x}) = e^{x_1 x_2}$, (b) $f(\mathbf{x}) = \log \sqrt{x_1^2 + x_2^2 - 1}$,

(c) $f(\mathbf{x}) = \exp (\tan (x_1/x_2))$, (d) $f(\mathbf{x}) = \log \left(\dfrac{x_1 + 1}{x_2 - 3} \right)$.

10. Explain how the theorems and the definitions of this section are used in the evaluation of the following limits.

(a) $\lim_{\mathbf{x}\to(2,1)} \dfrac{3x_1^2 - 5x_2 + 7}{x_1 + 4x_1 x_2}$, (b) $\lim_{\mathbf{x}\to(1,\pi)} \dfrac{x_1^2 \cos x_2}{x_2 + 3}$,

(c) $\lim_{\mathbf{x}\to(e,2)} (e^{-x_2} + \log x_1)$.

11. It is sometimes desirable to show that a function is discontinuous at a point in question. If the point is not in the domain of the function, then clearly, by the definition, it is not a point of continuity. However, if the point is in the domain and is an accumulation point of the domain, then for the function to be discontinuous at the point, the function must fail to have a limit. If either of the two following conditions hold, then the function is discontinuous at the point in question. Suppose f is the function and $\mathbf{x}_0$ is the point.

(a) There exists a sequence $\{\mathbf{x}_n\}$ such that

$$\lim_{n\to\infty} \mathbf{x}_n = \mathbf{x}_0 \quad \text{and} \quad \lim_{n\to\infty} f(\mathbf{x}_n) = \infty \ (\text{or} \ -\infty).$$

(b) There exist two sequences $\{\mathbf{x}_n\}$ and $\{\mathbf{y}_n\}$ such that

$$\lim_{n\to\infty} \mathbf{x}_n = \lim_{n\to\infty} \mathbf{y}_n = \mathbf{x}_0, \qquad \lim_{n\to\infty} f(\mathbf{x}_n) = a,$$

$$\lim_{n\to\infty} f(\mathbf{y}_n) = b, \quad \text{and } a \neq b.$$

Show that $\lim_{\mathbf{x}\to\mathbf{x}_0} f(\mathbf{x})$ fails to exist if either condition holds.

12. In the following three functions defined, show that the origin is a point of discontinuity.

(a) $$f(\mathbf{x}) = \begin{cases} \dfrac{1}{x_1^2 + x_2^2}, & \mathbf{x} \neq \mathbf{0}; \\ 0, & \mathbf{x} = \mathbf{0}. \end{cases}$$

(b) $$f(\mathbf{x}) = \begin{cases} \dfrac{x_1 x_2}{x_1^2 + x_2^2}, & \mathbf{x} \neq 0; \\ 0, & \mathbf{x} = \mathbf{0}. \end{cases}$$

(c) $$f(\mathbf{x}) = \begin{cases} 1, & x_1 \text{ and } x_2 \text{ are rational}, \\ 0, & \text{otherwise}. \end{cases}$$

13. In this section the function f was assumed to have its domain in E^n. Suppose that the domain of f is a subset of R^n. Restate Definitions 2–3.1, 3–1.1, and 3–1.2 for this weaker circumstance. If the domain of f is a subset of the space of ordered real n-tuples, then again restate these definitions for this still weaker circumstance (cf. Section 2–2). Show that if f is continuous at a point $\mathbf{a}$ with respect to any one of these definitions, it is continuous with respect to the other two.

14. Prove that if a function f defined by $z = f(x, y)$ is continuous on an interval $[\mathbf{a}, \mathbf{b}]$ in E^2, then $f(\ \ , y)$ is continuous on $[a_1, b_1]$.

Answers

6. (a) $\text{Do}(f) = \{(x_1, x_2): x_1^2 + x_2^2 > 0\}$,
   (b) $\{(x_1, x_2): x_1 \neq 4 \text{ and } x_1 \neq -2\}$,
   (c) $\{(x_1, x_2): x_2 = \pm x_1\}$, (d) $\{(x_1, x_2): x_1^2 + x_2^2 > 1\}$.
   Each function is continuous on its domain.
7. (a) $\text{Do}(f) = \{(x_1, x_2): 4x_1^2 + 9x_2^2 \leq 36\}$,
   (b) $\{(x_1, x_2): 4x_1^2 + 9x_2^2 \leq 36 \text{ and } x_1 \neq 0\}$,
   (c) $\{(x_1, x_2): x_2^2 - x_1^2 \leq 1\}$, (d) E^3.
   Each function is continuous on its domain.
8. (a) $\text{Do}(f): \{(x_1, x_2): x_1^2 + x_2^2 > 0\}$,
   (b) E^2, (c) $\{(x_1, x_2): x_1 \neq (\pi/2) + n\pi, n \text{ an integer}\}$,
   (d) $\{(x_1, x_2): (x_1, x_2) \neq (0, 0)\}$.
   Each function is continuous on its domain.
9. (a) $\text{Do}(f) = E^2$, (b) $\{(x_1, x_2): x_1^2 + x_2^2 > 1\}$,
   (c) $\{(x_1, x_2): x_1 \neq ((\pi/2) + n\pi)x_2; n \text{ an integer}, x_2 \neq 0\}$,
   (d) $\{(x_1, x_2): x_1 > -1 \text{ and } x_2 > 3 \text{ or } x_1 < -1 \text{ and } x_2 < 3\}$.
   Each function is continuous on its domain.

10. (a) $\dfrac{7}{5}$, (b) $\dfrac{-1}{\pi + 3}$, (c) $e^{-2} + 1$.

3–2. Partial Differentiation

Before considering a function of n variables and defining a partial derivative in general, let us first consider a function f of two variables: $f\colon E^2 \to R$. Suppose, furthermore, that the domain of f is an open set $S \subset E^2$. Let

$$\mathbf{x}_0 = (x_{01}, x_{02}) \in S.$$

Then there exists a neighborhood $N(\mathbf{x}_0\,;\,\delta) \subset S$. If x_1 is chosen such that $|x_1 - x_{01}| < \delta$, then $(x_1, x_{02}) \in N(\mathbf{x}_0\,;\,\delta) \subset S$, and the ratio

$$\frac{f(x_1, x_{02}) - f(x_{01}, x_{02})}{x_1 - x_{01}}$$

is defined for all $0 < |x_1 - x_{01}| < \delta$. This ratio is really a function of the single real variable x_1 whose domain contains the set $(x_{01} - \delta, x_{01}) \cup (x_{01}, x_{01} + \delta)$. Thus x_{01} is an accumulation point of this domain; hence it is meaningful to consider the limit of this ratio as $x_1 \to x_{01}$; that is,

$$\lim_{x_1 \to x_{01}} \frac{f(x_1, x_{02}) - f(x_{01}, x_{02})}{x_1 - x_{01}}.$$

If this limit exists, then the value of the limit is the *partial derivative of f with respect to x_1 at $\mathbf{x}_0$*, and its value is denoted by any one of the following three symbols:

$$\frac{\partial}{\partial x_1} f(\mathbf{x}_0), \quad f_{x_1}(\mathbf{x}_0) \quad \text{or} \quad D_{x_1} f(\mathbf{x}_0).$$

Similarly, if

$$\lim_{x_2 \to x_{02}} \frac{f(x_{01}, x_2) - f(x_{01}, x_{02})}{x_2 - x_{02}}$$

exists, then the value of the limit is the *partial derivative of f with respect to x_2 at $\mathbf{x}_0$*, and its value is denoted by any of the following three symbols:

$$\frac{\partial}{\partial x_2} f(\mathbf{x}_0), \quad f_{x_2}(\mathbf{x}_0) \quad \text{or} \quad D_{x_2} f(\mathbf{x}_0).$$

Figure 3–2.1 illustrates geometrically the meaning of the above concepts. The point A in the figure is $\mathbf{x}_0$, B is the point (x_1, x_{02}) and D is the point (x_{01}, x_2). The length of the line segment with end points A and E, $\overline{AE}$, is $f(\mathbf{x}_0)$. Similarly

$$\overline{BP} = f(x_1, x_{02}) \quad \text{and} \quad \overline{DR} = f(x_{01}, x_2).$$

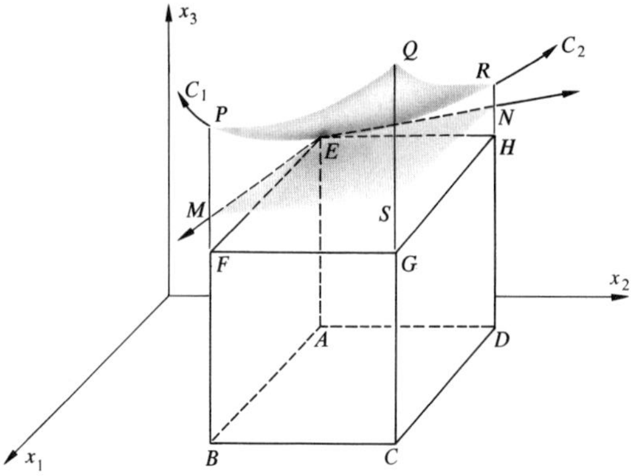

Figure 3-2.1

Thus

$$\frac{f(x_1, x_{02}) - f(x_{01}, x_{02})}{x_1 - x_{01}} = \frac{\overline{FP}}{\overline{EF}},$$

and

$$\frac{f(x_{01}, x_2) - f(x_{01}, x_{02})}{x_2 - x_{02}} = \frac{\overline{HR}}{\overline{EH}}.$$

These ratios are clearly the slope of the secant lines determined by E and P and by E and R, respectively. From the figure, it is intuitively easy to see that the limits of these ratios will be $\overline{FM}/\overline{EF}$ and $\overline{HN}/\overline{EH}$, respectively. That is, the slopes of the tangent lines EM, the lines determined by E and M and by EN are the partial derivatives $f_{x_1}(\mathbf{x}_0)$ and $f_{x_2}(\mathbf{x}_0)$, respectively.

If $f(\mathbf{x})$ is the magnitude of some physical quantity at $\mathbf{x}$ and if x_1 and x_2 are distances, then $f_{x_1}(\mathbf{x}_0)$ is the *instantaneous rate of change of* $f(\mathbf{x})$ *with respect to a change in* x_1 *at* x_0. A similar statement can be made for $f_{x_2}(\mathbf{x}_0)$.

Notice that in each of the previous limits one of the two variables was held fixed or constant. The limit of this ratio is precisely the required limit and ratio for the derivative if the function is considered a function of a single variable with all other variables held constant. Hence the differentiation formulas for calculating derivatives of a function of single-variable functions can be used to calculate partial derivatives. One simply regards as constant all other variables except the one with which the derivative is taken. This is illustrated in the following two examples.

Example 3-2.1. Consider $f(x_1, x_2) = x_1^2 + 3x_1 x_2^3$. Then

$$\frac{\partial}{\partial x_1} f(x_{01}, x_{02}) = 2x_{01} + 3x_{02}^3,$$

and

$$\frac{\partial}{\partial x_2} f(x_{01}, x_{02}) = 9x_{01} x_{02}^2.$$

Consider

$$g(x_1, x_2) = x_1 \sin x_2^2.$$

Then

$$\frac{\partial}{\partial x_1} g(x_{01}, x_{02}) = \sin x_{02}^2,$$

and

$$\frac{\partial}{\partial x_2} g(x_{01}, x_{02}) = 2x_{01} x_{02} \cos x_{02}^2.$$

Next is the definition of a partial derivative in general.

Definition 3–2.1. *Let $f: S \to R$ where S is an open subset of E^n and let $\mathbf{x}_0 \in S$. If for a fixed i, $1 \leq i \leq n$,*

$$\lim_{x_i \to x_{0i}} \frac{f(x_{01}, \cdots, x_i, \cdots, x_{0n}) - f(x_{01}, \cdots, x_{0n})}{x_i - x_{0i}}$$

exists, then the value of this limit is the partial derivative of f with respect to x_i at $\mathbf{x}_0$, and this value is denoted by any one of the following three symbols:

$$\frac{\partial}{\partial x_i} f(\mathbf{x}_0), \quad f_{x_i}(\mathbf{x}_0), \quad \text{or} \quad D_{x_i} f(\mathbf{x}_0).$$

The function consisting of these ordered pairs $(\mathbf{x}_0, f_{x_i}(\mathbf{x}_0))$ for all $\mathbf{x}_0 \in S$ where the limit exists is called the partial derivative of f with respect to x_i, and it is denoted by any one of the following three symbols:

$$\frac{\partial f}{\partial x_i}, \quad f_{x_i}, \quad \text{or} \quad D_{x_i} f.$$

Note that a function of n variables may have n partial derivatives, but all of these partials are functions of n variables, and each of them may have n partial derivatives, depending upon whether the required limits exist or not.

For example, suppose $f: S \to R$ and $S \subset E^2$ and f has both partial derivatives f_{x_1} and f_{x_2}. If f_{x_1} has a partial derivative with respect to x_1 and x_2, then they would be denoted by $f_{x_1x_1}$ and $f_{x_2x_1}$, respectively. Similarly, the partial derivatives of f_{x_2} with respect to x_1 and x_2 are denoted by $f_{x_1x_2}$ and $f_{x_2x_2}$, respectively. The partial derivatives f_{x_1} and f_{x_2} are, for emphasis, called *first partials* of f (or *first partial derivatives* of f). The derivatives $f_{x_1x_1}$, $f_{x_2x_1}$, $f_{x_1x_2}$, and $f_{x_2x_2}$ are called *second partials*. It is entirely possible that each of the

functions that are second partial derivatives may have partial derivatives—that is, the *third partials* of f. In fact, it is possible to have a function that has partial derivatives of any order.

The symbol $f_{x_2x_2x_1}$ means a third partial of f. It is taken first with respect to x_1 and on the second and third times with respect to x_2. The notation may appear to be backward, since we read from left to right, but the convention arose from the use of one of the other symbols for partial derivatives, namely $\partial/\partial x_1$. The partial derivative of $\partial f/\partial x_1$ with respect to x_2 would quite naturally be written

$$\frac{\partial}{\partial x_2}\left(\frac{\partial f}{\partial x_1}\right)$$

or, for short,

$$\frac{\partial^2 f}{\partial x_2\,\partial x_1}.$$

From this, the order of the differentiation, which is first with respect to x_1 and then with respect to x_2, reads backwards, or right to left. In keeping with this convention the other symbols for a partial are handled similarly, so that

$$\frac{\partial^2 f}{\partial x_2\,\partial x_1} = f_{x_2x_1} = D_{x_2x_1} f.$$

The following example illustrates the calculation of these higher partial derivatives.

Example 3–2.2. Consider f given by $f(\mathbf{x}) = 5x_1^2x_2 - x_2^3$. Then

$$\begin{aligned} f_{x_1}(\mathbf{x}) &= 10x_1 x_2, & f_{x_2}(\mathbf{x}) &= 5x_1^2 - 3x_2^2,\\ f_{x_1x_1}(\mathbf{x}) &= 10x_2, & f_{x_2x_1}(\mathbf{x}) &= 10x_1,\\ f_{x_1x_2}(\mathbf{x}) &= 10x_1, & f_{x_2x_2}(\mathbf{x}) &= -6x_2. \end{aligned}$$

Notice that in this example $f_{x_1x_2}(\mathbf{x}) = f_{x_2x_1}(\mathbf{x})$. This happens very often, but there are examples of functions for which it is not true. The example that follows is a case in point.

Example 3–2.3. Consider f given by

$$f(x) = \begin{cases} x_1 x_2 \dfrac{x_1^2 - x_2^2}{x_1^2 + x_2^2}, & \mathbf{x} \neq \mathbf{0},\\[2mm] 0, \quad \mathbf{x} = \mathbf{0}. \end{cases}$$

Since the origin is an exceptional point, it is necessary to use the definition to calculate the partial derivatives at the origin. Elsewhere the partials can be calculated by use of the differentiation formulas. In this way we get

$$f_{x_1}(\mathbf{x}) = x_2 \frac{x_1^4 + 4x_1^2x_2^2 - x_2^4}{(x_1^2 + x_2^2)^2},$$

and

$$f_{x_2}(\mathbf{x}) = x_1 \frac{x_1^4 - 4x_1^2x_2^2 - x_2^4}{(x_1^2 + x_2^2)^2},$$

when $\mathbf{x} \neq \mathbf{0}$. Also

$$f_{x_1}(\mathbf{0}) = \lim_{x_1 \to 0} \frac{f(x_1, 0) - f(0, 0)}{x_1 - 0}$$

$$= \lim_{x_1 \to 0} \frac{0 - 0}{x_1 - 0} = 0,$$

and similarly $f_{x_2}(\mathbf{0}) = 0$.

Again the origin is an exceptional point. Thus following the definition we have

$$f_{x_2x_1}(\mathbf{0}) = \lim_{x_2 \to 0} \frac{f_{x_1}(0, x_2) - f_{x_1}(0, 0)}{x_2 - 0}$$

$$= \lim_{x_2 \to 0} \frac{x_2 \dfrac{-x_2^4}{x_2^4} - 0}{x_2} = -1,$$

and

$$f_{x_1x_2}(\mathbf{0}) = \lim_{x_1 \to 0} \frac{f_{x_2}(x_1, 0) - f_{x_2}(0, 0)}{x_1 - 0}$$

$$= \lim_{x_1 \to 0} \frac{x_1 \dfrac{x_1^4}{x_1^4} - 0}{x_1} = 1.$$

Thus $f_{x_1x_2}(\mathbf{0}) \neq f_{x_2x_1}(\mathbf{0})$.

This example is very significant. It serves as a warning to the reader that the order of differentiation is important. It is only in certain circumstances that $f_{x_1x_2} = f_{x_2x_1}$. A sufficient criterion for equality follows.

Proposition 3–2.1. *Let* $f\colon S \to R$ *where* S *is an open set in* E^2. *Let* $\mathbf{x}_0 \in S$ *and let* $f_{x_2x_1}$ *be continuous in some neighborhood of* $\mathbf{x}_0$, $N(\mathbf{x}_0\,;\delta)$, *and suppose* $f_{x_1x_2}(\mathbf{x}_0)$ *exists; then*

$$f_{x_1x_2}(\mathbf{x}_0) = f_{x_2x_1}(\mathbf{x}_0).$$

Clearly the roles of $f_{x_1x_2}$ and $f_{x_2x_1}$ in the hypothesis could be interchanged. See Problem 7 for the proof.

Problems

1. For the functions defined by the following equations, calculate f_{x_1} by first using only the definition; then check your answer by differentiating formally.

   (a) $f(\mathbf{x}) = 2x_1^2 + x_2^3$, (b) $f(\mathbf{x}) = \dfrac{1}{2x_1 - 3x_2}$,

   (c) $f(\mathbf{x}) = \sqrt{x_1^2 - x_1 x_2}$, (d) $f(\mathbf{x}) = \sin x_1 \cos x_2$.

2. Cut a piece of light cardboard (a manila folder is excellent for this) in the shape of the sketch in Figure 3–2.2. Fold along each of the vertical lines

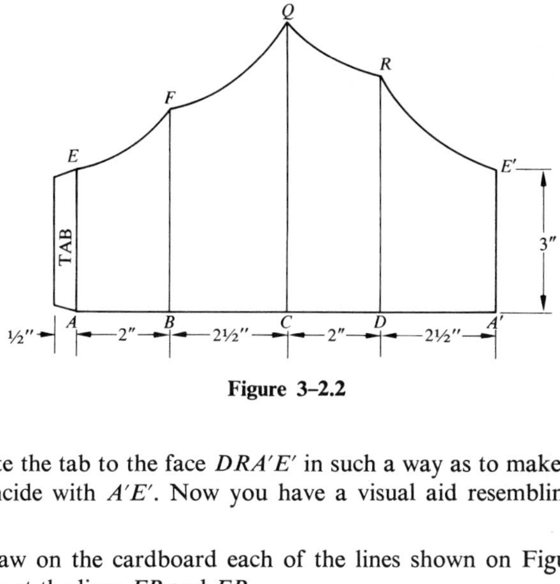

Figure 3–2.2

and paste the tab to the face $DRA'E'$ in such a way as to make the edge AE coincide with $A'E'$. Now you have a visual aid resembling Figure 3–2.1.

(a) Draw on the cardboard each of the lines shown on Figure 3–2.1 except the lines EP and ER.

(b) Label the coordinates of each of the points: $A = (x_{01}, x_{02}, 0)$, $E = (x_{01}, x_{02}, f(x_{01}, x_{02}))$, etc.

(c) Locate and label the lengths equal to $f(x_1, x_{02}) - f(x_{01}, x_{02})$, $x_1 - x_{01}, f(x_{01}, x_2) - f(x_{01}, x_{02})$, and $x_2 - x_{02}$.

(d) Label the angles whose slopes are $f_{x_1}(\mathbf{x}_0)$ and $f_{x_2}(\mathbf{x}_0)$.

(e) The points, E, M, N, and S, lie on a tangent plane to the surface. How is the point S located if it is granted that M and N have been located correctly?

3. Find the first partials of the functions defined by the following equations.

(a) $f(\mathbf{x}) = \sin^3(x_1 - x_2^2)$, (b) $f(\mathbf{x}) = \dfrac{e^{x_1x_2}}{x_1^2 - x_2^2}$,

(c) $f(\mathbf{x}) = \log\sqrt{x_1^2 + x_2 + 1}$, (d) $f(\mathbf{x}) = \tan^{-1}(x_2/x_1)$,

(e) $f(\mathbf{x}) = \dfrac{x_2}{\sqrt{x_1 + x_3^2}}$, (f) $f(\mathbf{x}) = x_1\, e^{(x_3/x_2)}$.

Find the domains of definition of these partial derivatives.

4. Find the first and second partials of the functions defined by the following equations.

(a) $f(\mathbf{x}) = e^{x_1x_2^2}$, (b) $f(\mathbf{x}) = x_1^2\, x_2^3\, x_3$,

(c) $f(\mathbf{x}) = x_1 \cos x_2$.

5. If the function f is defined by $f(\mathbf{x}) = x_1^2 \tan x_2$, find the values of the following:

(a) $f_{x_1}(0, \pi/4)$, (b) $f_{x_2}(3, 0)$, (c) $f_{x_2x_1}(1, -\pi/4)$.

6. Show that on certain sets of points the functions defined by the following equations satisfy identically the given partial differential equation.

(a) $f(\mathbf{x}) = \log\sqrt{x_1^2 + x_2^2}$, $f_{x_1x_1} + f_{x_2x_2} = 0$,

(b) $f(\mathbf{x}) = x_1 + 2x_2 + x_1 \tan\dfrac{x_1}{x_2}$, $x_1\, f_{x_1} + x_2\, f_{x_2} = f$,

(c) $f(\mathbf{x}) = e^{x_1 + ax_2} + \cos(x_1 - ax_2)$, $f_{x_2x_2} = a^2 f_{x_1x_1}$,

(d) $f(\mathbf{x}) = x_1 \tan^{-1}\dfrac{x_2}{x_1} + e^{(x_2/x_1)}$, $x_1^2 f_{x_1x_1} + 2x_2\, x_2 f_{x_1x_2}$

$+ x_2^2 f_{x_2x_2} = 0$.

7. Suppose that f satisfies the hypothesis of Proposition 3–2.1. Show that $f_{x_1x_2}(\mathbf{x}_0)$ can be written

$$\lim_{x_1 \to x_{01}} \lim_{x_2 \to x_{02}} \frac{[f(x_1, x_2) - f(x_1, x_{02})] - [f(x_{01}, x_2) - f(x_{01}, x_{02})]}{(x_1 - x_{01})(x_2 - x_{02})}.$$

Notice that the first bracket in the numerator of the above iterated limit is the same as the second bracket, only evaluated at a different value of x_1. Assume $(x_1, x_2) \in N(\mathbf{x}_0\,;\,\delta)$ and, by use of the mean-value theorem, show that for some $0 < \theta_1 < 1$

$$f_{x_1x_2}(\mathbf{x}_0) = \lim_{x_1 \to x_{01}} \lim_{x_2 \to x_{02}} \frac{f_{x_1}(x_{01} + \theta_1(x_1 - x_{01}), x_2) - f_{x_1}(x_{01} + \theta_1(x_1 - x_{01}), x_{02})}{x_2 - x_{02}}.$$

By use of the mean-value theorem, again on the numerator, show that for some $0 < \theta_2 < 1$,

$$f_{x_1x_2}(\mathbf{x}_0) = \lim_{x_1 \to x_{01}} \lim_{x_2 \to x_{02}} f_{x_2x_1}(x_{01} + \theta_1(x_1 - x_{01}), x_{02} + \theta_2(x_2 - x_{02}))$$
$$= f_{x_2x_1}(\mathbf{x}_0).$$

Answers

1. (a) $4x_1$, (b) $-2(2x_1 - 3x_2)^{-2}$,
   (c) $\frac{1}{2}(2x_1 - x_2)(x_1^2 - x_1 x_2)^{-1/2}$, (d) $\cos x_1 \cos x_2$.
3. (a) $f_{x_1}(\mathbf{x}) = 3\sin^2(x_1 - x_2^2)\cos(x_1 - x_2^2)$,
   $f_{x_2}(x) = -6x_2 \sin^2(x_1 - x_2^2)\cos(x_1 - x_2^2)$.
   (b) $e^{x_1x_2}(x_1^2 x_2 - x_2^3 - 2x_1)(x_1^2 - x_2^2)^{-2}$,
   $e^{x_1x_2}(x_1^3 - x_1 x_2^2 + 2x_2)(x_1^2 - x_2^2)^{-2}$.
   (c) $x_1(x_1^2 + x_2 + 1)^{-1}$, $\frac{1}{2}(x_1^2 + x_2 + 1)^{-1}$.
   (d) $-x_2(x_1^2 + x_2^2)^{-1}$, $x_1(x_1^2 + x_2^2)^{-1}$.
   (e) $f_{x_1}(\mathbf{x}) = (-\frac{1}{2})x_2(x_1 + x_3^2)^{-(3/2)}$, $f_{x_2}(\mathbf{x}) = (x_1 + x_3^2)^{-1/2}$,
   $f_{x_3}(\mathbf{x}) = -x_2 x_3(x_1 + x_3^2)^{-(3/2)}$.
   (f) $e^{(x_3/x_2)}$, $-(x_1 x_3 x_2^{-2})e^{(x_3/x_2)}$, $x_1 x_2^{-1}e^{(x_3/x_2)}$.
4. (a) $f_{x_1}(\mathbf{x}) = x_2^2 e^{x_1x_2^2}$, $f_{x_2}(\mathbf{x}) = 2x_1 x_2 e^{x_1x_2^2}$,
   $f_{x_1x_1}(\mathbf{x}) = x_2^4 e^{x_1x_2^2}$, $f_{x_1x_2}(\mathbf{x}) = 2x_2 e^{x_1x_2^2}(1 + x_1 x_2^2)$,
   $f_{x_2x_2}(\mathbf{x}) = 2x_1 e^{x_1x_2^2}(1 + 2x_1 x_2^2)$.
   (b) $f_{x_1}(\mathbf{x}) = 2x_1 x_2^3 x_3$, $f_{x_2}(\mathbf{x}) = 3x_1^2 x_2^2 x_3$, $f_{x_3}(\mathbf{x}) = x_1^2 x_2^3$,
   $f_{x_1x_1}(\mathbf{x}) = 2x_2^3 x_3$, $f_{x_1x_2}(\mathbf{x}) = 6x_1 x_2^2 x_3$, $f_{x_1x_3}(\mathbf{x}) = 2x_1 x_2^3$,
   $f_{x_2x_2}(\mathbf{x}) = 6x_1^2 x_2 x_3$, $f_{x_2x_3}(\mathbf{x}) = 3x_1^2 x_2^2$, $f_{x_3x_3}(\mathbf{x}) = 0$.
   (c) $f_{x_1}(\mathbf{x}) = \cos x_2$, $f_{x_2}(\mathbf{x}) = -x_1 \sin x_2$,
   $f_{x_1x_1}(\mathbf{x}) = 0$, $f_{x_1x_2}(\mathbf{x}) = -\sin x_2$, $f_{x_2x_2}(\mathbf{x}) = -x_1 \cos x_2$.
5. (a) 0, (b) 9, (c) 4.

3-3. Differentials, Tangent Planes, and Differentiability

If $f: S \to R$ where S is an open subset of $E^1 = R$, the reals, then by definition the differential of f at x_{01}, where $x_{01} \in S$, is $df = f'(x_{01})\, dx_1$ (cf. Figure 3-3.1). The length of the segment BD, $\overline{BD}$, is df. The differential of f

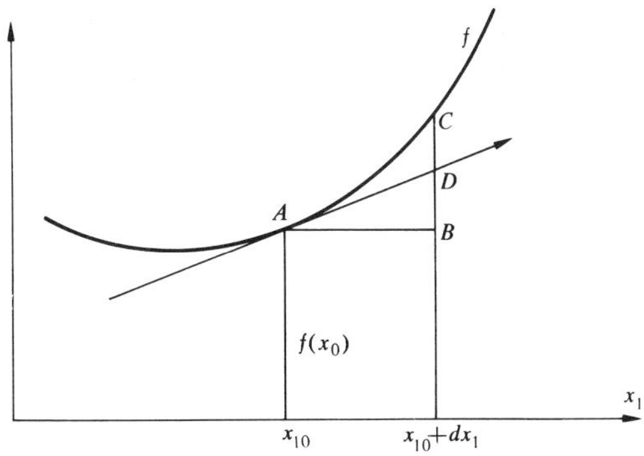

Figure 3-3.1

at x_{01} depends upon two variables—the point x_{01} and the value of dx_1. To emphasize this relationship, it would be better to write $df = df(x_{01}\,; dx_1)$. From the figure it is clear that for dx_1 small, $\overline{BD} = df(x_{01}\,; dx_1)$ is approximately equal to $\overline{BC}$, the increment in $f(x)$ at x_{01} due to the increment dx_1, which is written $\Delta f(x_{01}\,; dx_1)$. Another way to say this is that the graph of f is approximately the tangent line to f at $(x_{01}, f(x_{01}))$ in a neighborhood of x_{01}.

Let us be more precise about the manner of this approximation. Since f has a derivative at x_{01} it follows that given $\varepsilon > 0$ there exists $\delta > 0$ such that

$$\left|\frac{f(x_1) - f(x_{01})}{x_1 - x_{01}} - f'(x_{01})\right| < \varepsilon$$

for $0 < |x_1 - x_{01}| < \delta$. If $x_1 = x_{01} + dx_1$, then

$$\left|\frac{f(x_{01} + dx_1) - f(x_{01})}{dx_1} - f'(x_{01})\right| < \varepsilon.$$

Since $dx_1 \neq 0$,

$$|f(x_{01} + dx_1) - f(x_{01}) - f'(x_{01})\, dx_1| < \varepsilon\, |dx_1|,$$

or

$$|\Delta f(x_{01}\,; dx_1) - df(x_{01}\,; dx_1)| < \varepsilon\, |dx_1|.$$

That is, given ε arbitrarily small, there exists a neighborhood $N(x_{01}\,; \delta)$ such that the increment of f and the differential of f differ by an amount less than ε times the magnitude of the increment where $x_{01} + dx_1 \in N'(x_{01}\,; \delta)$.

The related discussion for functions of several variables is very similar. The main difference is that more hypothesis is required of f, as we shall see. Let f: $S \to R$ where S is an open subset of E^2; let $\mathbf{x}_0 \in S$; and suppose *there is some neighborhood of* $\mathbf{x}_0$, $N(\mathbf{x}_0\,; \delta)$ *such that the first partials of f exist in* $N(\mathbf{x}_0\,; \delta)$ *and at* $\mathbf{x}_0$ *they are continuous*. The tangent plane to f at $(\mathbf{x}_0, f(\mathbf{x}_0))$, which will play a role analogous to the tangent line, is given by the equation

$$x_3 - f(\mathbf{x}_0) = f_{x_1}(\mathbf{x}_0)(x_1 - x_{01}) + f_{x_2}(\mathbf{x}_0)(x_2 - x_{02}),$$

which can be written using the dot-product notation as follows:

$$(f_{x_1}(\mathbf{x}_0), f_{x_2}(\mathbf{x}_0), -1) \cdot (x_1 - x_{01}, x_2 - x_{02}, x_3 - f(\mathbf{x}_0)) = 0.$$

Let us take a moment to motivate the last equation as the equation of the tangent plane. In Figure 3–2.1 (page 83), a nice smooth surface, the graph determined by a function f, is pictured. The parametric representation of the curve C_1 is

$$x_1 = x_1,$$
$$x_2 = x_{02},$$
$$x_3 = f(x_1, x_{02}).$$

Thus a tangent vector to C_1 at E parallel to the vector $\overrightarrow{EM}$ is $(1, 0, f_{x_1}(\mathbf{x}_0))$. Similarly, a tangent vector to C_2 at E parallel to $\overrightarrow{EN}$ is $(0, 1, f_{x_2}(\mathbf{x}_0))$. A tangent plane to the surface given by f at E would be determined by these two lines, EM and EN. Thus the vector

$$(0, 1, f_{x_2}(\mathbf{x}_0)) \times (1, 0, f_{x_1}(\mathbf{x}_0)) = (f_{x_1}(\mathbf{x}_0), f_{x_2}(\mathbf{x}_0), -1)$$

is perpendicular to the tangent plane. Therefore, the equation of the plane through $E = (x_{01}, x_{02}, f(\mathbf{x}_0))$ perpendicular to $(f_{x_1}(\mathbf{x}_0), f_{x_2}(\mathbf{x}_0). -1)$ is precisely the preceding dot product, which was given as the equation of the tangent plane.

The differential of f at $\mathbf{x}_0 = (x_{01}, x_{02})$ due to increments (dx_1, dx_2) is

$$df(x_{01}, x_{02}; dx_1, dx_2) = f_{x_1}(\mathbf{x}_0)\, dx_1 + f_{x_2}(\mathbf{x}_0)\, dx_2.$$

If on Figure 3–2.1

$$A = \mathbf{x}_0, \qquad \overline{AB} = dx_1, \quad \text{and} \quad \overline{AD} = dx_2,$$

then

$$\overline{FM} = f_{x_1}(\mathbf{x}_0)\, dx_1 \quad \text{and} \quad \overline{HN} = f_{x_2}(\mathbf{x}_0)\, dx_2.$$

Since the line MS is parallel to EN, $\overline{GS} = \overline{FM} + \overline{HN}$, and thus

$$\overline{GS} = df(x_{01}, x_{02}; dx_1, dx_2).$$

The increment in f at (x_{01}, x_{02}) due to the increments (dx_1, dx_2) is

$$\Delta f(x_{01}, x_{02}; dx_1, dx_2) = f(x_{01} + dx_1, x_{02} + dx_2) - f(x_{01}, x_{02});$$

that is, $\Delta f(x_{01}, x_{02}; dx_1, dx_2) = \overline{GQ}$ in Figure 3–2.1. With the above hypothesis on f, it only remains to show that locally (that is, in a neighborhood of $\mathbf{x}_0$) df approximates Δf in a manner similar to the single variable case. Next, this is formally stated as a proposition, and the proof is given.

Proposition 3–3.1. *Let $f\colon S \to R$ where S is an open set in E^2, $\mathbf{x}_0 \in S$, and let $N(\mathbf{x}_0; \delta)$ be a neighborhood of $\mathbf{x}_0$ contained in S such that the first partials of f exist in $N(\mathbf{x}_0; \delta)$ and are continuous at $\mathbf{x}_0$. Then, given $\varepsilon > 0$, there exists $N(\mathbf{x}_0; \delta_1)$ such that if $\mathbf{x}_0 + \mathbf{dx} \in N(\mathbf{x}_0; \delta_1)$, then*

$$|\Delta f(x_{01}, x_{02}; dx_1, dx_2) - df(x_{01}, x_{02}; dx_1, dx_2)| < \varepsilon\, \|\mathbf{dx}\|.$$

Proof. Given $\varepsilon > 0$. Write

$$\begin{aligned}\Delta f(x_{01}, x_{02}; dx_1, dx_2) &= f(x_{01} + dx_1, x_{02} + dx_2) - f(x_{01}, x_{02}),\\ &= f(x_{01} + dx_1, x_{02} + dx_2) - f(x_{01}, x_{02} + dx_2)\\ &\quad + f(x_{01}, x_{02} + dx_2) - f(x_{01}, x_{02}).\end{aligned}$$

If x_{02} and dx_2 are fixed with $\|\mathbf{dx}\| < \delta$, then $f(x_1, x_{02} + dx_2)$ is a function of a single variable that satisfies the mean-value theorem on the interval with end points x_{01} and $x_{01} + dx_1$. The existence of f_{x_1} in $N(\mathbf{x}_0; \delta)$ implies that on the interval with end points $x_{01}, x_{01} + dx_1, f(x_1, x_{01} + dx_1)$ as a function of the single variable, x_1 is continuous on this interval and has a derivative in the interior of this interval. Therefore, there exists θ_1, $0 < \theta_1 < 1$, such that

$$f(x_{01} + dx_1, x_{02} + dx_2) - f(x_{01}, x_{02} + dx_2) = f_{x_1}(x_{01} + \theta_1\, dx_1, x_{02} + dx_2)\, dx_1.$$

Similarly, there exists θ_2, $0 < \theta_2 < 1$ such that

$$f(x_{01}, x_{02} + dx_2) - f(x_{01}, x_{02}) = f_{x_2}(x_{01}, x_{02} + \theta_2\, dx_2)\, dx_2.$$

Then

$$\begin{aligned}&|\Delta f(x_{01}, x_{02}; dx_1, dx_2) - df(x_{01}, x_{02}; dx_1, dx_2)|\\ &\qquad = |f_{x_1}(x_{01} + \theta_1\, dx_1, x_{02} + dx_2)\, dx_1 + f_{x_2}(x_{01}, x_{02} + \theta_2\, dx_2)\, dx_2\\ &\qquad\quad - (f_{x_1}(x_{01}, x_{02})\, dx_1 + f_{x_2}(x_{01}, x_{02})\, dx_2)|\\ &\qquad \le |f_{x_1}(x_{01} + \theta_1\, dx_1, x_{02} + dx_2) - f_{x_1}(x_{01}, x_{02})|\, |dx_1|\\ &\qquad\quad + |f_{x_2}(x_{01}, x_{02} + \theta_2\, dx_2) - f_{x_2}(x_{01}, x_{02})|\, |dx_2|.\end{aligned}$$

Since the first partials are continuous at $\mathbf{x}_0$, there exists $\delta_1 > 0$ such that, for any $\mathbf{x} \in N(\mathbf{x}_0; \delta_1)$,

$$|f_{x_1}(\mathbf{x}) - f_{x_1}(\mathbf{x}_0)| < \varepsilon/2$$

and

$$|f_{x_2}(\mathbf{x}) - f_{x_2}(\mathbf{x}_0)| < \varepsilon/2.$$

If $\mathbf{x}_0 + \mathbf{dx} \in N(\mathbf{x}_0; \delta_1)$, then $(x_{01} + \theta_1\, dx_1, x_{02} + dx_2)$ and $(x_{01}, x_{02} + \theta_2\, dx_2) \in N(\mathbf{x}_0; \delta_1)$, and both $|dx_1|$ and $|dx_2| \le \|\mathbf{dx}\|$. Hence

$$\begin{aligned}|\Delta f(x_{01}, x_{02}; dx_1, dx_2) - df(x_{01}, x_{02}; dx_1, dx_2)| &< \frac{\varepsilon}{2}\|\mathbf{dx}\| + \frac{\varepsilon}{2}\|\mathbf{dx}\|\\ &= \varepsilon\, \|\mathbf{dx}\|. \quad \blacksquare\end{aligned}$$

The requirement in the proposition that f has partial derivatives in a neighborhood of $\mathbf{x}_0$ is essential even though f is continuous at $\mathbf{x}_0$. The following example illustrates this.

Example 3–3.1. Consider the function f given by the following equation:

$$f(x_1, x_2) = \begin{cases} x_1^{2/3}, & x_1 = x_2, \quad x_1 > 0, \\ 0, & \text{elsewhere}, \end{cases}$$

(cf. Figure 3–3.2).

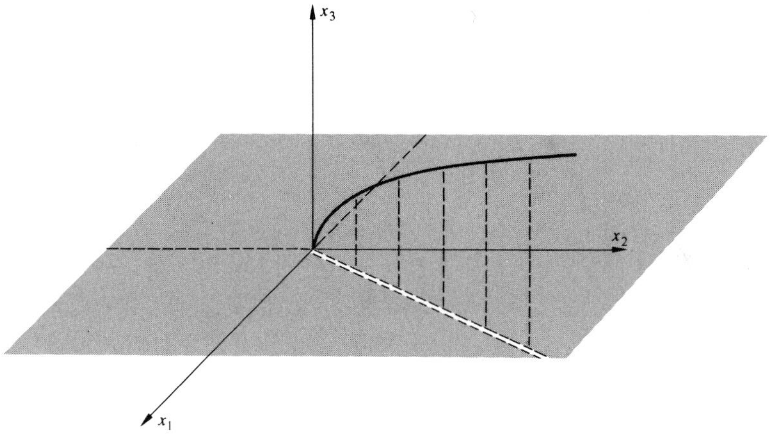

Figure 3–3.2

Let $\mathbf{x}_0$ be the origin. Then $S = E^2$, an open set. Both f_{x_1} and f_{x_2} are identically zero everywhere except on $\{(x_1, x_2) \colon x_1 = x_2, x_1 > 0\}$, where they are not defined. Thus both partials are continuous at the origin. Furthermore, $df(0, 0; dx_1, dx_2) = 0$ for all $\mathbf{dx}$, but for $dx_1 = dx_2$ and $dx_1 > 0$,

$$\Delta f(0, 0; dx_1, dx_2) = (dx_1)^{2/3}.$$

Hence, take $\varepsilon = 1$ and, given $\delta_1 > 0$, there exists $0 < dx_1 < \delta_1$ such that

$$|\Delta f(0, 0; dx_1, dx_1) - df(0, 0; dx_1, dx_1)| = (dx_1)^{2/3} > \sqrt{2}\, dx_1 = \varepsilon\, \|\mathbf{dx}\|.$$

For example, take $dx_1 = \frac{1}{2} \min(\delta_1, 1, 2^{-(3/8)})$.

The requirement that the partial derivatives of f are continuous at $\mathbf{x}_0$ is essential also. In the following example, all of the first partial derivatives **exist in a** neighborhood of $\mathbf{x}_0$.

Example 3–3.2. Consider the function f given by the following equation:

$$f(x_1, x_2) = \begin{cases} 4\dfrac{x_1^{\,2}x_2^{\,2}}{(x_1^{\,2} + x_2^{\,2})^{5/3}}, & (x_1, x_2) \neq (0, 0), \\ 0, \quad (x_1, x_2) = (0, 0), \end{cases}$$

(cf. Figure 3–3.3).

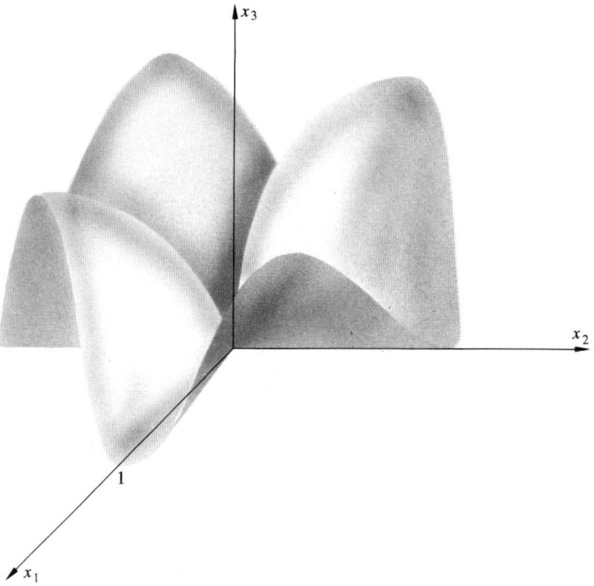

Figure 3–3.3

Again $\mathbf{x}_0 = \mathbf{0}$. The continuity is clear everywhere except possibly at the origin. Changing to polar coordinates $f(r\cos\theta, r\sin\theta) = r^{2/3}\sin^2 2\theta$. In this form it is easy to see that f is continuous at the origin. The function f_{x_1} is given by the following equation:

$$f_{x_1}(x_1, x_2) = \begin{cases} x_1\dfrac{6x_2^{\,2} - 4x_1^{\,2}}{3(x_1^{\,2} + x_2^{\,2})^{(8/3)}}, & (x_1, x_2) \neq (0, 0), \\ 0, \quad (x_1, x_2) = (0, 0). \end{cases}$$

Due to the symmetry in the definition of f, f_{x_2} is defined by a similar equation. In order to see that f_{x_1} is discontinuous at the origin, observe that $f(x_1, x_1)$ approaches ∞ as $x_1 \to 0$. Since f_{x_1} and f_{x_2} are zero at the origin,

$$df(0, 0; dx_1, dx_2) = 0$$

for all increments $\mathbf{dx}$. Again f is such that the approximation inequality fails to hold. The argument for the previous example could be used again here, since the function values of both examples differ only by a positive constant on $\{(x_1, x_2): x_1 = x_2, x_1 > 0\}$ and in both examples $df(0, 0; dx_1, dx_2) = 0$.

With this background, it is now possible for the reader to understand and appreciate the following general definitions.

Definition 3–3.1. *Let $f: S \to R$ where S is an open subset of E^n. Let $\mathbf{x}_0 \in S$ and let all first partials of f exist in some neighborhood of $\mathbf{x}_0$ and suppose each of these partials is continuous at $\mathbf{x}_0$. Then the surface given by $f(\mathbf{x}) = z$, the graph of f, has a tangent plane at $(\mathbf{x}_0, f(\mathbf{x}_0))$, and the equation of this tangent plane is given by*

$$x_{n+1} - f(\mathbf{x}_0) = f_{x_1}(\mathbf{x}_0)(x_1 - x_{01}) + \cdots + f_{x_n}(\mathbf{x}_0)(x_n - x_{0n}),$$

which can be written

$$(f_{x_1}(\mathbf{x}_0), \cdots, f_{x_n}(\mathbf{x}_0), -1) \cdot (x_1 - x_{01}, \cdots, x_n - x_{0n}, x_{n+1} - f(\mathbf{x}_0)) = 0.$$

The vector $(f_{x_1}(\mathbf{x}_0), \cdots, f_x(\mathbf{x}_0), -1)$ is normal at $\mathbf{x}_0$ to the surface given by $f(\mathbf{x}) = z$. The function f, which has a differential at $\mathbf{x}_0$ with increment $\mathbf{dx}$ is a function of $2n$ real variables—defined by the following equation:

$$df(x_{01}, \cdots, x_{0n}; dx_1, \cdots, dx_n) = f_x(\mathbf{x}_0)\, dx_1 + \cdots + f_{x_n}(\mathbf{x}_0)\, dx_n. \tag{3–3.1}$$

It also can be written

$$df(\mathbf{x}_0; \mathbf{dx}) = (f_{x_1}(\mathbf{x}_0), \cdots, f_{n_n}(\mathbf{x}_0)) \cdot (dx_1, \cdots, dx_n). \tag{3–3.2}$$

The function f is differentiable at $\mathbf{x}_0$. If S_1 is some subset of S and f is differentiable at each point of S_1, then f is differentiable on S_1.

The increment in $f(\mathbf{x})$ at $\mathbf{x}_0$ due to the increment $\mathbf{dx}$ is a function of $2n$ real variables, defined by the following equation:

$$\Delta f(x_{01}, \cdots, x_{0n}; dx_1, \cdots, dx_n) = f(x_{01} + dx_1, \cdots, x_{0n} + dx_n) - f(x_{01}, \cdots, x_{0n}),$$

which can be written

$$\Delta f(\mathbf{x}_0; \mathbf{dx}) = f(\mathbf{x}_0 + \mathbf{dx}) - f(\mathbf{x}_0).$$

The reader should note that the definition of tangent plane and differential for $n > 1$ would agree with the earlier definition of tangent line and differential for $n = 1$ on functions that satisfy the extra conditions required in this definition.

Proposition 3–3.1 can now be stated more generally as follows:

Proposition 3–3.2. *Let* $f\colon S \to R$ *where* S *is an open set in* E^n, $n > 1$. *If* f *is differentiable at* $\mathbf{x}_0$, *then, given* $\varepsilon > 0$, *there exists* $N(\mathbf{x}_0\,;\,\delta)$ *such that when* $\mathbf{x}_0 + \mathbf{dx} \in N(\mathbf{x}_0\,;\,\delta)$,

$$|\Delta f(\mathbf{x}_0\,;\,\mathbf{dx}) - df(\mathbf{x}_0\,;\,\mathbf{dx})| < \varepsilon\,\|\mathbf{dx}\|.$$

A proof for this proposition could be given in a manner exactly analogous to the proof of Proposition 3–3.1.

The reader may have noticed that the discussion has said nothing about the continuity of f. This is quite intentional. The motivating idea is the following: To say that a function of a single variable has a derivative at a point is sufficient to imply that the function is continuous at the point. Thus we try to use similar reasoning with functions of several variables. The following proposition shows that the above condition of differentiability implies continuity.

Proposition 3–3.3. *Let* $f\colon S \to R$ *where* S *is an open subset of* E^n. *If* f *is differentiable at* $\mathbf{x}_0$, *a point of* S, *then* f *is continuous at* $\mathbf{x}_0$.

Proof. By Proposition 3–3.2, with $\varepsilon = 1$, it follows that there exists $N(\mathbf{x}_0\,;\,\delta)$ such that if $\mathbf{x} + \mathbf{dx} \in N(\mathbf{x}_0\,;\,\delta)$,

$$|\Delta f(\mathbf{x}_0\,;\,\mathbf{dx}) - df(\mathbf{x}_0\,;\,\mathbf{dx})| < \|\mathbf{dx}\|.$$

Let $\mathbf{y} \in N(\mathbf{x}_0\,;\,\delta)$ and let $\mathbf{dx} = \mathbf{y} - \mathbf{x}_0$; then the above inequality becomes

$$|\Delta f(\mathbf{x}_0\,;\,\mathbf{y} - \mathbf{x}_0) - df(\mathbf{x}_0\,;\,\mathbf{y} - \mathbf{x}_0)| < \|\mathbf{y} - \mathbf{x}_0\|.$$

Then

$$|\Delta f(\mathbf{x}_0\,;\,\mathbf{y} - \mathbf{x}_0)| - |df(\mathbf{x}_0\,;\,\mathbf{y} - \mathbf{x}_0)| \le |\Delta f(\mathbf{x}_0\,;\,\mathbf{y} - \mathbf{x}_0) - df(\mathbf{x}_0\,;\,\mathbf{y} - \mathbf{x}_0)|.$$

By combining these last two inequalities, we have

$$|\Delta f(\mathbf{x}_0\,;\,\mathbf{y} - \mathbf{x}_0)| \le |df(\mathbf{x}_0\,;\,\mathbf{y} - \mathbf{x}_0)| + \|\mathbf{y} - \mathbf{x}_0\|.$$

However,

$$df(\mathbf{x}_0\,;\,\mathbf{y} - \mathbf{x}_0) = (f_{x_1}(\mathbf{x}_0), \cdots, f_{x_n}(\mathbf{x}_0)) \cdot (y_1 - x_{01}, \cdots, y_n - x_{0n}),$$

and, by the Cauchy-Schwarz inequality,

$$|df(\mathbf{x}_0\,;\,\mathbf{y} - \mathbf{x}_0)| \le \|(f_{x_1}(\mathbf{x}_0), \cdots, f_{x_n}(\mathbf{x}_0))\|\ \|\mathbf{y} - \mathbf{x}_0\|.$$

Thus

$$|\Delta f(\mathbf{x}_0\,;\,\mathbf{y} - \mathbf{x}_0)| \le \|(f_{x_1}(\mathbf{x}_0), \cdots, f_{x_n}(\mathbf{x}_0))\|\ \|\mathbf{y} - \mathbf{x}_0\| + \|\mathbf{y} - \mathbf{x}_0\|$$

or

$$|f(\mathbf{y}) - f(\mathbf{x}_0)| \leq M \|\mathbf{y} - \mathbf{x}_0\|$$

where $M = \|(f_{x_1}(\mathbf{x}_0), \cdots, f_{x_n}(\mathbf{x}_0))\| + 1$.
Therefore, given $\varepsilon > 0$, we shall take $\delta_1 = \min(\varepsilon/M, \delta)$ and then

$$|f(\mathbf{y}) - f(\mathbf{x}_0)| \leq M\frac{\varepsilon}{M} = \varepsilon$$

when $\|\mathbf{y} - \mathbf{x}_0\| < \delta_1$. Hence f is continuous at $\mathbf{x}_0$. ▌

Let $f: S \to R$ where S is a subset of E^n. If f is continuous on S_1, a subset of S, then $f \in C(S_1)$, the set of functions that are continuous on S_1. If $S_1 \subset S$ is an open subset of E^n and if each first partial is continuous at each point of S_1, written $f \in C^1(S_1)$, then f is differentiable on S_1 and, by Proposition 3–3.3, f is continuous on S_1. That is, $f \in C^1(S_1)$ implies $f \in C(S_1)$. Similarly, if $S_1 \subset S$ is an open subset of E^n and all of the kth order partials are continuous on S_1, written $f \in C^k(S_1)$, $k \geq 1$, then $f \in C^{k-1}(S_1)$. Thus $f \in C^k(S_1)$ implies $f \in C^{k-1}(S_1), \cdots, f \in C^1(S_1)$, and $f \in C(S_1)$. If $f \in C^k(S_1)$ for all positive integers k, then we write $f \in C^\infty(S_1)$.

Let S be a plane regular region of E^2 and let $f: S \to R$. The function f is *continuously differentiable on the plane regular region S iff $f \in C(S)$, $f \in C^1(S^o)$*, and each of the first order partials of f on S^o is extendable to a continuous function on S. The function f is *piecewise continuous on the finite plane region S* if S is a finite sum of plane regular regions $S_1, \cdots, S_m$ such that f is continuous on each S_i^o, $1 \leq i \leq m$, and if the restriction of f to each S_i^o can be extended to a continuous function on S_i, $1 \leq i \leq m$. The function f is *piecewise differentiable on the finite plane region S* if S is a finite sum of plane regular regions $S_1, \cdots, S_m$ such that f is continuously differentiable on each S_i, $1 \leq i \leq m$.

The functions f and g defined below are piecewise continuous and piecewise differentiable, respectively, on the plane regular region,

$$S = \{(x_1, x_2) \;\; |x_1| + |x_2| \leq 1\}:$$

$$f(x_1, x_2) = \frac{x_1}{|x_1|} + \frac{x_2}{|x_2|},$$

$$g(x_1, x_2) = |x_1| + |x_2|.$$

The graphs of these two functions are shown in Figure 3–3.4(a) and (b).

The function h such that

$$h(x_1, x_2) = \sqrt{4 - x_1^2 - x_2^2}$$

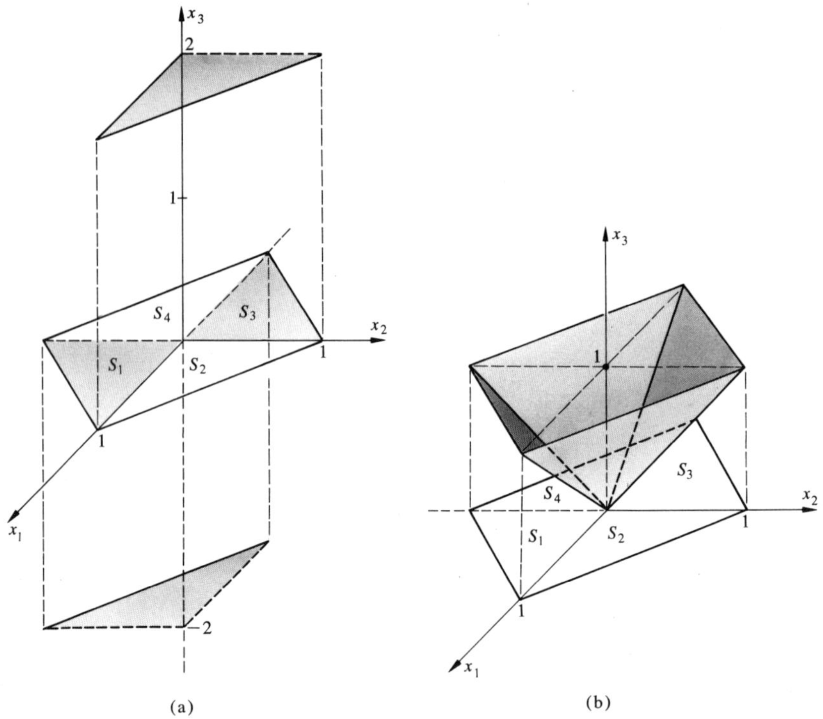

Figure 3–3.4

on the plane regular region $S = \{(x_1, x_2): x_1^2 + x_2^2 \leq 4\}$ is continuous on S, but it is not continuously differentiable on S, since the two first partials at $\mathbf{x}$, $h_{x_1}(\mathbf{x})$, and $h_{x_2}(\mathbf{x})$ approach infinity (either plus or minus) as $\mathbf{x}$ approaches points on the boundary of S.

Problems

1. For the functions f given by the following equations calculate $df(\mathbf{x}\,;\mathbf{dx})$.

   (a) $f(\mathbf{x}) = x_1^3 - 5x_1 x_2^2$, (b) $f(\mathbf{x}) = \dfrac{x_1 - 3x_2^3}{x_1^2 + 4}$,

   (c) $f(\mathbf{x}) = \sin^2(x_1 + x_2^2)$, (d) $f(\mathbf{x}) = [\exp(x_1^2 x_2)] + x_3$.

2. For each of the functions f given by the equations below find the equation of the tangent plane to the surface of the function f at the point indicated.

(a) $f(\mathbf{x}) = x_1 + x_1 x_2 + x_2^2$ at $(1, 1, 3)$,
(b) $f(\mathbf{x}) = (x_1 - x_2)^2$ at $(1, 0, 1)$,
(c) $f(\mathbf{x}) = \cos (x_1^2 + x_2^2)^{1/2}$ at $(\pi/\sqrt{2}, \pi/\sqrt{2}, -1)$,
(d) $f(\mathbf{x}) = x_1 + x_2 x_3^2$ at $(1, 0, 2, 1)$.

3. If $\mathbf{a}$ and $\mathbf{b}$ are vectors in E^n where $\mathbf{a} \neq \mathbf{0}$, then the set $\{\mathbf{a}t + \mathbf{b}: t \in R\}$ is a line through $\mathbf{b}$ in the direction of $\mathbf{a}$. The equation of this line is given parametrically as $\mathbf{x}(t) = \mathbf{a}t + \mathbf{b}$. The direction of

$$(f_{x_1}(\mathbf{x}_0), \cdots, f_{x_n}(\mathbf{x}_0), -1)$$

is *normal* to the surface given by f at $(\mathbf{x}_0, f(\mathbf{x}_0))$. Write parametrically the equation of the normal line to each of the surfaces in Problem 2 at the point indicated.

4. Let the function f be given by the equation $f(\mathbf{x}) = x_1^2 + x_1 x_2^3$. Find
(a) $df(1, 2; .1, -.02)$, (b) $\Delta f(1, 2; .1, -.02)$,
(c) $df(0, -1; -.03, .2)$, (d) $\Delta f(0, -1; -.03, .2)$.

5. There is a certain room of length 20 feet, width 15 feet, and height 8 feet. If the engineers wish to soundproof the room by placing on the floor carpet of thickness ¾ inch, acoustical tile on the ceiling of approximately ¾ inch thickness, and paneling on the walls of the thickness ½ inch, find the approximate loss in volume of the room.

6. Two sides of a triangle are measured to be 12 and 16 feet, respectively. The angle between these sides is measured to be 30 degrees. If these measurements are, respectively, .1, .01, and .2 degrees smaller than they should be, then what is the approximate error in the length of the third side if the third side is calculated by the law of cosines?

7. Prove that if the function f is given by the equation

$$f(\mathbf{x}) = \begin{cases} \dfrac{4x_1^2x_2^2}{(x_1^2 + x_2^2)^{(9/5)}}, & \mathbf{x} \neq \mathbf{0}, \\ 0, \quad \mathbf{x} = \mathbf{0}, \end{cases}$$

f is not differentiable at $\mathbf{0}$.

8. Suppose the function f is given by the following equation:

$$f(\mathbf{x}) = \begin{cases} \dfrac{4x_1^2x_2^2}{x_1^2 + x_2^2}, & \mathbf{x} \neq \mathbf{0}, \\ 0, \quad \mathbf{x} = \mathbf{0}. \end{cases}$$

Prove or disprove that f is differentiable at $\mathbf{0}$.

9. Given f and g mapping an open set S in E^n into R. If f and g are differentiable at $\mathbf{x}_0 \in S$, prove that $f + g$ and $f \cdot g$ are differentiable at $\mathbf{x}_0$. What extra condition must be true in order for f/g to be differentiable at $\mathbf{x}_0$? Find the formulas for $d(f + g)$, $d(f \cdot g)$, and $d(f/g)$.

10. Notice that for a given plane regular region S, {cont. diff.} $\subset$ {piece. diff.} $\subset$ {cont.} $\subset$ {piece. cont.}; that is, the class of continuously differentiable functions is contained in the class of piecewise differentiable functions, etc. For each of the following functions f on the indicated plane regular region S, find the most restrictive of the above classes to which it belongs.
Sketch the graph of each f.
   (a) $f(x_1, x_2) = [x_1] + [x_2]$, $S = \{(x_1, x_2): 0 \le x_1 \le 2, 0 \le x_2 \le 2\}$
   where $[x]$ means the largest integer less than or equal to x.
   (b) $f(x_1, x_2) = [x_1 + x_2]$, $S = \{(x_1, x_2): |x_1| + |x_2| \le 2\}$.
   (c) $f(x_1, x_2) = |x_1| - |x_2|$, $S = \{x_1, x_2): |x_1| \le 1, |x_2| \le 1\}$.
   (d) $f(x_1, x_2) = |(x_1 - x_2) - [x_1 - x_2] - \frac{1}{2}|$,
   $S = \{(x_1, x_2): |x_1| + |x_2| \le 2\}$.
11. Give the proof to Proposition 3–3.2.

Answers

1. (a) $(3x_1^2 - 5x_2^2)\,dx_1 - 10x_1x_2\,dx_2$,
   (b) $\left[\dfrac{6x_1x_2^3 - x_2^2 + 4}{(x_1^2 + 4)^2}\right] dx_1 - \left[\dfrac{9x_2^2}{(x_1^2 + 4)^2}\right] dx_2$,
   (c) $2 \sin (x_1 + x_2^2) \cos (x_1 + x_2^2)\,dx_1 + 4x_2 \sin (x_1 + x_2^2) \cos (x_1 + x_2^2)\,dx_2$,
   (d) $2x_1x_2 \exp (x_1^2x_2)\,dx_1 + x_1^2 \exp (x_1^2x_2)\,dx_2 + dx_3$.
2. (a) $2(x_1 - 1) + 3(x_2 - 1) = (x_3 - 3)$,
   (b) $2(x_1 - 1) - 2x_2 = x_3 - 1$,
   (c) $x_3 = -1$,
   (d) $(x_1 - 1) + 4x_2 = x_4 - 1$.
3. (a) $\mathbf{x}(t) = (2, 3, -1)\,t + (1, 1, 3)$,
   (b) $\mathbf{x}(t) = t(2, -2, -1) + (1, 0, 1)$,
   (c) $\mathbf{x}(t) = (0, 0, -1)\,t + (\pi/\sqrt{2}, \pi/\sqrt{2}, -1)$,
   (d) $\mathbf{x}(t) = t(1, 4, 0, -1) + (1, 0, 2, 1)$.
4. (a) .76, (b) .7486312, (c) .03, (d) .01626.
5. -60.83 ft^3.
6. $-.025$ ft.
10. (a) Piece. cont. (b) Piece. cont.
    (c) Piece. diff. (d) Piece. diff.

3–4. Chain Rule

The use of the chain rule is in the taking of partial derivatives of composite functions. Let $\mathbf{f}$ be a vector-valued function which maps a set S in

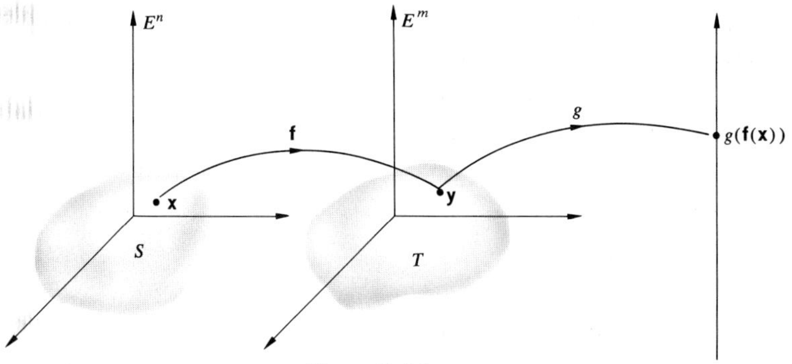

Figure 3–4.1

E^n into a set $\mathbf{f}(S)$ in E^m, and let g be a real-valued function that maps T into R where $\mathbf{f}(S) \subset T \subset E^m$ (cf. Figure 3–4.1). Let the function $\mathbf{f}$ be given by the vector equation $\mathbf{y} = \mathbf{f}(\mathbf{x})$ where $\mathbf{x} \in S$ and $\mathbf{y} \in T$ or by the equivalent m equations

$$\begin{aligned} y_1 &= f_1(x_1, \cdots, x_n), \\ y_2 &= f_2(x_1, \cdots, x_n), \\ &\cdots\cdots\cdots\cdots \\ y_m &= f_m(x_1, \cdots, x_n), \end{aligned}$$

which are obtained from the vector equation by equating each of the m components. Let the function g be defined by an equation of the form $z = g(y_1, \cdots, y_m)$. The variables $x_1, \cdots, x_n$ are the independent variables; $y_1, \cdots, y_m$ are the intermediate variables, and z is the dependent variable. The composite function $g\mathbf{f}$ is then defined by the equation $z = g(\mathbf{f}(\mathbf{x}))$. Figure 3–4.2, which might be described as a diagram of the relationship of the

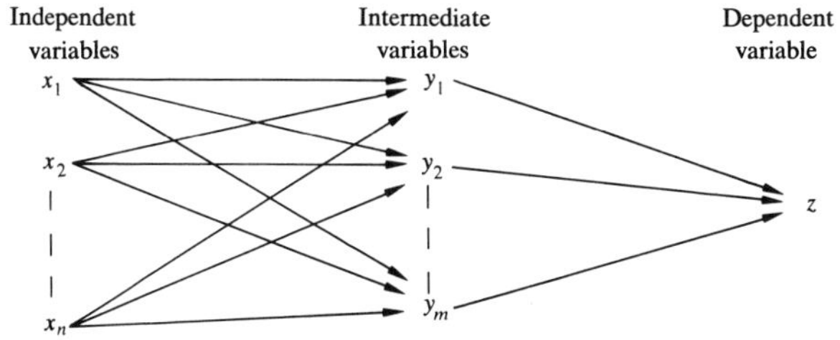

Figure 3–4.2

variables, indicates that each of the variables $x_1, \cdots, x_n$ contributes to the value of each of the variables $y_1, \cdots, y_m$, which in turn contribute to the value of z. The chain rule gives the formula for taking the derivative of the dependent variable with respect to an independent variable.

Let us for the moment put aside the question of existence of the derivatives involved, the justification of the chain rule, and even the statement of the chain rule. Instead we shall illustrate the chain rule by examples. We begin with the very simplest case.

Example 3–4.1. Let the diagram of the relationship of the variables be

$$x_1 \longrightarrow y_1 \longrightarrow z.$$

The equations of the functions are

$$y_1 = f(x_1), \qquad z = g(y_1), \qquad z = g(f(x_1)).$$

The chain rule gives the derivative of the composite function gf to be

$$(gf)'(x) = g'(y_1)f'(x_1),$$

or we can write

$$\frac{d}{dx_1}(gf)(x_1) = \frac{d}{dx_1}g(f(x_1)) = \frac{d}{dy_1}g(y_1)\cdot\frac{d}{dx_1}f(x_1).$$

If we wish to suppress the notation, we could write the chain rule as

$$\frac{dgf}{dx_1} = \frac{dg}{dy_1}\cdot\frac{df}{dx_1},$$

so that the independent variable of the function is shown on the bottom each time. Notice that this equation is an identity between the derivatives of the functions, whereas the first two equations are equalities of real numbers. However, none of the above forms of the chain rule compares in popularity to the following form:

$$\frac{dz}{dx_1} = \frac{dz}{dy_1}\cdot\frac{dy_1}{dx_1}.$$

This simple form, which better lends itself to exhibiting the pattern of the chain rule, has certain shortcomings. The symbol z in this notation is occupying the position of a symbol standing for a *function*, whereas z was defined above to be the *real number*—that is, the image of y_1 by the function g. Thus there are *two* very different concepts. This ambiguity is, of course, irksome, but there is a compelling need to familiarize the reader with this popular notation. This same notational convention will be followed not only in the subsequent examples, but frequently throughout the remainder of this book.

Example 3-4.2.
Diagram:

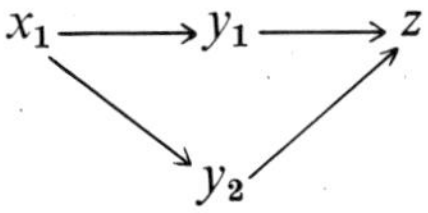

Equations:

$$\left.\begin{aligned} y_1 &= f_1(x_1) \\ y_2 &= f_2(x_1) \end{aligned}\right\}, \quad z = g(y_1, y_2), \quad z = g(f_1(x_1), f_2(x_1)).$$

Derivative:

$$\frac{dz}{dx_1} = \frac{\partial z}{\partial y_1}\frac{dy_1}{dx_1} + \frac{\partial z}{\partial y_2}\frac{dy_2}{dx_1}.$$

Example 3-4.3.
Diagram:

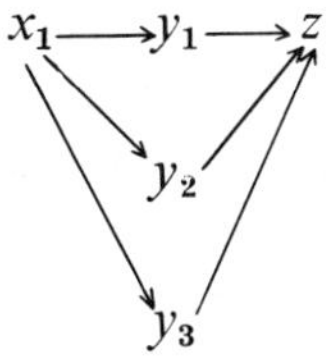

Equations:

$$\left.\begin{aligned} y_1 &= f_1(x_1) \\ y_2 &= f_2(x_1) \\ y_3 &= f_3(x_1) \end{aligned}\right\}, \quad z = g(y_1, y_2, y_3), \quad z = g(f_1(x_1), f_2(x_1), f_3(x_1)).$$

Derivative:

$$\frac{dz}{dx_1} = \frac{\partial z}{\partial y_1}\frac{dy_1}{dx_1} + \frac{\partial z}{\partial y_2}\frac{dy_2}{dx_1} + \frac{\partial z}{\partial y_3}\frac{dy_3}{dx_1}.$$

Example 3-4.4.
Diagram:

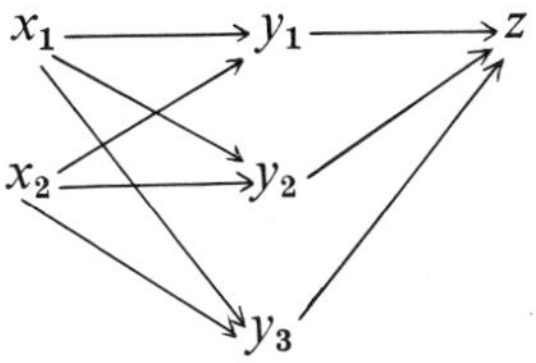

Equations:

$$\left.\begin{aligned} y_1 &= f_1(x_1, x_2) \\ y_2 &= f_2(x_1, x_2) \\ y_3 &= f_3(x_1, x_2) \end{aligned}\right\}, \quad z = g(y_1, y_2, y_3),$$

$$z = g(f_1(x_1, x_2), f_2(x_1, x_2), f_3(x_1, x_2)).$$

Derivatives:

$$\frac{\partial z}{\partial x_1} = \frac{\partial z}{\partial y_1}\frac{\partial y_1}{\partial x_1} + \frac{\partial z}{\partial y_2}\frac{\partial y_2}{\partial x_1} + \frac{\partial z}{\partial y_3}\frac{\partial y_3}{\partial x_1},$$

$$\frac{\partial z}{\partial x_2} = \frac{\partial z}{\partial y_1}\frac{\partial y_1}{\partial x_2} + \frac{\partial z}{\partial y_2}\frac{\partial y_2}{\partial x_2} + \frac{\partial z}{\partial y_3}\frac{\partial y_3}{\partial x_2}.$$

The first three of these examples illustrate the reason for the name "chain rule." As each successive intermediate variable is added, the derivative adds correspondingly another term, another chain link as it were. The number of terms in the derivative is determined by the number of intermediate variables. The number of derivatives is, of course, the same as the number of independent variables.

Notice the pattern of the symbols in the derivatives. In each term on the right side, the upper left symbol, ∂z, is the same as the upper symbol on the left side. Similarly, the lower right symbol, ∂x_1, for example, agrees with the lower symbol on the left. The other two symbols of each term on the right are the same, except possibly for the difference indicated by d, as compared with ∂. Notice that d is used when the function is a function of a single variable. Otherwise the symbol is ∂.

Now that the pattern for taking the partials of composite functions has become somewhat familiar, we shall take as our example the general circumstance introduced in the beginning. This will allow us to give a general statement of the chain rule. Figure 3–4.2 shows the diagram of the relationship of the variables. The composite function, $g\mathbf{f}$, is defined by the equation

$$z = g(f_1(x_1, \cdots, x_n), \cdots, f_m(x_1, \cdots, x_n)).$$

Then the derivative with respect to x_i is calculated by the *chain rule*, thus:

$$\frac{\partial z}{\partial x_i} = \frac{\partial z}{\partial y_1}\frac{\partial y_1}{\partial x_i} + \cdots + \frac{\partial z}{\partial y_m}\frac{\partial y_m}{\partial x_i} = \sum_{j=1}^{m}\frac{\partial z}{\partial y_j}\frac{\partial y_j}{\partial x_i},$$

$1 \leq i \leq n$. With indicial notation we could write simply

$$\frac{\partial z}{\partial x_i} = \frac{\partial z}{\partial y_j}\frac{\partial y_j}{\partial x_i}.$$

The following example will illustrate a use of the chain rule.

Example 3–4.5. In E^3 let the position of a particle at time t be given by $\mathbf{x}(t) = (t, t^2, t^3)$. Suppose that temperature T in E^3 varies inversely with respect to the distance from the origin. What is the rate of change in temperature at time $t = 1$? Thus $T = (x_1^2 + x_2^2 + x_3^2)^{-1/2}$. The quantity desired is

$$\left.\frac{dT}{dt}\right|_{t=1}.$$

By the chain rule,

$$\frac{dT}{dt} = \frac{\partial T}{\partial x_1}\frac{dx_1}{dt} + \frac{\partial T}{\partial x_2}\frac{dx_2}{dt} + \frac{\partial T}{\partial x_3}\frac{dx_3}{dt}$$

$$= -(x_1^2 + x_2^2 + x_3^2)^{-(3/2)}[x_1(1) + x_2(2t) + (x_3)(3t^2)].$$

At $t = 1$, $\mathbf{x}(t) = (1, 1, 1)$. Hence,

$$\left.\frac{dT}{dt}\right|_{t=1} = -3^{-(3/2)}(1 + 2 + 3) = -2\sqrt{3}/3.$$

Thus at time $t = 1$, the temperature is decreasing at the rate of $2\sqrt{3}/3$ units of temperature per unit of time.

As a check

$$T = (t^2 + (t^2)^2 + (t^3)^2)^{-(1/2)}$$

and

$$\frac{dT}{dt} = -\tfrac{1}{2}(t^2 + t^4 + t^6)^{-(3/2)}(2t + 4t^3 + 6t^5).$$

Again

$$\left.\frac{dT}{dt}\right|_{t=1} = -2\sqrt{3}/3.$$

In order to take second partials of composites, the chain rule is applied several times. Consider the following example.

Example 3–4.6. Let $u = f(x, y)$, $v = g(x, y)$, and $z = h(u, v)$, and suppose that it is desired to find z_{xx}. By the chain rule,

$$\frac{\partial z}{\partial x} = \frac{\partial z}{\partial u}\frac{\partial u}{\partial x} + \frac{\partial z}{\partial v}\frac{\partial v}{\partial x}$$

or

$$z_x = z_u u_x + z_v v_x.$$

Then it follows by using the differentiation formulas for sums and products that

$$z_{xx} = \frac{\partial z_x}{\partial x} = \frac{\partial}{\partial x}(z_u u_x) + \frac{\partial}{\partial x}(z_v v_x)$$

$$= \frac{\partial z_u}{\partial x} u_x + z_u \frac{\partial u_x}{\partial x} + \frac{\partial z_v}{\partial x} v_x + z_v \frac{\partial v_x}{\partial x}.$$

Now z_u (and z_v) are composite functions in that

$$z_u = \frac{\partial}{\partial u} z(u, v), \qquad u = f(x, y), \quad \text{and} \quad v = g(x, y).$$

Thus the derivatives in the first and third terms are again derivatives of composite functions. The derivative,

$$\frac{\partial u_x}{\partial x},$$

is merely the second partial of u with respect to x since u_x, as u, maps the (x, y)-space into R. Hence, using the chain rule on the first and third term, we get

$$z_{xx} = \left(\frac{\partial z_u}{\partial u}\frac{\partial u}{\partial x} + \frac{\partial z_u}{\partial v}\frac{\partial v}{\partial x}\right) u_x + z_u u_{xx} + \left(\frac{\partial z_v}{\partial u}\frac{\partial u}{\partial x} + \frac{\partial z_v}{\partial v}\frac{\partial v}{\partial x}\right) v_x + z_v v_{xx}.$$

After simplifying and rearranging we get

$$z_{xx} = z_{uu} u_x u_x + (z_{vu} + z_{uv}) u_x v_x + z_{vv} v_x v_x + z_u u_{xx} + z_v v_{xx}.$$

With this background on the chain rule and its use, we go now to a proposition that gives a sufficient condition for using the chain rule and proves, for a special case, that the chain rule is indeed the correct formula.

Proposition 3–4.1. *Let* $\mathbf{f}\colon S \to E^m$ *where* S *is an open set in* E^n, *and let* $g\colon T \to R$ *where* T *is an open set in* E^m *such that* $f(S) \subset T$. *Suppose that* $\mathbf{f}$ *is given by the* m *equations,* $y_i = f_i(\mathbf{x})$, $1 \le i \le m$, *and that* g *is given by the equation* $z = g(\mathbf{y})$. *Let each* f_i *be differentiable at* $\mathbf{x}_0 \in S$, $1 \le i \le m$, *and let* g *be differentiable at* $\mathbf{y}_0 = \mathbf{f}(\mathbf{x}_0)$. *Then the composite function* $g\mathbf{f}$ *is differentiable at* $\mathbf{x}_0$ *and*

$$\frac{\partial z(\mathbf{x}_0)}{\partial x_i} = \frac{\partial g\mathbf{f}(\mathbf{x}_0)}{\partial x_i} = \sum_{j=1}^{m} \frac{\partial g(\mathbf{f}(\mathbf{x}_0))}{\partial y_j}\frac{\partial f_j(\mathbf{x}_0)}{\partial x_i}, \quad 1 \le i \le m.$$

Proof for $n = 1$, $m = 2$. For this case $\mathbf{x}_0$ is x_{01}. Let

$$\mathbf{y} = (y_1, y_2) = (f_1(x_1), f_2(x_1)),$$

and let

$$\mathbf{y}_0 = (y_{01}, y_{02}) = (f_1(x_{01}), f_2(x_{02})).$$

Now, by adding and subtracting a term,

$$g(y_1, y_2) - g(y_{01}, y_{02}) = [g(y_1, y_2) - g(y_{01}, y_2)] + [g(y_{01}, y_2) - g(y_{01}, y_{02})].$$

Since g is differentiable at $\mathbf{y}_0$, there is a neighborhood $N(\mathbf{y}_0; \varepsilon)$ in which both partials of g exist. Since both f_1 and f_2 are differentiable at x_{01}, then they are continuous at x_{01}. Hence there exists a neighborhood $N(x_{01}; \delta)$ such that when $x_1 \in N(x_{01}; \delta)$, then $\mathbf{y} \in N(\mathbf{y}_0; \varepsilon)$. Suppose $x_1 \in N(x_{01}, \delta)$. Then $g(y_1, y_2)$, as a function of the single variable y_1, satisfies the mean value theorem, and thus

$$g(y_1, y_2) - g(y_{01}, y_2) = g_{y_1}(y_{01} + \theta_1(y_1 - y_{01}), y_2)(y_1 - y_{01})$$

where $0 < \theta_1 < 1$. A similar substitution can be made for the quantity in the second bracket. Thus

$$g(y_1, y_2) - g(y_{01}, y_{02}) = g_{y_1}(y_{01} + \theta_1(y_1 - y_{01}), y_2)[y_1 - y_{01}] + g_{y_2}(y_{01}, y_{02} + \theta_2(y_2 - y_{02}))[y_2 - y_{02}],$$

or, equivalently,

$$g(\mathbf{f}(x_1)) - g(\mathbf{f}(x_{01})) = g_{y_1}(y_{01} + \theta_1(y_1 - y_{01}), y_2)[f_1(x_1) - f_1(x_{01})] + g_{y_2}(y_{01}, y_{02} + \theta_2(y_2 - y_{02}))[f_2(x_1) - f_2(x_{01})]$$

where $0 < \theta_1, \theta_2 < 1$.

Suppose $x_1 \neq x_{01}$. Divide both sides by $x_1 - x_{01}$; then

$$\frac{g(\mathbf{f}(x_1)) - g(\mathbf{f}(x_{01}))}{x_1 - x_0} = g_{y_1}(y_{01} + \theta_1(y_1 - y_{01}), y_2) \cdot \frac{f_1(x_1) - f_1(x_{01})}{x_1 - x_{01}} + g_{y_2}(y_{01}, y_{02} + \theta_2(y_2 - y_{02})) \cdot \frac{f_2(x_1) - f_2(x_{01})}{x_1 - x_{01}}.$$

The chain rule follows by taking the limit at x_{01} on both sides because of the existence of $f_1'(x_{01})$, $f_2'(x_{01})$, since the continuity of f_1 and f_2 at x_{01} implies

$$\lim_{x_1 \to x_{01}} g_{y_1}(y_{01} + \theta_1(y_1 - y_{01}), y_2) = \lim_{y \to y_0} g_{y_1}(y_{01} + \theta_1(y_1 - y_{01}), y_2)$$

(and similarly for g_{y_2}), and since g_{y_1} and g_{y_2} are continuous at $\mathbf{y}_0$. ∎

Consider the problem of finding differentials of composite functions. In particular, let us use the circumstances of Example 3–4.4. By definition,

$$dh(\mathbf{x}_0; \mathbf{dx}) = h_{x_1}(\mathbf{x}_0)\, dx_1 + h_{x_2}(\mathbf{x}_0)\, dx_2, \qquad (3\text{–}4.1)$$

and also

$$dg(\mathbf{y}_0; \mathbf{dy}) = g_{y_1}(\mathbf{y}_0)\, dy_1 + g_{y_2}(\mathbf{y}_0)\, dy_2 + g_{y_3}(\mathbf{y}_0)\, dy_3$$

where $z = h(\mathbf{x}) = g(\mathbf{f}(\mathbf{x}))$. The question here is: If $\mathbf{y}_0 = \mathbf{f}(\mathbf{x}_0)$ and $dy_i = df_i(\mathbf{x}_0; \mathbf{dx})$, $i = 1, 2$, then does $dh(\mathbf{x}_0; \mathbf{dx}) = dg(\mathbf{y}_0; \mathbf{dy})$? That is, if $\mathbf{y}_0$ is the point corresponding to $\mathbf{x}_0$ and $\mathbf{dy}$ the increment corresponding to $\mathbf{dx}$, are the differentials equal? The answer is yes, for by the chain rule,

$$h_{x_i}(\mathbf{x}_0) = g_{y_1}(\mathbf{y}_0)\frac{\partial y_1}{\partial x_i} + g_{y_2}(\mathbf{y}_0)\frac{\partial y_2}{\partial x_i} + g_{y_3}(\mathbf{y}_0)\frac{\partial y_3}{\partial x_i}, \quad i = 1, 2.$$

Substituting above in equation (3–4.1), we get

$$dh(\mathbf{x}_0\,;\mathbf{dx}) = \left(g_{y_1}(\mathbf{y}_0)\frac{\partial y_1}{\partial x_1} + g_{y_2}(\mathbf{y}_0)\frac{\partial y_2}{\partial x_1} + g_{y_3}(\mathbf{y}_0)\frac{\partial y_3}{\partial x_1}\right)dx_1 + \left(g_{y_1}(\mathbf{y}_0)\frac{\partial y_1}{\partial x_2} + g_{y_2}(\mathbf{y}_0)\frac{\partial y_2}{\partial x_2} + g_{y_3}(\mathbf{y}_0)\frac{\partial y_3}{\partial x_2}\right)dx_2.$$

Rearranging,

$$dh(\mathbf{x}_0\,;\mathbf{dx}) = g_{y_1}(\mathbf{y}_0)\left(\frac{\partial y_1}{\partial x_1}dx_1 + \frac{\partial y_1}{\partial x_2}dx_2\right) + g_{y_2}(\mathbf{y}_0)\left(\frac{\partial y_2}{\partial x_1}dx_1 + \frac{\partial y_2}{\partial x_2}dx_2\right) + g_{y_3}(\mathbf{y}_0)\left(\frac{\partial y_3}{\partial x_1}dx_1 + \frac{\partial y_3}{\partial x_2}dx_2\right),$$

where the partials, $\partial y_j/\partial x_i$, are evaluated at $\mathbf{x}_0$. Thus

$$\begin{aligned} dh(\mathbf{x}_0\,;\mathbf{dx}) &= g_{y_1}(\mathbf{y}_0)\,dy_1(\mathbf{y}_0\,;\mathbf{dx}) + g_{y_2}(\mathbf{y}_0)\,dy_2(\mathbf{y}_0\,;\mathbf{dx}) + g_{y_3}(\mathbf{y}_0)\,dy_3(\mathbf{y}_0\,;\mathbf{dx}) \\ &= dg(\mathbf{y}_0\,;\mathbf{dy}). \end{aligned}$$

Clearly, the proof could be carried out the same way in general circumstances—that is, with n and m arbitrary. The equality of these two differentials under the above circumstances is known as *Cauchy's Invariant Rule*. As a summary, this rule is stated below in complete generality.

Proposition 3–4.2. *Under the same hypothesis as Proposition* 3–4.1, *if* $\mathbf{dy}$ *is such that*

$$dy_i = df_i(\mathbf{x}_0\,;\mathbf{dx}), \quad 1 \le i \le m,$$

then

$$dg(\mathbf{y}_0\,;\mathbf{dy}) = dg\mathbf{f}(\mathbf{x}_0\,;\mathbf{dx})$$

where $g\mathbf{f}$ *is the composite function.*

Problems

1. Let the function $\mathbf{f}$ and the function g be defined by the equations below. Let h be the composite function $g\mathbf{f}$. At the point indicated calculate the **derivatives** indicated in each of the following cases.

(a) $\left.\begin{aligned} f_1(x_1) &= 3x_1 \\ f_2(x_1) &= x_1^2 + 1 \end{aligned}\right\}, \quad g(y_1, y_2) = y_1 + y_1^2 y_2.$

Find $\dfrac{dh}{dx_1}$ at $x_{01} = -1$.

(b) $\left.\begin{aligned} f_1(x_1) &= x_1^2 \\ f_2(x_1) &= (\pi/2)(x_1 - 1) \\ f_3(x_1) &= 2x_1 - 3 \end{aligned}\right\}, \quad g(y_1, y_2, y_3) = y_1^2 \cos(y_2 y_3).$

Find $\dfrac{dh}{dx_1}$ at $x_{01} = 2$.

(c) $f_1(x_1, x_2) = x_1^2 x_2 - 3x_2^3, \quad g(y_1) = e^{y_1^2}.$

Find $\dfrac{\partial h}{\partial x_1}$ and $\dfrac{\partial h}{\partial x_2}$ at (2, 1).

(d) $\left.\begin{aligned} f_1(x_1, x_2, x_3) &= 2x_1 + 3x_2 + x_3 \\ f_2(x_1, x_2, x_3) &= x_1 x_2 \log x_3 \end{aligned}\right\}, \quad g(y_1, y_2) = y_1 \tan^{-1} y_2.$

Find $\dfrac{\partial h}{\partial x_i}$, $i = 1, 2, 3$, at $(1, 1, e)$.

2. Find the derivatives of Problem 1 by first finding the equation $h(\mathbf{x}) = g(\mathbf{f}(\mathbf{x}))$; in other words, substitute for $\mathbf{y}$, $\mathbf{f}(\mathbf{x})$ and then calculate the derivatives directly.

3. Let $\mathbf{f}$ be given by equations indicated below, and let g be an arbitrary differentiable function of m variables as indicated. Calculate each of the derivatives asked for below in terms of $\mathbf{x}$ and derivatives of g.

(a) $f_1(\mathbf{x}) = x_1 + x_1 x_3 + x_2^2, \quad m = 1.$

Find $\dfrac{\partial h}{\partial x_i}$, $\quad i = 1, 2, 3$.

(b) $\left.\begin{aligned} f_1(\mathbf{x}) &= x_1^2 + x_2^3 \\ f_2(\mathbf{x}) &= x_1 \sec x_2 \end{aligned}\right\}, \quad m = 2.$

Find $\dfrac{dh}{dx_i}$, $\quad i = 1, 2$.

(c) $\left.\begin{aligned} f_1(\mathbf{x}) &= x_1 x_2 \\ f_2(\mathbf{x}) &= 2x_1 - x_2 \\ f_3(\mathbf{x}) &= x_2^{2^{x_1}} \end{aligned}\right\}, \quad m = 3.$

Find $\dfrac{\partial h}{\partial x_i}$, $\quad i = 1, 2$.

(d) $\left.\begin{aligned} f_1(\mathbf{x}) &= x_3 \sinh x_1x_2 \\ f_2(\mathbf{x}) &= x_1x_2 + x_3 \\ f_3(\mathbf{x}) &= (x_1x_3)/x_2 \end{aligned}\right\}, \quad m = 3.$

Find $\dfrac{\partial h}{\partial x_i}$, $i = 1, 2, 3.$

4. The variables x, y, z, r, and θ are related by the equations

$$\left.\begin{aligned} x &= r \cos \theta \\ y &= r \sin \theta \end{aligned}\right\}, \quad z = f(x, y).$$

Assuming that the required derivatives exist, calculate the partials of the composite function with respect to r and θ, written f_r and f_θ, in terms of f_x, f_y, r, and θ. Show that

$$(f_x)^2 + (f_y)^2 = (f_r)^2 + \frac{1}{r^2}(f_\theta)^2$$

where $r \neq 0$. The equality here is an equality of functions with the meaning that if the function on the right is evaluated at (r_0, θ_0), then it is equal to the function on the left evaluated at (x_0, y_0) where $x_0 = r_0 \cos \theta_0$, $y_0 = r_0 \sin \theta_0$.

5. Use the same technique as shown in the above problem to show

$$f_x^2 + f_y^2 + f_z^2 = f_r^2 + \frac{1}{r^2} f_\varphi^2 + \frac{1}{r^2 \sin^2 \varphi} f_\theta^2$$

where $r \sin \varphi \neq 0$ if

$$x = r \sin \varphi \cos \theta,$$
$$y = r \sin \varphi \sin \theta,$$
$$z = r \cos \varphi,$$

and f is a real-valued function whose domain is (x, y, z)-space.

6. Suppose the functions f_1, f_2, and g are defined by the following equations: $u = f_1(x, y)$, $v = f_2(x, y)$, and $z = g(u, v)$. Use the chain rule to calculate the following derivatives:

(a) $\dfrac{\partial^2 z}{\partial y^2}$, (b) $\dfrac{\partial^2 z}{\partial y\, \partial x}$.

7. Suppose f and g are real-valued functions of a single variable such that their second derivatives exist on a certain open set S of E^1. Show that on a certain open set in the E^2 space given by x and y, the function F defined by

$$F(x, y) = f(y + ax) + g(y - ax)$$

is such that

$$F_{xx} = a^2 F_{yy}.$$

8. Use the technique of Problem 4 to show the following equalities when the variables are related as indicated.

(a) $f_{xx} + f_{yy} = f_{rr} + \frac{1}{r} f_r + \frac{1}{r^2} f_{\theta\theta}$,

where $x = r \cos \theta$, $y = r \sin \theta$.

(b) $f_{xx} + f_{yy} + f_{zz} = f_{rr} + \frac{1}{r^2} f_{\varphi\varphi} + \frac{1}{r^2 \sin^2 \varphi} f_{\theta\theta} + \frac{2}{r} f_r + \frac{\cot \varphi}{r^2} f_\varphi$

where $x = r \sin \varphi \cos \theta$,

$y = r \sin \varphi \sin \theta$,

$z = r \cos \varphi$.

9. Let the functions f_1, f_2, and g be defined by the following equations:

$$u = f_1(x, y) = x^2 - y^2, \qquad z = g(u, v) = 3u - v^2,$$
$$v = f_2(x, y) = 2xy.$$

(a) Find $dh(3, 1; .1, -.2)$ where $h = g\mathbf{f}$, the composite function. If $(x_0, y_0) = (3, 1)$, then

$$(u_0, v_0) = (f_1(3, 1), f_2(3, 1)) = (8, 6).$$

Find $df_i(3, 1; .1, -.2)$, $i = 1, 2$. Use these two answers in dg to demonstrate Proposition 3–4.2; that is, show

$$dg(8, 6; df_1, df_2) = dh(3, 1; .1, -.2)$$

with the above values df_1 and df_2.

(b) Change the given data of part (a) to $(x_0, y_0) = (1, -1)$ and $(dx, dy) = (-.03, .01)$, and again demonstrate Proposition 3–4.2.

Answers

1. (a) -51. (b) -24π. (c) $8e$, $-10e$.

(d) $\frac{\pi}{2} - \frac{5 + e}{2}, \frac{3\pi}{4} + \frac{5 + e}{2}, \frac{\pi}{4} + \frac{5 + e}{2e}$.

3. (a) $\frac{dg}{dy_1}(1 + x_3), \frac{dg}{dy_1}(2x_2), \frac{dg}{dy_1}(x_1)$.

(b) $\frac{\partial g}{\partial y_1}(2x_1) + \frac{\partial g}{\partial y_2}(\sec x_2), \frac{\partial g}{\partial y_1}(3x_2^2) + \frac{\partial g}{\partial y_2}(x_1 \sec x_2 \tan x_2)$.

(c) $\frac{\partial g}{\partial y_1}(x_2) + \frac{\partial g}{\partial y_2}(2) + \frac{\partial g}{\partial y_3}(x_2^{2^{x_1}} (\log x_2)\, 2^{x_1}(\log 2))$,

$\frac{\partial g}{\partial y_1}(x_1) + \frac{\partial g}{\partial y_2}(-1) + \frac{\partial g}{\partial y_3}(2^{x_1} x_2^{2^{x_1}-1})$.

(d) $\dfrac{\partial g}{\partial y_1}(x_2 x_3 \cosh(x_1 x_2)) + \dfrac{\partial g}{\partial y_2}(x_2) + \dfrac{\partial g}{\partial y_3}\left(\dfrac{x_3}{x_2}\right),$

$\dfrac{\partial g}{\partial y_1}(x_1 x_3 \cosh(x_1 x_2)) + \dfrac{\partial g}{\partial y_2}(x_1) + \dfrac{\partial g}{\partial y_3}\left(-\dfrac{x_1 x_3}{x_2^2}\right),$

$\dfrac{\partial g}{\partial y_1}(\sinh(x_1 x_2)) + \dfrac{\partial g}{\partial y_2} + \dfrac{\partial g}{\partial y_3}\left(\dfrac{x_1}{x_2}\right).$

9. $df_1(3, 1; .1, -.2) = 1,$
$df_2(3, 1; .1, -.2) = -1,$
$dg(8, 6; df_1, df_2) = 15.$

3–5. Functions Defined Implicitly

The functions for which we have been finding derivatives and differentials have been defined by an equation. If the function $f\colon S \to R$ where $S \subset E^n$, then the equation that defined f was always of the form $y = f(x_1, \cdots, x_n)$; that is, f is defined *explicitly* in that the image of $\mathbf{x}$ under the mapping f is explicitly known when the components of $\mathbf{x}$ are substituted on the right and the required calculations are made. Functions may be defined in other ways by equations, for consider the following equation:

$$F(x_1, x_2, y_1) = x_1^2 + x_2^2 + y_1^2 - 6 = 0. \qquad (3\text{–}5.1)$$

In (x_1, x_2, y_1)-space, the set of points that satisfy this equation lies on the surface of a sphere of radius $\sqrt{6}$ with center at the origin. The point $(1, 1, 2)$ is one such point. It is possible to solve this equation explicitly for y_1 in terms of x_1 and x_2, by finding an equation involving y_1 and only y_1 on one side of the equation and x_1 and x_2 on the other, namely,

$$y_1 = \pm\sqrt{6 - x_1^2 - x_2^2}.$$

If we wish $(1, 1, 2)$ to satisfy this equation, we take the plus sign in front of the radical. The function defined by the equation

$$y_1 = \sqrt{6 - x_1^2 - x_2^2} = f(x_1, x_2) \qquad (3\text{–}5.2)$$

is a closed hemispherical surface in (x_1, x_2, y_1)-space that coincides with the above sphere everywhere that $y_1 \geq 0$ (cf. Figure 3–5.1). Thus (3–5.1) defines a function f such that f maps (x_1, x_2)-space into y_1-space, $f(1, 1) = 2$, and in some neighborhood of $(1, 1)$ $F(x_1, x_2, f(x_1, x_2)) = 0$ for all (x_1, x_2) in the neighborhood. In this example any neighborhood about $(1, 1)$ that is contained in the disk $\{(x_1, x_2)\colon x_1^2 + x_2^2 \leq 6\}$ would be satisfactory.

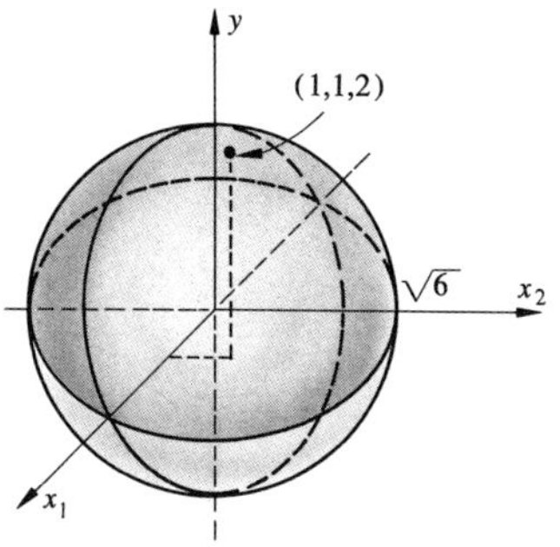

Figure 3–5.1

This function f defined by (3–5.1) is said to be defined *implicitly*. It was easy in this example to find an equation that defined f explicitly, but it is not always easy; for example, consider the equation

$$x_1 x_2 + y_1 + \sin y_1 = 0.$$

Clearly there is difficulty here in solving for y_1.

Functions may be defined implicitly in quite a variety of circumstances. In the example that follows, the equation on the left is the given equation $F = 0$ that presumably defines implicitly a function f. On the right is shown the form of an equation, if such could be found, that would define f explicitly.

Example 3–5.1.

Case 1. $F_1(x_1, y_1) = 0$ $\quad y_1 = f_1(x_1)$

Case 2. $F_1(x_1, x_2, y_1) = 0$ $\quad y_1 = f_1(x_1, x_2)$

Case 3. $F_1(x_1, \cdots, x_n, y_1) = 0$ $\quad y_1 = f_1(x_1, \cdots, x_n)$

It is possible that two equations may be given, in which case two functions could be defined implicitly. For example,

$$F_1(x_1, y_1, y_2) = 2x_1 - 3y_1 + y_2 - 2 = 0,$$

$$F_2(x_1, y_1, y_2) = x_1 + 6y_1 + y_2 + 3 = 0$$

define explicitly the two functions f_1 and f_2 defined by the following equations:

$$y_1 = f_1(x_1) = \frac{1}{9}(x_1 - 5),$$

$$y_2 = f_2(x_1) = \frac{1}{3}(-5x_1 + 1).$$

Following the pattern of Example 3–5.1, two other cases are indicated, the latter being the general case.

Example 3–5.2.

Case 1. $F_1(x_1, x_2, y_1, y_2, y_3) = 0, \qquad y_1 = f_1(x_1, x_2),$
$F_2(x_1, x_2, y_1, y_2, y_3) = 0, \qquad y_2 = f_2(x_1, x_2),$
$F_3(x_1, x_2, y_1, y_2, y_3) = 0, \qquad y_3 = f_3(x_1, x_2),$

Case 2. $F_1(x_1, \cdots, x_n, y_1, \cdots, y_m) = 0, \quad y_1 = f_1(x_1, \cdots, x_n),$
. . . .
$F_m(x_1, \cdots, x_n, y_1, \cdots, y_m) = 0, \quad y_m = f_m(x_1, \cdots, x_n).$

It is the purpose of this section to consider certain equations in order to (a) *give sufficient conditions for the equations to define a function implicitly that has certain partial derivatives* and (b) *determine how partial derivatives of a function defined implicitly can be calculated* when they exist. We shall consider the second problem first.

Consider Case 1 of Example 3–5.2 and look at the composite function $F_1 \, \mathbf{f}$. Here x_1 and x_2 are the independent variables x_1, x_2; y_1, y_2, and y_3 are the intermediate variables, and $z_1 = F_1(\mathbf{f}(x_1, x_2, y_1, y_2, y_3))$ is the dependent variable. Suppose the chain rule is applicable. Then, taking the derivative of both sides of the equation $F_1 = 0$ with respect to x_1 and holding x_2 constant, we have

$$0 = \left.\frac{\partial z_1}{\partial x_1}\right)_{x_2} = \left.\frac{\partial F_1}{\partial x_1}\right)_{x_2,y_1,y_2,y_3} \cdot \left.\frac{\partial x_1}{\partial x_1}\right)_{x_2} + \left.\frac{\partial F_1}{\partial x_2}\right)_{x_1,y_1,y_2,y_3} \cdot \left.\frac{\partial x_2}{\partial x_1}\right)_{x_2}$$
$$+ \left.\frac{\partial F_1}{\partial y_1}\right)_{x_1,x_1,y_2,y_3} \cdot \left.\frac{\partial y_1}{\partial x_1}\right)_{x_2} + \left.\frac{\partial F_1}{\partial y_2}\right)_{x_1,x_2,y_1,y_3} \cdot \left.\frac{\partial y_2}{\partial x_1}\right)_{x_2}$$
$$+ \left.\frac{\partial F_1}{\partial y_3}\right)_{x_1,x_2,y_1,y_2} \cdot \left.\frac{\partial y_3}{\partial x_1}\right)_{x_2}$$

where

$$\left.\frac{\partial F_1}{\partial x_1}\right)_{x_2,y_1,y_2,y_3}$$

means the partial of F_1 with respect to x_1 holding x_2, y_1, y_2, and y_3 constant. This extra notation, which was understood always in the discussion of the chain rule, is used here to help us to understand clearly the meaning of the symbols in that some of the variables are both independent variables and intermediate variables. The difficulty becomes apparent if the extra notation is omitted, and on the left F_1 is written, as it very well may be, in place of z_1.

Since

$$\left.\frac{\partial x_1}{\partial x_1}\right)_{x_2} = 1, \qquad \left.\frac{\partial x_2}{\partial x_1}\right)_{x_2} = 0,$$

we have

$$\frac{\partial F_1}{\partial x_1} + \frac{\partial F_1}{\partial y_1}\frac{\partial y_1}{\partial x_1} + \frac{\partial F_1}{\partial y_2}\frac{\partial y_2}{\partial x_1} + \frac{\partial F_1}{\partial y_3}\frac{\partial y_3}{\partial x_1} = 0$$

or

$$\frac{\partial F_1}{\partial y_1}\frac{\partial y_1}{\partial x_1} + \frac{\partial F_1}{\partial y_2}\frac{\partial y_2}{\partial x_1} + \frac{\partial F_1}{\partial y_3}\frac{\partial y_3}{\partial x_1} = -\frac{\partial F_1}{\partial x_1}. \tag{3–5.3}$$

Treating F_2 and F_3 similarly, we get

$$\frac{\partial F_2}{\partial y_1}\frac{\partial y_1}{\partial x_1} + \frac{\partial F_2}{\partial y_2}\frac{\partial y_2}{\partial x_1} + \frac{\partial F_2}{\partial y_3}\frac{\partial y_3}{\partial x_1} = -\frac{\partial F_2}{\partial x_1}, \tag{3–5.4}$$

$$\frac{\partial F_3}{\partial y_1}\frac{\partial y_1}{\partial x_1} + \frac{\partial F_3}{\partial y_2}\frac{\partial y_2}{\partial x_1} + \frac{\partial F_3}{\partial y_3}\frac{\partial y_3}{\partial x_1} = -\frac{\partial F_3}{\partial x_1}. \tag{3–5.5}$$

Solving (3–5.3), (3–5.4), and (3–5.5) for $\partial y_1/\partial x_1$ by Cramer's rule, we get

$$\frac{\partial y_1}{\partial x_1} = -\left.\begin{vmatrix} \frac{\partial F_1}{\partial x_1} & \frac{\partial F_1}{\partial y_2} & \frac{\partial F_1}{\partial y_3} \\ \frac{\partial F_2}{\partial x_1} & \frac{\partial F_2}{\partial y_2} & \frac{\partial F_2}{\partial y_3} \\ \frac{\partial F_3}{\partial x_1} & \frac{\partial F_3}{\partial y_2} & \frac{\partial F_3}{\partial y_3} \end{vmatrix} \middle/ \begin{vmatrix} \frac{\partial F_1}{\partial y_1} & \frac{\partial F_1}{\partial y_2} & \frac{\partial F_1}{\partial y_3} \\ \frac{\partial F_2}{\partial y_1} & \frac{\partial F_2}{\partial y_2} & \frac{\partial F_2}{\partial y_3} \\ \frac{\partial F_3}{\partial y_1} & \frac{\partial F_3}{\partial y_2} & \frac{\partial F_3}{\partial y_3} \end{vmatrix}\right.,$$

if the determinant in the denominator is not zero. This method of calculating the partial derivative is known as *differentiating implicitly*.

The notation here is quite cumbersome. To remedy this, these determinants are written more simply. For example, the determinant of the denominator is frequently written

$$\frac{\partial(F_1, F_2, F_3)}{\partial(y_1, y_2, y_3)}$$

and is read, "*the Jacobian of* (F_1, F_2, F_3) *with respect to* (y_1, y_2, y_3)." Note that the order of writing the variables is important in that an interchange of two of the variables amounts to a change in sign on the determinant.

Following the Jacobian notation,

$$\frac{\partial y_1}{\partial x_1} = -\frac{\dfrac{\partial(F_1, F_2, F_3)}{\partial(x_1, y_2, y_3)}}{\dfrac{\partial(F_1, F_2, F_3)}{\partial(y_1, y_2, y_3)}}.$$

Notice the repetition in the numerator and denominator of the symbols F_1, F_2, F_3, y_2, and y_3. Note that x_1, the variable that the derivative is taken with respect to, on the left appears in the top symbol in the position corresponding to y_1 in the bottom. Thus

$$\frac{\partial y_2}{\partial x_2} = -\frac{\dfrac{\partial(F_1, F_2, F_3)}{\partial(y_1, x_2, y_3)}}{\dfrac{\partial(F_1, F_2, F_3)}{\partial(y_1, y_2, y_3)}}, \tag{3-5.6}$$

or, in general, for Example 3–5.2, Case 2

$$\frac{\partial y_j}{\partial x_i} = -\frac{\dfrac{\partial(F_1, \cdots, F_j, \cdots, F_m)}{\partial(y_1, \cdots, x_i, \cdots, y_m)}}{\dfrac{\partial(F_1, \cdots, F_j, \cdots, F_m)}{\partial(y_1, \cdots, y_j, \cdots, y_m)}}, \tag{3-5.7}$$

$1 \leq i \leq n$, $1 \leq j \leq m$. In Case 2 there are mn first partial derivatives that can be considered.

Let us now illustrate the use of this formula for taking derivatives with specific functions.

Example 3–5.3. Find $\partial y_1/\partial x_1$ at $(x_1, x_2, y_1, y_2) = (1, 1, -1, 2)$ if

$$x_1^2 + x_2^2 + y_1^2 + y_2^2 - 7 = 0,$$

$$x_1 x_2 y_1 y_2 + 2 = 0.$$

By (3–5.7),

$$\frac{\partial y_1}{\partial x_1} = -\frac{\dfrac{\partial(F_1, F_2)}{\partial(x_1, y_2)}}{\dfrac{\partial(F_1, F_2)}{\partial(y_1, y_2)}}$$

where

$$F_1(x_1, x_2, y_1, y_2) = x_1^2 + x_2^2 + y_1^2 + y_2^2 - 7,$$

$$F_2(x_1, x_2, y_1, y_2) = x_1 x_2 y_1 y_2 + 2.$$

Then

$$\frac{\partial y_1}{\partial x_1} = -\frac{\begin{vmatrix} 2x_1 & 2y_2 \\ x_2\,y_1\,y_2 & x_1\,x_2\,y_1 \end{vmatrix}}{\begin{vmatrix} 2y_1 & 2y_2 \\ x_1\,x_2\,y_2 & x_1\,x_2\,y_1 \end{vmatrix}},$$

and $\partial y_1/\partial x_1 = 1$ at $(1, 1, -1, 2)$.

As a means of checking, solve the two equations for the equation $y_1 = f_1(x_1, x_2)$:

$$f_1(x_1, x_2) = -\frac{1}{\sqrt{2}}\sqrt{-(x_1^2 + x_2^2 - 7) - \sqrt{(x_1^2 + x_2^2 - 7)^2 - (16)/(x_1^2 x_2^2)}}.$$

Differentiating with respect to x_1 and evaluating at (1, 1) give again

$$\frac{\partial y_1}{\partial x_1} = 1.$$

Notice that in finding the function f_1 there are two choices of signs of the radicals that must be made. We are guided here by the desire that $f_1(1, 1) = -1$. If we had solved for f_2, we would have wanted $f_2(1, 1) = 2$; that is,

$$f_2(x_1, x_2) = \frac{1}{\sqrt{2}}\sqrt{-(x_1^2 + x_2^2 - 7) + \sqrt{(x_1^2 + x_2^2 - 7)^2 - (16)/(x_1^2 x_2^2)}}.$$

Notice that

$$F_i(x_1, x_2\,; f_1(x_1, x_2), f_2(x_1, x_2)) = 0,$$

$i = 1, 2$ in some neighborhood of (1, 1).

This example shows very nicely that the partial derivatives of functions defined implicitly can be found without first finding an equation that defines the function explicitly. This is indeed a very pleasant situation, since finding an equation that defines the function explicitly may be difficult or even impossible.

Until now we have ignored the possibility that m given equations,

$$F_i(x_1, \cdots, x_n\,; y_1, \cdots, y_m) = F_i(\mathbf{x}\,; \mathbf{y}) = 0,$$

$1 \leq i \leq m$, *may not* define a function implicitly. In case they do not, then any thought of taking partial derivatives is meaningless. What we need is a set of conditions, sufficiency conditions, such that if the functions F_i satisfy these conditions, then a certain function f_j is defined implicitly, and the function f_j has the partial derivative in which we are interested. We now seek these sufficiency conditions as the second objective of this discussion.

Some conditions that must be met are readily apparent. The point $(\mathbf{x}_0 ; \mathbf{y}_0)$, which plays the role of $(1, 1, -1, 2)$ in Example 3–5.3, needs to satisfy each equation; that is, $F_i(\mathbf{x}_0 ; \mathbf{y}_0) = 0$. Also from (3–5.7), it is clear that the Jacobian of the denominator must not be zero at $(\mathbf{x}_0 ; \mathbf{y}_0)$. Lastly, if the partial derivatives are to be obtained by the method given above, it is necessary that the chain rule can be applied to the composite functions defined by

$$z_i = F_i(\mathbf{x} ; f_1(\mathbf{x}), \cdots, f_m(\mathbf{x})),$$

$1 \leq i \leq m$. By Proposition 3–4.1 a sufficient condition for this would be F_i and f_j differentiable at $(\mathbf{x}_0 ; \mathbf{y}_0)$ and $\mathbf{x}_0$, respectively, $1 \leq i \leq m, 1 \leq j \leq m$. If each function F_i has continuous first partials in a neighborhood of $(\mathbf{x}_0 ; \mathbf{y}_0)$, then we have a little more than enough to insure that each F_i is differentiable at $(\mathbf{x}_0 ; \mathbf{y}_0)$. We can only hope that these conditions on each F_i make each f_j differentiable at $\mathbf{x}_0$. The *implicit function theorem* next states that these several conditions on the functions F_i, $1 \leq i \leq m$, are sufficient to insure that each f_j, $1 \leq j \leq m$, exists and that the partial derivatives can be calculated as indicated above.

Theorem 3–5.1. *Let F_j, $1 \leq j \leq m$, be functions such that $F_j : S \to R$ where S is an open subset of E^{n+m}. Let*

$$(\mathbf{x}_0 ; \mathbf{y}_0) = (x_{01}, \cdots, x_{0n}, y_{01}, \cdots, y_{0m}) \in S, \qquad F_j(\mathbf{x}_0 ; \mathbf{y}_0) = 0$$

and for some $\delta_1 > 0$,

$$F_j \in C^1(N((\mathbf{x}_0 ; \mathbf{y}_0); \delta_1)).$$

If at $(\mathbf{x}_0, \mathbf{y}_0)$

$$\frac{\partial(F_1, \cdots, F_m)}{\partial(y_1, \cdots, y_m)} \neq 0,$$

then there exists a neighborhood of $\mathbf{x}_0$, $N(\mathbf{x}_0 ; \delta)$, and there exists uniquely function f_i, $1 \leq i \leq m$, defined on $N(\mathbf{x}_0 ; \delta)$ such that

(a) $f_i(\mathbf{x}_0) = y_{0i}$, $\quad 1 \leq i \leq m$,

(b) $f_j(x_1, \cdots, x_n, f_1(x_1, \cdots, x_n), \cdots, f_m(x_1, \cdots, x_n)) = 0, 1 \leq j \leq m$, *for all* $(x_1, \cdots, x_n) \in N(\mathbf{x}_0 ; \delta)$,

(c) *at* $(\mathbf{x}, \mathbf{y}) = (x_1, \cdots, x_n ; f_1(\mathbf{x}), \cdots, f_m(\mathbf{x}))$ *where* $\mathbf{x} \in N(\mathbf{x}_0 ; \delta)$,

$$\frac{\partial y_j}{\partial x_i} = -\frac{\dfrac{\partial(F_1, \cdots, F_j, \cdots, F_m)}{\partial(y_1, \cdots, x_i, \cdots, y_m)}}{\dfrac{\partial(F_1, \cdots, F_j, \cdots, F_m)}{\partial(y_1, \cdots, y_j, \cdots, y_m)}},$$

(d) $f_i \in C^1(N(\mathbf{x}_0 ; \delta))$, $\quad 1 \leq i \leq m$.

The complete proof is given by induction on m, the number of given equations. Thus the proof is first given for $m = 1$, then assume the theorem is true for the positive integer $m - 1$ and show that this implies that the theorem is true for m. Rather than do what is necessary to give a complete proof, we shall give the proof for $m = 1$ and then for $m = 2$. From this the reader will have a chance to see how the general induction step can be taken without the extra notational burden. Since there is nothing gained in understanding the proof by taking n arbitrary, we shall take $n = 1$.

Proof for $m = 1$. In this case $F_1(x_{01}, y_{01}) = 0$, and $\partial F_1/\partial y_1 \neq 0$ at (x_{01}, y_{01}). Assume $\partial F_1/\partial y_1 > 0$ at (x_{01}, y_{01}). There is no loss in assuming this, for if $\partial F_1/\partial y_1 < 0$ at (x_{01}, y_{01}), we could consider instead $-F_1$, which would satisfy the hypothesis of the theorem, and at (x_{01}, y_{01}) the partial with respect to y_1 would be positive. Since $\partial F_1/\partial y_1$ is continuous in $N((x_{01}, y_{01}); \delta_1)$, there exists δ_2, $0 < \delta_2 \leq \delta_1$, such that $(\partial/\partial y_1)F_1(x_1, y_1) > 0$ for all $(x_1, y_1) \in N((x_{01}, y_{01}); \delta_2)$ (cf. Figure 3–5.2). The restriction of F_1 to

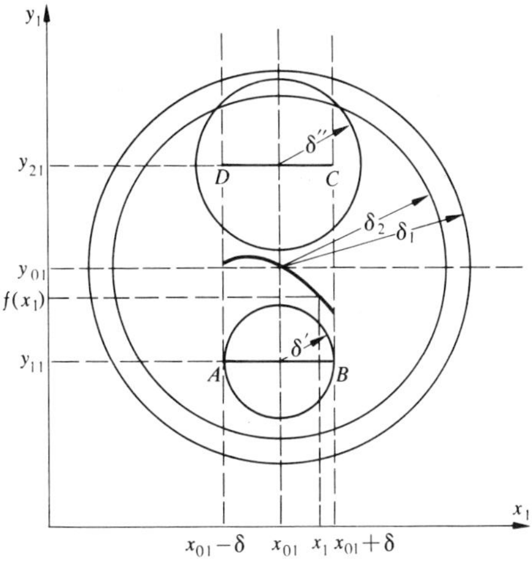

Figure 3–5.2

the line $x_1 = x_{01}$, so that $F_1(x_{01},\ \ )$, is a strictly increasing function on the interval $(x_{01} - \delta_2, x_{01} + \delta_2)$. Thus there exist y_{11} and y_{21} such that

$$y_{01} - \delta_2 < y_{11} < y_{01} < y_{21} < y_{01} + \delta_2$$

and

$$F_1(x_{01}, y_{11}) < 0,$$

$$F_1(x_{01}, y_{21}) > 0.$$

Since $F_1 \in C^1(N((x_{01}, y_{01}); \delta_1))$, then F_1 is continuous at (x_{01}, y_{11}) and (x_{01}, y_{21}). Thus there exists $\delta', \delta'' > 0$, such that

$$F_1(x_1, y_1) < 0$$

for all $(x_1, y_1) \in N((x_{11}, y_{01}); \delta')$, and

$$F_1(x_1, y_1) > 0$$

for all $(x_1, y_1) \in N((x_{21}, y_{01}); \delta'')$. Let $\delta = \min(\delta', \delta'')$. Then for $x_1 \in N(x_{01}; \delta)$,

$$F_1(x_1; y_{11}) < 0, \qquad F_1(x_1; y_{21}) > 0,$$

and the restriction of F_1 to the line $x_1 = x_1$, $F_1(x_1, \quad)$ is strictly increasing. Thus there exists a *unique* y_1 depending on x_1, which we shall call $f_1(x_1)$, such that

$$F_1(x_1, f_1(x_1)) = 0.$$

Let f_1 be the set of ordered pairs $\{(x_1, f_1(x_1)) : x_1 \in N(x_{01}; \delta)\}$. Then clearly, from the construction, (b) holds and f_1 is unique. From the uniqueness it follows that (a) holds.

Next let us demonstrate that f_1 is continuous on $N(x_{01}; \delta)$ by first checking the continuity of f_1 at x_{01}. Given $\varepsilon > 0$. Then in the proof above pick y_{11} and y_{21} such that in addition they are within ε of y_{01}. This is possible since $F_1(x_{01}, \quad)$ is strictly increasing. Continue the proof as above and let the new δ that results be δ_ε, since it will now depend upon ε. Then $0 < \delta_\varepsilon \leq \delta$. For $x_1 \in N(x_{01}; \delta_\varepsilon)$,

$$y_{11} < f_1(x_1) < y_{21}$$

as before, but now this has the added meaning that

$$|f_1(x_1) - f_1(x_{01})| < \varepsilon,$$

and hence f_1 is continuous at x_{01}. For x_{01}', some other point of $N(x_{01}, \delta)$ than x_{01}, take $\eta > 0$ but small enough for $N((x_{01}', f_1(x_{01}')); \eta)$ to be contained in the rectangle with vertices A, B, C, and D (cf. Figure 3–5.2). Repeat the above proof now from the beginning letting $N((x_{01}', f_1(x_{01}')); \eta)$ play the role of $N((x_{01}, y_{01}); \varepsilon)$. By the uniqueness feature, the new function f_1 defined in a neighborhood of x_{01}' must be identical to the original function f_1. The above continuity argument applied at x_{01}' instead of x_{01} tells us that f_1 is continuous at x_{01}', and hence $f_1 \in C(N(x_{01}; \delta))$. Next we demonstrate that conclusion (c) holds.

Let x_1 and $x_{01}' \in N(x_{01}; \delta)$. Then

$$0 = F_1(x_1, f_1(x_1)) - F_1(x_{01}', f_1(x_{01}')),$$

since each term is zero. Adding and subtracting a term we have

$$0 = (F_1(x_1, f_1(x_1)) - F_1(x_{01}', f_1(x_1))) + (F_1(x_{01}', f_1(x_1)) - F_1(x_{01}', f_1(x_{01}'))).$$

Each parenthesis can be replaced by applying the mean-value theorem successively to $F_1(\ \ , f_1(x_1))$ and $F_1(x_{01}',\ \ )$, obtaining

$$0 = \left[\frac{\partial F_1}{\partial x_1}(x_{01}' + \theta_1(x_1 - x_{01}'), f_1(x_1))\right](x_1 - x_{01}')$$
$$+ \left[\frac{\partial F_1}{\partial y_1}(x_{01}', f_1(x_{01}') + \theta_2(f_1(x_1) - f_1(x_{01}'))\right](f_1(x_1') - f_1(x_{01}'))$$

where $0 < \theta_1, \theta_2 < 1$. Suppose $x_1 \neq x_{01}'$; divide both sides by

$$(x_1 - x_{01}')\left[\frac{\partial F_1}{\partial y_1}(x_{01}', f_1(x_{01}') + \theta_2(f_1(x_1) - f_1(x_{01}'))\right],$$

transpose the first term across the equality, and then take the limit of both sides as $x_1 \to x_{01}'$. We get

$$\frac{df_1}{dx_1} f_1(x_{01}') = -\frac{\dfrac{\partial}{\partial x_1} F_1(x_{01}', f_1(x_{01}'))}{\dfrac{\partial}{\partial y_1} F_1(x_{01}', f_1(x_{01}'))}. \tag{3–5.8}$$

Notice that both the continuity of f_1 at x_{01}' and the continuity of $\partial F_1/\partial x_1$, $\partial F_1/\partial y_1$ in $N((x_{01}, y_{01}); \delta_1)$ are used here. There is no problem about the denominator going to zero, since we are operating in $N((x_{01}, y_{01}); \delta_2)$. Since the composite functions appearing on the right of (3–5.8) are continuous at each point $x_{01}' \in N(x_{01}, \delta)$, df_1/dx_1 is continuous on $N(x_{01}, \delta)$, and thus (d) is satisfied. ∎

Proof for $m = 2$. By hypothesis we have two functions F_1 and F_2 such that $F_i(x_{01}, y_{01}, y_{02}) = 0$, $i = 1, 2$, and at (x_{01}, y_{01}, y_{02}),

$$\frac{\partial(F_1, F_2)}{\partial(y_1, y_2)} \neq 0.$$

This Jacobian not being equal to zero at (x_{01}, y_{01}, y_{02}) means that

$$\frac{\partial F_1}{\partial y_1}\frac{\partial F_2}{\partial y_2} - \frac{\partial F_2}{\partial y_1}\frac{\partial F_1}{\partial y_2} \neq 0$$

at (x_{01}, y_{01}, y_{02}). However, this means that $\partial F_2/\partial y_2$ and $\partial F_1/\partial y_2$ are not both zero at (x_{01}, y_{01}, y_{02}). Suppose $\partial F_2/\partial y_2 \neq 0$ at (x_{01}, y_{01}, y_{02}). Then we have the hypothesis of this theorem satisfied where $m = 1$. Therefore, there exists $N((x_{01}, y_{01}); \delta_1)$, and there exists a unique function g_2 defined on

$N((x_{01}, y_{01}); \delta_1)$ such that

$$y_{02} = g_2(x_{01}, y_{02})$$

for all $(x_1, y_1) \in N((x_{01}, y_{01}); \delta_1)$,

$$F_2(x_1, y_1, g_2(x_1, y_1)) = 0, \tag{3–5.9}$$

at $(x_1, y_1, g_2(x_1, y_1))$ where $(x_1, y_1) \in N((x_{01}, y_{01}); \delta_1)$

$$\frac{\partial y_2}{\partial y_1} = \frac{\partial g_2}{\partial y_1} = -\frac{\dfrac{\partial F_2}{\partial y_1}}{\dfrac{\partial F_2}{\partial y_2}}, \tag{3–5.10}$$

and $g_2 \in C^1(N((x_{01}, y_{01}); \delta_1))$.

We have sufficient conditions to apply the chain rule to the composite function defined by $z_1 = F_1(x_1, y_1, g_2(x_1, y_1))$. Thus

$$\left.\frac{\partial F_1}{\partial y_1}\right)_{x_1} = \left.\frac{\partial F_1}{\partial y_1}\right)_{x_1,y_2} + \left.\frac{\partial F_1}{\partial y_2}\right)_{x_1,y_1} \left.\frac{\partial y_2}{\partial y_1}\right)_{x_1}.$$

Substituting from (3–5.10), we have

$$\left.\frac{\partial F_1}{\partial y_1}\right)_{x_1} = \left.\frac{\partial F_1}{\partial y_1}\right)_{x_1,y_2} + \left.\frac{\partial F_1}{\partial y_2}\right)_{x_1,y_1} \left(-\frac{\left.\dfrac{\partial F_2}{\partial y_1}\right)_{x_1,y_2}}{\left.\dfrac{\partial F_2}{\partial y_2}\right)_{x_1,y_1}} \right);$$

or, by adding fractions, we have

$$\frac{\partial F_1}{\partial y_1} = \frac{\dfrac{\partial F_1}{\partial y_1}\dfrac{\partial F_2}{\partial y_2} - \dfrac{\partial F_2}{\partial y_1}\dfrac{\partial F_1}{\partial y_2}}{\dfrac{\partial F_1}{\partial y_2}} = \frac{\dfrac{\partial(F_1, F_2)}{\partial(y_1, y_2)}}{\dfrac{\partial F_1}{\partial y_2}}$$

where some of the notation has been suppressed. Since the Jacobian on the right is not zero at (x_{01}, y_{01}, y_{02}), then $\partial F_1/\partial y_1$ is not zero at (x_{01}, y_{01}). Again the hypothesis of Theorem 3–5.1 is satisfied for $m = 1$, and thus there exists a neighborhood $N(x_{01}; \delta_2)$ and a unique function g_1 defined on $N(x_{01}; \delta_2)$ such that

$$y_{01} = g_1(x_{01})$$

for all $x_1 \in N(x_{01}; \delta_2)$,

$$F_1(x_1, g_1(x_1), g_2(x_1, g_1(x_1))) = 0,$$

and $g_1 \in C^1(N(x_{01}; \delta_2))$.

Since g_1 is continuous at x_{01}, there exists δ_3, $0 < \delta_3 \le (1/\sqrt{2})\delta_1$, such that if $x_1 \in N(x_{01}\,;\,\delta_3)$,

$$g_1(x_1) \in N\left(y_{01}\,;\,\frac{1}{\sqrt{2}}\,\delta_1\right),$$

and thus

$$(x_1, g_1(x_1)) \in N((x_{01}, y_{01});\, \delta_1).$$

Let f_1 and f_2 be functions defined on $N(x_{01}, \delta)$, $\delta = \min(\delta_2, \delta_3)$, as follows:

$$f_1(x_1) = g_1(x_1),$$

$$f_2(x_1) = g_2(x_1, g_1(x_1)),$$

for $x_1 \in N(x_{01}\,;\,\delta)$. Then f_1 and f_2 satisfy (a), (b), (c), and (d). ▮

Problems

1. (a) Consider the equation $x\,y^2\,z = 6$. The point $(x, y, z) = (2, 1, 3)$ satisfies the equation. Find a function f such that

$$y = f(x, z), \qquad f(2, 3) = 1, \quad \text{and} \quad x(f(x, z))^2 z = 6$$

in some neighborhood of (2, 3). Find the radius of the largest neighborhood at (2, 3) such that $x(f(x, z))^2 z = 6$ for all (x, z) in the neighborhood. Find

$$\left.\frac{\partial f}{\partial x}\right|_{(2,\,1,\,3)}$$

by directly differentiating the function f and by use of (3–5.7) as a check.

(b) Consider the equation $x^2 - y^2\,z^2 = -5$ at $(x, y, z) = (2, -3, -1)$. Proceed as in (a).

2. Given the two equations $F(w, x, y, z) = 0$ and $G(w, x, y, z) = 0$. It is possible in certain circumstances that functions f and g are defined implicitly where explicitly f and g would be defined by equations of the form $w = f(x, y)$, $z = g(x, y)$. Conceivably how many functions could the above two equations define implicitly? In the above setting $\partial w/\partial x$ means $(\partial w/\partial x)_y$. What other meaning could it have?

3. Suppose that in each circumstance indicated below the hypothesis of the implicit function theorem is satisfied. In each case calculate the

derivatives required by using the chain rule and then Cramer's rule, that is, differentiate implicitly. Check by using (3–5.7).

(a) $F(x, y) = 0$; find dy/dx.
(b) $F(x, y, z) = 0$, $G(x, y, z) = 0$; find dy/dx, dz/dx.
(c) $F(x, y, u, v) = 0$, $G(x, y, u, v) = 0$; find $\partial u/\partial x$ and $\partial u/\partial y$.
(d) $F(x, y, z, u, v, w) = 0$, $G(x, y, z, u, v, w) = 0$,
$H(x, y, z, u, v, w) = 0$;
find

$$\left.\frac{\partial w}{\partial y}\right)_{x,z} \text{ and } \left.\frac{\partial w}{\partial z}\right)_{x,y}.$$

4. Demonstrate that

$$\frac{\partial(F_1, F_2, F_3)}{\partial(y_1, y_2, y_3)} = \frac{\partial(F_3, F_1, F_2)}{\partial(y_2, y_3, y_1)}.$$

How many other similar Jacobians are equal to the Jacobian on the left?

5. Use (3–5.7) to find the derivatives indicated and check your answers by finding the derivatives by differentiating implicitly. In each problem there are some obvious points where the derivatives are not defined. Be sure each time to indicate this restriction.

(a) $x \sin y + z - 3 = 0, \quad x \log y + y \log z + z^3 - 1 = 0$;
find dz/dx and dy/dx.
(b) $x^2 + y^2 + u + v = 5, \quad xy + uv = 3$;

$$\text{find } \left.\frac{\partial u}{\partial x}\right)_y \text{ and } \left.\frac{\partial v}{\partial y}\right)_x.$$

(c) $x + y + z + w = 2$,
$x^2 + y + z^2 + w = 4$,
$x + y^2 + z + w^2 = 4$;
find dy/dx, dz/dx, dw/dx at $(x, y, z, w) = (2, 0, 1, -1)$.

Solve explicitly for f where $z = f(x)$. Calculate dz/dx at $x = 2$ and check the above answer for dz/dx.

6. Consider the two equations:

$$x^2 + y^2 + u + \sin v = 5, \qquad xy + uv = 0.$$

(a) Find the functions f and g, where $x = f(u, v)$ and $y = g(u, v)$, that are defined implicitly in a neighborhood of $(x, y, u, v) = (1, 0, 4, 0)$ by the above two equations.
(b) Find $\dfrac{\partial x}{\partial u}, \dfrac{\partial x}{\partial v}, \dfrac{\partial y}{\partial u}$, and $\dfrac{\partial y}{\partial v}$ at $(x, y, u, v) = (1, 0, 4, 0)$ by using (3–5.7).
(c) Check the results of (b) by using the results of (a) to find f_u, f_v, g_u, and g_v at $(u, v) = (4, 0)$.

7. In each of the cases shown below answer the following two questions: (i) Does the function F satisfy the hypothesis of the implicit function theorem in a neighborhood of $(0, 0)$ and if not, why not? (ii) Does the equation $F = 0$ define implicitly a function f where f maps a neighborhood of $x = 0$ in x-space into y-space?

   (a) $F(x, y) = y - |x| = 0$, (b) $F(x, y) = y^2 - x = 0$,
   (c) $F(x, y) = y^3 - x = 0$.

8. Given the functions f_1, f_2, g_1, and g_2 defined by the following equations:

$$x = g_1(u, v), \qquad y = g_2(u, v), \qquad u = f_1(r, s), \qquad v = f_2(r, s).$$

   Assume that the chain rule can be applied to these two composite functions and prove the chain rule for Jacobians—that is,

$$\frac{\partial(x, y)}{\partial(r, s)} = \frac{\partial(x, y)}{\partial(u, v)} \cdot \frac{\partial(u, v)}{\partial(r, s)} .$$

9. Suppose that from the equation $F(x, y) = 0$ we have found

$$\frac{dy}{dx} = -\frac{F_x}{F_y}.$$

   To find d^2y/dx^2, differentiate on the right recalling that there exists a function f such that $y = f(x)$. Show that in terms of partials of F

$$\frac{d^2y}{dx^2} = \frac{(F_{xy} + F_{yx})\, F_x F_y - F_{xx}(F_y)^2 - F_{yy}(F_x)^2}{F_y^{\,3}} .$$

   From the equation $x^3 + xy^2 + y^3 = 1$ find d^2y/dx^2 at $(1, -1)$.

10. Given the equations $u = f(x, y)$ and $v = g(v, x, y)$, express

$$\text{(a)}\ \left.\frac{\partial u}{\partial x}\right)_v, \qquad \text{(b)}\ \left.\frac{\partial y}{\partial x}\right)_v, \qquad \text{(c)}\ \left.\frac{\partial^2 u}{\partial x^2}\right)_v,$$

    in terms of partials of f and g. Give sufficient conditions on f and g for this to be true.

11. In thermodynamics there are equations of the form $F(p, v, T) = 0$ where p is pressure, v is volume, and T is temperature. From this equation the following equation may be written

$$\frac{\partial p}{\partial T}\frac{\partial T}{\partial v}\frac{\partial v}{\partial p} = -1.$$

    What is the precise meaning of these partials? Give sufficient conditions on F for this last equation to hold and demonstrate your reasoning.

12. Consider the implicit function theorem for $m = 3$. The hypothesis would say that at $(\mathbf{x}_0 ; \mathbf{y}_0)$,

$$\frac{\partial(F_1, F_2, F_3)}{\partial(y_1, y_2, y_3)} \neq 0.$$

Expand the determinant about the first row and show that not all of the second-order determinants involving partials of F_2 and F_3 are zero. Using this hint, write out in detail the proof of the implicit function theorem for $m = 3$.

Answers

1. (a) $y = f(x, z) = \sqrt{6/xz}$, the radius of the largest neighborhood $= 2$,

$$f_x(2, 3) = -\tfrac{1}{4}, \quad \left.\frac{\partial y}{\partial x}\right|_{(2,3)} = -\frac{F_x(2, 1, 3)}{F_y(2, 1, 3)} = -\left.\frac{y}{2x}\right|_{(2,1,3)} = -\tfrac{1}{4}.$$

(b) $y = f(x, z) = -\sqrt{(x^2 + 5)/z^2}$, the radius of the largest neighborhood $= 1$,

$$f_x(2, -1) = -\tfrac{2}{3}, \quad \left.\frac{\partial y}{dx}\right|_{(2,-1)} = -\frac{F_x(2, -3, -1)}{F_y(2, -3, -1)}$$

$$= \left.\frac{x}{yz^2}\right|_{(2,-3,-1)} = -\tfrac{2}{3}.$$

2. $\dbinom{4}{2} = 6$ pairs of functions, $\left.\dfrac{\partial w}{\partial x}\right)$.

3. (a) $\dfrac{dy}{dx} = -\dfrac{F_x(x, y)}{F_y(x, y)}$, (b) $\dfrac{dy}{dx} = -\dfrac{\dfrac{\partial(F, G)}{\partial(x, z)}}{\dfrac{\partial(F, G)}{\partial(y, z)}}$, $\dfrac{dz}{dx} = -\dfrac{\dfrac{\partial(F, G)}{\partial(y, x)}}{\dfrac{\partial(F, G)}{\partial(y, z)}}$,

(c) $\dfrac{\partial u}{dx} = -\dfrac{\dfrac{\partial(F, G)}{\partial(x, v)}}{\dfrac{\partial(F, G)}{\partial(u, v)}}$, $\dfrac{\partial u}{\partial y} = -\dfrac{\dfrac{\partial(F, G)}{\partial(y, v)}}{\dfrac{\partial(F, G)}{\partial(u, v)}}$,

(d) $\left.\dfrac{\partial w}{\partial y}\right)_{x,z} = -\dfrac{\dfrac{\partial(F, G, H)}{\partial(u, v, y)}}{\dfrac{\partial(F, G, H)}{\partial(u, v, w)}}$, $\left.\dfrac{\partial w}{\partial z}\right)_{x,y} = -\dfrac{\dfrac{\partial(F, G, H)}{\partial(u, v, z)}}{\dfrac{\partial(F, G, H)}{\partial(u, v, w)}}$.

5. (a) $\dfrac{dz}{dx} = \dfrac{-xz \sin y - yz \sin y \log z + zxy(\cos y) \log y}{xz + zy \log z - y^2 x \cos y - 3\, xyz^3 \cos y}$,

when the denominator $\neq 0$, $y > 0$, and $z > 0$,

$$\frac{dy}{dx} = \frac{-y^2 \sin y - 3\, yz^3 \sin y + zy \log y}{xy^2 \cos y + 3\, xyz^3 \cos y - xz - yz \log z},$$

(b) $\left.\dfrac{\partial u}{\partial x}\right)_y = \dfrac{y - 2xu}{u - x}$, $\left.\dfrac{\partial v}{\partial y}\right)_x = \dfrac{2vy - x}{u - v}$,

(c) 3, −3, −1.

6. (a) $f(u, v) = [-(u + \sin v - 5) + ((u + \sin v - 5)^2 - 4u^2v^2)^{1/2}]^{1/2}/\sqrt{2},$
$g(u, v) = -[-(u + \sin v - 5) - ((u + \sin v - 5)^2 - 4u^2v^2)^{1/2}]^{1/2}/\sqrt{2}, \quad v \geq 0,$
$g(u, v) = -g(u, -v), \quad v < 0.$
(b) At $(x, y, u, v) = (1, 0, 4, 0)$, $x_u = -\frac{1}{2}$, $x_v = -\frac{1}{2}$, $y_u = 0$, $y_v = -4$.

7. (i) No, $F \notin C^1(N((0, 0); \delta))$ for any $\delta > 0$,
No, $F_y(0, 0) = 0$,
No, $F_y(0, 0) = 0$,
(ii) No. No not on $(-\delta, \delta)$, $\delta > 0$. Yes, namely $y = x^{1/3}$.

9. 42.

10. (a) $(f_x g_y - f_y g_x)/g_y, g_y(v, x, y) \neq 0,$
(b) $-g_x/g_y, g_y(v, x, y) \neq 0,$
(c) $[-f_y g_y^2 g_{xx} - 2g_x g_y^2 f_{xy} + 2f_y g_x g_y g_{xy} + g_y^3 f_{xx} + g_x^2 g_y f_{yy} - f_y g_x^2 g_{yy}]/g_y^3.$

3–6. Directional Derivative and Gradient

The partial derivatives of a function f are derivatives in a particular direction. That is, in the case where f is a function of two variables

$$\frac{\partial}{\partial x_1} f(\mathbf{x}_0) = \lim_{x_1 \to x_{01}} \frac{f(x_1, x_{02}) - f(x_{01}, x_{02})}{x_1 - x_{01}}.$$

Here the limit is taken with x_2 remaining constant. The quantity $(\partial/\partial x_1) f(\mathbf{x}_0)$ interpreted geometrically is the instantaneous rate of change in $f(\mathbf{x})$ at $\mathbf{x}_0$ as $\mathbf{x}$ moves along the line $x_2 = x_{02}$ in the (x_1, x_2)-plane. Similarly, $(\partial/\partial x_2) f(\mathbf{x}_0)$ is the instantaneous rate of change of $f(\mathbf{x})$ at $\mathbf{x}_0$ as $\mathbf{x}$ moves along the line $x_1 = x_{01}$ in the (x_1, x_2)-plane. The question could very well be asked, why not consider the instantaneous rate of change in $f(\mathbf{x})$ at $\mathbf{x}_0$ along other lines through $\mathbf{x}_0$? This is precisely the role filled by the directional derivative.

Let $f: S \to R$ where S is an open subset of E^2 and let $\mathbf{x}_0 \in S$. Consider a unit vector $\mathbf{u} = (u_1, u_2)$. Then for t any real number, $\mathbf{x}_0 + t\mathbf{u}$ is a point on the line through $\mathbf{x}_0$ in the *direction* $\mathbf{u}$ (cf. Figure 3–6.1). Forming a ratio and taking a limit in a manner exactly analogous to the way partial derivatives were obtained give us

$$\lim_{t \to 0} \frac{f(\mathbf{x}_0 + t\mathbf{u}) - f(\mathbf{x}_0)}{t}.$$

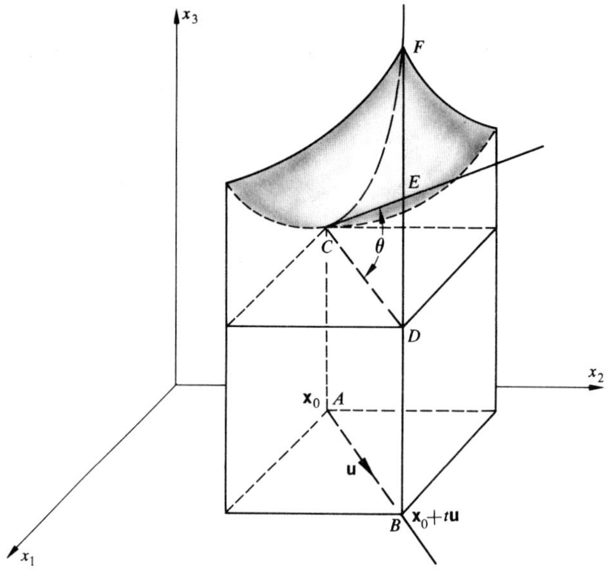

Figure 3–6.1

If this limit exists, then the value of this limit is denoted by $(d/ds)f(\mathbf{x}_0\,;\mathbf{u})$, and it is called the *directional derivative of* f *at* $\mathbf{x}_0$ *in the direction* $\mathbf{u}$.

Notice from the figure that

$$f(\mathbf{x}_0 + t\mathbf{u}) - f(\mathbf{x}_0) = \overline{BF} - \overline{BD} = \overline{DF}$$

and that $\overline{AB} = \overline{CD} = t$. Thus

$$\frac{f(\mathbf{x}_0 + t\mathbf{u}) - f(\mathbf{x}_0)}{t} = \frac{\overline{DF}}{\overline{CD}}$$

is the slope of the secant line, the line determined by C and F, and

$$\frac{d}{ds}f(\mathbf{x}_0\,;\mathbf{u}) = \frac{\overline{DE}}{\overline{CD}} = \tan\theta$$

is the slope of the tangent line, the line determined by C and E.

Notice from the figure that if $\mathbf{u} = (1, 0)$, a unit vector in the direction of the x_1-axis, then the directional derivative is the partial of f with respect to x_1; or in other words,

$$\frac{d}{ds}f(x_1, x_2\,; 1, 0) = \frac{\partial}{\partial x_1}f(x_1, x_2).$$

Similarly,

$$\frac{d}{ds}f(x_1, x_2; 0, 1) = \frac{\partial}{\partial x_2}f(x_1, x_2).$$

Example 3–6.1. Consider the function f defined by

$$f(x, y) = x^2 - 2yx + 1.$$

Find

$$\frac{d}{ds}f\left(2, 1; \frac{3}{5}, \frac{4}{5}\right).$$

Now

$$f\left(2 + \frac{3t}{5}, 1 + \frac{4t}{5}\right) - f(2, 1) = -2t - \frac{3t^2}{5}$$

and thus

$$\frac{d}{ds}f\left(2, 1; \frac{3}{5}, \frac{4}{5}\right) = \lim_{t\to 0}\frac{f\left(2 + \frac{3t}{5}, 1 + \frac{4t}{5}\right) - f(2, 1)}{t} = -2.$$

Below, the definition of the directional derivative is given for the general circumstance.

Definition 3–6.1. *Let $f\colon S \to R$ where S is an open set in E^n. Let $\mathbf{x}_0 \in S$ and let $\mathbf{u}$ be a unit vector in E^n. If*

$$\lim_{t\to 0}\frac{f(\mathbf{x}_0 + t\mathbf{u}) - f(\mathbf{x}_0)}{t}$$

exists, then the value of this limit is the directional derivative of f at $\mathbf{x}_0$ in the direction $\mathbf{u}$, and it is denoted by

$$\frac{d}{ds}f(\mathbf{x}_0; \mathbf{u}).$$

Whereas the definition is clear and precise, it is cumbersome to apply. We desire a formula for finding the directional derivative that will allow us to make use of our standard differentiation formulas. The following proposition gives us such a formula which is useful for a certain general class of functions.

Proposition 3–6.1. *Let $f: S \to R$ where S is an open set in E^n. Let f be differentiable at $\mathbf{x}_0 \in S$; then the directional derivative of f at $\mathbf{x}_0$ in the direction $\mathbf{u}$ exists and is given by*

$$\frac{d}{ds} f(\mathbf{x}_0 ; \mathbf{u}) = \frac{\partial}{\partial x_1} f(\mathbf{x}_0) u_1 + \cdots + \frac{\partial}{\partial x_n} f(\mathbf{x}_0) u_n = df(\mathbf{x}_0 ; \mathbf{u}). \quad (3\text{–}6.1)$$

Proof. Since f is differentiable at $\mathbf{x}_0$, it follows by Proposition 3–3.1 that, given $\varepsilon > 0$, there exists a neighborhood of $\mathbf{x}_0$, $N(\mathbf{x}_0 ; \delta)$ such that when $\mathbf{x}_0 + \mathbf{dx} \in N(\mathbf{x}_0 ; \delta)$,

$$|\Delta f(\mathbf{x}_0 ; \mathbf{dx}) - df(\mathbf{x}_0 ; \mathbf{dx})| < \varepsilon \, \|\mathbf{dx}\|. \quad (3\text{–}6.2)$$

Take $\mathbf{dx} = t\mathbf{u}$ in (3–6.2). Then for $|t| < \delta$

$$|f(\mathbf{x}_0 + t\mathbf{u}) - f(\mathbf{x}_0) - df(\mathbf{x}_0 ; t\mathbf{u})| < \varepsilon \, \|\mathbf{dx}\| = \varepsilon \, |t|.$$

Assume $t \neq 0$ and divide by $|t|$, obtaining

$$\left| \frac{f(\mathbf{x}_0 + t\mathbf{u}) - f(\mathbf{x}_0)}{t} - df(\mathbf{x}_0 ; \mathbf{u}) \right| < \varepsilon.$$

Therefore,

$$\frac{d}{ds} f(\mathbf{x}_0 ; \mathbf{u}) = \lim_{t \to 0} \frac{f(\mathbf{x}_0 + t\mathbf{u}) - f(\mathbf{x}_0)}{t} = df(\mathbf{x}_0 ; \mathbf{u}),$$

but

$$df(\mathbf{x}_0 ; \mathbf{u}) = \frac{\partial}{\partial x_1} f(\mathbf{x}_0) u_1 + \cdots + \frac{\partial}{\partial x_n} f(\mathbf{x}_0) u_n . \quad \blacksquare$$

Thus, applying (3–6.1) to the problem of Example 3–6.1, we have

$$\frac{d}{ds} f(2, 1; 3/5, 4/5) = (2(2) - 2(1))(3/5) + (-2(2))(4/5) = -2.$$

Since the direction cosines of a directed line are the components of a unit vector, or

$$\cos^2 \alpha_1 + \cdots + \cos^2 \alpha_n = 1,$$

it is possible to write (3–6.1) as follows:

$$\frac{d}{ds} f(\mathbf{x}_0 ; \mathbf{u}) = \frac{\partial}{\partial x_1} f(\mathbf{x}_0) \cos \alpha_1 + \cdots + \frac{\partial}{\partial x_n} f(\mathbf{x}_0) \cos \alpha_n . \quad (3\text{–}6.3)$$

In particular, if $n = 2$, (3–6.1) is often written

$$\frac{d}{ds} (\mathbf{x}_0 ; \mathbf{u}_0) = \frac{\partial}{\partial x_1} (\mathbf{x}_0) \cos \alpha_1 + \frac{\partial}{\partial x_2} f(\mathbf{x}_0) \sin \alpha_1 \quad (3\text{–}6.4)$$

where α_1 is understood to be an angle in standard position and

$$\mathbf{u} = (\cos \alpha_1, \sin \alpha_1).$$

Sometimes it is desired that *the directional derivative be found along some curve*. What is meant is that there is some curve

$$\Gamma = \{\mathbf{x}(s): s \in (a, b)\}$$

such that $\mathbf{x} \in C^1((a, b))$, $\mathbf{x}(s_0) = \mathbf{x}_0$ for a unique s_0, $a < s_0 < b$, and $\mathbf{x}'(s_0) \neq 0$. Then the directional derivative of f at $\mathbf{x}_0$ along Γ is the instantaneous rate of change of $f(\mathbf{x}(s))$ with respect to s at $s = s_0$; that is,

$$\frac{d}{ds} f(\mathbf{x}(s_0)).$$

Assuming that f is, as in Proposition 3–6.1, differentiable at $\mathbf{x}_0$, the chain rule is applicable, and

$$\frac{d}{ds} f(\mathbf{x}(s_0)) = \frac{\partial}{\partial x_1} f(\mathbf{x}_0) \frac{d}{ds_1} x_1(s_0) + \cdots + \frac{\partial}{\partial x_n} f(\mathbf{x}_0) \frac{d}{ds} x_n(s_0).$$

Since s is arc length, the vector $\left(\frac{d}{ds} x_1(s_0), \cdots, \frac{d}{ds} x_n(s_0)\right)$ is the unit tangent vector to the curve Γ at $\mathbf{x}_0$ in the direction of increasing s (cf. (1–5.6)). Thus

$$\frac{d}{ds} f(\mathbf{x}(s_0)) = \frac{d}{ds} f(\mathbf{x}_0; \mathbf{t}_0) \tag{3–6.5}$$

where $\mathbf{t}_0$ is the unit tangent vector mentioned. Equation (3–6.5) shows several things. It somewhat explains the presence of the s in the notation of the directional derivative. Other notations are in use that suppress the s, namely

$$\frac{d}{ds}(\mathbf{x}_0; \mathbf{u}) = D_{\mathbf{u}} f(\mathbf{x}_0) = \nabla_{\mathbf{u}} f(\mathbf{x}_0),$$

but the notations that are probably used the most widely are the simple, but somewhat ambiguous, df/ds or $\partial f/\partial s$. Secondly, the equation shows that the directional derivative along a curve at a given point is the same for all curves through the point having the same unit tangent.

There is a vector that is deeply related to the idea of a directional derivative, the *gradient*. The following definition defines this vector explicitly.

Definition 3–6.2. *Let $f: S \to R$ where S is an open subset of E^n. Let $\mathbf{x}_0 \in S$, and let all the first partials of f exist at $\mathbf{x}_0$. The gradient of f at $\mathbf{x}_0$ is the vector $(f_{x_1}(\mathbf{x}_0), \cdots, f_{x_n}(\mathbf{x}_0))$. It is denoted by* **grad** $f(\mathbf{x}_0)$, *which is read, "gradient of f at $\mathbf{x}_0$" or by $\nabla f(\mathbf{x}_0)$, which is read, "del of f at $\mathbf{x}_0$."*

In Example 3–6.1, $\mathbf{grad} f((2, 1)) = (2, -4)$.

The gradient vector offers immediately some new ways to write some earlier formulas. For instance, (3–3.2) becomes

$$df(\mathbf{x}_0 ; \mathbf{dx}) = \mathbf{grad} f(\mathbf{x}_0) \cdot \mathbf{dx} \tag{3–6.6}$$

and (3–6.1) becomes

$$\frac{d}{ds} f(\mathbf{x}_0 ; \mathbf{u}) = \mathbf{grad} f(\mathbf{x}_0) \cdot \mathbf{u}. \tag{3–6.7}$$

The following proposition summarizes the relationship between the directional derivative and the gradient.

Proposition 3–6.2. *Let $f: S \to R$ where S is an open set in E^n, and let f be differentiable at $\mathbf{x}_0 \in S$. Then the directional derivative at $\mathbf{x}_0$ is bounded as follows for all choices of $\mathbf{u}$:*

(a) $-\|\mathbf{grad} f(\mathbf{x}_0)\| \leq \dfrac{d}{ds} f(\mathbf{x}_0 ; \mathbf{u}) \leq \|\mathbf{grad} f(\mathbf{x}_0)\|$,

(b) *if* $\mathbf{grad} f(\mathbf{x}_0) \neq \mathbf{0}$,

$$\frac{d}{ds} f(\mathbf{x}_0 ; \pm\mathbf{u}_0) = \pm\|\mathbf{grad} f(\mathbf{x}_0)\|$$

where

$$\mathbf{u}_0 = \frac{\mathbf{grad} f(\mathbf{x}_0)}{\|\mathbf{grad} f(\mathbf{x}_0)\|},$$

(c) *if* $\mathbf{grad} f(\mathbf{x}_0) \neq \mathbf{0}$ *and* $\mathbf{r}$ *is a function defined by* $\mathbf{r} = \mathbf{r}(s)$, $a < s < b$ *where* s *is arc length,* $\mathbf{r}(s_0) = \mathbf{x}_0$, $\mathbf{r} \in C^1((a, b))$, $f(\mathbf{r}(s)) = f(\mathbf{x}_0)$ *for all* $s \in (a, b)$, *then for* $\mathbf{u}_1 = \mathbf{r}'(s_0)$,

$$\frac{d}{ds} f(\mathbf{x}_0 ; \mathbf{u}_1) = 0.$$

Proof. From (3–6.7) and by (1–2.15)

$$\left| \frac{d}{ds} f(\mathbf{x}_0 ; \mathbf{u}) \right| = |\mathbf{grad} f(\mathbf{x}_0) \cdot \mathbf{u}| \leq \|\mathbf{grad} f(\mathbf{x}_0)\|,$$

which establishes (a). Also from (3–6.7) we have on substituting the above vector, $\mathbf{u}_0$,

$$\frac{d}{ds} f(\mathbf{x}_0 ; \mathbf{u}_0) = \frac{\mathbf{grad} f(\mathbf{x}_0) \cdot \mathbf{grad} f(\mathbf{x}_0)}{\|\mathbf{grad} f(\mathbf{x}_0)\|} = \|\mathbf{grad} f(\mathbf{x}_0)\|.$$

Clearly, if the sign of $\mathbf{u}_0$ is changed, the sign of the answer is changed, and thus (b) is established.

Differentiating both sides of $f(\mathbf{r}(s)) = f(\mathbf{x}_0)$ with respect to s and evaluating at s_0 gives

$$\frac{\partial}{\partial x_1} f(\mathbf{x}_0) \frac{d}{ds} r_1(s_0) + \cdots + \frac{\partial}{\partial x_n} f(\mathbf{x}_0) \frac{d}{ds} r_n(s_0) = 0,$$

$$\mathbf{grad}\, f(\mathbf{x}_0) \cdot \mathbf{r}'(s_0) = 0,$$

$$\frac{d}{ds} f(\mathbf{x}_0\,;\mathbf{u}_1) = 0. \quad \blacksquare$$

This proposition tells us that the magnitude of the directional derivative is largest when the direction is the direction of the gradient vector. Its value in that direction is the magnitude of the gradient. Thus the direction of the gradient vector is the *direction of steepest ascent.* Furthermore, by (b), the direction of the negative of the gradient vector, $-\mathbf{u}_0$, is the *direction of steepest descent.*

Part (c) tells us that if the equation $\mathbf{r} = \mathbf{r}(s)$ gives a curve in the *level surface* $\{\mathbf{x}: f(\mathbf{x}) = f(\mathbf{x}_0)\}$ *that passes through* $\mathbf{x}_0$ and if $\mathbf{u}_1$ is the unit tangent to this curve, the directional derivative in this direction is zero. Note from the proof of (c) that $\mathbf{u}_1$ is orthogonal to the gradient; that is, the gradient is orthogonal to every curve in the surface through the point $\mathbf{x}_0$. In other words, $\mathbf{grad}\, f(\mathbf{x}_0)$ *is a vector normal to the surface defined by* $f(\mathbf{x}) = f(\mathbf{x}_0)$. This agrees with Definition 3–3.1. For let us write

$$F(\mathbf{x}) = f(\mathbf{x}) - f(\mathbf{x}_0) = 0.$$

Then $\mathbf{grad}\, f(\mathbf{x}_0) \neq \mathbf{0}$ implies that at least one of the first partials of F is not zero at $\mathbf{x}_0$. Suppose

$$F_{x_n}(\mathbf{x}_0) = f_n(\mathbf{x}_0) \neq 0.$$

Then by the implicit function theorem there exists a function g such that

$$\{\mathbf{x}: x_n = g(x_1, \cdots, x_{n-1})\} = \{\mathbf{x}: f(\mathbf{x}) = f(\mathbf{x}_0)\}$$

in a neighborhood of $\mathbf{x}_0$. Following Definition 3–3.1, the vector

$$(g_{x_1}(\mathbf{x}_0), \cdots, g_{x_{n-1}}(\mathbf{x}_0), -1)$$

is normal to the surface given by g, $\{\mathbf{x}: x_n = g(x_1, \cdots, x_{n-1})\}$.

However, by the implicit function theorem,

$$g_{x_i}(\mathbf{x}_0) = -\frac{F_{x_i}(\mathbf{x}_0)}{F_{x_n}(\mathbf{x}_0)} = -\frac{f_{x_i}(\mathbf{x}_0)}{f_{x_n}(\mathbf{x}_0)}.$$

Hence

$$-f_{x_n}(\mathbf{x}_0)(g_{x_1}(\mathbf{x}_0), \cdots, g_{x_{n-1}}(\mathbf{x}_0), -1) = (f_{x_1}(\mathbf{x}_0), \cdots, f_{x_n}(\mathbf{x}_0)),$$

and the two vectors are in the same direction.

Thus since the direction of the gradient vector, **grad** $h(\mathbf{x}_0)$, is normal to the level surface, S, that passes through $(\mathbf{x}_0, h(\mathbf{x}_0))$, $\{\mathbf{x}: h(\mathbf{x}) = h(\mathbf{x}_0)\}$, the directional derivative of f in that direction is denoted by

$$\frac{d}{ds} f(\mathbf{x}_0\, ; \mathbf{n})$$

where

$$\mathbf{n} = \frac{\mathbf{grad}\ h(\mathbf{x}_0)}{\|\mathbf{grad}\ h(\mathbf{x}_0)\|} .$$

The symbol,

$$\frac{\partial}{\partial n} f(\mathbf{x}_0),$$

the *normal derivative* of f to S at $\mathbf{x}_0$, is often used to refer to

$$\frac{d}{ds} f(\mathbf{x}_0\, ; \pm\mathbf{n})$$

where the correct sign for $\mathbf{n}$ and the surface S is clear from the context.

The following example will serve to illustrate the concepts just discussed.

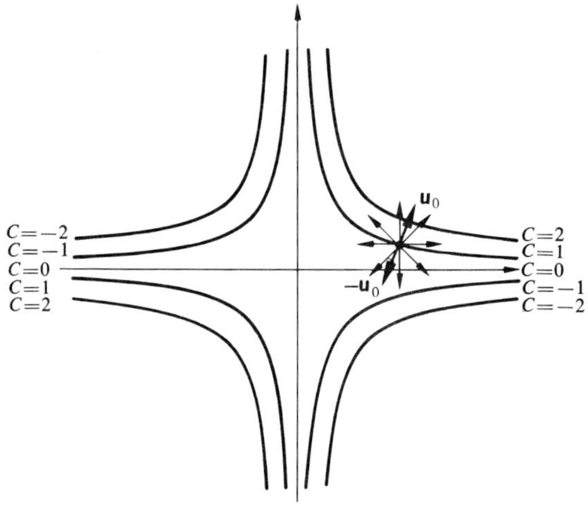

Figure 3–6.2

Example 3–6.2. Given the equation $f(x, y) = xy = c$ where c is a constant. The *level curves* for $c = -2, -1, 0, 1$, and 2, $\{(x, y): xy = c\}$, are plotted in Figure 3–6.2. The surface of which these are the level curves is a hyperbolic-paraboloid—that is, a saddle-shaped surface. Table 3–6.1 tabulates the directional derivative at $(2, \frac{1}{2})$ for different values of $\mathbf{u}$.

$\mathbf{u}$	$\frac{d}{ds} f(2, \frac{1}{2}; \mathbf{u})$	$\mathbf{u}$	$\frac{d}{ds} f(2, \frac{1}{2}; \mathbf{u})$
$(1, 0)$	.5	$(-1, 0)$	−.5
$\left(\frac{1}{\sqrt{2}}, \frac{1}{\sqrt{2}}\right)$	1.76	$\left(\frac{-1}{\sqrt{2}}, \frac{-1}{\sqrt{2}}\right)$	−1.76
$\left(\frac{1}{\sqrt{17}}, \frac{4}{\sqrt{17}}\right)$	2.06	$\left(\frac{-1}{\sqrt{17}}, \frac{-4}{\sqrt{17}}\right)$	−2.06
$(0, 1)$	2.	$(0, -1)$	−2.
$\left(\frac{-1}{\sqrt{2}}, \frac{1}{\sqrt{2}}\right)$	1.06	$\left(\frac{1}{\sqrt{2}}, \frac{-1}{\sqrt{2}}\right)$	−1.06
$\left(\frac{-4}{\sqrt{17}}, \frac{1}{\sqrt{17}}\right)$	0	$\left(\frac{4}{\sqrt{17}}, \frac{-1}{\sqrt{17}}\right)$	0

Table 3–6.1

Calculating, we have $\mathbf{grad}\, f((2, \frac{1}{2})) = (\frac{1}{2}, 2)$ and hence

$$\|\mathbf{grad}\, f((2, \tfrac{1}{2}))\| = \frac{\sqrt{17}}{2} = 2.06.$$

Then

$$\mathbf{u}_0 = \frac{\mathbf{grad}\, f(2, \frac{1}{2})}{\|\mathbf{grad}\, f(2, \frac{1}{2})\|} = \left(\frac{1}{\sqrt{17}}, \frac{4}{\sqrt{17}}\right).$$

Notice that the directional derivative at $(2, \frac{1}{2})$ is greatest in the table in the direction $\mathbf{u}_0$ and least in the direction $-\mathbf{u}_0$. The direction of any curve in the level curve $xy = 1$ through $(2, \frac{1}{2})$ has as unit tangent $\mathbf{u}_1$ either

$$(-4/\sqrt{17}, 1/\sqrt{17}) \qquad \text{or} \qquad (4/\sqrt{17}, -1/\sqrt{17}).$$

In both cases

$$\frac{d}{ds} f(\mathbf{x}_0\,; \mathbf{u}_1) = 0.$$

If $f(x, y)$ represents the height above sea level at the point (x, y), then a drop of water at $(2, \frac{1}{2})$ would flow in the direction of $-\mathbf{grad}\, f(2, \frac{1}{2})$, or perpendicular to the level curve through $(2, \frac{1}{2})$. Similarly, if $f(x, y)$ is the temperature of a plane lamina at (x, y), then heat, which flows from hot toward cold, would flow in the direction of $-\mathbf{grad}\, f(x, y)$, which would again be perpendicular to the level, or the isothermal, curves. The "lines of flow" form a one-parameter family of curves, the *orthogonal trajectories* of the family of level curves. The orthogonal trajectories in this example are the curves:

$$\{(x, y)\colon x^2 - y^2 = c_1\},$$

where c_1 is a constant.

Problems

1. Use Definition 3–6.1 to find $(d/ds)\, f(\mathbf{x}_0\,; \mathbf{u})$ for the conditions given below.

   (a) $f(x_1, x_2) = x_1^3 - 2x_1 x_2 + x_2^2$, at (2, 1) where $\mathbf{u} = (3/\sqrt{10}, 1/\sqrt{10})$
   (b) $f(x_1, x_2, x_3) = x_2^2 - x_1 x_3 + 2x_2 x_1$ at (1, −2, 1) where

   $$\mathbf{u} = (-1/\sqrt{2}, 0, 1/\sqrt{2}).$$

   (c) Plot several of the level curves for (a) near (2, 1).

2. Consider the function f defined by

   $$f(\mathbf{x}) = \begin{cases} \dfrac{x_1 x_2^3}{x_1^3 + x_2^9}, & \mathbf{x} \neq \mathbf{0}, \\ 0, \quad \mathbf{x} = \mathbf{0}. \end{cases}$$

   Find $(d/ds) f(\mathbf{0}; \mathbf{u})$ for

   $$\mathbf{u} = \left(\frac{1}{\sqrt{1 + m^2}}, \frac{m}{\sqrt{1 + m^2}}\right)$$

   where m is any real constant, and also find $(d/ds) f(\mathbf{0}; (0, 1))$ showing thus that $(d/ds) f(\mathbf{0}; \mathbf{u})$ exists for all possible $\mathbf{u}$. Show f is not continuous at $\mathbf{0}$ by considering $f(\mathbf{x})$ where $x_1 = x_2^3$.

3. Let $f(x, y) = x^2 + y^2 = 25$. Find

   $$\frac{d}{ds} f((4, 3); \mathbf{u})$$

   where $\mathbf{u} = (\cos \alpha, \sin \alpha)$. Make a table like Table 3–6.1 for $\alpha = 0°, 45°, 90°$, etc. Find $\mathbf{grad}\, f(4, 3)$ and then find $\mathbf{u}_0$. Find the vectors $\mathbf{u}_1$ such that

   $$\frac{d}{ds} f((4, 3); \mathbf{u}_1) = 0.$$

   Plot several of the level curves of the function f near (4, 3).

4. For $n = 2$ the directional derivative at $\mathbf{x}_0$ which is given by Equation (3–6.4) is

   $$\frac{d}{ds} f(\mathbf{x}_0\,; \cos \alpha, \sin \alpha) = f_{x_1}(\mathbf{x}_0) \cos \alpha + f_{x_2}(\mathbf{x}_0) \sin \alpha.$$

That is, it has a value $F(\alpha)$ for each choice of α. By differentiating the function F so defined, find the values α such that F has a maximum and a minimum. Prove that the maximum and minimum values of the directional derivative at $\mathbf{x}_0$ are, indeed, $\|\mathbf{grad}\, f(\mathbf{x}_0)\|$ and $-\|\mathbf{grad}\, f(\mathbf{x}_0)\|$, respectively.

5. Let $f(\mathbf{x}) = x_1 - x_2 x_3 + x_3^2$. At the point indicated find $(d/ds) f(\mathbf{x}_0; \mathbf{u})$ in the direction indicated.

   (a) At $(2, -1, 2)$ in the direction of the point $(3, -2, -1)$.
   (b) At $(1, 1, -1)$ in the direction of the curve $\mathbf{x}(t) = (t, t^2, -t^3)$, $0 \leq t \leq 2$.
   (c) At $(2, 0, 2)$ in the direction of the line segment from $(3, -2, 5)$ to $(-1, -2, 1)$.
   (d) At $(1, 2, -2)$ in the direction of the gradient of f at the same point.

6. The two surfaces given by

$$\{(x, y, z): f(x, y, z) = 0\} \quad \text{and} \quad \{(x, y, z): g(x, y, z) = 0\}$$

intersect in a curve. The coordinates of a point on this curve (x_0, y_0, z_0) satisfy both equations. The vectors $\nabla f(x_0, y_0, z_0)$ and $\nabla g(x_0, y_0, z_0)$ are perpendicular to the surfaces given by the f and g equations, respectively, and hence they are perpendicular to the curve of intersection. Thus the vectors

$$\pm \left. \frac{\nabla f \times \nabla g}{\|\nabla f \times \nabla g\|} \right|_{(x_0, y_0, z_0)}$$

are unit tangents of the curve if $\nabla f \times \nabla g \neq \mathbf{0}$.

Find the unit tangents of the curve of intersection of the two surfaces given by the following equations at the point indicated, and each time make a sketch of the level surfaces through the given point.

   (a) $f(x, y, z) = x^2 + y^2 - z^2 - 4 = 0$, $g(x, y, z) = xyz - 2 = 0$ at $(1, 2, 1)$,
   (b) $f(x, y, z) = (x + y)^2 + z^2 - 4 = 0$, $g(x, y, z) = x^2y^2 - 2z + 3 = 0$ at $(1, -1, 2)$.

7. Let $\theta(x, y, z) = y^2 + 2yz - x$ be the temperature at the point (x, y, z). At $(3, -2, 1)$ find the direction that the temperature is increasing the most rapidly. Find the direction that heat is flowing. Find

$$\max \left\{ \frac{d}{ds} \theta(3, -2, 1) \right\}.$$

Find both normal derivatives of θ to S if

$$S = \{(x, y, z): 2x^2 + y^2 - z^2 = 21\}.$$

8. Show that if f and g are defined in an open set S in E^2, if f and $g \in C^1(S)$, and if on S,

$$f_{x_1} = g_{x_2} \quad \text{and} \quad f_{x_2} = -g_{x_1},$$

then the level curves $\{(x_1, x_2): g(x_1, x_2) = c_2\}$ are orthogonal trajectories of the level curves $\{(x_1, x_2): f(x_1, x_2) = c_1\}$.

9. If $f(r, \theta)$ is the function value of a function f at a point (r, θ) in the plane where r and θ are the polar coordinates, then show

$$\frac{d}{ds} f(r, \theta; \cos(\theta + \psi), \sin(\theta + \psi)) = f_r(r, \theta) \cos \psi + \frac{1}{r} f_\theta(r, \theta) \sin \psi.$$

Use this formula to find the rate of change of $f(r, \theta)$ at $(1, \pi/4)$ in the direction of the unit tangent to the circle $\{(r, \theta): r = 1\}$ such that motion in the direction of this unit tangent would increase θ.

Answers

1. (a) $28/\sqrt{10}$, (b) $2\sqrt{2}$.

2. $\frac{d}{ds} f(\mathbf{0}; (1/\sqrt{1 + m^2}, m/\sqrt{1 + m^2})) = m^3/\sqrt{1 + m^2}$,

$\frac{d}{ds} f(\mathbf{0}; (0, 1)) = 0$,

$$f(x_2^3, x_2) = \frac{x_2^3 \cdot x_2^3}{x_2^9 + x_2^9} = \frac{1}{2x_2^3}, \quad x_2 \neq 0,$$

$$\lim_{x_2 \to 0} f(x_2^3, x_2) = \lim_{x_2 \to 0} \frac{1}{2x_2^2} = \infty$$

and hence $\lim_{\mathbf{x} \to \mathbf{0}} f(\mathbf{x}) \neq f(\mathbf{0})$.

3. $\frac{d}{ds} f((4, 3); \mathbf{u}) = 8 \cos \alpha + 6 \sin \alpha$,

grad $f(4, 3) = (8, 6)$.

$\mathbf{u}$	$\frac{d}{ds} f((4, 3); \mathbf{u})$	$\mathbf{u}$	$\frac{d}{ds} f((4, 3); \mathbf{u})$
$(1, 0)$	8	$(-1, 0)$	-8
$\left(\frac{4}{5}, \frac{3}{5}\right)$	10	$\left(\frac{-4}{5}, \frac{-3}{5}\right)$	-10
$\left(\frac{1}{\sqrt{2}}, \frac{1}{\sqrt{2}}\right)$	$\frac{14}{\sqrt{2}} \approx 9.8$	$\left(\frac{-1}{\sqrt{2}}, \frac{-1}{\sqrt{2}}\right)$	$\frac{-14}{\sqrt{2}} \approx -9.8$
$(0, 1)$	6	$(-1, 0)$	-8
$\left(\frac{-3}{5}, \frac{4}{5}\right)$	0	$\left(\frac{3}{5}, \frac{-4}{5}\right)$	0
$\left(\frac{-1}{\sqrt{2}}, \frac{1}{\sqrt{2}}\right)$	$\frac{-2}{\sqrt{2}} \approx -1.41$	$\left(\frac{1}{\sqrt{2}}, \frac{-1}{\sqrt{2}}\right)$	$\frac{2}{\sqrt{2}} \approx 1.41$

5. (a) $-12/\sqrt{11}$, (b) $12/\sqrt{14}$, (c) $-5\sqrt{2}/2$, (d) $\sqrt{41}$.
6. (a) $\pm(1/\sqrt{2}, -4/5\sqrt{2}, -3/5\sqrt{2})$. (b) $\pm(1/\sqrt{2}, 1/\sqrt{2}, 0)$.
7. $(1, 2, 4)$.

4

Applications of Differential Calculus

4–1. Taylor's Series of Several Variables

Suppose that f is a real-valued function whose domain is S, an open set in R^n. Let $f \in C^{m+1}(S)$ and let $\mathbf{a} \in S$. It is ultimately our goal to consider Taylor's series of functions f in this general setting of n variables. To build up to this, it is better to start out in the simple setting of $n = 2$. To further simplify the presentation we shall use (x, y)-space, that is, E^2 where the general point is written (x, y) instead of (x_1, x_2), and we shall use (a, b) instead of (a_1, a_2) for our point in S.

The general real polynomial of degree m in the variables $(x - a)$ and $(y - b)$ is

$$\begin{aligned} C_{00} &+ [C_{10}(x-a) + C_{01}(y-b)] \\ &+ [C_{20}(x-a)^2 + C_{11}(x-a)(y-b) + C_{02}(y-b)^2] \\ &\cdots\cdots\cdots\cdots\cdots\cdots\cdots\cdots \\ &+ [C_{m0}(x-a)^m + C_{(m-1)1}(x-a)^{m-1}(y-b) + \cdots + C_{0m}(y-b)^m] \end{aligned} \tag{4–1.1}$$

where C_{ij} ($i = 0, 1, \cdots, m$ and $j = 0, 1, \cdots, i$) are arbitrary real numbers. If

$$C_{ij} = \frac{1}{(i+j)!}\binom{i+j}{j}\frac{\partial^{i+j}}{\partial x^i \partial y^j} f(a, b) \qquad (4\text{–}1.2)$$

for $0 \leq i \leq m$, $0 \leq j \leq m$ where $\binom{k}{j}$ is the coefficient of $s^{k-j}t^j$ in the expansion of $(s+t)^k$, which means

$$\binom{k}{j} = \frac{k!}{j!\,(k-j)!},$$

then the polynomial of (4–1.1) is the *Taylor's polynomial of degree m at (a, b) relative to f*, written

$$P_m(a, b; x, y).$$

Notice that there is sufficient hypothesis to calculate the partial derivatives of (4–1.2) in any order (cf. Proposition 3–2.1).

Writing out $P_m(a, b; x, y)$, we have

$$P_m(a, b; x, y) = f(a, b) + \frac{1}{1!}\left[\frac{\partial}{\partial x} f(a, b)(x-a) + \frac{\partial}{\partial y} f(a, b)(y-b)\right]$$
$$+ \frac{1}{2!}\left[\frac{\partial^2}{\partial x^2} f(a,b)(x-a)^2 + 2\frac{\partial^2}{\partial x \partial y} f(a,b)(x-a)(y-b) + \frac{\partial^2}{\partial y^2} f(a,b)(y-b)^2\right]$$
$$\cdots\cdots\cdots\cdots\cdots\cdots\cdots\cdots\cdots$$
$$+ \frac{1}{m!}\left[\frac{\partial^m}{\partial x^m} f(a,b)(x-a)^m + m\frac{\partial^m}{\partial x^{m-1}\partial y} f(a,b)(x-a)^{m-1}(y-b) + \cdots \right.$$
$$\left. + \frac{\partial^m}{\partial y^m} f(a,b)(y-b)^m\right]. \qquad (4\text{–}1.3)$$

To say the least, this polynomial is cumbersome to write. Of course it would be much worse if f were a function of more variables. Notice that the terms in each bracket somewhat resemble a binomial expansion. The coefficients are exactly the binomial coefficients. An operator notation that plays upon this binomial expansion idea is used to simplify the writing of (4–1.3). That is,

$$P_m(a, b; x, y) = f(a, b) + \frac{1}{1!}\left[(x-a)\frac{\partial}{\partial x} + (y-b)\frac{\partial}{\partial y}\right] f(a, b)$$
$$+ \frac{1}{2!}\left[(x-a)\frac{\partial}{\partial x} + (y-b)\frac{\partial}{\partial y}\right]^2 f(a, b)$$
$$\cdots\cdots\cdots\cdots\cdots\cdots\cdots\cdots\cdots$$
$$+ \frac{1}{m!}\left[(x-a)\frac{\partial}{\partial x} + (y-b)\frac{\partial}{\partial y}\right]^m f(a, b) \qquad (4\text{–}1.4)$$

where the meaning of each term of (4–1.4) is precisely the corresponding term of (4–1.3). Using the sigma symbol, (4–1.4) could be written

$$P_m(a, b; x, y) = \sum_{i=0}^{m} \frac{1}{i!}\left[(x-a)\frac{\partial}{\partial x} + (y-b)\frac{\partial}{\partial y}\right]^i f(a, b)$$

since $0! = 1$ and since it is natural to understand that

$$\left[(x-a)\frac{\partial}{\partial x} + (y-b)\frac{\partial}{\partial y}\right]^0 f(a, b) = f(a, b).$$

The motivation for choosing these particular coefficients is that this choice produces *exact agreement between f and $P_m(a, b;\ )$ and their corresponding partial derivatives at* (a, b). That is,

$$\frac{\partial^i}{\partial x^{i-j}\,\partial y^j} f(a, b) = \frac{\partial^i}{\partial x^{i-j}\,\partial y^j} P_m(a, b; a, b) \tag{4–1.5}$$

for all i and j, $0 \le i \le m$ and $0 \le j \le i$.

The following example illustrates a Taylor's polynomial.

Example 4–1.1. Let the function f be given by

$$f(x, y) = \frac{y+3}{x+1}.$$

Find $P_m(1, 2; x, y)$ for $m = 0$, 1, and 2. Then

$$f(1, 2) = \frac{5}{2},$$

$$\frac{\partial}{\partial x} f(x, y) = -\frac{y+3}{(x+1)^2}, \qquad \frac{1}{1!}\binom{1}{0}\frac{\partial}{\partial x} f(1, 2) = -\frac{5}{4},$$

$$\frac{\partial}{\partial y} f(x, y) = \frac{1}{x+1}, \qquad \frac{1}{1!}\binom{1}{0}\frac{\partial}{\partial y} f(1, 2) = \frac{1}{2},$$

$$\frac{\partial^2}{\partial x^2} f(x, y) = 2\frac{y+3}{(x+1)^3}, \qquad \frac{1}{2!}\binom{2}{0}\frac{\partial^2}{\partial x^2} f(1, 2) = \frac{5}{8},$$

$$\frac{\partial^2}{\partial x\,\partial y} f(x, y) = -\frac{1}{(x+1)^2}, \qquad \frac{1}{2!}\binom{2}{1}\frac{\partial^2}{\partial x\,\partial y} f(1, 2) = -\frac{1}{4},$$

and

$$\frac{\partial^2}{\partial y^2} f(x, y) = 0, \qquad \frac{1}{2!}\binom{2}{2}\frac{\partial^2}{\partial y^2} f(1, 2) = 0.$$

Hence

$$P_0(1, 2; x, y) = \frac{5}{2},$$

$$P_1(1, 2; x, y) = \frac{5}{2} + \left[-\frac{5}{4}(x - 1) + \frac{1}{2}(y - 2)\right],$$

and

$$P_2(1, 2; x, y) = \frac{5}{2} + \left[-\frac{5}{4}(x - 1) + \frac{1}{2}(y - 2)\right]$$

$$+ \left[\frac{5}{8}(x - 1)^2 - \frac{1}{4}(x - 1)(y - 2)\right].$$

Figure 4–1.1(a), (b), and (c) show how the graph of $P_0(1, 2;\quad)$, $P_1(1, 2;\quad)$, and $P_2(1, 2;\quad)$, respectively, are related to the graph of f in a neighborhood of (1, 2).

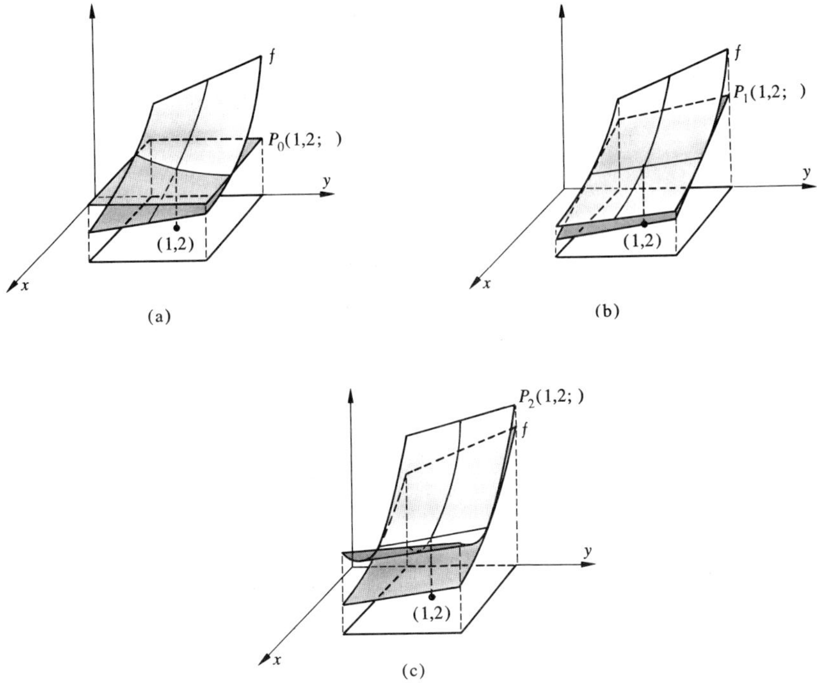

Figure 4–1.1

Because of the way that $P_m(a, b;\quad)$ and its partial derivatives are related to f at (a, b) as seen in (4–1.5), it would seem that $P_m(a, b; x, y)$ and $f(x, y)$ are approximately equal for (x, y) near (a, b). For many of the elementary functions that are often used, this is the case. Thus Taylor's polynomial plays a role relative to a function of several variables that is analogous to the role of a Taylor's polynomial to a function of a single variable. The difference between the two functions is called the *remainder function*, written $R_m(a, b;\quad)$; that is,

$$R_m(a, b; x, y) = f(x, y) - P_m(a, b; x, y). \tag{4–1.6}$$

Notice that the domain of the remainder function is always the domain of f since $P_m(a, b;\quad)$ is a polynomial and has E^2 as its domain. Thus the domain of $R_m(a, b;\quad)$ is S, an open set in (x, y)-space. In the case of Example 4–1.1,

$$\text{Do}\,(f) = \{(x, y): x \neq -1\},$$

an open set in E^2, and hence $R_m(1, 2;\quad)$ is also not defined where $x = -1$.

The following theorem, which is Taylor's theorem for the case of two variables, gives Taylor's formula, which shows the remainder function in a form that is much more useful than (4–1.6).

Theorem 4–1.1. *Let f be a function defined on an open set S in (x, y)-space. If $f \in C^{m+1}(N((a, b); \delta))$ where $N((a, b); \delta) \subset S$, then for*

$$(x, y) \in N'((a, b); \delta)$$

$$f(x, y) = \textstyle\sum_{i=0}^{m} \dfrac{1}{i!}\left[(x - a)\dfrac{\partial}{\partial x} + (y - b)\dfrac{\partial}{\partial y}\right]^i f(a, b)$$

$$+ \frac{1}{(m + 1)!}\left[(x - a)\frac{\partial}{\partial x} + (y - b)\frac{\partial}{\partial y}\right]^{m+1} f(x_\theta, y_\theta) \tag{4–1.7}$$

where $(x_\theta, y_\theta) = (a + \theta(x - a), b + \theta(y - b))$ for some θ, $0 < \theta < 1$.

Proof. Let $h = x - a$ and $k = y - b$; let the function F be defined by

$$F(t) = f(a + th, b + tk) \tag{4–1.8}$$

for $-t_0 \leq t \leq t_0$ where $t_0 = \delta/\|(h, k)\|$. Notice that $t_0 > 1$. Then

$$F \in C^{m+1}((0, t_0))^2$$

by virtue of the hypothesis and Proposition 3–4.1. Then by Taylor's theorem for a function of a single variable applied to F on the interval $(-t_0, t_0)$, it follows that

$$F(t) = F(0) + \frac{F'(0)}{1!}t + \frac{F''(0)}{2!}t^2 + \cdots + \frac{F^{(m)}(0)}{m!}t^m + \frac{F^{(m+1)}(\theta t)}{(m + 1)!}t^{m+1}, \tag{4–1.9}$$

where $0 < \theta < 1$ and $0 < |t| < t_0$. Take $t = 1$ in (4–1.9); calculate by the chain rule the derivatives $F^{(k)}$, $k = 0, 1, \cdots, m + 1$, of F given by (4–1.8); substitute back $x = a + h$, $y = b + k$, and thus obtain (4–1.7). ▎

Equation (4–1.7) is *Taylor's formula* for the function f about (a, b). Comparing (4–1.6) and (4–1.7) gives

$$R_m(a, b; x, y) = \frac{1}{(m+1)!}\left[(x-a)\frac{\partial}{\partial x} + (y-b)\frac{\partial}{\partial y}\right]^{m+1} f(x_\theta, y_\theta). \tag{4–1.10}$$

The point (x_θ, y_θ) is a point between (x, y) and (a, b) as is illustrated in Figure 1–2.4 in a similar situation.

Let f be a function defined on an open set S in (x, y)-space and suppose $f \in C^\infty(S)$ where $N((a, b); \delta) \subset S$. Then the infinite series

$$\textstyle\sum_{i=0}^{\infty} \frac{1}{i!}\left[(x-a)\frac{\partial}{\partial x} + (y-b)\frac{\partial}{\partial y}\right]^{i} f(a, b) \tag{4–1.11}$$

is the *Taylor's series of* f *about* (a, b). There is no assurance that this series converges for any (x, y) other than (a, b). Even if the series does converge for a particular (x, y), then it still may be that the series (4–1.11) does not converge to the value $f(x, y)$. However, for a particular (x, y),

$$\lim_{m\to\infty} R_m(a, b; x, y) = 0, \tag{4–1.12}$$

if and only if the series (4–1.11) converges to $f(x, y)$, so that

$$f(x, y) = \textstyle\sum_{i=0}^{\infty} \frac{1}{i!}\left[(x-a)\frac{\partial}{\partial x} + (y-b)\frac{\partial}{\partial y}\right]^{i} f(a, b). \tag{4–1.13}$$

Since the main object of a Taylor's series is to represent a function by a series, or to have (4–1.13) hold for a given function on a given set of points (x, y), it is then very important to be able to show that (4–1.12) holds. The following proposition does this.

Proposition 4–1.1. *Let* $f \in C^\infty(S)$ *where* $\text{Do}(f) = S$ *is an open set in* (x, y)*-space. Let* $N((a, b); \delta)$ *be a neighborhood contained in* S. *If there exists a positive number* M *such that*

$$\left|\frac{\partial^m}{\partial x^{m-i}\,\partial y^i} f(x, y)\right| \le M \tag{4–1.14}$$

for all $(x, y) \in N((a, b); \delta)$, *for all integers* $m \geq m_0$, *for some non-negative integer* m_0, *and for all integers* $0 \leq i \leq m$, *then for all* $(x, y) \in N((a, b); \delta)$

$$\lim_{m \to \infty} R_m(a, b; x, y) = 0$$

and

$$f(x, y) = \sum_{i=0}^{\infty} \frac{1}{i!}\left[(x - a)\frac{\partial}{\partial x} + (y - b)\frac{\partial}{\partial y}\right]^i f(a, b).$$

The following example will illustrate a Taylor's series and the use of the foregoing proposition.

Example 4–1.2. Let the function f be given by $f(x, y) = e^{x+y}$, and let $(a, b) = (0, 0)$. Then

$$\frac{\partial^m}{\partial x^{m-i}\,\partial y^i} f(x, y) = e^{x+y}$$

and hence

$$\frac{\partial^m}{\partial x^{m-i}\,\partial y^i} f(0, 0) = 1$$

for $m \geq 0$, $0 \leq i \leq m$. Thus the Taylor's series for f about $(0, 0)$ is

$$1 + \frac{1}{1!}[x \cdot 1 + y \cdot 1] + \frac{1}{2!}[x^2 \cdot 1 + 2(xy)(1) + y^2 \cdot (1)] + \cdots$$

$$= 1 + \frac{1}{1!}(x + y) + \frac{1}{2!}(x + y)^2 + \cdots$$

$$= \sum_{i=0}^{\infty} \frac{1}{i!}(x + y)^i. \qquad (4\text{–}1.15)$$

Since the domain of f is all of (x, y)-space, then for any fixed $\delta > 0$, $N((a, b); \delta) \subset S$, the domain of f. Then for $(x, y) \in N((a, b); \delta)$,

$$\left|\frac{\partial^m}{\partial x^{m-i}\,\partial y^i} f(x, y)\right| = |e^{x+y}| \leq e^{2\delta} = M$$

and thus by Proposition 4–1.1 we have

$$e^{x+y} = \sum_{i=0}^{\infty} \frac{1}{i!}(x + y)^i. \qquad (4\text{–}1.16)$$

Theorem 4–1.1 is really a mean-value theorem, because in equation (4–1.7) we find that $f(x, y) - P_m(a, b; x, y)$ is equal to the remainder term where the point (x_θ, y_θ) is between (a, b) and (x, y). A particular case of Theorem 4–1.1, when $m = 0$, is actually called "the mean-value theorem" for functions of several variables (in our case, two). Since it is a special case of Theorem 4–1.1, no proof will be necessary, and we shall call it a corollary.

Corollary 4–1.1. *Let f be a function defined on an open set S in (x, y)-space. If $f \in C^1(N((a, b); \delta))$ where $N((a, b); \delta) \subset S$, then for $(x, y) \in N((a, b); \delta)$,*

$$f(x, y) - f(a, b) = (x - a)\frac{\partial}{\partial x} f(x_\theta, y_\theta) + (y - b)\frac{\partial}{\partial y} f(x_\theta, y_\theta) \quad (4\text{–}1.17)$$

where $(x_\theta, y_\theta) = (a + \theta(x - a), b + \theta(y - b))$ for some θ, $0 < \theta < 1$.

In order to refer to it later, we give below the general statement of Taylor's theorem.

Theorem 4–1.2. *Let f be a function defined on an open set S in E^n. If $f \in C^{m+1}(N(\mathbf{a}; \delta))$ where $N(\mathbf{a}; \delta) \subset S$, then for $\mathbf{x} \in N'(\mathbf{a}; \delta)$*

$$f(\mathbf{x}) = \sum_{i=0}^{m} \frac{1}{i!}\left[\sum_{j=1}^{n} (x_j - a_j)\frac{\partial}{\partial x_j}\right]^i f(\mathbf{a})$$
$$+ \frac{1}{(m + 1)!}\left[\sum_{j=1}^{n} (x_j - a_j)\frac{\partial}{\partial x_j}\right]^{m+1} f(\mathbf{x}_\theta) \quad (4\text{–}1.18)$$

where $\mathbf{x}_\theta = \mathbf{a} + \theta(\mathbf{x} - \mathbf{a})$ for some θ, $0 < \theta < 1$.

Problems

1. Write the general real polynomial of degree m in the variables $(x - a)$, $(y - b)$, and $(z - c)$ using triple subscripts for the coefficients.

2. Calculate $\binom{8}{j}$ for $j = 0, 1, \cdots, 8$.

3. Show that the trinomial coefficient of $r^{i_1}s^{i_2}t^{i_3}$ in the expansion of $(r + s + t)^m$ is

$$\frac{m!}{(i_1)!\,(i_2)!\,(i_3)!}$$

where $0 \leq i_1, i_2$, and $i_3 \leq m$ and $i_1 + i_2 + i_3 = m$.

4. Define the Taylor's polynomial of degree m at (a, b, c) relative to f if f is a function whose domain is an open subset S in (x, y, z)-space. Write this Taylor's polynomial using an operator notation similar to that of (4–1.4).

5. Find the indicated Taylor's polynomial of each of the following functions f:

   (a) $f(x, y) = 2x^2y + y^3$, $P_3(1, 1; x, y)$,
   (b) $f(x, y) = x^2y - y^3$, $P_3(1, 0; x, y)$,

(c) $f(x, y) = (x - 3y + 2)^3$, $P_3(1, -1; x, y)$,
(d) $f(x, y) = e^y \cos x$, $P_3(\pi/2, 0; x, y)$,
(e) $f(x, y) = (x + 1)^2(y - 2)^{-1}$, $P_3(0, 1; x, y)$.

6. By using (4–1.6) show that
   (a) $R_3(1, 1; x, y) = 0$ for Problem 5(a),
   (b) $R_3(1, -1; x, y) = 0$ for Problem 5(b).

7. Find the Taylor's polynomial $P_3(1, -1, 1; x, y, z)$ of the function f defined by
$$f(x, y, z) = x^3 + 2xy - y^2 + yz^2 - 4z + xz + 5.$$

8. Find Taylor's series of the following functions f defined below about the indicated point. That is, the problem is complete when an equation like (4–1.16) has been found.
   (a) $f(x, y) = e^{xy}$, $(0, 0)$, (b) $f(x, y) = \sin x \cos y$, $(0, 0)$,
   (c) $f(x, y) = (1 + x + y)^{1/2}$, $(0, 0)$.

9. Show that (4–1.12) holds by using Proposition 4–1.1 for the following functions f in a neighborhood of the indicated point.
   (a) $f(x, y) = \cos x \cos y$, $(0, 0)$, (b) $f(x, y) = ye^x$, $(0, 1)$.

10. For $m = 0$ in Taylor's formula, (4–1.7), find the point (x_θ, y_θ) for each of the following circumstances:
    (a) $f(x, y) = x^3 - 4xy^2 + y$, $(x, y) = (1, 1)$, $(a, b) = (0, 0)$,
    (b) $f(x, y) = \dfrac{2x + 1}{y - 3}$, $(x, y) = (0, 5)$, $(a, b) = (-1, 4)$.

11. Prove Proposition 4–1.1.

12. Consider the function f defined by
$$f(x, y) = \begin{cases} e^{-(1/x^2)}, & x \neq 0, \\ 0, & x = 0, \end{cases} + \begin{cases} e^{-(1/y^2)}, & y \neq 0, \\ 0, & y = 0. \end{cases}$$
Show
$$\frac{\partial}{\partial x} f(x, y) = \begin{cases} 2x^{-3}e^{-(1/x^2)}, & (x, y) \neq (0, 0), \\ 0, & (x, y) = (0, 0), \end{cases}$$
and
$$\frac{\partial^2}{\partial x^2} f(x, y) = \begin{cases} \dfrac{4 - 6x^2}{x^6} e^{-(1/x^2)}, & (x, y) \neq (0, 0), \\ 0, & (x, y) = (0, 0). \end{cases}$$
Calculate the other first and second partials of f. Show by induction that for k, a non-negative integer,
$$\frac{\partial^k}{\partial x^k} f(x, y) = \begin{cases} \dfrac{g_k(x)}{x^{3k}} e^{-(1/x^2)}, & (x, y) = (0, 0), \\ 0, & (x, y) = (0, 0), \end{cases}$$

where g_k is a polynomial in x of degree less than or equal to $3(k-1)$. Show that all partials of f involving differentiation with respect to x and y are zero. Show $P_m(0, 0; x, y) \equiv 0$, and hence show (4–1.12) does not hold.

13. State a corollary to Theorem 4–1.2 that generalizes Corollary 4–1.1 to functions of n-variables.

Answers

1. $$c_{000} + [c_{100}(x-a) + c_{010}(y-b) + c_{001}(z-c)] + [c_{200}(x-a)^2 + c_{110}(x-a)(y-b) + c_{101}(x-a)(z-c) + c_{020}(y-b)^2 + c_{011}(y-b)(z-c) + c_{002}(z-c)^2] + \cdots + \left[\sum_{\substack{i,j,k\geq 0\\ i+j+k=m}} c_{ijk}(x-a)^i(y-b)^j(z-c)^k\right].$$

2. $$\binom{8}{0} = \binom{8}{8} = 1, \quad \binom{8}{1} = \binom{8}{7} = 8, \quad \binom{8}{2} = \binom{8}{6} = 28,$$
$$\binom{8}{3} = \binom{8}{5} = 56, \quad \binom{8}{4} = 70.$$

4. $$P_m(a, b, c; x, y, z) = \sum_{k=0}^{m} \frac{1}{k!}\left[(x-a)\frac{\partial}{\partial x} + (y-b)\frac{\partial}{\partial y} + (z-c)\frac{\partial}{\partial z}\right]^k f(a, b, c).$$

5. (a) $$P_3(1, 1; x, y) = 3 + \frac{1}{1!}[4(x-1) + 5(y-1)] + \frac{1}{2!}[4(x-1)^2 + 8(x-1)(y-1) + 6(y-1)^2] + \frac{1}{3!}[12(x-1)^2(y-1) + 6(y-1)^3].$$

(b) $$P_3(1, 0; x, y) = y + \frac{1}{2!}[4(x-1)y] + \frac{1}{3!}[6(x-1)^2y - 6y^3].$$

(c) $$P_3(1, -1; x, y) = 216 + \frac{1}{1!}[108(x-1) - 324(y+1)] + \frac{1}{2!}[36(x-1)^2 - 216\,(x-1)(y+1) + 324(y+1)^2] + \frac{1}{3!}[6(x-1)^3 - 54(x-1)^2(y+1) + 162(x-1)(y+1)^2 - 162(y+1)^3].$$

(d) $$P_3\left(\frac{\pi}{2}, 0; x, y\right) = -\frac{1}{1!}\left(x-\frac{\pi}{2}\right) + \frac{1}{2!}\left[-2\left(x-\frac{\pi}{2}\right)y\right] + \frac{1}{3!}\left[\left(x-\frac{\pi}{2}\right)^3 - 3\left(x-\frac{\pi}{2}\right)y^2\right].$$

(e) $$P_3(0, 1; x, y) = -1 + \frac{1}{1!}[-2x - (y-1)] + \frac{1}{2!}[-2x^2 - 4x(y-1) - 2(y-1)^2] + \frac{1}{3!}[-6x^2(y-1) - 12x(y-1)^2 - 6(y-1)^3].$$

7. $$P_3(1, -1, 1; x, y, z) = -1 + \frac{1}{1!}[2(x-1) + 5(y+1) - 5(z-1)] + \frac{1}{2!}\begin{bmatrix} 6(x-1)^2 + 4(x-1)(y+1) + 2(x-1)(z-1) \\ -2(y+1)^2 + 4(y+1)(z-1) \\ -2(z-1)^2 \end{bmatrix} + \frac{1}{3!}[6(x-1)^3 + 6(y+1)(z-1)^2].$$

8. (a) $f(x, y) = \sum_{k=0}^{\infty} \frac{1}{k!} x^k y^k$,

(b) $$f(x, y) = \sum_{k=0}^{\infty} (-1)^k \left[\frac{x^{2k+1}}{(2k+1)!} + \frac{x^{2k-1}}{(2k-1)!}\frac{y^2}{2!} + \cdots + \frac{xy^{2k}}{(2k)!}\right],$$

(c) $$f(x, y) = \sum_{k=0}^{\infty} \frac{(-1)^k(1 \cdot 3 \cdot 5 \cdot \cdots \cdot (2k-3))}{2^k k!}(x+y)^k.$$

10. (a) $\left(\frac{\sqrt{3}}{3}, \frac{\sqrt{3}}{3}\right)$ (b) $(\sqrt{2} - 2, \sqrt{2} + 3)$.

4–2. Transformation of Coordinates

A *transformation of coordinates* (or simply *transformation*) is a mapping (a function) from R^n space into R^n space. Thus, if $\mathbf{x} \in \text{Do}(\mathbf{f})$ where $\text{Do}(\mathbf{f}) \subset R^n$ then $\mathbf{y} = \mathbf{f}(\mathbf{x}) \in R^n$. That is, $\mathbf{f}: R^n \to R^n$. A transformation of coordinates is usually defined by n functions $f_1, \cdots, f_n$ such that each of them maps a set S in R^n into R. Thus, if $\mathbf{x} = (x_1, \cdots, x_n) \in S$, then

$$y_i = f_i(x_1, \cdots, x_n), \tag{4–2.1}$$

$1 \leq i \leq n$, and $\mathbf{y} = (y_1, \cdots, y_n)$ is the image of $\mathbf{x}$.

A common example of a transformation of coordinates defined in this way is:

$$x = r\cos\theta, \quad y = r\sin\theta, \tag{4–2.2}$$

which transforms (r, θ)-space into (x, y)-space. The usual understanding here is that r and θ are the variables of polar coordinates and hence the (r, θ)-space would not normally have any of the vector operations defined on it. That is, the space would be an R^2 space without vector addition, (1–1.1), and scalar multiplication, (1–1.6). Thus we wish to consider some examples in which, initially, the domain space does not have all of the structure that our definition of a transformation indicates it should have. However, this deficiency can easily be remedied by defining on (r, θ)-space vector addition and scalar multiplication according to (1–1.1) and (1–1.6), respectively. It may be that in this additional structure something is lost or confused—a possibility of which the reader should be aware. For example, according to (1–1.1),

$$\left(1, \frac{\pi}{3}\right) + \left(1, \frac{2\pi}{3}\right) = (2, \pi).$$

If the points $(1, \pi/3)$, $(1, 2\pi/3)$, and $(2, \pi)$ are plotted in the usual manner of polar coordinates, the point $(2, \pi)$ is not on the diagonal through $(0, 0)$ of the parallelogram that has vertices $(0, 0)$, $(1, \pi/3)$, and $(1, 2\pi/3)$. Thus the parallelogram law of addition (cf. Figure 1–2.3) does not hold.

The foregoing discussion shows how the structure of vector addition and scalar multiplication may be added to the domain space in order that the space can become an R^n space. Of course the method would work equally well if initially the range space were just the space of n-tuples.

The transformations we wish to consider are continuous. Intuitively, this means that points "near" one another in the domain space have images in the range space that are "near" one another. However, our definition is not precise until we assert clearly the meaning of "near." That is, a "neighborhood of a point" in both the domain and range spaces must be defined.

If the transformation is given in the form of the equations of (4–2.1), then the domain and range spaces are at least n-tuples of real numbers. In the same way, we earlier indicated how to add sufficient structure to make the spaces R^n spaces; we may also define on the spaces the dot product (1–2.4) and the norm (1–2.12), making the spaces E^n spaces. Then the spaces would automatically get the spherical neighborhoods $N(\mathbf{a}; \delta)$, and the topology would be the usual topology.

Adding the structure of a dot product and a norm according to (1–2.4) and (1–2.12) may bring conflict with something that is already understood about the domain or range spaces. For example, in the transformation given by equations of (4–2.2), the distance between two points (r_1, θ_1) and (r_2, θ_2)

would usually be taken to be the Euclidean distance induced from the (x, y)-space—namely,

$$\rho'((r_1, \theta_1), (r_2, \theta_2)) = \sqrt{(r_1 \cos \theta_1 - r_2 \cos \theta_2)^2 + (r_1 \sin \theta_1 - r_2 \sin \theta_2)^2}. \tag{4–2.3}$$

The distance between these same two points given by the usual distance formula for an E^n space would be

$$\|(r_1, \theta_1) - (r_2, \theta_2)\| = \sqrt{(r_1 - r_2)^2 + (\theta_1 - \theta_2)^2} \tag{4–2.4}$$

(cf. Problem 10, Section 1–2). The two measures of the distance between two points would be different for certain pairs of points—for example, $(1, 0)$ and $(1, \pi/2)$.

Although in this example there is conflict in the matter of distances between pairs of points, there are important areas of agreement. The topology generated by the neighborhoods, $N_M((r_0, \theta_0); \delta)$, $\delta > 0$, where

$$N_M((r_0, \theta_0); \delta) = \{(r, \theta) \colon \rho'((r, \theta), (r_0, \theta_0)) < \delta\}$$

and the topology generated by the neighborhood $N((r_0, \theta_0); \delta)$, $\delta > 0$, where

$$N((r_0, \theta_0); \delta) = \{(r, \theta) \colon \|(r, \theta) - (r_0, \theta_0)\| < \delta\}$$

are equivalent (cf. Section 2–2). That is, the topology that was understood for the space originally and the topology given by introducing the new structure are not in conflict; they are one and the same. The topologies on the respective spaces are of central importance in the definitions of continuous transformations.

To begin, let us suppose that we have given a transformation by the n equations of (4–2.1). If the domain space is not already an E^n space, then add the necessary structure to make it so. Thus the topology on the domain space is the *usual topology*. If $\mathbf{x}_0 \in S = \text{Do}(\mathbf{f})$, then $\mathbf{f}$ is *continuous* at $\mathbf{x}_0$ if each function f_i, $1 \le i \le n$, is continuous at $\mathbf{x}_0$. If S_1 is a subset of S, then $\mathbf{f}$ is *continuous on* S_1 if $\mathbf{f}$ is continuous at each point of S_1. For this we write $\mathbf{f} \in C(S_1)$. Similarly, if S_1 is an open set in E^n and each $f_i \in C^k(S_1)$, $1 \le i \le n$, $k = 1, 2, 3, \cdots$, then $\mathbf{f} \in C^k(S_1)$.

The definition of continuity at $\mathbf{x}_0$ just stated really says that $\mathbf{f}$ is continuous at $\mathbf{x}_0$ if $\mathbf{f}$ is continuous at $\mathbf{x}_0$ with respect to the usual topology. Thus, if we wished not to add to the domain space sufficient structure to make it into an E^n space, it follows by Problem 13 in Section 3–1 that the same transformations would be rendered continuous if we obtained this same topology on the domain space by simply generating the topology on the space with the square neighborhoods.

By the above definition the neighborhoods that are effectively being used in the range space are the rectangular neighborhoods. Since any one of the functions f_i, $1 \leq i \leq n$, is continuous at $\mathbf{x}_0$, it follows from Definitions 3–1.1 and 3–1.2 that the appropriate neighborhood of y_{0i} is of the form $N(y_{0i}, b_i)$, $b_i > 0$. Thus $y_i \in N(y_{0i}, b_i)$ when $\mathbf{x} \in N_i(\mathbf{x}_0 ; \delta_i) \cap S$. That is, if

$$\mathbf{x} \in N(\mathbf{x}_0 ; \delta') \cap S$$

where $\delta' = \min \{\delta_1, \cdots, \delta_n\}$, then $|y_i - y_{0i}| < b_i$, $1 \leq i \leq n$, and hence

$$\mathbf{y} \in N_R(\mathbf{y}_0 ; \mathbf{b})$$

where $\mathbf{b} = (b_1, \cdots, b_n)$. Thus the transformation $\mathbf{f}$ is continuous at $\mathbf{x}_0$ if, given $N_R(\mathbf{y}_0 ; \mathbf{b})$, $b_i > 0$, $1 \leq i \leq n$, there exists $N(\mathbf{x}_0 ; \delta)$, $\delta > 0$ such that if $\mathbf{x} \in N(\mathbf{x}_0 ; \delta) \cap S$ then $\mathbf{f}(\mathbf{x}) \in N_R(\mathbf{y}_0 ; \delta)$. Since for any given neighborhood $N(\mathbf{y}_0 ; \varepsilon)$ there exists a $\mathbf{b} = (b_1, \cdots, b_n)$, $b_i > 0$, such that

$$N_R(\mathbf{y}_0 ; b) \subset N(\mathbf{y}_0 ; \varepsilon),$$

it is possible to restate the definition using spherical neighborhoods in both spaces (cf. Problem 2, Section 2–1). Accordingly, both the domain and the range spaces have the usual topology. Other topologies could be taken, but the definitions of continuity would need to reflect these changes.

Transformations of coordinates occur mainly in two settings—namely, in the changing of the independent variables of a partial differential equation and in the changing of the variables of integration of a multiple integral. In both cases it is desirable that the transformation $\mathbf{f}$ have an inverse transformation, at least locally. That is, if $\mathbf{f}: S \to E^n$ where S is a subset of E^n and if $\mathbf{x}_0 \in S$, then there exists a neighborhood of $\mathbf{x}_0$, $N(\mathbf{x}_0 ; \delta)$ contained in S such that $\mathbf{f}$ is 1–1 on $N(\mathbf{x}_0 ; \delta)$. Also, if $\mathbf{y}_0 = f(\mathbf{x}_0)$ and if $\mathbf{g}$ is the inverse transformation of $\mathbf{f}$ relative to the neighborhood $N(\mathbf{x}_0 ; \delta)$, then it is also desired that $\mathbf{g} = (g_1, \cdots, g_n)$ be such that each function g_i have continuous first partials in a neighborhood of $\mathbf{y}_0$.

The following theorem, which is known as the *inverse function theorem*, gives sufficient conditions on $\mathbf{f}$ such that $\mathbf{f}$ has all of the desired properties mentioned above. Most of the proof is given by noticing that our theorem can be treated as a special case of the implicit function theorem.

Theorem 4–2.1. *Let* $\mathbf{f} = (f_1, \cdots, f_n)$ *be a transformation of coordinates from an open set* S *in* E^n *into* E^n *defined by the* n *equations*

$$y_i = f_i(x_1, \cdots, x_n),$$

$1 \leq i \leq n$. *Furthermore, let* $\mathbf{f} \in C^1(S)$, *and at* $\mathbf{x}_0 \in S$ *let*

$$\frac{\partial(f_1, \cdots, f_n)}{\partial(x_1, \cdots, x_n)} \neq 0. \tag{4–2.5}$$

Then there exist a neighborhood of $\mathbf{y}_0 = \mathbf{f}(\mathbf{x}_0)$, $N(\mathbf{y}_0 ; \delta)$ *and a unique transformation* $\mathbf{g} = (g_1, \cdots, g_n)$ *defined on* $N(\mathbf{y}_0 ; \delta)$ *such that*

(a) $\mathbf{g}(\mathbf{y}_0) = \mathbf{x}_0$,

(b) $\mathbf{f}(\mathbf{g}(\mathbf{y})) = \mathbf{y}$ *for all* $\mathbf{y} \in N(\mathbf{y}_0 ; \delta)$,

(c) *at* $(\mathbf{x}_0, \mathbf{y}_0)$

$$\frac{\partial x_j}{\partial y_i} = (-1)^{i+j} \frac{\dfrac{\partial(f_1, \cdots, f_{i-1}, f_{i+1}, \cdots, f_n)}{\partial(x_1, \cdots, x_{j-1}, x_{j+1}, \cdots, x_n)}}{\dfrac{\partial(f_1, \cdots, f_n)}{\partial(x_1, \cdots, x_n)}}, \tag{4–2.6}$$

(d) $\mathbf{g} \in C^1(N(\mathbf{y}_0 ; \delta))$,

(e) $\mathbf{f}$ *is* 1–1 *in a neighborhood of* $\mathbf{x}_0$, *and*

(f) $S_\varepsilon = \{\mathbf{g}(\mathbf{y}) : \mathbf{y} \in N(\mathbf{y}_0 ; \varepsilon)\}$ *is open for some* $\varepsilon > 0$.

Proof. Let F_i, $1 \leq i \leq n$, be functions from E^{2n} into R defined by the following equations:

$$F_i(x_1, \cdots, x_n, y_1, \cdots, y_n) = f_i(x_1, \cdots, x_n) - y_i,$$

$1 \leq i \leq n$. The domain of each F_i,

$$\text{Do}\,(F_i) = \{(\mathbf{x}, \mathbf{y}) : \mathbf{x} \in S\} = S',$$

is an open set in E^{2n}. If $(\mathbf{x}_1, \mathbf{y}_1) \in S'$, then $\mathbf{x}_1 \in S$. Since S is open in E^n, then there exists $N(\mathbf{x}_1 ; \delta_1) \subset S$. Then $N((\mathbf{x}_1, \mathbf{y}_1); \delta_1) \subset S'$ since

$$\|(\mathbf{x}, \mathbf{y}) - (\mathbf{x}_1, \mathbf{y}_1)\| < \delta_1$$

implies

$$\delta_1 > \sqrt{\textstyle\sum_{i=1}^n (x_i - x_{1i})^2 + \sum_{i=1}^n (y_i - y_{1i})^2} \geq \sqrt{\textstyle\sum_{i=1}^n (x_i - x_{1i})^2} = \|x - x_1\|.$$

Furthermore, $(\mathbf{x}_0, \mathbf{y}_0) \in S'$, and for each i, $1 \leq i \leq n$,

$$F_i \in C'(S') \quad \text{and} \quad F_i(\mathbf{x}_0, \mathbf{y}_0) = 0.$$

Lastly, at $(\mathbf{x}_0, \mathbf{y}_0)$,

$$\left.\frac{\partial(F_1, \cdots, F_n)}{\partial(x_1, \cdots, x_n)}\right|_{(\mathbf{x}_0, \mathbf{y}_0)} = \left.\frac{\partial(f_1, \cdots, f_n)}{\partial(x_1, \cdots, x_n)}\right|_{\mathbf{x}_0} \neq 0.$$

Thus the hypothesis of the implicit function theorem is satisfied. Then there exists a neighborhood of $\mathbf{y}_0$, $N(\mathbf{y}_0 ; \delta)$, and there exist uniquely functions

g_i, $1 \le i \le n$, defined on $N(\mathbf{y}_0\,;\delta)$ such that

(a) $g_i(\mathbf{y}_0) = x_{0i}$, $1 \le i \le n$,

(b) $f_i(g_1(\mathbf{y}), \cdots, g_n(\mathbf{y})) = y_i$, $1 \le i \le n$, for all $\mathbf{y} \in N(\mathbf{y}_0\,;\delta)$, and

(c) at $(\mathbf{x}_0, \mathbf{y}_0)$,

$$\frac{\partial x_j}{\partial y_i} = -\frac{\dfrac{\partial(F_1, \cdots, F_j, \cdots, F_n)}{\partial(x_1, \cdots, y_i, \cdots, x_n)}}{\dfrac{\partial(F_1, \cdots, F_n)}{\partial(x_1, \cdots, x_n)}}$$

$$= (-1)^{i+j}\frac{\dfrac{\partial(f_1, \cdots, f_{i-1}, f_{i+1}, \cdots, f_n)}{\partial(x_1, \cdots, x_{j-1}, x_{j+1}, \cdots, x_n)}}{\dfrac{\partial(f_1, \cdots, f_n)}{\partial(x_1, \cdots, x_n)}},$$

since the jth column in the determinant of the numerator involving the functions $F_1, \cdots, F_n$ are all zero except for -1 in the ith place. Furthermore, $\mathbf{g} \in C^1(N(\mathbf{y}_0\,;\delta))$.

Now let us show that $\mathbf{f}$ is 1–1 in a neighborhood of $\mathbf{x}_0$. Since $\mathbf{f} \in C^1(S)$, then the function h defined on

$$S^n = \overbrace{S \times \cdots \times S}^{n \text{ factors}}$$

in E^{n^2} by

$$h(\mathbf{x}_1\,; \cdots ; \mathbf{x}_n) = \begin{vmatrix} \dfrac{\partial}{\partial x_1} f_1(\mathbf{x}_1) & \cdots & \dfrac{\partial}{\partial x_n} f_1(\mathbf{x}_1) \\ \dfrac{\partial}{\partial x_1} f_2(\mathbf{x}_2) & \cdots & \dfrac{\partial}{\partial x_n} f_2(\mathbf{x}_2) \\ \cdots & \cdots & \cdots \\ \dfrac{\partial}{\partial x_n} f_n(\mathbf{x}_n) & \cdots & \dfrac{\partial}{\partial x_n} f_n(\mathbf{x}_n) \end{vmatrix}$$

belongs to $C(S^n)$. Furthermore, by (4–2.5) it follows that $h(\mathbf{x}_0\,; \cdots ; \mathbf{x}_0) \ne 0$. Hence there exists $\delta' > 0$ such that $h(\mathbf{x}_1\,; \cdots ; \mathbf{x}_n) \ne 0$ for $(\mathbf{x}_1\,; \cdots ; \mathbf{x}_n) \in N((\mathbf{x}_0, \cdots, \mathbf{x}_0); \delta')$. However, if $\mathbf{x}_i \in N(\mathbf{x}_0\,;\delta'')$, $1 \le i \le n$, where $\delta'' = \sqrt{\delta'/n}$, then $(\mathbf{x}_1\,; \cdots ; \mathbf{x}_n) \in N(\mathbf{x}_0\,; \cdots ; \mathbf{x}_0\,; \delta')$.

Let $\mathbf{x}$ and $\mathbf{a}$ be points of $N(\mathbf{x}_0\,;\delta'')$. Using Theorem 4–1.2 for $m = 0$, we have for each i, $1 \le i \le n$,

$$f_i(\mathbf{x}) = f_i(\mathbf{a}) + \textstyle\sum_{j=1}^{n} (x_j - a_j)\dfrac{\partial}{\partial x_j} f_i(\mathbf{a} + \theta_i(\mathbf{x} - \mathbf{a})) \qquad (4\text{–}2.7)$$

where θ_i is such that $0 < \theta_i < 1$. Since $N(\mathbf{x}_0\,;\delta'')$ is convex,

$$\mathbf{a} + \theta_i(\mathbf{x} - \mathbf{a}) = \mathbf{z}_i \in N(\mathbf{x}_0\,;\delta'')$$

for each i, $1 \leq i \leq n$ (cf. Problem 11(a) Section 1–2). If $\mathbf{x}$ and $\mathbf{a}$ both map into the same point, so that $\mathbf{f}(\mathbf{x}) = \mathbf{f}(\mathbf{a})$, then $f_i(\mathbf{x}) = f_i(\mathbf{a})$ for each i, $1 \leq i \leq n$. In this case the equations of (4–2.7) become

$$\sum_{j=1}^{n} (x_j - a_j) \frac{\partial}{\partial x_j} f_i(\mathbf{z}_i) = 0 \tag{4–2.8}$$

for $1 \leq i \leq n$. The determinant of this linear system (4–2.8) is

$$\left| \frac{\partial f_i}{\partial x_j}(\mathbf{z}_i) \right| = h(\mathbf{z}_1, \cdots, \mathbf{z}_n),$$

which is not zero since $\mathbf{z}_i \in N(\mathbf{x}_0\,; \delta'')$, $1 \leq i \leq n$. Therefore, $x_j - a_j = 0$, $1 \leq j \leq n$ or $\mathbf{x} = \mathbf{a}$. Therefore, $\mathbf{f}$ is 1–1 on $N(\mathbf{x}_0\,; \delta'')$.

Since $\mathbf{g}$ is continuous at $\mathbf{y}_0$, there exists $\varepsilon > 0$ (and such that $\varepsilon < \delta$) such that if $\mathbf{y} \in N(\mathbf{y}_0\,; \varepsilon)$, then $\mathbf{g}(\mathbf{y}) \in N(\mathbf{x}_0\,; \delta'')$. Let

$$S_\varepsilon = \{\mathbf{g}(\mathbf{y}) \colon \mathbf{y} \in N(\mathbf{y}_0\,; \varepsilon)\}.$$

We shall show S_ε is open. Let $\mathbf{x}_1 \in S_\varepsilon$. Then there exists $\mathbf{y}_1 \in N(\mathbf{y}_0\,; \varepsilon)$ such that $\mathbf{g}(\mathbf{y}_1) = \mathbf{x}_1$. There exists $\varepsilon_1 > 0$ such that $N(\mathbf{y}_1\,; \varepsilon_1) \subset N(\mathbf{y}_0\,; \varepsilon)$. Since $\mathbf{f}$ is continuous at $\mathbf{x}_1$, there exists $\eta > 0$ such that if $\mathbf{x} \in N(\mathbf{x}_1\,; \eta)$, then $\mathbf{f}(\mathbf{x}) \in N(\mathbf{y}_1\,; \varepsilon_1)$. Then $\mathbf{g}(\mathbf{f}(\mathbf{x})) = \mathbf{x}$, for otherwise we would have a contradiction of the fact that $\mathbf{f}$ is 1–1 on $N(\mathbf{x}_0\,; \delta'')$. Therefore, $\mathbf{x} \in S_\varepsilon$ and hence $N(\mathbf{x}_1\,; \eta) \subset S_\varepsilon$. Thus S_ε is open. ▌

The following examples will illustrate the meaning of this inverse-function theorem.

Example 4–2.1. Let us consider the transformation given by the equations (4–2.2). That is,

$$x = f_1(r, \theta) = r \cos \theta,$$
$$y = f_2(r, \theta) = r \sin \theta.$$

The domain of $\mathbf{f}$, Do $(\mathbf{f}) = E^2$, or all of (r, θ)-space. Since

$$\begin{aligned} \lim_{(r,\theta)\to(r_0,\theta_0)} r \cos \theta &= (\lim_{(r,\theta)\to(r_0,\theta_0)} r)(\lim_{(r,\theta)\to(r_0,\theta_0)} \cos \theta) \\ &= (\lim_{r\to r_0} r)(\lim_{\theta\to\theta_0} \cos \theta) \\ &= r_0 \cos \theta_0 \end{aligned}$$

and since a similar thing holds for the function f_2, then $\mathbf{f} \in C(E^2)$. The first partials of each of the functions, which can be treated similarly, are also continuous at each point of E^2. The Jacobian for this transformation is

$$\frac{\partial(f_1, f_2)}{\partial(r, \theta)} = \begin{vmatrix} \cos \theta & -r \sin \theta \\ \sin \theta & r \cos \theta \end{vmatrix} = r.$$

Thus, the inequality of (4–2.5) holds for all points except those where $r = 0$.

From our prior knowledge of polar coordinates we know that the regions

$$S_0^+ = \{(r, \theta): r > 0, 0 \leq \theta < 2\pi\}$$

and

$$S_0^- = \{(r, \theta): r < 0, 0 \leq \theta < 2\pi\}$$

map 1–1 onto the (x, y)-plane with the origin removed (cf. Figure 4–2.1). A similar statement is true for each of the regions $S_1^+, S_2^+, \cdots; S_{-1}^+, S_{-2}^+, \cdots; S_1^-, S_2^-, \cdots$, and $S_{-1}^-, S_{-2}^-, \cdots$.

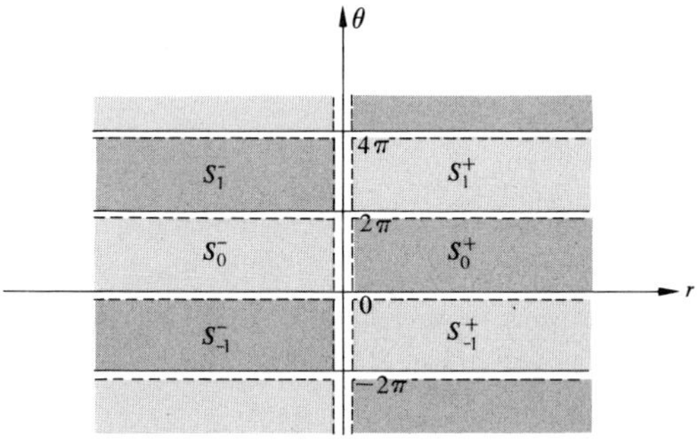

Figure 4–2.1

Let us now consider what the transformation **g** of Theorem 4–2.1 might be for the points (r_0, θ_0), $r_0 \neq 0$. If $r_0 > 0$, then take

$$r = g_1(x, y) = \sqrt{x^2 + y^2},$$

or if $r_0 < 0$, take

$$r = g_1(x, y) = -\sqrt{x^2 + y^2}.$$

Since the principal values of the arc tangent are on the interval $(-\pi/2, \pi/2)$, we may take

$$\theta = g_2(x, y) = \tan^{-1}\frac{y}{x}$$

if $-\pi/2 < \theta_0 < \pi/2$. Or, in general,

$$\theta = g_2(x, y) = n\pi + \tan^{-1}\frac{y}{x},$$

if $n\pi - (\pi/2) < \theta_0 < n\pi + (\pi/2)$, $n = \cdots, -2, -1, 0, 1, 2, \cdots$. If $\theta_0 = (\pi/2) + n\pi$, $n = \cdots, -2, -1, 0, 1, 2, \cdots$, then take

$$\theta = g_2(x, y) = n\pi + \cot^{-1}\frac{x}{y}.$$

Recall that the principal values of the arc cotangent are on the interval $(0, \pi)$. Now let us observe the conclusions of Theorem 4–2.1 for a particular point, (r_0, θ_0). Let

$$(r_0, \theta_0) = \left(1, \frac{\pi}{4}\right),$$

then

$$(x_0, y_0) = \left(\frac{1}{\sqrt{2}}, \frac{1}{\sqrt{2}}\right)$$

and

$$\mathbf{g}(x, y) = \left(\sqrt{x^2 + y^2}, \tan^{-1}\frac{y}{x}\right).$$

The domain of $\mathbf{g}$,

$$\text{Do}\,(\mathbf{g}) = \{(x, y): x > 0\}.$$

Hence, δ may be any positive real number less than $1/\sqrt{2}$. Clearly (a) checks; that is, $\mathbf{g}(1/\sqrt{2}, 1/\sqrt{2}) = (1, \pi/4)$. Furthermore, (b) checks also as follows:

$$\left(\sqrt{x^2 + y^2}\cos\left(\tan^{-1}\frac{y}{x}\right), \sqrt{x^2 + y^2}\sin\left(\tan^{-1}\frac{y}{x}\right)\right) = (x, y)$$

for all $(x, y) \in N((1/\sqrt{2}, 1/\sqrt{2}); \delta)$.

Equation (4–2.6) gives the value of four partial derivatives at

$$((1, \pi/4), (1/\sqrt{2}, 1/\sqrt{2})).$$

For example,

$$\frac{\partial r}{\partial x} = (-1)^{1+1}\frac{\dfrac{\partial y}{\partial \theta}}{\dfrac{\partial(x, y)}{\partial(r, \theta)}} = \frac{r\cos\theta}{\begin{vmatrix}\cos\theta & -r\sin\theta \\ \sin\theta & r\cos\theta\end{vmatrix}} = \cos\theta.$$

Then at the point in question

$$\frac{\partial r}{\partial x} = \frac{1}{\sqrt{2}}.$$

As a check, we may differentiate directly the inverse function, since in this case the inverse function $\mathbf{g}$ is known explicitly. Thus, from $r = \sqrt{x^2 + y^2}$,

$$\frac{\partial r}{\partial x} = \frac{x}{\sqrt{x^2 + y^2}},$$

and again we have

$$\frac{\partial r}{\partial x} = \frac{1}{\sqrt{2}}.$$

The other first partials of r and θ can be calculated directly, and it is easy to observe that they are continuous in $N((1/\sqrt{2}, 1/\sqrt{2}); \delta)$.

The transformation **f** is 1–1 in $N((1, \pi/4); \pi/4)$, since this neighborhood is a subset of S_0^+, a set on which **f** is 1–1 (cf. Figure 4–2.1).

Figure 4–2.2(a) shows an open disk of radius ε centered at $(1/\sqrt{2}, 1/\sqrt{2})$. In Figure 4–2.2(b), the image of the open disk S_ε is shown. The points A, B, B_1, C, C_1, etc. on the boundary of the disk go into the corresponding primed points.

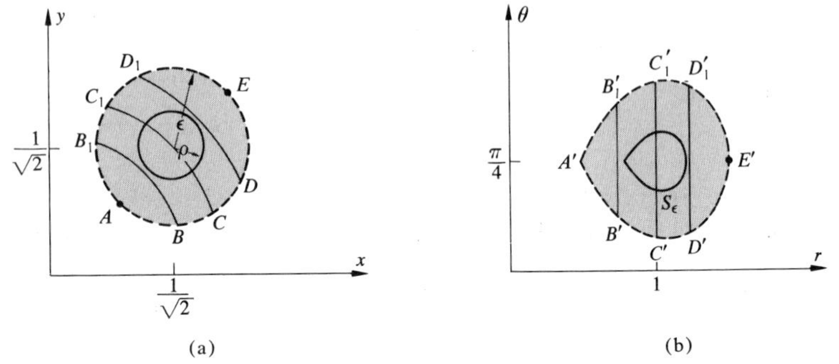

Figure 4–2.2

The sketch of S_ε can be found another way by first writing parametrically the equation of a circle of radius ρ centered at the same point, $(1/\sqrt{2}, 1/\sqrt{2})$. That is, write

$$x = \frac{1}{\sqrt{2}} + \rho \cos t,$$

$$y = \frac{1}{\sqrt{2}} + \rho \sin t$$

for $0 \leq t \leq 2\pi$, and then substitute these into

$$r = \sqrt{x^2 + y^2},$$

$$\theta = \tan^{-1} \frac{y}{x}.$$

The result will be a closed curve in the interior of S_ε, as shown in Figure 4–2.2(b).

It is often instructive, particularly in the case of $n = 2$, to draw the curves in the range space where the variables of the domain space are constant. In the transformation of (4–2.2), the curves where r is constant are circles centered at the origin, and the curves where θ is constant are lines through the origin (cf. Figure 4–2.3). For each positive constant r_0, the circles $r = r_0$ and $r = -r_0$ are one and the same. Thus each point away from the origin has two circles passing through it. However, if we require that $r > 0$ and require that $0 \le \theta < 2\pi$, then clearly there exist a unique circle (curve where r is constant) and a unique straight line (curve where θ

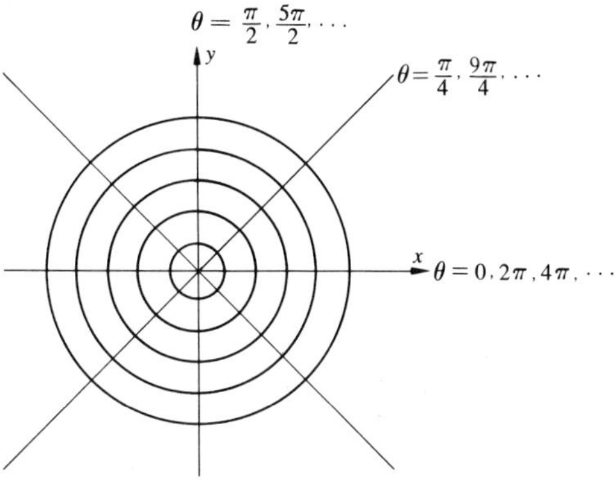

Figure 4–2.3

is constant) through each point except the origin. Thus, by drawing these two families of curves in the (x, y)-plane, we have been able to observe that the transformation (4–2.2) maps S_0^+ of Figure 4–2.1 in a 1–1 manner onto the (x, y)-plane less the point (0, 0). Thus we are observing geometrically the existence of an inverse transformation in the neighborhood of any point (r_0, θ_0), $r_0 \neq 0$.

At the point $(x, y) = (0, 0)$, the image of the line $r = 0$, there are many lines that pass through (0, 0). This is the point where the Jacobian of the transformation,

$$\frac{\partial(x, y)}{\partial(r, \theta)} = 0,$$

and hence the hypothesis of Theorem 4–2.1, is not satisfied. However, it is easy to observe that all of the conclusions of the theorem, in particular (e), do not hold in any neighborhood of a point $(0, \theta_0)$. In any neighborhood of $(0, \theta_0)$, there are always two points on the line $r = 0$ and in this neighborhood. Both points by this transformation will map into (0, 0), and hence the transformation is not 1–1 on this neighborhood.

Problems

1. For the transformation of (4–2.2) calculate the following derivatives at $r = 1$, $\theta = \pi/4$:
   (a) $\dfrac{\partial r}{\partial y}, \dfrac{\partial \theta}{\partial x}$, and $\dfrac{\partial \theta}{\partial y}$.
   (b) Check the above derivatives by differentiating directly the functions g_1 and g_2 and evaluating at $(1/\sqrt{2}, 1/\sqrt{2})$.
2. Consider the transformation given by
$$x = \frac{u}{u^2 + v^2}, \qquad y = -\frac{v}{u^2 + v^2}.$$
   (a) Find $\dfrac{\partial(x, y)}{\partial(u, v)}$.
   (b) What is the set of points on which the hypothesis of Theorem 4–2.1 is satisfied?
   (c) Find the inverse transformation.
   (d) Show that the transformation is a 1–1 onto map between the (x, y)-plane less the origin and (u, v)-plane less the origin.
   (e) Calculate in two ways the derivatives u_x, u_y, v_x, and v_y. Check your answers as you did in Problem 1.
   (f) Plot in the (x, y)-plane the following two families of curves: the family where u is constant and the family where v is constant.
3. Consider the transformation given by
$$x = u^2 + 2v, \qquad y = u - v.$$
   (a) and (b) Follow the directions of (a) and (b) of Problem 2.
   (c) Plot in the (x, y)-plane the family of curves where u is constant and the family of curves where v is constant, and thus show that the range of the transformation is $\{(x, y): x + 2y \geq -1\}$.
   (d) Find the inverse transformation that maps $\{(x, y): x + 2y > -1\}$ onto $\{(u, v): u > -1)\}$. Also find the inverse transformation that maps $\{(x, y): x + 2y > -1\}$ onto $\{(u, v); u < -1\}$.
   (e) Demonstrate that at least one of the conclusions is not true at a point $(-1, v_0)$ for any real constant v_0. In particular, show that $(-1 + \varepsilon, v_0 + \varepsilon)$ and $(-1 - \varepsilon, v_0 - \varepsilon)$ map into the same point for any $\varepsilon > 0$.
   (f) Calculate in two ways the partial derivatives u_x, u_y, v_x, and v_y at the point $(u, v) = (1, 2)$.
4. Discuss each of the following transformations in a manner similar to the way the transformations of Problems 2 and 3 were discussed.
   (a) $x = e^u \cos v, y = e^u \sin v$,
   (b) $x = \dfrac{u}{1 + u + v}, y = \dfrac{v}{1 + u + v}$,

(c) $x = u^2 - v^2, y = 2uv,$ (d) $x = v^2, y = \dfrac{u}{v}.$

(e) $x = au,\ y = a \log \tan \left(\dfrac{v}{2} + \dfrac{\pi}{4}\right)$, $u =$ longitude, $v =$ latitude of Mercator maps.

5. If the functions f_i of equation (4–2.1) are defined as

$$f_i(x_1, \cdots, x_n) = a_{i1} x_1 + \cdots + a_{in} x_n,$$

$1 \leq i \leq n$, then the transformation is, of course, a linear transformation. Show that the Jacobian of the transformation,

$$\frac{\partial(f_1, \cdots, f_n)}{\partial(x_1, \cdots, x_n)} = \left|\frac{\partial f_i}{\partial x_j}\right|,$$

is $|a_{ij}|$. Thus if the matrix (a_{ij}) is nonsingular, show that the hypothesis of Theorem 4–2.1 is satisfied at any point in E^n. Demonstrate that the inverse transformation is indeed the transformation **g** of Theorem 4–2.1.

6. Consider the transformation of spherical coordinates—that is,

$$x = \rho \cos \theta \sin \varphi,$$
$$y = \rho \sin \theta \sin \varphi,$$
$$z = \rho \cos \varphi.$$

(a) Calculate the Jacobian of this transformation from (ρ, φ, θ)-space to (x, y, z)-space.
(b) Find the set of points on which the hypothesis of Theorem 4–2.1 is satisfied.
(c) Demonstrate that the above transformation is 1–1 on

$$\{(\rho, \varphi, \theta)\colon \rho > 0, 0 \leq \theta < 2\pi, 0 \leq \varphi < \pi\}.$$

(d) For the point $(\rho, \varphi, \theta) = (1, \pi/2, \pi/4)$ find the **g** transformation of Theorem 4–2.1.
(e) At $(1, \pi/2, \pi/4)$, calculate in two ways the following derivatives: ρ_x, θ_y, and φ_z.

7. Suppose that **f** and **g** are transformations from E^2 into E^2 given as follows:

$$y_1 = f_1(x_1, x_2), \qquad y_2 = f_2(x_1, x_2),$$
$$z_1 = g_1(y_1, y_2), \qquad z_2 = g_2(y_1, y_2).$$

Suppose, furthermore, that Do (**g**) $\supset$ Ra (**f**), so that the composite transformation $\mathbf{h} = \mathbf{gf}$ is formed. If the chain rule is applicable at a certain point $\mathbf{x}_0$ for each function $g_i\,\mathbf{f}$, $1 \leq i \leq n$, show that

$$\frac{\partial(h_1, h_2)}{\partial(x_1, x_2)} = \frac{\partial(h_1, h_2)}{\partial(y_1, y_2)} \cdot \frac{\partial(f_1, f_2)}{\partial(x_1, x_2)}.$$

Generalize this result to transformations from E^n into E^n. What would be the result if **g** is locally the inverse of **f**?

8. Consider a transformation $x = f_1(u, v), y = f_2(u, v)$ that satisfies the hypothesis of Theorem 4–2.1 at (u_0, v_0). The curve in the (x, y)-plane where $u = u_0$ is given parametrically by considering v as the parameter in

$$x = f_1(u_0, v), \qquad y = f_2(u_0, v).$$

The direction of this curve at (x_0, y_0) is

$$\left(\frac{\partial}{\partial v} f_1(u_0, v_0), \frac{\partial}{\partial v} f_2(u_0, v_0)\right).$$

Similarly, the direction of the curve where $v = v_0$ is given by

$$\left(\frac{\partial}{\partial u} f_1(u_0, v_0), \frac{\partial}{\partial u} f_2(u_0, v_0)\right).$$

Suppose that neither of these vectors is the zero vector. Think of each of these as vectors in E^3 with the third component zero. Calculate the sine of the angle between these two vectors by use of the vector product. Show that the two curves cannot be tangent at (x_0, y_0).

9. Consider the transformation given by

$$x = \tfrac{1}{2}(u^3 + v^3), \qquad y = \tfrac{1}{2}(u^3 - v^3).$$

(a) Show that the hypothesis of Theorem 4–2.1 is not satisfied at (0, 0).
(b) Show that this transformation maps the (u, v)-plane in a 1–1 manner onto the (x, y)-plane.
(c) Find the inverse transformation.
(d) Do any of the conclusions of Theorem 4–2.1 fail to hold at (0, 0)?

Answers

1. At $r = 1, \theta = \frac{\pi}{4}: \frac{\partial r}{\partial \theta} = \frac{1}{\sqrt{2}}, \frac{\partial \theta}{\partial x} = -\frac{1}{\sqrt{2}}, \frac{\partial \theta}{\partial y} = \frac{1}{\sqrt{2}}.$

2. (a) $1/(u^2 + v^2)^2$, (b) $\{(u, v): (u, v) \neq (0, 0)\}$,

(c) $u = \dfrac{x}{x^2 + y^2}, v = \dfrac{-y}{x^2 + y^2}$,

(e) $\dfrac{\partial u}{\partial x} = \dfrac{y^2 - x^2}{(x^2 + y^2)^2} = v^2 - u^2$, etc.

3. (a) $\dfrac{\partial(x, y)}{\partial(u, v)} = -2u - 2.$ (b) $\{(u, v): u \neq -1\}$.

(d) $\left.\begin{aligned} u &= -1 + \sqrt{x + 2y + 1} \\ v &= -1 - y + \sqrt{x + 2y + 1}\end{aligned}\right\}, \quad \left.\begin{aligned} u &= -1 - \sqrt{x + 2y + 1} \\ v &= -1 - y - \sqrt{x + 2y + 1}\end{aligned}\right\}.$

(f) $\tfrac{1}{4}, \tfrac{1}{2}, \tfrac{1}{4}, -\tfrac{1}{2}$.

6. (a) $\rho^2 \sin \varphi$, (b) $\{(\rho, \varphi, \theta): \rho \sin \varphi = 0\}$,

(d) $\rho = \sqrt{x^2 + y^2 + z^2}$, $\varphi = \cos^{-1}\left(\frac{z}{\sqrt{x^2 + y^2 + z^2}}\right)$, $\theta = \tan^{-1}\left(\frac{y}{x}\right)$,

(e) $\sqrt{2/2}$, $\sqrt{2/2}$, -1.

9. (c) $u = (x + y)^{1/3}$, $v = (x - y)^{1/3}$.

4–3. Functional Dependence

There are some transformations that are unlike any of those considered in the previous section. The following examples illustrate various things that could happen.

Example 4–3.1. Consider the transformation given by

$$x = u + v + 2,$$
$$y = 3u^2 + 6uv + 3v^2 - 5.$$

The Jacobian of the transformation is identically equal to zero for all points (u, v), and thus the hypothesis of Theorem 4-2.1 is satisfied at no point. Notice that

$$x - 2 = u + v,$$
$$y + 5 = 3(u + v)^2.$$

Thus any point (u_0, v_0) on the line $u + v = t_0$ in the (u, v)-plane maps into the same point (x_0, y_0) in the (x, y)-plane (cf. Figure 4–3.1). Thus the whole (u, v)-plane maps into the parabola defined by

$$y + 5 = 3(x - 2)^2. \tag{4–3.1}$$

In this example a two-dimensional set maps onto a one-dimensional set. It is, of course, possible to have the whole (u, v)-plane map onto a single point, a zero-dimensional set. For example, $x = a$, $y = b$ maps the whole (u, v)-plane onto the point (a, b).

Notice the striking parallel that these more general transformations have with linear transformations. In Example 4–3.1, the matrix of the transformation

$$\begin{pmatrix} x_u & x_v \\ y_u & y_v \end{pmatrix}$$

has rank one and in the trivial case, $x = a$, $y = b$, the matrix of the transformation would have rank zero. In the first case the image is one-dimensional,

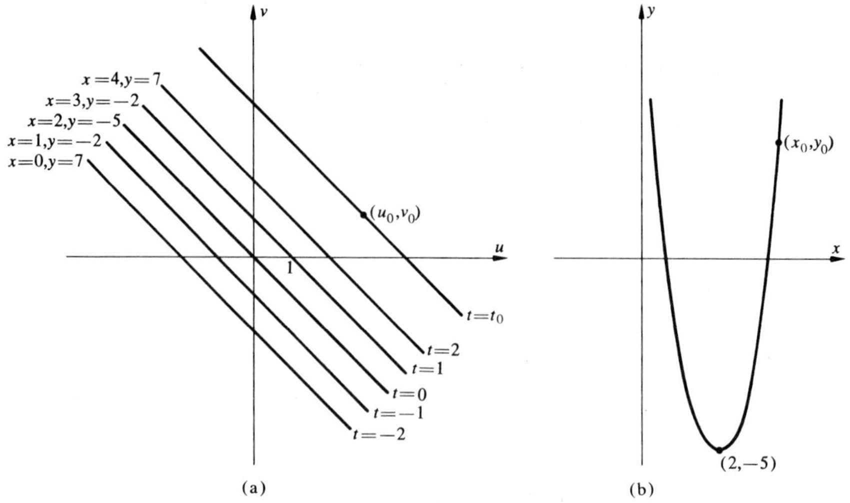

Figure 4–3.1

and in the second case it is zero-dimensional. This corresponds exactly with the relationship between the rank of the matrix and the dimension of the image set for linear transformations. The only difference is that in the case of linear transformations, the image set is always a subspace.

There is another feature of the transformation of Example 4–3.1 that we wish to observe. In Figure 4–3.1(a) the curves (in this case they are lines) where x is constant are plotted. The curves where y is constant are plotted also, but the two families of curves are one and the same. However, different constants give the same curve, as you can see on the figure. If these pairs of values are plotted, they all lie on the parabola of Figure 4–3.1. Thus, there is a relationship between x and y—namely, the relationship defined by (4–3.1). In the correct terminology, "the set of functions $\{f_1, f_2\}$ is functionally dependent on (u, v)-space" where

$$f_1(u, v) = u + v + 2, \qquad f_2(u, v) = 3(u + v)^2 - 5.$$

Before defining functional dependence precisely, let us look at some more examples of functional dependence.

Example 4–3.2. The set of functions $\{f_1, f_2\}$ where

$$f_1(x, y, z) = \sin(x + y - z), \qquad f_2(x, y, z) = \cos(x + y - z)$$

is functionally dependent on (x, y, z)-space, since

$$F(u, v) = u^2 + v^2 - 1 = 0$$

for all (x, y, z) where $u = f_1(x, y, z)$, $v = f_2(x, y, z)$. Notice that for all (x, y, z),

$$\frac{\partial(u, v)}{\partial(x, y)} = \frac{\partial(u, v)}{\partial(x, z)} = \frac{\partial(u, v)}{\partial(y, z)} = 0.$$

Example 4–3.3. Consider the functions f_1 and f_2 where

$$u = f_1(x, y) = \frac{x + y}{|x + y|},$$

$$v = f_2(x, y) = |x - y|.$$

Clearly, $f_1(x, y) = 1$ when $x + y > 0$ and -1 when $x + y < 0$, and $f_1(x, y)$ is not defined when $x + y = 0$. The function f_2 is such that $f_2(x, y) = x - y$ when $x - y \geq 0$ and $-(x - y)$ when $x - y < 0$. Along the line $x - y = 0$, f_2 fails to have its first partial derivatives. On each of the following four regions, the set of functions $\{f_1, f_2\}$ is functionally dependent. The appropriate function, F, for each region is indicated beside each of the regions:

$$S_1 = \{(x, y): x + y > 0, x - y < 0\}, \qquad F(u, v) = u - 1,$$
$$S_2 = \{(x, y): x + y > 0, x - y > 0\}, \qquad F(u, v) = u - 1,$$
$$S_3 = \{(x, y): x + y < 0, x - y > 0\}, \qquad F(u, v) = u + 1,$$
$$S_4 = \{(x, y): x + y < 0, x - y < 0\}, \qquad F(u, v) = u + 1.$$

In each case, $F(f_1(x, y), f_2(x, y)) = 0$ for all (x, y) in the corresponding region. Notice that on each of these four regions

$$\frac{\partial(u, v)}{(\partial x, y)} = 0.$$

We are now ready to proceed with a formal definition of functional dependence.

Definition 4–3.1. *Let $f_1, \cdots, f_m$ be m functions $m > 1$, such that $f_i : S \to R$, $1 \leq i \leq m$, where S is an open set in E^n and $m \leq n$. Suppose that $f_i \in C^1(S)$. If there exists a function F: $S_1 \to R$ where S_1 is an open set in E^m such that $S_1 \supset$ Ra* (**f**) *where* $\mathbf{f} = (f_1, \cdots, f_n)$ *such that $F \in C^1(S_1)$, $\nabla F \neq \mathbf{0}$ at any point of S_1 and such that $F(f_1(\mathbf{x}), \cdots, f_m(\mathbf{x})) = 0$ for all $\mathbf{x} \in S$, then the set of functions $\{f_1, \cdots, f_m\}$ is functionally dependent on S. Otherwise, the set of functions $\{f_1, \cdots, f_m\}$ is functionally independent.*

Direct use of this definition is limited in that we can conclude functional dependence or independence only as we are able to find a function F or demonstrate that it fails to exist. The examples indicate the possibility of the identical vanishing of certain Jacobians on S as being related to functional dependence. The following proposition demonstrates the necessity of certain Jacobians vanishing on S. For simplicity, the proposition shall be stated for the case $m = 2$, $n = 3$.

Proposition 4–3.1. *Let f_1 and f_2 be functions given by*

$$u = f_1(x, y, z), \qquad v = f_2(x, y, z).$$

If $\{f_1, f_2\}$ is functionally dependent on S, then

$$\frac{\partial(u, v)}{\partial(x, y)}, \quad \frac{\partial(u, v)}{\partial(x, z)}, \quad \text{and} \quad \frac{\partial(u, v)}{\partial(y, z)}$$

are identically zero on S.

Proof. Let F be the function such that

$$F(f_1(x, y, z), f_2(x, y, z)) = 0$$

for all $(x, y, z) \in S$. Since we have sufficient conditions to use the chain rule, it follows that:

$$\frac{\partial F}{\partial u}\frac{\partial u}{\partial x} + \frac{\partial F}{\partial v}\frac{\partial v}{\partial x} = 0,$$

$$\frac{\partial F}{\partial u}\frac{\partial u}{\partial y} + \frac{\partial F}{\partial v}\frac{\partial v}{\partial y} = 0,$$

$$\frac{\partial F}{\partial u}\frac{\partial u}{\partial z} + \frac{\partial F}{\partial v}\frac{\partial v}{\partial z} = 0,$$

for any $(x, y, z) \in S$. Taking the equations by pairs, it follows that since

$$\nabla F = \left(\frac{\partial F}{\partial u}, \frac{\partial F}{\partial v}\right) \neq \mathbf{0},$$

the determinant of the linear system must be zero. Thus each of the Jacobians is zero. ▮

The above proposition assures us that certain Jacobians vanish on certain open sets if a set of functions is functionally dependent. Thus, if the appropriate Jacobians are checked, and at least one fails to vanish on a certain open set, then we can conclude that the set of functions is functionally independent. This is a particularly useful result in finding a set of functions that are first integrals of a system of first-order ordinary differential equations.† However, it is interesting and desirable to know if the vanishing of certain Jacobians is a sufficient condition. The following proposition indicates that the answer is a qualified yes. The set of functions is functionally dependent locally. Again for the sake of simplicity the proposition is stated and proved for $n = 3$, $m = 2$.

Proposition 4–3.2. *Let f_1 and f_2 be real functions defined by*

$$u = f_1(x, y, z), \qquad v = f_2(x, y, z).$$

Suppose that both f_1 and f_2 have the same domain, an open set S in (x, y, z)-space. Suppose, furthermore, that $f_i \in C^1(S)$, $i = 1, 2$. Let $(x_0, y_0, z_0) \in S$, and map into (u_0, v_0) by the above functions f_1 and f_2. Suppose that at least one of the first partials of f_1 or of f_2 is not zero at (x_0, y_0, z_0). If the Jacobians

$$\frac{\partial(u, v)}{\partial(x, y)}, \quad \frac{\partial(u, v)}{\partial(x, z)}, \quad \text{and} \quad \frac{\partial(u, v)}{\partial(y, z)}$$

are identically zero on S, then there exists a neighborhood, $N((x_0, y_0, z_0); \delta)$ such that $\{f_1, f_2\}$ is functionally dependent on this neighborhood.

Proof. Suppose that

$$\frac{\partial}{\partial z} f_1(x_0, y_0, z_0) \neq 0,$$

then the function G, defined by

$$G(x, y, z, u) = f_1(x, y, z) - u,$$

satisfies the hypothesis of the implicit function theorem in a neighborhood of (x_0, y_0, z_0, u_0). Notice that

$$\frac{\partial}{\partial z} G(x_0, y_0, z_0, u_0) = \frac{\partial}{\partial z} f_1(x_0, y_0, z_0).$$

Thus there exists $N((x_0, y_0, u_0); \delta_1)$, and there exists a unique function g_1 defined on $N((x_0, y_0, u_0); \delta_1)$ such that

$$\begin{aligned} g_1(x_0, y_0, u_0) &= z_0, \\ G(x, y, g_1(x, y, u), u) &= f_1(x, y, g_1(x, y, u)) - u \\ &= 0 \end{aligned}$$

† Cf. G. F. D. Duff, *Partial Differential Equations*, pp. 26 ff.

for all $(x, y, u) \in N((x_0, y_0, u_0); \delta_1)$, and at $(x, y, u) \in N((x_0, y_0, u_0); \delta_1)$,

$$\left.\frac{\partial g_1}{\partial x}\right)_{y,u} = -\frac{\left.\frac{\partial G}{\partial x}\right)_{x,z,u}}{\left.\frac{\partial G}{\partial z}\right)_{x,y,u}} = -\frac{\left.\frac{\partial f_1}{\partial x}\right)_{y,z}}{\left.\frac{\partial f_1}{\partial z}\right)_{x,y}}, \tag{4–3.2}$$

$$\left.\frac{\partial g_1}{\partial y}\right)_{x,u} = -\frac{\left.\frac{\partial G}{\partial y}\right)_{x,z,u}}{\left.\frac{\partial G}{\partial z}\right)_{x,y,u}} = -\frac{\left.\frac{\partial f_1}{\partial y}\right)_{x,z}}{\left.\frac{\partial f_1}{\partial z}\right)_{x,y}}, \tag{4–3.3}$$

$$\left.\frac{\partial g_1}{\partial u}\right)_{x,y} = -\frac{\left.\frac{\partial G}{\partial u}\right)_{x,y,z}}{\left.\frac{\partial G}{\partial z}\right)_{x,y,u}} = -\frac{(-1)}{\left.\frac{\partial f_1}{\partial z}\right)_{x,y}} \tag{4–3.4}$$

where $z = g_1(x, y, u)$ and $g_1 \in C^1(N((x_0, y_0, u_0); \delta_1))$.

Since g_1 is continuous at (x_0, y_0, u_0), then, given $\varepsilon > 0$ such that $N((x_0, y_0, z_0); \varepsilon) \subset S$, there exists δ_1', where $0 < \delta_1' \leq \delta_1$, such that if $(x, y, u) \in N((x_0, y_0, u_0); \delta_1')$, then $g_1(x, y, u) \in N((x_0, y_0, z_0); \varepsilon)$. Therefore, the function ψ defined by

$$\psi(x, y, u) = f_2(x, y, g_1(x, y, u))$$

satisfies the conditions for using the chain rule in $N((x_0, y_0, u_0); \delta_1')$. Thus if $(x, y, u) \in N((x_0, y_0, z_0); \delta_1')$,

$$\left.\frac{\partial \psi}{\partial x}\right)_{y,u} = \left.\frac{\partial f_2}{\partial x}\right)_{y,z} + \left.\frac{\partial f_2}{\partial z}\right)_{x,y}\left.\frac{\partial g_1}{\partial x}\right)_{y,u}, \tag{4–3.5}$$

$$\left.\frac{\partial \psi}{\partial y}\right)_{x,u} = \left.\frac{\partial f_2}{\partial y}\right)_{x,z} + \left.\frac{\partial f_2}{\partial z}\right)_{x,y}\left.\frac{\partial g_1}{\partial y}\right)_{x,u}, \tag{4–3.6}$$

and

$$\left.\frac{\partial \psi}{\partial u}\right)_{x,y} = \left.\frac{\partial f_2}{\partial z}\right)_{x,y}\left.\frac{\partial g_1}{\partial u}\right)_{x,y}. \tag{4–3.7}$$

Substituting from (4–3.2) and (4–3.3) into (4–3.3) and (4–3.4), respectively, gives

$$\frac{\partial \psi}{\partial x} = \frac{\frac{\partial f_2}{\partial x}\frac{\partial f_1}{\partial z} - \frac{\partial f_2}{\partial z}\frac{\partial f_1}{\partial x}}{\frac{\partial f_1}{\partial z}} = -\frac{1}{\frac{\partial f_1}{\partial z}}\frac{\partial(f_1, f_2)}{\partial(x, z)},$$

$$\frac{\partial \psi}{\partial y} = \frac{\frac{\partial f_2}{\partial y}\frac{\partial f_1}{\partial z} - \frac{\partial f_2}{\partial z}\frac{\partial f_1}{\partial y}}{\frac{\partial f_1}{\partial z}} = -\frac{1}{\frac{\partial f_1}{\partial z}}\frac{\partial(f_1, f_2)}{\partial(y, z)}.$$

By the hypothesis, both Jacobians on the right are zero, and hence ψ is a function only of u in $N((x_0, y_0, z_0); \delta_1')$.

Let H be a function defined by

$$H(u) = f_2(x, y, g_1(x, y, u)),$$

where $u \in N(u_0; \delta_1')$. Then

$$H'(u) = \left.\frac{\partial f_2}{\partial z}\right)_{x,y} \left.\frac{\partial g_1}{\partial u}\right)_{x,y}. \tag{4–3.8}$$

Substituting (4–3.4) into (4–3.8) gives

$$H'(u) = \frac{\left.\dfrac{\partial f_2}{\partial z}\right)_{x,y}}{\left.\dfrac{\partial f_1}{\partial z}\right)_{x,y}} \tag{4–3.9}$$

for $u \in N(u_0; \delta_1')$, and hence $H \in C^1(N(u_0; \delta_1'))$.

Let F be a function defined by $F(u, v) = H(u) - v$ where (u, v) is any point such that $u \in N(u_0; \delta_1')$; that is, the domain of f is an open set S_1 in (u, v)-space. On S_1, F has continuous first partials since $H \in C^1(N(u_0; \delta_1'))$. Furthermore, $\nabla F = (H', -1) \neq \mathbf{0}$ on S_1. Since f_1 is continuous at (x_0, y_0, z_0), there exists δ such that if

$$(x, y, z) \in N((x_0, y_0, z_0); \delta),$$

then

$$u = f_1(x, y, z) \in N(u_0; \delta_1').$$

Thus for $(x, y, z) \in N((x_0, y_0, z_0); \delta)$,

$$\begin{aligned} F(f_1(x, y, z), f_2(x, y, z)) &= H(f_1(x, y, z)) - f_2(x, y, z) \\ &= f_2(x, y, g_1(x, y, f_1(x, y, z))) - f_2(x, y, z) \\ &= f_2(x, y, z) - f_2(x, y, z) \\ &= 0 \end{aligned}$$

and hence $\{f_1, f_2\}$ is functionally dependent on $N((x_0, y_0, z_0); \delta)$. ▮

The following example demonstrates the use of the above proposition.

Example 4–3.4. Let the functions f_1 and f_2 be defined as follows:

$$u = f_1(x, y, z) = \frac{y}{x + z}, \tag{4–3.10}$$

$$v = f_2(x, y, z) = \frac{x - y + z}{x + y + z}. \tag{4–3.11}$$

In this case, $S = (\{(x, y, z): x + y + z = 0\} \cup \{(x, y, z): x + z = 0\})^c$. Then

$$\frac{\partial(u, v)}{\partial(x, y)} = \begin{vmatrix} -\dfrac{y}{(x+z)^2} & \dfrac{1}{x+z} \\ \dfrac{2y}{(x+y+z)^2} & \dfrac{-2(x+z)}{(x+y+z)^2} \end{vmatrix} = 0$$

for all $(x, y, z) \in S$. Similarly,

$$\frac{\partial(u, v)}{\partial(x, z)} = \frac{\partial(u, v)}{\partial(y, z)} = 0$$

for all $(x, y, z) \in S$. Since

$$\frac{\partial u}{\partial y} = \frac{1}{x+z},$$

then $\partial u/\partial y \neq 0$ for any point $(x_0, y_0, z_0) \in S$. Thus there exists a neighborhood of (x_0, y_0, z_0) on which $\{f_1, f_2\}$ is functionally dependent. In order to find the function F try the technique used in the proof. Since $\partial f_1/\partial y \neq 0$, solve for y in (4–3.10):

$$y = (x + z)u.$$

Substitute into (4–3.11);

$$v = \frac{1-u}{1+u}.$$

Thus

$$H(u) = \frac{1-u}{1+u}, \qquad F(u, v) = \frac{1-u}{1+u} - v.$$

Notice that F is not unique in that $F_1(u, v) = 1 - u - v(1 + u)$ works equally as well.

Problems

1. Given sufficient conditions on the coefficients for the sets of functions defined below to be functionally dependent.
   (a) $f(x, y) = a_1 x + b_1 y + c_1, \qquad g(x, y) = a_2 x + b_2 y + c_2.$
   (b) $f(x, y, z) = a_1 x + b_1 y + c_1 z + d_1,$
   $$g(x, y, z) = a_2 x + b_2 y + c_2 z + d_2.$$
   (c) $f_i(x, y, z) = a_i x + b_i y + c_i z + d_i, \qquad i = 1, 2, 3.$
   Give a geometrical interpretation to each of the above circumstances.
2. Test the sets of functions defined for functional dependence. When they are functionally dependent in a neighborhood of the indicated point, find a function F that demonstrates this functional dependence.

(a) $f(x, y) = \tan^{1/3}(x + y^2)$, $g(x, y) = \sec(x + y^2)$, $(x_0, y_0) = (0, 0)$.

(b) $f(x, y) = y + \log\cos x$, $g(x, y) = e^y \cos x$, $(x_0, y_0) = (0, 1)$.

(c) $f(x, y, z) = \dfrac{1 + x + y + z}{1 + x + y - z}$, $g(x, y, z) = \dfrac{z}{1 + x + y}$, $(1, 1, -1)$.

(d) $f(x, y, z, t) = \dfrac{tx}{yz} + \dfrac{yz}{tx}$, $g(x, y, z, t) = \dfrac{tx}{yz} - \dfrac{yz}{tx}$, $(1, 1, 1, 1)$.

(e) $f_1(x, y, z) = 9x^2 - 9y^2 + 6xz - 6yz + 12z^2$,
$f_2(x, y, z) = x - y + 2z$,
$f_3(x, y, z) = x + 2y - z$, $(1, 0, 1)$.

(f) $f_1(x, y, z) = \dfrac{y}{x + 2z}$,

$f_2(x, y, z) = \dfrac{z}{y + 2x}$,

$f_3(x, y, z) = \dfrac{x}{z + 2y}$, $(1, 1, 1)$.

3. State and prove Proposition 4–3.2 for the case $n = 3$, $m = 3$.

4. A special case of Proposition 4–3.2 would be $n \geq 1$ and $m = 1$. Notice that this case is ruled out in Definition 4–3.1. The hypothesis for this circumstance could be as follows: Let f be a function such that $f: S \to R$ where S is an open region in E^n. Let $f \in C^1(S)$. If

$$\nabla f = \left(\frac{\partial f}{\partial x_1}, \ldots, \frac{\partial f}{\partial x_n}\right) = \mathbf{0}$$

at points $\mathbf{x} \in S$, then $f(\mathbf{x}) = c$, a constant for all $\mathbf{x} \in S$.

Note that since it is an open region, S is polygonally connected. Show by using the directional derivative that if $\mathbf{x}_0$ and $\mathbf{x}_1$ are two points in S such that the points between them are in S, then $f(\mathbf{x}_1) = f(\mathbf{x}_0)$. Extend this result to show that $f(\mathbf{x}) = f(\mathbf{x}_0)$ for all $\mathbf{x} \in S$.

5. Use the result of Problem 4 to show that Proposition 4–3.2 is trivially true even in the general case of m and n arbitrary, $1 < m \leq n$, if all first partials are identically zero in S where S is an open region in E^n.

Answers

1. (a) $\begin{vmatrix} a_1 & b_1 \\ a_2 & b_2 \end{vmatrix} = 0.$ (b) $\begin{vmatrix} a_1 & b_1 \\ a_2 & b_2 \end{vmatrix} = \begin{vmatrix} a_1 & c_1 \\ a_2 & c_2 \end{vmatrix} = \begin{vmatrix} b_1 & c_1 \\ b_2 & c_2 \end{vmatrix} = 0.$

(c) $\begin{vmatrix} a_1 & b_1 & c_1 \\ a_2 & b_2 & c_2 \\ a_3 & b_3 & c_3 \end{vmatrix} = 0.$

2. (a) Yes, $F(u, v) = u^6 - v^2 + 1 = 0$.
   (b) Yes, $F(u, v) = \log v - u = 0$.
   (c) Yes, $F(u, v) = uv - u + v + 1 = 0$.
   (d) Yes, $F(u, v) = u^2 - v^2 - 4 = 0$.
   (e) No.
   (f) Yes, $F(u, v, w) = 9\,uvw + 2uv + 2uw + 2vw - 1 = 0$.

4–4. Maxima and Minima Theory

There are many problems in which a maximum or a minimum value of a certain physical quantity is of interest. In some cases the quantity is a function of two variables—for example, the elevation or height above sea level at a point (x, y) where x and y are the coordinates of a point of a certain piece of land relative to a coordinate system. In this case if $h(x, y)$ represents the height at (x, y), then we have a function h defined in a very natural way. Usually there is a certain set on which we wish to conduct our investigation of maxima and minima. For instance, in this example, we may wish to find the highest point and the lowest point on a given piece of land. Thus it would seem that a natural domain for our functions in general would be a region.

It is necessary first to make some definitions. The definitions will be stated for a function of n variables, since it is just as easy to do so.

Definition 4–4.1. *Let $f: S \to R$ where S is a subset of E^n. The function f has an absolute maximum (absolute minimum) on S at $\mathbf{x}_0 \in S$ if $f(\mathbf{x}_0) \geq f(\mathbf{x})$ $(f(\mathbf{x}_0) \leq f(\mathbf{x}))$ for all $\mathbf{x} \in S$. The function f has a relative maximum (relative minimum) on S at $\mathbf{x}_0 \in S$ if there exists a neighborhood, $N(\mathbf{x}_0\,;\,\delta)$, such that $f(\mathbf{x}_0) \geq f(\mathbf{x})$ $(f(\mathbf{x}_0) \leq f(\mathbf{x}))$ for all $\mathbf{x} \in S \cap N(\mathbf{x}_0\,;\,\delta)$. A function f has a relative extremum on S at $\mathbf{x}_0 \in S$ if f has a relative maximum or a relative minimum at $\mathbf{x}_0 \in S$.*

A function f, even a continuous function, may have many points of S such that f has an absolute maximum at each of them. For example, if $f(x, y) = c$, a constant, on a set S, then f has an absolute maximum at each point of S. Thus there is not necessarily a unique point where f has an absolute maximum.

The continuous function f defined by

$$f(x, y) = x + y$$

has no absolute maximum on S_1 if $S_1 = \{(x, y): x^2 + y^2 < 1\}$, a bounded set. The same function has no absolute maximum on $S_2 = \{(x, y): x \geq 0, y \geq 0\}$, a closed set. However, if the set is $S_3 = \overline{S_1}$, a closed and bounded set, or a compact set, then f has an absolute maximum at $(1/\sqrt{2}, 1/\sqrt{2})$. Figure 4–4.1(a), (b), and (c) illustrate the level curves of f on the three sets S_1, S_2, and S_3, respectively.

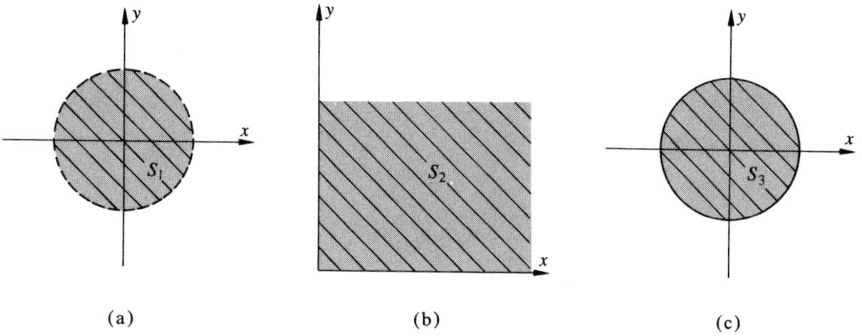

Figure 4–4.1

This function, as considered in these three circumstances, leads us to suspect the following proposition.

Proposition 4–4.1. *Let f be a function such that $f: S \to R$ where S is a compact subset of E^n. If f is continuous on S, there exists a point $\mathbf{x}_0 \in S$ ($\mathbf{x}_0' \in S$) such that f has an absolute maximum (absolute minimum) on S at $\mathbf{x}_0$ (at $\mathbf{x}_0'$).*

The proof of this proposition, which gives us sufficient conditions for the existence of an absolute maximum and an absolute minimum, is sketched in Problem 9 of this section. A continuous function can have an absolute maximum and an absolute minimum on a set that is not compact, as the following example shows.

Example 4–4.1. Consider the function f defined by

$$f(x, y) = \begin{cases} \dfrac{2x^2}{(1 + x^2)\sqrt{x^2 + y^2}}, & (x, y) \neq (0, 0), \\ 0, & (x, y) = 0. \end{cases}$$

On either the set $\{(x, y): x^2 + y^2 < 4\}$, which is bounded but not closed, or on the set consisting of the whole (x, y)-plane, which is closed but not

(f) (2, 6), rel. min.; (0, 0), saddle point; (−2, −6), rel. min.;

(g) $\left(\frac{-\sqrt{3}}{2}, \frac{\sqrt{3}}{6}\right)$, rel. max., $\left(\frac{\sqrt{3}}{2}, \frac{-\sqrt{3}}{6}\right)$, rel. min., $\left(\frac{-\sqrt{3}}{6}, \frac{\sqrt{3}}{2}\right)$ and $\left(\frac{\sqrt{3}}{6}, \frac{-\sqrt{3}}{2}\right)$, saddle points,

(h) (1, −2), rel. max., (0, −1), (0, −3), (2, −1), and (2, −3), saddle points.

3. (a) Abs. max. of 12 at (3, 0), abs. min. of −5 at (1, 1).
   (b) Abs. max. of 4 at (1, −1), abs. min. of −71 at (−2, 3).
   (c) Abs. max. at $(\sqrt{(16 - \sqrt{112})/8}, (4 + \sqrt{112})/8)$,
   Abs. min. of $7 - \sqrt{2}$ at $(\sqrt{2}, -\sqrt{2})$.
   (d) Abs. max. of $11\sqrt{22} + 27 \approx 78.59$ at $(-\sqrt{11/2}, 3 - \sqrt{11/2})$,
   Abs. min. of $-7\sqrt{14} - 27 \approx -53.18$ at $(-\sqrt{7/2}, -3 + \sqrt{7/2})$.

4. (a) (0, 0), saddle point, (b) (1, −1), rel. max.,
   (c) (0, 0), saddle point, (d) (0, 0), rel. min.

5. $x = y = \left(\frac{V_0}{2}\right)^{(1/3)}$, $z = (4V_0)^{(1/3)}$, i.e. $z = 2x = 2y$.

6. $x = y = z = V_0^{(1/3)}$.

7. $\dfrac{32\sqrt{3}}{9}$.

4–5. Extensions of Extrema Theory

The sufficiency conditions of Proposition 4–4.3 for classifying the critical points of a function f on a region S were limited to functions of two variables. It is the object of this section to discuss sufficient conditions for classifying critical points of functions having more than two variables. In order to see how the general sufficiency conditions are a natural generalization of the sufficiency conditions for $n = 2$, it will be necessary to look at Proposition 4–4.3 from another viewpoint.

Since $f \in C^2(S^0)$, then at the critical point (x_0, y_0) there is always the following symmetric matrix:

$$\begin{pmatrix} f_{xx}(x_0, y_0) & f_{xy}(x_0, y_0) \\ f_{yx}(x_0, y_0) & f_{yy}(x_0, y_0) \end{pmatrix}.$$

This matrix determines a quadratic form,

$$Q(x, y) = (x - x_0 \quad y - y_0)\begin{pmatrix} f_{xx}(x_0, y_0) & f_{xy}(x_0, y_0) \\ f_{yx}(x_0, y_0) & f_{yy}(x_0, y_0) \end{pmatrix}\begin{pmatrix} x - x_0 \\ y - y_0 \end{pmatrix} \qquad (4\text{–}5.1)$$

which, when multiplied out, becomes

$$Q(x, y) = f_{xx}(x_0, y_0)(x - x_0)^2 + 2f_{xy}(x_0, y_0)(x - x_0)(y - y_0) + f_{yy}(x_0, y_0)(y - y_0)^2.$$

Notice that this form of $Q(x, y)$ closely resembles the bracket term of (4–4.3). *The difference is in the point where the partials are evaluated.*

In general, a real quadratic form Q on E^n has the form

$$Q(\mathbf{x}) = (x_1 \cdots x_n)(a_{ij})\begin{pmatrix} x_1 \\ \cdot \\ \cdot \\ \cdot \\ x_n \end{pmatrix} = (\mathbf{x})^T A(\mathbf{x}) \tag{4–5.2}$$

where $A = (a_{ij})$ is a real symmetric $n \times n$ matrix and $(\mathbf{x})^T$ is the transpose of $(\mathbf{x})$. If we multiply out the matrices, we obtain

$$Q(\mathbf{x}) = \textstyle\sum_{i,j=1}^{n} a_{ij} x_i x_j.$$

The quadratic form Q of (4.5.2) is *positive* (*negative*) definite if $Q(\mathbf{x}) > 0$ (<0) for all $\mathbf{x} \neq \mathbf{0}$. If Q is not positive (negative) definite, but $Q(\mathbf{x}) \geq 0$ (≤ 0) for all $\mathbf{x} \neq \mathbf{0}$, then Q is *positive* (*negative*) *semidefinite*. Otherwise, Q is an *indefinite* quadratic form.

There is a very simple test known as Gundelfinger's rule that can be used to determine whether Q of (4–5.2) is positive definite or not.† Namely, the quadratic form Q is positive definite if and only if the n subdeterminants of A,

$$|a_{11}|, \quad \begin{vmatrix} a_{11} & a_{12} \\ a_{21} & a_{22} \end{vmatrix}, \quad \begin{vmatrix} a_{11} & a_{12} & a_{13} \\ a_{21} & a_{22} & a_{23} \\ a_{31} & a_{32} & a_{33} \end{vmatrix}, \ldots, \begin{vmatrix} a_{11} & \cdots & a_{1n} \\ \cdots & \cdots & \cdots \\ a_{n1} & \cdots & a_{nn} \end{vmatrix}, \tag{4–5.3}$$

are all *strictly* positive. Since Q is negative definite if and only if $-Q$ is positive definite, it follows that Q is negative definite if and only if all of the above odd order subdeterminants are strictly negative and all of the even order subdeterminants are strictly positive.

Upon this basis we can see that Proposition 4–4.3 says that f has a relative minimum (relative maximum) at (x_0, y_0), if the quadratic form of (4–5.1) is positive definite (negative definite). Although it is not now apparent, it is, furthermore, true that f has a saddle point if the quadratic form of 4–5.1 is indefinite. The critical point may be of any type when the quadratic form

† See Franz E. Hohn, *Elementary Matrix Algebra*, Theorem 9.15.3, p. 260.

of (4–5.1) is semidefinite, either positive or negative. Thus it appears that sufficient conditions for classifying critical points can be stated in terms of the type of quadratic form that (4–5.1) happens to be. This is exactly how the extension of the theory to functions of more than two variables is made.

We have already mentioned tests for checking positive and negative definite quadratic forms for n variables—that is, Gundelfinger's rule—but this leaves open the question of further classification when these tests fail. *Any real symmetric quadratic form may be classified by transforming it to a canonical form.* A method for changing any quadratic form to canonical form is illustrated below for $n = 3$.

If the element $a_{11} \neq 0$ in the A matrix, then take

$$P_1 = \begin{pmatrix} 1 & -\dfrac{a_{12}}{a_{11}} & -\dfrac{a_{13}}{a_{11}} \\ 0 & 1 & 0 \\ 0 & 0 & 1 \end{pmatrix}. \tag{4–5.4}$$

Then

$$P_1^T A P_1 = \begin{pmatrix} a_{11} & 0 & 0 \\ 0 & b_{11} & b_{12} \\ 0 & b_{21} & b_{22} \end{pmatrix} \tag{4–5.5}$$

where

$$b_{11} = \frac{a_{11}\, a_{22} - (a_{12})^2}{a_{11}},$$

$$b_{12} = b_{21} = \frac{a_{11}\, a_{23} - a_{12}\, a_{13}}{a_{11}},$$

and

$$b_{22} = \frac{a_{11}\, a_{23} - (a_{13})^2}{a_{11}}.$$

If $b_{11} \neq 0$, then take

$$P_2 = \begin{pmatrix} 1 & 0 & 0 \\ 0 & 1 & -\dfrac{-b_{12}}{b_{11}} \\ 0 & 0 & 1 \end{pmatrix}.$$

Then

$$P_2^T P_1^T A P_1 P_2 = \begin{pmatrix} a_{11} & 0 & 0 \\ 0 & b_{11} & 0 \\ 0 & 0 & c_{11} \end{pmatrix} \tag{4–5.6}$$

where $c_{11} = [b_{11}\, b_{22} - (b_{12})^2]/b_{11}$. If $c_{11} \neq 0$, then take

$$P_3 = \begin{pmatrix} \dfrac{1}{\sqrt{|a_{11}|}} & 0 & 0 \\ 0 & \dfrac{1}{\sqrt{|b_{11}|}} & 0 \\ 0 & 0 & \dfrac{1}{\sqrt{|c_{11}|}} \end{pmatrix}. \tag{4–5.7}$$

If $P = P_1\, P_2\, P_3$, then, since $P_3{}^T P_2{}^T P_1{}^T = (P_1 \;\; P_2 \;\; P_3)^T$,

$$P^T AP = \begin{pmatrix} \varepsilon_1 & 0 & 0 \\ 0 & \varepsilon_2 & 0 \\ 0 & 0 & \varepsilon_3 \end{pmatrix} \tag{4–5.8}$$

where $\varepsilon_i = \pm 1$, $i = 1, 2, 3$.

If $a_{11} = 0$, but a_{22} or $a_{33} \neq 0$, then take P_1 as follows if $a_{22} \neq 0$:

$$P_1 = \begin{pmatrix} 0 & 0 & 0 \\ \dfrac{-a_{21}}{a_{22}} & 1 & \dfrac{-a_{23}}{a_{22}} \\ 0 & 0 & 1 \end{pmatrix}.$$

A similar adjustment may be necessary for P_2 if $b_{11} = 0$ and $b_{22} \neq 0$.

If $a_{11} = a_{22} = a_{33} = 0$, then we have a trivial quadratic form unless a_{12}, a_{13}, or $a_{23} \neq 0$. If all of the diagonal elements of A are zero and if for example $a_{12} \neq 0$, then take

$$P_1 = \begin{pmatrix} 1 & -1 & 0 \\ 1 & 1 & 0 \\ 0 & 0 & 1 \end{pmatrix}, \tag{4–5.9}$$

that is, -1 is in the position of the nonzero element and $+1$ is in the symmetric position. By so doing

$$P_1^T AP_1 = \begin{pmatrix} 2a_{12} & 0 & a_{13} + a_{23} \\ 0 & -2a_{12} & -a_{13} + a_{23} \\ a_{13} + a_{23} & -a_{13} + a_{23} & 0 \end{pmatrix},$$

a matrix with some nonzero elements on the diagonal. We can now proceed as before. It may be that after obtaining the matrix of (4–5.5) we have $b_{11} = b_{12} = 0$, but $b_{12} \neq 0$. In such a case it would be necessary to multiply by a matrix like (4–5.9) to get nonzero elements back on the diagonal before proceeding.

Once we have reduced the matrix to the form of (4–5.6), then the last step is to use a matrix similar to the matrix of (4–5.7), in which all nondiagonal elements are zero, the first diagonal element is $1/\sqrt{|a_{11}|}$ if $a_{11} \neq 0$ and 1 otherwise, and similarly for the second and third diagonal elements.

Thus in this method we successively find matrices $P_1, P_2, \cdots, P_s$ such that if $P = P_1 P_2 \cdots P_s$, then $P^T AP = \text{diag}\,(\varepsilon_1, \varepsilon_2, \cdots, \varepsilon_n)$ where $\varepsilon_i = +1$, -1, or 0, $1 \leq i \leq n$. If $\mathbf{x} = P\mathbf{y}$ and we substitute into (4–5.2), we get

$$Q(P\mathbf{y}) = (P\mathbf{y})^T A(P\mathbf{y}) = (\mathbf{y})^T P^T AP(\mathbf{y})$$

or

$$Q(P\mathbf{y}) = \varepsilon_1 {y_1}^2 + \cdots + \varepsilon_n {y_n}^2,$$

a canonical form of the quadratic form. Table 4–5.1 summarizes the relationship between the type of quadratic form and the coefficients ε_i of its canonical

Type of quadratic form	*Coefficients ε_i of a canonical form*
positive definite	$\varepsilon_i = 1$ for all i.
negative definite	$\varepsilon_i = -1$ for all i.
positive semi-definite	$\varepsilon_i = 1$ or 0 for all i, and 1 and 0 each appear at least once.
negative semi-definite	$\varepsilon_i = -1$ or 0 for all i, and -1 and 0 each appear at least once
indefinite	$\varepsilon_i = 1$ for at least one i, and $\epsilon_i = -1$ for at least one i.

Table 4–5.1

forms. By Sylvester's law of inertia the number of plus ones, minus ones, and zeros is the same in any canonical form for a given quadratic form.†

The following proposition will give us sufficient conditions for classifying critical points of functions of two or more variables. It is thus an extension of Proposition 4–4.3.

Proposition 4–5.1. *Let S be a region in E^n, $n \geq 2$, and let f be a function such that f: $S \to R$. Let $f \in C^2(S^0)$ and let $\mathbf{x}_0$, a point of S^0, be a critical point of f on S. Let Q be the quadratic form defined by*

$$Q(\mathbf{x}) = (\mathbf{x} - \mathbf{x}_0)^T \left(\frac{\partial^2}{\partial x_i \, \partial x_j} f(\mathbf{x}_0) \right) (\mathbf{x} - \mathbf{x}_0). \tag{4–5.10}$$

† Cf. Hohn, pp. 249–50.

If Q is positive definite, f has a relative minimum at $\mathbf{x}_0$ on S; if Q is negative definite, f has a relative maximum at $\mathbf{x}_0$ on S; if Q is indefinite, f has a saddle point at $\mathbf{x}_0$ on S, and if Q is positive or negative semi-definite, the test fails.

Proof. By Taylor's theorem (Theorem 4–1.2), if $N(\mathbf{x}_0\,;\delta) \subset S$, then it follows that for $\mathbf{x} \in N'(\mathbf{x}_0\,;\delta)$,

$$f(\mathbf{x}) - f(\mathbf{x}_0) = \frac{1}{2!}\left[\sum_{j=1}^{n}(x_j - x_{0j})\frac{\partial}{\partial x_j}\right]^2 f(\mathbf{x}_\theta)$$

$$= \frac{1}{2}\sum_{i,j=1}^{n}\frac{\partial^2}{\partial x_i\,\partial x_j}f(\mathbf{x}_\theta)(x_i - x_{0i})(x_j - x_{0j})$$

where $\mathbf{x}_\theta = \mathbf{x}_0 + \theta(\mathbf{x} - \mathbf{x}_0)$, $0 < \theta < 1$. Hence when $\mathbf{x} \in N(\mathbf{x}_0\,;\delta)$,

$$\Delta f(\mathbf{x}_0\,;\mathbf{x} - \mathbf{x}_0) = \frac{1}{2}\sum_{i,j=1}^{n}\frac{\partial^2}{\partial x_i\,\partial x_j}f(\mathbf{x}_\theta)(x_i - x_{0i})(x_j - x_{0j}) \quad (4\text{–}5.11)$$

if $\mathbf{x} \neq \mathbf{x}_0$ and zero if $\mathbf{x} = \mathbf{x}_0$.

If the quadratic form of (4–5.10) is positive definite, then the n principal minor determinants of (4–5.3) are all strictly positive. Since $f \in C^2(S^0)$, there is a neighborhood of $\mathbf{x}_0$, $N(\mathbf{x}_0\,;\delta_1)$, $0 < \delta_1 \leq \delta$, such that if $\mathbf{x} \in N(\mathbf{x}_0\,;\delta_1)$ the corresponding n principal minor determinants of

$$\left(\frac{\partial^2}{\partial x_i\,\partial x_j}f(\mathbf{x})\right)$$

are strictly positive. Since $N(\mathbf{x}_0\,;\delta_1)$ is convex, $\mathbf{x}_\theta \in N'(\mathbf{x}_0'\,;\delta_1)$ when $\mathbf{x} \in N'(\mathbf{x}_0\,;\delta_1)$, and hence the same principal minor determinants of

$$\left(\frac{\partial^2}{\partial x_i\,\partial x_j}f(\mathbf{x}_\theta)\right)$$

are also all strictly positive.

Upon this information we wish to show f has a relative minimum at $\mathbf{x}_0$. That is, we will show that

$$\Delta f(\mathbf{x}_0;\mathbf{x} - \mathbf{x}_0) = \sum_{i,j=1}^{n}\frac{\partial^2}{\partial x_i\,\partial x_j}f(\mathbf{x}_\theta)(x_i - x_{0i})(x_j - x_{0j})$$

$$= (\mathbf{x} - \mathbf{x}_0)^T\left(\frac{\partial^2}{\partial x_i\,\partial x_j}f(\mathbf{x}_\theta)\right)(\mathbf{x} - \mathbf{x}_0)$$

is non-negative for any $\mathbf{x} \in N'(\mathbf{x}_0\,;\delta_1)$.

Let $\mathbf{x}^*$ be an arbitrary temporary fixed point of $N'(\mathbf{x}_0\,;\delta_1)$. Then

$$\Delta f(\mathbf{x}_0;\mathbf{x}^* - \mathbf{x}_0) = (\mathbf{x}^* - \mathbf{x}_0)^T\left(\frac{\partial^2}{\partial x_i\,\partial x_j}f(\mathbf{x}_\theta{}^*)\right)(\mathbf{x}^* - \mathbf{x}_0). \quad (4\text{–}5.12)$$

With $\mathbf{x}^*$ fixed, then

$$\frac{\partial^2}{\partial x_i\, \partial x_j} f(\mathbf{x}_\theta^*), \quad 1 \le i, \quad j \le n,$$

are constants, and we can consider the following quadratic form:

$$Q_{\mathbf{x}_\theta}{}^*(\mathbf{y}) = (\mathbf{y})^T \left(\frac{\partial^2}{\partial x_i\, \partial x_j} f(\mathbf{x}_\theta^*)\right)(\mathbf{y}). \tag{4–5.13}$$

Since the n principal minor determinants in the upper left of

$$\left(\frac{\partial^2}{\partial x_i\, \partial x_j} f(\mathbf{x}_\theta^*)\right)$$

are strictly positive, if follows by Gundelfinger's rule that $Q_{\mathbf{x}_\theta}*$ is positive definite and, therefore, $Q_{\mathbf{x}_\theta}*(\mathbf{y}) > 0$ when $\mathbf{y} \neq \mathbf{0}$. In particular, take $\mathbf{y} = \mathbf{x}^* - \mathbf{x}_0 \neq 0$, and hence,

$$\Delta f(\mathbf{x}_0\,;\, \mathbf{x}^* - \mathbf{x}_0) = Q_{\mathbf{x}_\theta}{}^*(\mathbf{y}) > 0.$$

Thus $\Delta f(\mathbf{x}; \mathbf{x} - \mathbf{x}_0) \geq 0$ in $N(\mathbf{x}_0\,;\, \delta_1)$, and f has a relative minimum at $\mathbf{x}_0$.

If Q of (4–5.10) is negative definite, then $-Q$ would be positive definite. If in (4–5.10) $-f$ replaces f, the resulting quadratic form would be this same $-Q$. Thus by the above argument $-f$ has a relative minimum at $\mathbf{x}_0$. Consequently, f has a relative maximum at $\mathbf{x}_0$.

If Q of (4–5.10) is an indefinite quadratic form, then there exists $\mathbf{x}'(\mathbf{x}'')$ different from $\mathbf{x}_0$ such that $Q(\mathbf{x}') > 0$ $(Q(\mathbf{x}'')<0)$. Let

$$t\boldsymbol{\lambda} = \mathbf{x} - \mathbf{x}_0$$

where $\mathbf{x} \in N(\mathbf{x}_0\,;\, \delta)$ and $\lambda_1{}^2 + \cdots + \lambda_n{}^2 > 0$. Substituting into (4–5.11) gives

$$\Delta f(\mathbf{x}_0\,;\, t\boldsymbol{\lambda}) = \tfrac{1}{2}\,(t^2) \textstyle\sum_{i,j=1}^{n} \dfrac{\partial^2}{\partial x_i\, \partial x_j} f(\mathbf{x}_t)\lambda_i\, \lambda_j \tag{4–5.14}$$

where $\mathbf{x}_t = \mathbf{x}_0 + \theta(t\boldsymbol{\lambda})$ when $t \neq 0$ and where $\mathbf{x}_t = \mathbf{x}_0$ when $t = 0$. Take $\boldsymbol{\lambda}' = \mathbf{x}' - \mathbf{x}_0$. Then it follows that

$$M = \textstyle\sum_{i,j=1}^{n} \dfrac{\partial^2}{\partial x_i\, \partial x_j} f(\mathbf{x}_0)\lambda_i{}'\lambda_j{}' > 0.$$

Since $f \in C^2(N(\mathbf{x}_0\,;\, \delta))$, it follows that

$$\lim_{t\to 0} \textstyle\sum_{i,j=1}^{n} \dfrac{\partial^2}{\partial x_i\, \partial x_j} (\mathbf{x}_t)\lambda_i{}'\lambda_j{}' = M.$$

Thus there exists $\eta > 0$ such that

$$\textstyle\sum_{i,j=1}^{n} \dfrac{\partial^2}{\partial x_i\, \partial x_j} f(\mathbf{x}_t)\lambda_i{}'\lambda_j{}' \geq \dfrac{M}{2} > 0$$

for $|t| < \eta$. Therefore, $\Delta f(\mathbf{x}_0\,;t\boldsymbol{\lambda}') > 0$ for $0 < |t| < \eta$. Similarly, we could show $\Delta f(\mathbf{x}_0\,;\mathbf{x} - \mathbf{x}_0) < 0$ for points arbitrarily close to $\mathbf{x}_0$. Hence f has a saddle point at $\mathbf{x}_0$.

The functions of Example 4–4.4 show that when Q is semidefinite, it is possible at $\mathbf{x}_0$ for f to have a relative minimum, relative maximum or a saddle point. ∎

The following example illustrates the method of classifying critical points of functions of more than two variables.

Example 4–5.1. Consider the function f defined by

$$f(x, y, z) = x^2 + 3y^2 + z^2 - 4xy + 2xz - 8yz - 2x + 4y - 2z.$$

Then

$$f_x(x, y, z) = 2x - 4y + 2z - 2,$$
$$f_y(x, y, z) = -4x + 6y - 8z + 4,$$
$$f_z(x, y, z) = 2x - 8y + 2z - 2.$$

Thus there is one critical point, namely, $(\tfrac{2}{3}, 0, \tfrac{1}{3})$. Therefore, the matrix $\left(\dfrac{\partial^2}{\partial x_i\,\partial x_j} f(\mathbf{x}_0)\right)$ of (4–5.10) becomes:

$$A = \begin{pmatrix} 2 & -4 & 2 \\ -4 & 6 & -8 \\ 2 & -8 & 2 \end{pmatrix}.$$

Thus take

$$P_1 = \begin{pmatrix} 1 & 2 & -1 \\ 0 & 1 & 0 \\ 0 & 0 & 1 \end{pmatrix}, \quad P_1^T A P_1 = \begin{pmatrix} 2 & 0 & 0 \\ 0 & -2 & -4 \\ 0 & -4 & 0 \end{pmatrix}.$$

Then take

$$P_2 = \begin{pmatrix} 1 & 0 & 0 \\ 0 & 1 & -2 \\ 0 & 0 & 1 \end{pmatrix}, \quad P_2^T P_1^T A P_1 P_2 = \begin{pmatrix} 2 & 0 & 0 \\ 0 & -2 & 0 \\ 0 & 0 & 8 \end{pmatrix}.$$

Lastly, take

$$P_3 = \begin{pmatrix} \dfrac{1}{\sqrt{2}} & 0 & 0 \\ 0 & \dfrac{1}{\sqrt{2}} & 0 \\ 0 & 0 & \dfrac{1}{\sqrt{8}} \end{pmatrix}, \quad P^T A P = \begin{pmatrix} 1 & 0 & 0 \\ 0 & -1 & 0 \\ 0 & 0 & 1 \end{pmatrix}$$

where $P = P_1 P_2 P_3$. Thus from Table 4–5.1, Q of (4–5.10) is indefinite, and hence f has a saddle point at $(\tfrac{2}{3}, 0, \tfrac{1}{3})$.

Problems

1. Use the results of Proposition 4–5.1 to characterize the behavior of each of the following functions at each of the function's critical points.
   (a) $f(x, y, z) = -5x^2 - 5y^2 - 14z^2 + 4xy + 16xz - 10yz + 2x - 4z$,
   (b) $f(x, y, z) = 7x^2 + 5y^2 - 6xy - xz + 2yz - 2y - z$.
   (c) $f(x, y, z) = 14x^2 + 26y^2 + 13z^2 + 26xy - 18xz - 28yz - y$.
   (d) $f(x, y, z) = 5x^2 + 5y^2 + 9z^2 - 6xz - 12yz + 2x + 5$.

2. Follow the same instructions as for Problem 1 for the following functions.
   (a) $f(w, x, y, z) = 4w^2 + 3x^2 + 6y^2 + 7z^2 + 2wx - 2wz + 2xy - 2xz - 2yz - 2x$.
   (b) $f(w, x, y, z) = -6w^2 - 7x^2 - 3y^2 - 4z^2 + 2wx - 6wy + 4wz + 2xy + 6xz + 2yz + 2y$.

3. Follow the instructions of Problem 1 for each of the functions defined below:
   (a) $f(x, y, z) = 3x^4 - 12xy + 6y^2 + 4z^3 - 12z + 1$.
   (b) $f(x, y, z) = 2x^3 - 6xy + 3y^2 + 2z^3 - 6z$.
   (c) $f(x, y, z) = 6z^2x - 6x - 3x^2 + 3y^2 + 2z^3 - 6z$.
   (d) $f(x, y, z) = 2x^3 + 3y^2z - 3z^2 - 3x^2 - 6y$.

4. On the indicated region S find a point (x', y', z') $((x'', y'', z''))$ where each of the following functions f defined below has an absolute maximum (absolute minimum).
   (a) $f(x, y, z) = (x^2 + y^2 + z^2)^{-1/2}$,
   $S = \{(x, y, z): |x - 2| \le 1, |y - 2| \le 1, |z - 1| \le 1\}$.
   (b) $f(x, y, z) = (x^2 - 4)(3 - y^2)(e^z - z)$,
   $S = \{(x, y, z): |x| \le 1, |y| \le 1, |z| \le 1\}$.
   (c) $f(x, y, z) = 6x^2 + 2y^2 + 2z^2 + 2xy - 2xz - 2yz$,
   $S = \{(x, y, z): x^2 + y^2 + z^2 \le 1\}$.

5. In the functions f defined below, the tests of Proposition 4–5.1 fail to give information about the behavior of f in a neighborhood of at least one of the critical points. Use other means to classify the critical point.
   (a) $f(x, y, z) = xy^2 + yz^2 + zx^2$,
   (b) $f(x, y, z) = (x + y + z)^4 + (y - 1)^4 + z^4$,
   (c) $f(x, y, z) = (x - y + z)^4 + (y + z)^3 + z^2$,
   (d) $f(x, y, z) = x^4 + y^2 + 2yz^2 + z^4$.

6. Consider the plane in (x, y, z)-space given by $z = ax + by + c$. Use the technique of least squares to "pass this plane through the n points," $\{(x_i, y_i, z_i): i = 1, 2, \cdots, n\}$. That is, if

$$F(a, b, c) = \sum_{i=1}^{n} (ax_i + by_i + c - z_i)^2,$$

find (a, b, c) such that $F(a, b, c)$ is a minimum for all (a, b, c). You may assume the critical point that you find is a point where F is a minimum. Demonstrate this in particular for the set of points

$$\{(1, 1, 3), (-1, 2, -3), (0, 0, 3), (-1, -1, 1), (-2, 2, 7)\}.$$

7. Let $A = (a_{ij})$, $1 \leq i, j \leq 3$, and suppose that the principal minor determinants in the upper left of A of order 1, 2, and 3 are all strictly positive. Let

$$Q(\mathbf{x}) = \sum_{i,j=1}^{3} a_{ij}\, x_i\, x_j .$$

Calculate

$$Q_1(\mathbf{x}) = Q(\mathbf{x}) - \left(\frac{1}{a_{11}}\right)(a_{11}\, x_1 + a_{12}\, x_2 + a_{13}\, x_3)^2$$

and show Q_1 is of the following form:

$$Q_1(\mathbf{x}) = \sum_{i,j=2}^{n} b_{ij}\, x_i\, x_j .$$

Show $b_{22} > 0$. Repeat the same procedure on Q_1 obtaining

$$Q(\mathbf{x}) = \left(\frac{1}{a_{11}}\right)(a_{11}\, x_1 + a_{12}\, x_2 + a_{13}\, x_3)^2 + \left(\frac{1}{b_{22}}\right)(b_{22}\, x_2 + b_{23}\, x_3)^2 + \left(\frac{1}{c_{33}}\right)x_3^{\,2}.$$

Show $c_{33} > 0$. Without using Gundelfinger's rule, give reasons why Q is a positive definite quadratic form.

8. Use the technique outlined in Problem 7 to prove, without using Gundelfinger's rule, that a function of three variables has a relative minimum at a critical point $\mathbf{x}_0$, if the three principal minors in the upper left of $\left(\dfrac{\partial^2}{\partial x_i\, \partial x_j} f(\mathbf{x}_0)\right)$ are strictly positive.

Answers

1. Each of the problems has exactly one critical point.
   (a) Rel. max. at $(35/3, -10/3, 8)$. (b) Saddle pt.
   (c) Rel. min. (d) Test fails.
2. (a) Rel. min., (b) Rel. max.
3. (a) Rel. min. at $(1, 1, 1)$ and $(-1, -1, 1)$ and saddle point at $(1, 1, -1)$ and $(-1, -1, -1)$,
   (b) Rel. min. at $(1, 1, 1)$; saddle point at $(1, 1, -1)$, $(0, 0, 1)$, and $(0, 0, -1)$.

(c) Saddle point at (0, 0, 1), (0, 0, −1), and $\left(-\frac{3}{4}, 0, -\frac{1}{2}\right)$.

(d) Saddle point at $(0, 2^{(1/3)}, 2^{-(1/3)})$ and $(1, 2^{(1/3)}, 2^{-(1/3)})$.

4. (a) Abs. max. at $(x'', xy'', z'') = (1, 1, 0)$ of $\frac{1}{\sqrt{2}}$, abs. min. at (x', y', z) $= (3, 2, 2)$ of $\frac{1}{\sqrt{22}}$.

(b) Abs. max. at $(x'', y'', z'') = (0, 0, -1)$ of $12(1 - e^{-1}) \approx 7.58$; abs. min. at $(x', y', z') = (0, 0, 1)$ of $12(1 - e) \approx -20.52$.

(c) Abs. max. at

$$(x'', y'', z'') = \frac{(-3 - \sqrt{17}, -2, 2)}{\sqrt{34 + 6\sqrt{17}}} \text{ of } \frac{69 + 16\sqrt{17}}{17 + 3\sqrt{17}} \approx 4.59,$$

abs. min. at $(x', y', z') = (0, 0, 0)$ of 0.

5. (a) (0, 0, 0) is a saddle point, (b) (−1, 1, 0) is a rel. min., (c) (0, 0, 0) is a saddle point, (d) (0, 0, 0) is a rel. min.

6. $a = -\frac{18}{41}, \quad b = \frac{3}{41}, \quad c = \frac{77}{41}.$

4–6. Lagrange Multipliers

There is another type of maxima and minima problem that has a great number of applications. A typical example would be as follows: Suppose that f and g_1 are functions defined on all of (x, y, z)-space. Suppose, furthermore, that the points

$$S = \{(x, y, z): g_1(x, y, z) = 0\}$$

is a surface, as for example

$$\{(x, y, z): x^2 + y^2 + z^2 - 1 = 0\}$$

is a spherical surface. The problem is that we desire to find, if they exist, the maximum and minimum values of

$$\{f(x, y, z): g_1(x, y, z) = 0\}.$$

The condition that (x, y, z) must be such that $g_1(x, y, z) = 0$ is a *side condition* or *constraint*.

If, for example, $f(x, y, z)$ is the temperature at the point (x, y, z) and $g_1(x, y, z) = x^2 + y^2 + z^2 - 1$, then we would be desiring the maximum and minimum temperatures on this spherical surface.

Now it may well be that there are two constraints. For example, if $g_2(x, y, z) = x + y + z - 1$, then

$$\{(x, y, z): g_2(x, y, z) = 0\}$$

is a plane surface that intersects the spherical surface in a circular curve. We could very well be interested in finding the maximum and minimum temperatures on this circular curve—that is, the maximum and minimum of

$$\{f(x, y, z): g_1(x, y, z) = 0; g_2(x, y, z) = 0\}.$$

In this case there are *two* constraints in that the point (x, y, z) is confined to be on two surfaces, both the spherical surface and the plane surface.

For the same three-variable function f we will not be interested in finding the maximum or minimum value of $f(x, y, z)$ subject to three or more constraints. In general, the intersection of three or more surfaces in (x, y, z)-space would be a set of isolated points. In such a case the techniques of the calculus are not applicable. Thus in our consideration *the number of constraints will be strictly less than the number of independent variables in the function for which we wish to find maximum and minimum values.*

The following example, which is in the simplest possible circumstance, is a function of two variables subject to one constraint.

Example 4–6.1. Consider the problem of finding the point on the parabola $y = (1/\sqrt{2})x^2$ nearest to $(2, 0)$ (cf. Figure 4–6.1). In this case we

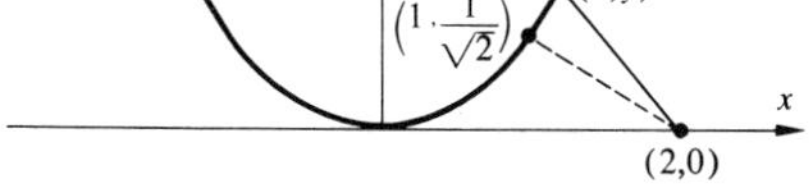

Figure 4–6.1

wish to find a minimum value for

$$\sqrt{(x - 2)^2 + y^2},$$

the distance from an arbitrary point (x, y) on the parabola to the point $(2, 0)$. Since a point (x_0, y_0) on the parabola such that

$$(x_0 - 2)^2 + y_0^2 \leq (x - 2)^2 + y^2$$

is also a point such that

$$\sqrt{(x_0 - 2)^2 + y_0^2} \le \sqrt{(x - 2)^2 + y^2}$$

where (x, y) is a point on the parabola, we may alternately seek a minimum value for

$$f(x, y) = (x - 2)^2 + y^2.$$

The constraint in this case is

$$g_1(x, y) = \frac{x^2}{\sqrt{2}} - y = 0.$$

Solving for y in $g_1(x, y) = 0$ and substituting into the f function gives

$$f\left(x, \frac{x^2}{\sqrt{2}}\right) = (x - 2)^2 + \tfrac{1}{2}x^4,$$

a function of a single variable. Differentiating with respect to x we get

$$\frac{d}{dx} f\left(x, \frac{x^2}{\sqrt{2}}\right) = 2(x - 2) + 2x^3.$$

The critical point, the only point for real x such that $2x^3 + 2x - 4 = 0$, is the point such that $x = 1$. By taking the second derivative with respect to x and evaluating at $x = 1$, we get a positive value, and hence a minimum occurs at $x = 1$. The corresponding point on the parabola is $(1, 1/\sqrt{2})$ and the minimum distance is $\sqrt{3/2}$.

Another way of working this problem is by the method of Lagrange multipliers. In the Lagrange multipliers technique, we define a new function F as follows:

$$\begin{aligned} F(x, y, \lambda_1) &= f(x, y) + \lambda_1 g_1(x, y) \\ &= (x - 2)^2 + y^2 + \lambda_1\left(\frac{1}{\sqrt{2}}x^2 - y\right). \end{aligned}$$

The new parameter that is introduced, λ_1, appears as a multiplier, giving the motivation for calling this technique "Lagrange multipliers."

From this function F consider the points (x, y, λ) such that

$$\begin{aligned} F_x(x, y, \lambda_1) &= 2(x - 2) + \lambda_1\left(\frac{2}{\sqrt{2}}\right) x = 0, \\ F_y(x, y, \lambda_1) &= 2y - \lambda_1 = 0, \\ g_1(x, y) &= \frac{1}{\sqrt{2}}x^2 - y = 0. \end{aligned} \tag{4–6.1}$$

Eliminating λ_1 between the first two equations gives us

$$2(x-2)+2y\left(\frac{2}{\sqrt{2}}\right)x=0.$$

Substituting for y from the last equation gives

$$2x^3+2x-4=0.$$

Again, we get $x=1$, and the only solution to the system of equations in (4–6.1) is

$$(x, y, \lambda_1)=(1, 1/\sqrt{2}, 2/\sqrt{2}).$$

Let us now consider the following example, which was suggested at the outset.

Example 4–6.2. Let $f(x, y, z)=((x+\tfrac{1}{2})^2+y^2+z^2)^{-1/2}$ be the temperature at the point (x, y, z). (a) Find by the method of Lagrange multipliers the maximum and the minimum values of the temperature on the spherical surface defined by $x^2+y^2+z^2=1$. (b) Do the same problem for the intersection of the above surface and the plane surface defined by $x+y+z=1$. For (a), by the Lagrange multipliers technique, let

$$F(x, y, z, \lambda_1)=[(x+\tfrac{1}{2})^2+y^2+x^2]^{-1/2}+\lambda_1(x^2+y^2+z^2-1).$$

Then consider the system

$$\begin{aligned}
F_x(x, y, z, \lambda_1)&=-[(x+\tfrac{1}{2})^2+y^2+z^2]^{-3/2}(x+\tfrac{1}{2})+2x\lambda_1=0,\\
F_y(x, y, z, \lambda_1)&=-[(x+\tfrac{1}{2})^2+y^2+z^2]^{-3/2}y+2y\lambda_1=0,\\
F_z(x, y, z, \lambda_1)&=-[(x+\tfrac{1}{2})^2+y^2+z^2]^{-3/2}z+2z\lambda_1=0,\\
g_1(x, y, z, \lambda_1)&=x^2+y^2+z^2-1=0.
\end{aligned} \qquad (4\text{–}6.2)$$

The only solutions of this system are $(x, y, z, \lambda_1)=(1, 0, 0, \tfrac{2}{9})$ and $(-1, 0, 0, 2)$. Thus $f(1, 0, 0)=\tfrac{2}{3}$ and $f(-1, 0, 0)=2$. Since the temperature is inversely proportional to the distance from $(-\tfrac{1}{2}, 0, 0)$, it is clear from the geometry that $\tfrac{2}{3}$ and 2 are the minimum and maximum values of the temperature on the spherical surface (cf. Figure 4–6.2 (a)).

For (b), by the method of Lagrange multipliers, let

$$\begin{aligned}
F(x, y, z, \lambda_1, \lambda_2)=[(x+\tfrac{1}{2})^2+y^2+z^2]^{-1/2}&+\lambda_1(x^2+y^2+z^2-1)\\
&+\lambda_2(x+y+z-1);
\end{aligned}$$

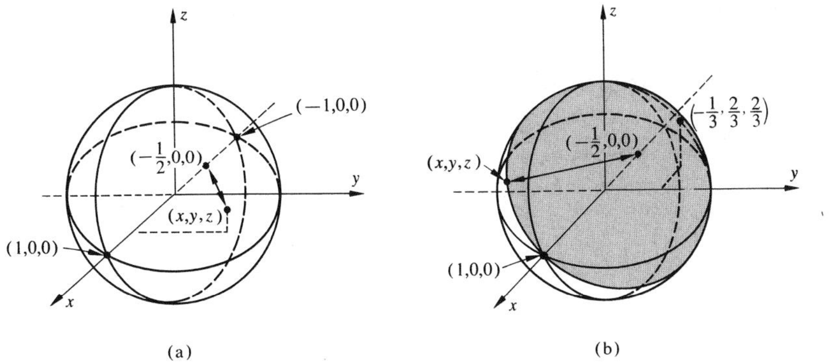

Figure 4–6.2

that is, two new parameters are introduced, one for each constraint. Consider the solutions to the following system:

$$\begin{aligned}
F_x(x, y, z, \lambda_1, \lambda_2) &= -[(x + \tfrac{1}{2})^2 + y^2 + z^2]^{-3/2}(x + \tfrac{1}{2}) + 2x\lambda_1 + \lambda_2 = 0,\\
F_y(x, y, z, \lambda_1, \lambda_2) &= -[(x + \tfrac{1}{2})^2 + y^2 + z^2]^{-3/2}(y) + 2y\lambda_1 + \lambda_2 = 0,\\
F_z(x, y, z, \lambda_1, \lambda_2) &= -[(x + \tfrac{1}{2})^2 + y^2 + z^2]^{-3/2}(z) + 2y\lambda_1 + \lambda_2 = 0,\\
g_1(x, y, z) &= x^2 + y^2 + z^2 - 1 = 0, \qquad (4\text{–}6.3)\\
g_2(x, y, z) &= x + y + z - 1 = 0.
\end{aligned}$$

By subtracting the F_z equation from the F_y equation, we obtain

$$-[(x + \tfrac{1}{2})^2 + y^2 + z^2]^{-3/2}(y - z) + 2(y - z)\lambda_1 = 0.$$

Thus letting $y = z$ in the last two equations of (4–6.3) gives $(x, y, z) = (1, 0, 0)$ and $(-\frac{1}{3}, \frac{2}{3}, \frac{2}{3})$. The only solutions of the entire system, (4–6.3), are $(1, 0, 0, \frac{2}{9}, 0)$ and $(-\frac{1}{3}, \frac{2}{3}, \frac{2}{3}, 54/(33)^{3/2}, 72/(33)^{3/2})$. Note that $f(1, 0, 0) = \frac{2}{3}$ and $f(-\frac{1}{3}, \frac{2}{3}, \frac{2}{3}) = 6/(33)^{1/2}$. Again the geometry makes it clear that $\frac{2}{3}$ and $6/(33)^{1/2}$ are the minimum and maximum values, respectively, of $f(x, y, z)$ on this circular curve (cf. Figure 4–6.2(b)).

In general, on some set, we wish to maximize or minimize $f(x_1, \cdots, x_n)$ subject to the m constraints

$$g_1(x_1, \cdots, x_n) = 0, \cdots, g_m(x_1, \cdots, x_n) = 0. \qquad (4\text{–}6.4)$$

That is, if

$$\{f(x_1, \cdots, x_n) : g_1(x_1, \cdots, x_n) = 0, \cdots, g_m(x_1, \cdots, x_n) = 0\} \qquad (4\text{–}6.5)$$

has a maximum value (minimum value) at $\mathbf{x}_0$ on a certain set of points, then we wish to find this extreme value.

By the method of Example 4–6.1, we would solve the set of equations (4–6.4) for m of the variables in terms of the other $n - m$ variables. Then substitute for these m variables into $f(x_1, \cdots, x_n)$ obtaining a function of $n - m$ variables. With this new function we can apply earlier techniques for finding the maximum and minimum values.

In the Lagrange multipliers technique for doing this, we first define a new function

$$F(x_1, \cdots, x_n; \lambda_1, \cdots, \lambda_m) = f(x_1, \cdots, x_n) + \textstyle\sum_{i=1}^{m} \lambda_i g_i(x_1, \cdots, x_n). \tag{4–6.6}$$

Assuming that the first partials of F with respect to x_i, $1 \leq i \leq n$, exist, then we obtain the following $m + n$ system of equations:

$$\begin{aligned} F_{x_i}(x_1, \cdots, x_n; \lambda_1, \cdots, \lambda_m) &= 0, \ 1 \leq i \leq n, \\ g_i(x_1, \cdots, x_n) &= 0, \ 1 \leq i \leq m. \end{aligned} \tag{4–6.7}$$

Among the solutions $(\mathbf{x}; \boldsymbol{\lambda})$ of the system (4–6.7) will be the solution $(\mathbf{x}_0; \boldsymbol{\lambda}_0)$ where $f(\mathbf{x}_0)$ is the maximum value (minimum value) of the set (4–6.5).

One of the advantages of the Lagrange multipliers method is that we do not have to solve the set of equations (4–6.4) for m of the variables. This could be difficult or impossible. One disadvantage of the Lagrange multipliers technique lies in not easily being able to decide whether a solution, $(\mathbf{x}'; \boldsymbol{\lambda}')$, of the system of equations (4–6.7) gives $\mathbf{x}'$ such that $f(\mathbf{x}')$ is a maximum value, a minimum value, or neither for the set (4–6.5). However, it is sometimes possible to use physical or geometrical reasoning as was done in both examples to make confident decisions on these questions.

We now state and prove the proposition that justifies the method of Lagrange multipliers. In order that the idea of the argument may be more easily followed, the proposition will be stated and proved for the case of a function of three variables subject to two constraints.

Proposition 4–6.1. *Let f, g_1, and g_2 be functions defined on an open set S in (x, y, z)-space. Let f, g_1, and $g_2 \in C^1(S)$ and suppose that at (x_0, y_0, z_0) in S,*

$$\frac{\partial(g_1, g_2)}{\partial(x, y)} \neq 0.$$

If there exists a neighborhood $N((x_0, y_0, z_0); \delta)$ such that $f(x_0, y_0, z_0) \geq f(x, y, z)$ (or $f(x_0, y_0, z_0) \leq f(x, y, z)$) for all (x, y, z) such that $g_1(x, y, z) = 0$, $g_2(x, y, z) = 0$, and $(x, y, z) \in S$, then there exist real numbers $\{\lambda_{01}, \lambda_{02}\}$

such that $(x_0, y_0, z_0; \lambda_{01}, \lambda_{02})$ *is a solution of the system*

$$\begin{aligned} F_x(x, y, z, \lambda_1, \lambda_2) &= 0, \\ F_y(x, y, z; \lambda_1, \lambda_2) &= 0, \\ F_z(x, y, z; \lambda_1, \lambda_2) &= 0, \\ g_1(x, y, z) &= 0, \\ g_2(x, y, z) &= 0, \end{aligned} \tag{4–6.8}$$

where

$$F(x, y, z; \lambda_1, \lambda_2) = f(x, y, z) + \lambda_1 g_1(x, y, z) + \lambda_2 g_2(x, y, z).$$

Proof. By the implicit function theorem, Theorem 3–5.1, since g_1 and $g_2 \in C^1(S)$, $g_i(x_0, y_0, z_0) = 0$, $i = 1, 2$, and since at (x_0, y_0, z_0)

$$\frac{\partial(g_1, g_2)}{\partial(x, y)} \neq 0,$$

there exists a neighborhood of z_0, $N(z_0; \delta_1)$, and there exist unique functions h_1 and h_2 defined on $N(z_0; \delta_1)$ such that

$$x_0 = h_1(z_0), \qquad y_0 = h_2(z_0)$$

and such that, for $z \in N(z_0; \delta_1)$,

$$g_i(h_1(z), h_2(z), z) = 0, i = 1, 2 \tag{4–6.9}$$

and

$$h_1' = \frac{dx}{dz} = -\frac{\dfrac{\partial(g_1, g_2)}{\partial(z, y)}}{\dfrac{\partial(g_1, g_2)}{\partial(x, y)}}, \qquad h_2' = \frac{dy}{dz} = -\frac{\dfrac{\partial(g_1, g_2)}{\partial(x, z)}}{\dfrac{\partial(g_1, g_2)}{\partial(x, y)}}. \tag{4–6.10}$$

Furthermore, $h_i \in C^1(N(z_0; \delta_1))$, $i = 1, 2$.

By continuity we may take δ_1 such that

$$(h_1(z), h_2(z), z) \in N((x_0, y_0, z_0); \delta)$$

when $z \in N(z_0; \delta_1)$. Then by hypothesis

$$f(h_1(z_0), h_2(z_0), z_0) \geq f(h_1(z), h_2(z), z)$$

(or the sense of the inequality is reversed) for all $z \in N(z_0; \delta_1)$. In either case at z_0,

$$\frac{d}{dz} f(h_1(z), h_2(z), z) = 0.$$

Thus by the chain rule it follows that at z_0

$$\frac{\partial f}{\partial x}\frac{dx}{dz} + \frac{\partial f}{\partial y}\frac{dy}{dz} + \frac{\partial f}{\partial z} = 0. \tag{4–6.11}$$

Substituting into (4–6.11) from (4–6.10), we get at z_0

$$\frac{\partial f}{\partial x}\cdot\frac{\dfrac{\partial(g_1, g_2)}{\partial(y, z)}}{\dfrac{\partial(g_1, g_2)}{\partial(x, y)}} - \frac{\partial f}{\partial y}\cdot\frac{\dfrac{\partial(g_1, g_2)}{\partial(x, z)}}{\dfrac{\partial(g_1, g_2)}{\partial(x, y)}} + \frac{\partial f}{\partial z} = 0$$

since

$$-\frac{\partial(g_1, g_2)}{\partial(z, y)} = \frac{\partial(g_1, g_2)}{\partial(y, z)}.$$

Hence at $(x_0, y_0, z_0) = (h_1(z_0), h_2(z_0), z_0)$,

$$\frac{\partial(f, g_1, g_2)}{\partial(x, y, z)} = 0. \tag{4–6.12}$$

Upon the basis of this Jacobian being zero at (x_0, y_0, z_0), we can obtain our desired conclusion.

From the first two equations of the system (4–6.8),

$$F_x(x, y, z; \lambda_1, \lambda_2) = f_x(x, y, z) + \lambda_1 g_{1x}(x, y, z) + \lambda_2 g_{2x}(x, y, z) = 0$$

and

$$F_y(x, y, z; \lambda_1, \lambda_2) = f_y(x, y, z) + \lambda_1 g_{1y}(x, y, z) + \lambda_2 g_{2y}(x, y, z) = 0,$$

it follows by Cramer's rule that at (x_0, y_0, z_0)

$$\lambda_{01} = -\frac{\dfrac{\partial(f, g_2)}{\partial(x, y)}}{\dfrac{\partial(g_1, g_2)}{\partial(x, y)}}, \qquad \lambda_{02} = -\frac{\dfrac{\partial(g_1, f)}{\partial(x, y)}}{\dfrac{\partial(g_1, g_2)}{\partial(x, y)}}. \tag{4–6.13}$$

Now we will have our desired conclusion if we can show that $(x_0, y_0, z_0; \lambda_{01}, \lambda_{02})$ satisfy the third equation of (4–6.8). Substituting from (4–16.13) into that equation gives

$$f_z - (g_{1z})\cdot\frac{\dfrac{\partial(f, g_2)}{\partial(x, y)}}{\dfrac{\partial(g_1, g_2)}{\partial(x, y)}} + (g_{2z})\cdot\frac{\dfrac{\partial(f, g_1)}{\partial(x, y)}}{\dfrac{\partial(g_1, g_2)}{\partial(x, y)}} = \frac{\dfrac{\partial(f, g_1, g_2)}{\partial(x, y, z)}}{\dfrac{\partial(g_1, g_2)}{\partial(x, y)}},$$

which is zero by virtue of (4–6.12). ∎

Problems

1. Find the point on the curve defined by $x = y^3$ that is nearest to the point (0, 4). Work the problem by the method of Example 4–6.1 and by the method of Lagrange multipliers.
2. Find the length of the semimajor and semiminor axes of the ellipse defined by $52x^2 - 72xy + 73y^2 = 400$.
3. Find the minimum distance of the point (0, 0) to the curve defined by $-381x^2 + 624xy - 209y^2 = 625$.
4. Use the method of Lagrange multipliers to find the points on the surface defined by $xyz = 1$ that are nearest the origin. What is the minimum distance?
5. Find the dimensions of a rectangular box of maximum volume if the box is made of thin sheets of material such that the bottom is to be made of four thicknesses of the material, the sides of two thicknesses and the top of one thickness where s^2 is the number of square units of material available.
6. Work Problem 5 under the conditions that the box has no top and the bottom requires only three thicknesses.
7. A rectangular grain storage bin with a square bottom of side length x and height y has a hipped roof. The highest point of the roof, which lies directly over the center of the bottom, is z units higher than the top of the sides. Find (x_0, y_0, z_0) such that the volume, including even up to the roof, is a maximum for a surface area of s^2 square units where the surface includes the bottom.
8. Find the point on the line given by the intersection of the two planes, $2x + y + z = 1$, $x - y - z = 4$, that is nearest to the origin.
9. Work both parts of Example 4–6.2 by the method of Example 4–6.1.
10. Find the points on the curve given by the intersection of

$$x^2 + y^2 + z - 4 = 0 \quad \text{and} \quad 4x + 2y + z - 6 = 0$$

that are a maximum and a minimum distance from (0, 0, 0). Work the problem both by the use of Lagrange multipliers and by the method of Example 4–6.1. What are the maximum and the minimum distances?

11. Find the minimum distance in E^n from the hyperplane $\mathbf{a} \cdot \mathbf{x} = 1$ to $\mathbf{0}$ if $\mathbf{a} = (a_1, \cdots, a_n)$ is such that $a_i = b > 0$ for $1 \leq i \leq n$.
12. Prove Proposition 4–6.1 if at (x_0, y_0, z_0) in S we assume instead that

$$\frac{\partial(g_1, g_2)}{\partial(x, z)} \neq 0.$$

13. State and prove Proposition 4–6.1 for the general case of f being a function of n variables and f being subject to m constraints where $1 \leq m < n$.

14. Use the technique of Lagrange multipliers to find the maximum and minimum values of $f(x, y)$ on the boundary of S for Problem 4–4.3(a), (b), (c), and (d).

Answers

1. (1, 1). 2. 4, 2 3. 5.

4. (1, 1, 1), (1, −1, −1), (−1, 1, −1), (−1, −1, 1); min. dist. = $\sqrt{3}$.
5. Length = width = $s/\sqrt{15}$; height = $5s/4\sqrt{15}$.
6. Length = width = $s/3$; height = $s/4$.

7. $x_0 = [40 + \sqrt{1600 + 15(55 - 8\sqrt{5})\, s^2}\,(55 - 8\sqrt{5})^{-1}$,

$$y_0 = \frac{15 - 4\sqrt{5}}{30}\, x_0 - \frac{4}{3}, \quad z_0 = \frac{1}{\sqrt{5}}\, x_0 .$$

8. (5/4, −9/10, −3/5).

10. Min. pt. $\left(\frac{11}{10} - \frac{\sqrt{159}}{20}, \frac{11}{10} + \frac{\sqrt{159}}{10}, \frac{1}{2}\right)$.

Max. pt. $\left(2 + \frac{2\sqrt{15}}{5}, 1 + \frac{\sqrt{15}}{5}, -4 - 2\sqrt{15}\right)$,

Max. dist. $\approx$ 4.10, min. dist. $\approx$ 1.14.

11. $\dfrac{\sqrt{n}}{nb}$.

5

Integral Calculus of Functions of Several Variables

5–1. Line Integrals

Before considering line integrals we shall first review the meaning of an integral of a real function of a single real variable. For example, if f is a real function defined on the interval $[a, b]$ by $y = f(x)$, if $P = \{x_0, x_1, \cdots, x_n\}$ where $a = x_0 < x_1 < \cdots < x_n = b$, and if $\bar{x}_i \in [x_i, x_{i-1}]$ for $1 \leq i = n$, then f is *integrable* on $[a, b]$ if

$$\lim_{|P| \to 0} \sum_{i=1}^{n} f(\bar{x}_i)(x_i - x_{i-1}) \qquad (5\text{–}1.1)$$

exists where $|P| = \max \{x_i - x_{i-1}: 1 \leq i \leq n\}$. The *Riemann integral* of f on $[a, b]$ is the value of the limit of (5–1.1), written $\int_a^b f(x)\, dx$ or simply $\int_a^b f\, dx$. The set P is a *partition* of $[a, b]$, and $|P|$ is the *mesh* of the partition P. The sum, $\sum_{i=1}^{n} f(\bar{x}_i)(x_i - x_{i-1})$, is a *Riemann sum* of f on $[a, b]$ for the partition P.

A sufficient condition for the limit (5–1.1) to exist is that f is continuous on $[a, b]$; that is, $f \in C([a, b])$.† It is in fact sufficient that f is bounded on $[a, b]$; that is, $f([a, b])$ is bounded, and f is continuous on all of $[a, b]$ except a set of points I^* such that I^* can be contained in a collection of closed intervals the sum of whose lengths is arbitrarily small.

If f is continuous on $[a, b]$, then by the Fundamental Theorem of Calculus,

$$\int_a^b f(x)\,dx = F(b) - F(a)$$

where F is any function such that $F'(x) = f(x)$ for all $x \in (a, b)$ and

$$F(a) = \lim_{x \to a^+} F(x), \qquad F(b) = \lim_{x \to b^-} F(x).‡$$

Thus, if f is a continuous function on $[a, b]$ and if F can be found, we have a simple way of calculating the limit of (5–1.1). If $f(x) \geq 0$ for all $x \in [a, b]$, then the *area* of

$$\{(x, y) : a \leq x \leq b, 0 \leq y \leq f(x)\}$$

is $\int_a^b f(x)\,dx$, and it is said to be the "area under the curve given by f between a and b."

The line integral is defined as follows:

Definition 5–1.1. *Let Γ be a piecewise smooth curve in E^n that joins $\boldsymbol{\alpha}$ to $\boldsymbol{\beta}$. Let $\mathbf{r}$ be a parametric representation of Γ, so that $\Gamma = \{\mathbf{r}(t) : a \leq t \leq b\}$ where $\boldsymbol{\alpha} = \mathbf{r}(a)$ and $\boldsymbol{\beta} = \mathbf{r}(b)$. Let $\mathrm{P} = \{t_0, t_1, \cdots, t_m\}$ be a partition of $[a, b]$ and take $\bar{t}_i \in [t_{i-1}, t_i]$, $1 \leq i \leq m$. For $\mathbf{f}$, a vector-valued function defined on Γ the (Riemann) line integral of $\mathbf{f}$ along Γ from $\boldsymbol{\alpha}$ to $\boldsymbol{\beta}$ is*

$$\int_\Gamma f \cdot \mathbf{dr} = {}_\Gamma\!\int_{\boldsymbol{\alpha}}^{\boldsymbol{\beta}} \mathbf{f} \cdot \mathbf{dr} = \lim_{|\mathrm{P}| \to 0} \sum_{i=1}^m \mathbf{f}(\mathbf{r}(\bar{t}_i)) \cdot (\mathbf{r}(t_i) - \mathbf{r}(t_{i-1})) \quad (5\text{–}1.2)$$

if this limit exists. The summation on the right is a Riemann sum of $\mathbf{f}$ *on* Γ *with respect to the partition* P.

Clearly, if $n = 1$ the limit of (5–1.2) would become (5–1.1), and Γ would be a "line segment" on the x_1-axis. Thus, a line integral is a generalization of an integral of a real-valued function of a single variable. It is possible to define a line integral in a more general setting than a piecewise smooth curve, but this is sufficient for our purposes.

---

† Cf. David V. Widder, *Advanced Calculus*, 2nd ed, Theorem A, p. 150.
‡ Cf. Robert C. James, *University Mathematics*, Theorem 7–6.2, p. 411.

If Γ is given parametrically as in the above definition, then a "direction" on Γ is given by this parametric representation. However, a curve is a set of points and no "direction" need be specified. For example,

$$\Gamma = \{(x, y): x^2 + y^2 = 1, x \geq 0, y \geq 0\}$$

is a curve. If we have given the line integral

$$_{\Gamma}\int_{(1,0)}^{(0,1)} \mathbf{f} \cdot \mathbf{dr},$$

then this notation implies a parametric representation $\mathbf{r}$ must be given such that if $[a, b]$ is the interval of the parameter then $\mathbf{r}(a) = (1, 0)$ and $\mathbf{r}(b) = (0, 1)$. For example,

$$\Gamma = \left\{(\cos t, \sin t): 0 \leq t \leq \frac{\pi}{2}\right\}$$

is such a representation. The points $\boldsymbol{\alpha}$ and $\boldsymbol{\beta}$ are the lower and upper limits, respectively, of the integration. When the meaning of Γ is very apparent from the context, the $\boldsymbol{\alpha}$ and $\boldsymbol{\beta}$ are suppressed in the notation and the briefer symbol, $\int \mathbf{f} \cdot \mathbf{dr}$, is used.

If Γ is not a simple piecewise smooth curve, or even if it is simple and is a closed curve, for which no parametric representation has been prescribed, the notation

$$_{\Gamma}\int_{\boldsymbol{\alpha}}^{\boldsymbol{\beta}} \mathbf{f} \cdot \mathbf{dr}$$

is not sufficient to indicate a precise meaning. Figure 5–1.1(a) and (b) indicate

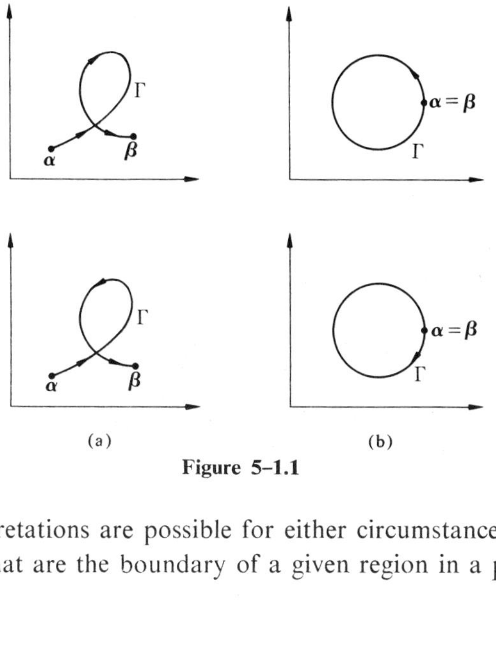

Figure 5–1.1

how two interpretations are possible for either circumstance. However, for closed curves that are the boundary of a given region in a plane there is a different integral notation that does clearly indicate the "direction" of the parametrization of Γ. For example, if $S = \{(x, y): x^2 + y^2 \leq 1\}$, then $\Gamma = \{(x, y): x^2 + y^2 = 1\}$ is the boundary of S, and $\mathbf{r}(t) = \{(\cos t, \sin t), 0 \leq t \leq 2\pi\}$ is a parametrization of Γ in the positive direction. The *positive direction* along Γ with respect to S means the direction along Γ such that the region S is to the left. In Figure 5–1.2, (a) ,(b), and (c) indicate the positive

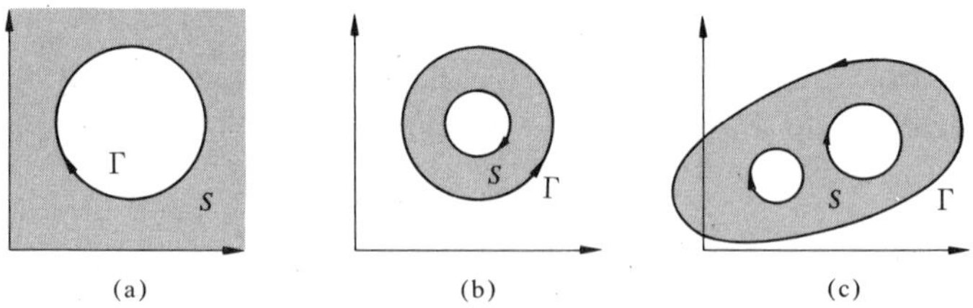

Figure 5–1.2

direction with respect to the shaded regions S. Notice that sometimes this is clockwise, and sometimes it is not. The symbol $\oint_{\Gamma}$ means that for the integration the parametrization of Γ is in the positive direction along Γ with respect to S. The negative direction, which is exactly the opposite direction to the positive direction, is indicated by the symbol $\ointclockwise_{\Gamma}$.

It can easily be seen from equation (5–1.2) of Definition 5–1.1 that *a reversal in the "direction" of* Γ *will result in a change in sign of the line integral.* Thus,

$$_{\Gamma}\int_{\alpha}^{\beta} \mathbf{f} \cdot \mathbf{dr} = -\, _{\Gamma}\int_{\beta}^{\alpha} \mathbf{f} \cdot \mathbf{dr} \tag{5–1.3}$$

and

$$\oint_{\Gamma} \mathbf{f} \cdot \mathbf{dr} = -\ointclockwise_{\Gamma} \mathbf{f} \cdot \mathbf{dr}. \tag{5–1.4}$$

The following proposition gives us a way of evaluating a line integral.

Proposition 5–1.1. *Let* Γ *be a smooth curve and let* **f** *be a vector-valued function that is continuous on* Γ. *Then*

$$_{\Gamma}\int_{\alpha}^{\beta} \mathbf{f} \cdot \mathbf{dr} = \sum_{i=1}^{n} \int_{a}^{b} f_i(\mathbf{r}(t))\mathbf{r}_i'(t)\, dt. \tag{5–1.5}$$

This theorem, whose proof is sketched in Problem 9, gives rise to an alternate notation, which follows, for

$$_{\Gamma}\int_{\alpha}^{\beta} \mathbf{f} \cdot \mathbf{dr}.$$

That is,

$$\int_{\Gamma\,\alpha}^{\beta} f_1\,dr_1 + \cdots + f_n\,dr_n, \tag{5–1.6}$$

or, since $\mathbf{x} = \mathbf{r}(t)$,

$$\int_{\Gamma\,\alpha}^{\beta} f_1\,dx_1 + \cdots + f_n\,dx_n, \tag{5–1.7}$$

or

$$\int_{\Gamma\,\alpha}^{\beta} f_1(\mathbf{x})\,dx_1 + \cdots + f_n(\mathbf{x})\,dx_n, \tag{5–1.8}$$

if you wish to exhibit the variables of the functions f_i. With the notation of (5–1.8) the change to the right side of (5–1.5) resembles very much the rule for changing the variables in an integral of a real-valued function on a closed interval $[a, b]$. That is, for $\mathbf{x} = \mathbf{r}(t)$, granting the ith parts are equal, we have

$$\int_{\Gamma\,\alpha}^{\beta} \textstyle\sum_{i=1}^{n} f_i(\mathbf{x})\,dx_i = \sum_{i=1}^{n} \displaystyle\int_a^b f_i(\mathbf{r}(t))r_i'(t)\,dt \tag{5–1.9}$$

(cf. Section 5–5).

Below is a simple example in (x, y)-space of a line integral and its evaluation.

Example 5–1.1. Consider the line integral

$$\int_{\Gamma\,(1,2)}^{(4,4)} f(x, y)\,dx$$

where $f(x, y) = (\frac{1}{3})y^2 - (\frac{1}{48})x^3$ and $\Gamma = \{(x, y): y^2 = 4x\}$. This is an example of (5–1.6) where $f_1 = f$, $f_2 = 0$, and $n = 2$. For a parametric representation of Γ we could take $\Gamma = \{(t, 2t^{(1/2)}): 1 \le t \le 4\}$. If, in particular, $\mathrm{P} = \{1, 2, 3, 4\}$ and $\bar{t}_i$ is the midpoint of $[t_i, t_{i-1}]$, $i = 2, 3, 4$, the Riemann sum of (5–1.2) becomes

$$\textstyle\sum_{i=1}^{3} f(\bar{t}_i, 2(\bar{t}_i)^{1/2})(t_i - t_{i-1}) = 8.7109 \cdots.$$

Making use of Proposition 5–1.1, we can evaluate our line integral as follows:

$$\begin{aligned}
\int_{\Gamma\,(1,2)}^{(4,4)} f(x, y)\,dx &= \int_1^4 f(t, 2t^{(1/2)})(1)\,dt \\
&= \int_1^4 \left[\frac{4}{3}\,t - \frac{1}{48}\,t^3\right] dt \\
&= \left[\frac{2}{3}\,t^2 - \frac{1}{192}\,t^4\right]_1^4 \\
&= 8.6770 \cdots.
\end{aligned}$$

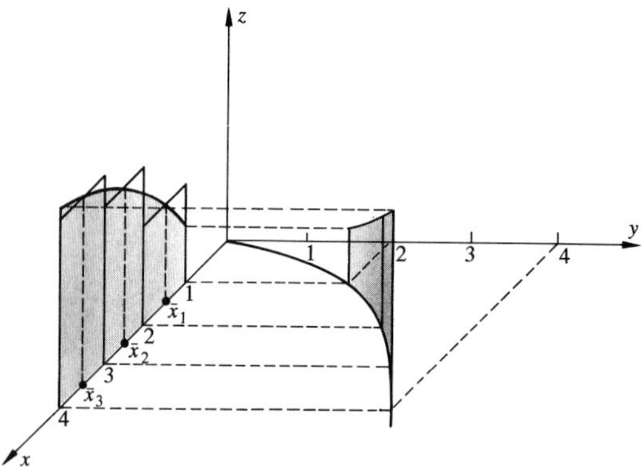

Figure 5–1.3

Thus, the Riemann sum is a good approximation of the line integral. Figure 5–1.3 illustrates that the value of this line integral is indeed the area of the projection of

$$\{(x, y, z): 4x = y^2, 0 \leq z \leq (1/3)y^2 - (1/48)x^3, 1 \leq x \leq 4\}$$

on the (x, z)-plane.

In the evaluation of this integral, $t = x$, and hence this evaluation is usually done without introducing a new variable t, which would simply amount to substituting for y in terms of x from the equation $y^2 = 4x$.

The following proposition gives some of the main properties of line integrals that we shall use.

Proposition 5–1.2. *If*

$$\Gamma\int_\beta^\alpha \mathbf{f} \cdot \mathbf{dr} \quad \textit{and} \quad \Gamma\int_\alpha^\beta \mathbf{g} \cdot \mathbf{dr}$$

exist, then

$$\Gamma\int_\alpha^\beta \lambda\mathbf{f} \cdot \mathbf{dr},$$

or λ real, and

$$\Gamma\int_\alpha^\beta (\mathbf{f} + \mathbf{g}) \cdot \mathbf{dr}$$

exist; and

$$\Gamma\int_\alpha^\beta \lambda\mathbf{f} \cdot \mathbf{dr} = \lambda \, \Gamma\int_\alpha^\beta \mathbf{f} \cdot \mathbf{dr}, \tag{5–1.10}$$

and

$$\Gamma\int_\alpha^\beta (\mathbf{f} + \mathbf{g}) \cdot \mathbf{dr} = \Gamma\int_\alpha^\beta \mathbf{f} \cdot \mathbf{dr} + \Gamma\int_\alpha^\beta \mathbf{g} \cdot \mathbf{dr}. \tag{5–1.11}$$

Furthermore, if $\Gamma = \{\mathbf{r}(t): a \le t \le b\}$ *is a piecewise smooth curve, then for any* c, $a < c < b$, *it follows that if*

$$\int_{\Gamma_1} \int_{\alpha}^{\gamma} \mathbf{f} \cdot \mathbf{dr} \quad \textit{and} \quad \int_{\Gamma_2} \int_{\gamma}^{\beta} \mathbf{f} \cdot \mathbf{dr}$$

exist where $\Gamma_1 = \{\mathbf{r}(t): a \le t \le c\}$ *and* $\Gamma_2 = \{\mathbf{r}(t): c \le t \le b\}$ *and where* $\boldsymbol{\alpha} = \mathbf{r}(a)$, $\boldsymbol{\beta} = \mathbf{r}(b)$, *and* $\boldsymbol{\gamma} = \mathbf{r}(c)$, *then*

$$\int_{\Gamma} \int_{\alpha}^{\beta} \mathbf{f} \cdot \mathbf{dr}$$

exists, and

$$\int_{\Gamma} \int_{\alpha}^{\beta} \mathbf{f} \cdot \mathbf{dr} = \int_{\Gamma_1} \int_{\alpha}^{\gamma} \mathbf{f} \cdot \mathbf{dr} + \int_{\Gamma_2} \int_{\gamma}^{\beta} \mathbf{f} \cdot \mathbf{dr}. \tag{5–1.12}$$

These three properties are quite easy to establish from Definition 5–1.1 (cf. Problem 10). Using (5–1.12) of Proposition 5–1.2 allows us to make Proposition 5–1.1 more generally applicable. That is, it is easy to see that (5–1.5) is true for piecewise smooth curves (cf. Problem 9).

In the following example a line integral is evaluated over three different curves between the same two points.

Example 5–1.2. Consider the three line integrals

$$\int_{\Gamma} \int_{(0,0)}^{(1,2)} (x + y^2)\, dx + (2xy^2)\, dy, \quad i = 1, 2, 3 \tag{5–1.13}$$

where $\Gamma_1 = \{(x, y): y = 2x\}$, $\Gamma_2 = \{(x, y): y = 2x^2\}$, and Γ_3 is made up of the line segment from (0, 0) to (1, 0), or Γ_3', and the line segment from (1, 0) to (1, 2), or Γ_3''.

For $i = 1$, we find by use of Proposition 5–1.1 that (5–1.13) becomes

$$\int_0^1 (x + 4x^2)\, dx + \int_0^2 y^3\, dy = \frac{35}{6},$$

and for $i = 2$ we see that (5–1.13) becomes

$$\int_0^1 (x + 4x^4)\, dx + \int_0^2 (\sqrt{2}y^{(5/2)})\, dy = \frac{411}{70}.$$

For $i = 3$, we get

$$\int_{\Gamma_3'} \int_{(0,0)}^{(1,0)} (x + y^2)\, dx + (2xy^2)\, dy = \int_0^1 x\, dx = \frac{1}{2}$$

since $y = 0$ on Γ_3', and

$$\int_{\Gamma_3''} \int_{(1,0)}^{(1,2)} (x + y^2)\, dx + (2xy^2)\, dy = \int_0^2 (2y^2)\, dy = \frac{16}{3}$$

since $x = 1$ on Γ_3''. Hence, by (5–1.12) of Proposition 5–1.2, it follows that

$$_{\Gamma_3}\int_{(0,0)}^{(1,2)} (x + y^2)\, dx + (2xy^2)\, dy = \frac{35}{6}.$$

A particular notation that is frequently used needs special attention here—namely,

$$_{\Gamma}\int_{\alpha}^{\beta} g(x, y, z)\, ds \tag{5–1.14}$$

where s is understood to be arc length. The intended interpretation is that

$$_{\Gamma}\int_{\alpha}^{\beta} g(x, y, z)\, ds = \int_a^b g(x(s), y(s), z(s))\, ds \tag{5–1.15}$$

where $\Gamma = \{(x(s), y(s), z(s)) \colon a \leq s \leq b\}$. Such an integral is frequently referred to as a line integral since, by (5–1.5),

$$_{\Gamma}\int_{\alpha}^{\beta} \mathbf{f} \cdot \mathbf{dr}$$

would equal the right side of (5–1.15) if a function $\mathbf{f}$ exists such that

$$\mathbf{f}(\mathbf{r}(s)) \cdot \mathbf{r}'(s) = g(\mathbf{r}(s))$$

where $\mathbf{r}(s) = (x(s), y(s), z(s))$. This convention is also followed if a parametric representation is given for Γ; for example, $\mathbf{r}(t) = (x(t), y(t), z(t))$, $a \leq t \leq b$. Again, as in (5–1.15), we would interpret the symbol

$$_{\Gamma}\int_{\alpha}^{\beta} g(x, y, z)\, dt$$

as

$$\int_a^b g(x(t), y(t), z(t))\, dt. \tag{5–1.16}$$

The following example illustrates this particular interpretation.

Example 5–1.3. Consider

$$_{\Gamma}\int_{(0,0)}^{(1,1)} g(x, y)\, ds$$

where $g(x, y) = x + y$ and $\Gamma = \{(x, y) \colon y = x^2\}$. By (5–1.15) we would have

$$_{\Gamma}\int_{(0,0)}^{(1,1)} g(x, y)\, ds = \int_a^b (x(s) + y(s))\, ds \tag{5–1.17}$$

where $\Gamma = \{(x(s), y(s)) \colon a \leq s \leq b\}$. In order to determine the functions x and y of the parametric representation of Γ, write Γ in any convenient parametric representation; for example,

$$\Gamma = \{(x(t), y(t)): x(t) = t, y(t) = t^2, 0 \le t \le 1\}. \tag{5–1.18}$$

Then s and t are related by

$$\frac{ds}{dt} = \sqrt{\left(\frac{dx}{dt}\right)^2 + \left(\frac{dy}{dt}\right)^2},$$

and hence

$$s = \int_0^t \sqrt{1 + 4\tau^2}\, d\tau = h(t). \tag{5–1.19}$$

By solving (5–1.19) for t in terms of s and substituting into (5–1.18), we could then proceed with the evaluation of the integral of (5–1.17). However, it is easier to change the variable of integration of this integral. For example, integrate on x. Then

$$ds = \sqrt{1 + \left(\frac{dy}{dx}\right)^2}\, dx,$$

and

$$\begin{aligned}\int_a^b (x(s) + y(s))\, ds &= \int_0^1 (x + x^2)\sqrt{1 + (2x)^2}\, dx \\ &= \frac{67}{96}\sqrt{5} - \frac{1}{12} - \frac{1}{16}\log(2 + \sqrt{5}).\end{aligned}$$

Similarly, if we were considering instead the integral

$$_\Gamma\int_{(0,0)}^{(1,1)} (x + y)\, dt$$

where Γ is given by (5–1.18), then the interpretation would be as in (5–1.16), thus:

$$_\Gamma\int_{(0,0)}^{(1,1)} (x + y)\, dt = \int_0^1 (t + t^2)\, dt = \frac{5}{6}.$$

As a conclusion to the discussion of line integrals a few remarks are in order about their dependence on the parametric representation and the sufficient conditions for their existence (cf. Definition 5–1.1).

It may appear from the definition that the value of the line integral could depend upon the parametric representations (parametrizations) of Γ. However, the value of the line integral will be the same among "properly equivalent" functions $\mathbf{r}_1$ and $\mathbf{r}_2$ that give parametric representations of Γ.†

† Cf. T. M. Apostal, *Mathematical Analysis*, pp. 171 ff., and Theorem 10–35, p. 278.

For simple piecewise smooth curves, either closed or not, any two parametrizations $\mathbf{r}_1$ and $\mathbf{r}_2$ of Γ are *properly equivalent* if $\mathbf{r}_1(t)$ and $\mathbf{r}_2(t)$ "move" around Γ in the same "direction" while t increases. Since most of the curves of interest to us will be either simple curves or curves that can be broken up into "parts" that are simple, it is sufficient for us to realize that the parametric representation must render the curve piecewise smooth, must have the correct limits (end points) if it is not a closed curve, and must have the correct "direction" in order to give the correct value of the line integral. This really leaves a great deal of freedom in choosing the parametric representation for evaluating a line integral (cf. Problem 4(f)).

In this connection it should be noted that a line integral like

$$_\Gamma\int_\alpha^\beta g(x, y, z)\, dt$$

as interpreted by (5–1.16) does depend on the parametric representation. For consider the integral of Example 5–1.3,

$$_\Gamma\int_{(0,0)}^{(1,1)} (x + y)\, dt$$

where $\Gamma = \{(t^2, t^4): 0 \leq t \leq 1\}$ is the parametric representation instead of the one given in (5–1.18). In this case,

$$_\Gamma\int_{(0,0)}^{(1,1)} (x + y)\, dt = \int_0^1 (t^2 + t^4)\, dt = \frac{8}{15},$$

instead of $5/6$. Yet these two parametric representations are properly equivalent.

The line integral of (5–1.2) will exist if $\mathbf{f}$ is such that $f_i([a, b])$ is bounded for each i, $1 \leq i \leq n$, and if in addition the composite function $\mathbf{fr}$ is continuous at all points of $[a, b]$ except a set of points I^* where I^* can be contained in a collection of closed intervals, the sum of whose length is arbitrarily small.

Problems

1. Consider the line integral

$$\int_{(1,2)}^{(4,4)} f(x, y)\, dy$$

where f and Γ are the same as in Example 5–1.1.

(a) Find the value of the Riemann sum pertinent to this line integral relative to the partition $P = \{2, 3, 4\}$ where $\bar{y}_i$ is a midpoint of the interval $[y_i, y_{i-1}]$.

(b) Find the value of this line integral by a method analogous to the method used in the example.

(c) Draw a figure like Figure 5–1.3 illustrating the areas that are equal to the Riemann sum of (a) and the value of the line integral of (b).

2. In Problem 1(b), the line integral is evaluated by integrating with respect to y, so that

$$\int_{(1,2)}^{(4,4)} f(x, y)\, dy = \int_2^4 f\left(\frac{y^2}{4}, y\right) dy.$$

(a) Evaluate this same line integral by integrating with respect to x. Thus change the variable of integration from y to x in the integral on the right where x and y are related by the equation $4x = y^2$.

(b) Evaluate this same line integral by integrating with respect to t where $x = t^2$ and $y = 2t$.

3. Consider the line integral $\int_\Gamma f(x, y)\, ds$ where f and Γ are the same as in Example 5–1.1 and where $s = h(t)$ is the arc length along Γ from $(1, 2)$ to $(x(t), y(t))$. If we find a parametrical representation of Γ such that the parameter is s and, therefore,

$$\Gamma = \{(x, y): x = g_1(s), y = g_2(s), h(1) \le s \le h(2)\},$$

then

$$\int_\Gamma f(x, y)\, ds = \int_{h(1)}^{h(2)} f(g_1(s), g_2(s))\, ds$$

(cf. 5–1.15).

(a) Find a set in Figure 5–1.3 whose area is equal to the value of this line integral.

(b) Find the value of this line integral by integrating with respect to t. In this case $ds = \sqrt{(x'(t))^2 + (y'(t))^2}\, dt$ where $x(t) = t^2$, $y(t) = 2t$.

(c) Find the value of this line integral by integrating with respect to x. In this case

$$ds = \sqrt{1 + \left(\frac{dy}{dx}\right)^2}\, dx, \quad \text{and} \quad 4x = y^2.$$

Note that ds must always be positive. Hence in some other cases it may be necessary to take

$$ds = -\sqrt{1 + \left(\frac{dy}{dx}\right)^2}\, dx.$$

4. Evaluate the line integrals. The meaning of the variable s is arc length as in Problem 3.

(a) $$\Gamma\int_{(0,1)}^{(2,0)} (y^2 - 3x)\, dx, \quad \Gamma\int_{(0,1)}^{(2,0)} (y^2 - 3x)\, dy \quad \text{and} \quad \Gamma\int_{(0,1)}^{(2,0)} (y^2 - 3x)\, ds$$

along Γ the line defined by $\{(x, y): x + 2y = 2, 0 \le x \le z\}$.

(b) $\int_\Gamma x^{1/2}y\,ds$ and $\int_\Gamma \frac{x}{1+y}\,dx + \frac{2y}{1+x}\,dy$

along the curve $\Gamma = \{(x, y): x = t^2, y = t^3, 0 \le t \le 2\}$.

(c) $\int_{\Gamma\,(1,2,1)}^{(3,0,2)} xyz\,ds$ and $\int_{\Gamma\,(1,2,1)}^{(3,0,2)} (xy + z)\,dx + (y^2 + xz)\,dz$

along the line Γ, joining (1, 2, 1) to (3, 0, 2).

(d) $\int_{\Gamma\,(0,0,0)}^{(1,1,2)} (x + z)\,dx + (y - x)\,dy + yz\,dz$

along the curve $\Gamma = \{(x, y, z): x = y^2, z = 2y^3, 0 \le x \le 1\}$.

(e) $\int_{\Gamma\,(0,1)}^{(1,c)} \sinh x\,ds$ and $\int_{\Gamma\,(0,1)}^{(1,c)} ye^x\,dx + dy$, $c = \dfrac{e^2+1}{2e}$,

along the curve $\Gamma = \{(x, y): y = \cosh x, 0 \le x \le 1\}$.

(f) $\int_{\Gamma\,(0,0,0)}^{(1,2,3)} yz\,dx + xz\,dy + xy\,dz$ and $\int_{\Gamma\,(0,0,0)}^{(1,2,3)} \frac{x^5 + yz}{1 + xy}\,dt$ along the curve $\Gamma = \{(x, y, z): x = t, y = 2t^2, z = 3t^3, 0 \le t \le 1\}$

Evaluate these two integrals again where

$$\Gamma = \{(x, y, t): x = t^2, y = 2t^4, z = 3t^6, 0 \le t \le 1\}.$$

5. Evaluate the following line integrals around the boundary Γ of the following regions S.

(a) $\oint_\Gamma (x + y)\,dx + 2xy\,dy$, $\quad S = \{(x, y): 4x^2 + y^2 \le 1\}$.

(b) $\oint_\Gamma (x^2 + y^2)\,ds$, S is the triangular region with vertices (0, 0), (2, 0), and (2, 3).

(c) $\oint_\Gamma yx\,dy + y^2\,dx$, $\quad S = \{(x, y): |x| + |y| \ge 1\}$.

(d) $\oint_\Gamma y\,dx - x\,dy$, $\quad S = \{(x, y): 1 \le x^2 + y^2 \le 4\}$.

6. If Γ is the boundary of a convex three-sided region of the plane, prove that the area of S is given by

$$A = \frac{1}{2}\oint_\Gamma x\,dy - y\,dx.$$

Extend this result to any convex region of the plane.

7. Consider a thin wire occupying the position of the points of the curve

$$\Gamma = \{(x, y, z): x = g_1(s), y = g_2(s), z = g_3(s), a \leq s \leq b\}$$

where s is the arc length as in Problem 3. If $\rho(x, y, z)$ is the linear density of the wire at the point $(x, y, z) \in \Gamma$, where the function ρ defined by $\rho = \rho(x, y, z)$ is continuous on Γ, then the mass of the wire is

$$M(\Gamma) = \int_\Gamma \rho(x, y, z)\, ds.$$

The moment of the wire about the (y, z)-plane is

$$M_{yz}(\Gamma) = \int_\Gamma x\rho(x, y, z)\, ds.$$

Similarly the moment of the wire about the (x, z)-plane (and about the (x, y)-plane) can be defined. The center of mass of the wire, $(\bar{x}, \bar{y}, \bar{z})$ is given by $M(\Gamma)\bar{x} = M_{yz}(\Gamma)$, etc. For each of the following circumstances find the mass of the wire, each of the three moments about each of the coordinate planes, and the center of mass.

(a) $\Gamma = \{(x, y, z): x^2 + y^2 = 1, |y| \leq x\}$, $\rho(x, y, z) = y^2$.

(b) $$\Gamma = \left\{(x, y, z): x = \cos\frac{s}{\sqrt{2}}, y = \sin\frac{s}{\sqrt{2}}, z = \frac{s}{\sqrt{2}}, \quad 0 \leq s \leq \frac{\sqrt{2}\pi}{2}\right\}, \quad \rho(x, y, z) = y.$$

(c) $\Gamma = \{(x, y, z): x = 1 - t, y = 2 + t, z = 3t, 0 \leq t \leq 1\}$,

$$\rho(x, y, z) = (x^2 + y^2 + z^2)^{-1/2}.$$

8. Consider the same wire discussed in Problem 7. The moment of inertia of the wire about the (x, y)-plane is

$$I_{xy}(\Gamma) = \int_\Gamma z^2\rho(x, y, z)\, ds.$$

Similarly, $I_{yz}(\Gamma)$ and $I_{xz}(\Gamma)$, the moments of inertia about the (y, z)- and (x, z)-planes, respectively, can be defined. The moment of inertia about the x-axis, $I_x(\Gamma)$, is given by

$$I_x(\Gamma) = I_{xz}(\Gamma) + I_{xy}(\Gamma),$$

and the moment of inertia about the origin, $I_0(\Gamma)$, is given by

$$I_0(\Gamma) = I_{xy}(\Gamma) + I_{xz}(\Gamma) + I_{yz}(\Gamma).$$

For the wire in the positions of Γ with density function ρ as defined in (a), (b), and (c) of Problem 7, find the moment of inertia about each of the three coordinate planes, about each of the three axes, and about the origin.

If $\bar{\bar{z}}$ is defined by the equation $M(\Gamma)(\bar{\bar{z}})^2 = I_{xy}(\Gamma)$ and x and $\bar{\bar{y}}$ are defined analogously, find $(\bar{\bar{x}}, \bar{\bar{y}}, \bar{\bar{z}})$. The number $\bar{\bar{z}}$ is referred to as the *radius of gyration* of the wire about the (x, y)-plane. The quantities $\bar{\bar{x}}$ and $\bar{\bar{y}}$ have similar names.

9. Prove Proposition 5–1.1 by writing out the meaning of each of the integrals of (5–1.5) and showing that the limits of these Riemann sums give (5–1.5). Extend this proposition to include piecewise smooth curves.

10. Prove Proposition 5–1.2.

11. Let Γ be a piecewise smooth curve in E^n and let $\mathbf{f}$ be a continuous function on an open region containing Γ. Show that

$$\int_\Gamma (\mathbf{f} \cdot \mathbf{t})\, ds = \int_\Gamma \mathbf{f} \cdot d\mathbf{r}$$

where s is arc length and where $\mathbf{t}$ is the unit tangent to Γ.

12. Given $\Gamma = \{\mathbf{r}(t) : a \le t \le b\}$, a piecewise smooth curve in (x, y, z)-space, take $\Gamma_1 = \{(r_1(t), r_2(t)) : a \le t \le b\}$, the piecewise smooth curve that is the projection of Γ on the (x, y)-plane. Show that if F is a continuous function on an open region in (x, y, z)-space that contains Γ and $\varphi \in C^1(S)$, where S is an open region in the (x, y)-plane that contains Γ_1, then

$$\int_\Gamma F(x, y, z)\, dz = \int_{\Gamma_1} F(x, y, \varphi(x, y))\varphi_x(x, y)\, dx + F(x, y, \varphi(x, y))\varphi_y(x, y)\, dy.$$

Answers

1. (a) 5.4888 ⋯. (b) 5.4663 ⋯.

3. (b) and (c) $\dfrac{24461}{4608}\sqrt{5} - \dfrac{9155}{9216}\sqrt{2} + \dfrac{429}{1152}\log\left(\dfrac{1+\sqrt{2}}{2+\sqrt{5}}\right)$.

4. (a) $-16/3,\ 8/3,\ -8\sqrt{5}/3$.
   (b) $(285440/15309)\sqrt{10},\ 16 + \log 500 + \pi\sqrt{3}/3$
   (c) $7,\ 65/6$. (d) $334/105$.
   (e) $\dfrac{(e^2-1)^2}{8e^2},\ \dfrac{e^3 + 2e^2 - 3e + 2}{4e}$. (f) $6,\ \dfrac{7}{6} - \dfrac{7}{12}\log 3$.

5. (a) $-\pi/2$. (b) $\dfrac{13\sqrt{13} + 71}{3}$ (c) 0. (d) -6π.

7. (a) $M(\Gamma) = (\pi - 2)/4$, $(\bar{x}, \bar{y}, \bar{z}) = (2\sqrt{2}(\pi - 2), 0, 0)$.
   (b) $M(\Gamma) = \sqrt{2}$, $(\bar{x}, \bar{y}, \bar{z}) = \left(\tfrac{1}{2}, \dfrac{\pi}{4}, 1\right)$.

(c) $M(\Gamma) = \log[(\sqrt{198} + 12)/(\sqrt{55} + 1)]$.

$$M_{yz}(\Gamma) = \frac{\sqrt{55} - \sqrt{198}}{11} + \frac{12}{11}\log\left(\frac{\sqrt{198} + 12}{\sqrt{55} + 1}\right),$$

$$M_{xz}(\Gamma) = \frac{\sqrt{198} - \sqrt{55}}{11} + \frac{21}{11}\log\left(\frac{\sqrt{198} + 12}{\sqrt{55} + 1}\right),$$

$$M_{xy}(\Gamma) = 3\,\frac{\sqrt{198} - \sqrt{55}}{11} - \frac{3}{11}\log\left(\frac{\sqrt{198} + 12}{\sqrt{55} + 1}\right).$$

8. (a) $I_{xy}(\Gamma) = 0, \quad I_{yz} = \dfrac{\pi}{16}, \quad I_{xz}(\Gamma) = \dfrac{3\pi + 8}{16},$

$$I_x(\Gamma) = \frac{3\pi + 8}{16}, \quad I_y(\Gamma) = \frac{\pi}{16}, \quad I_z(\Gamma) = \frac{\pi + 2}{4},$$

$$I_0(\Gamma) = \frac{\pi + 2}{4},$$

$$\bar{\bar{x}} = \tfrac{1}{2}\sqrt{\pi/(\pi - 2)}, \quad \bar{\bar{y}} = \tfrac{1}{2}\sqrt{(3\pi + 8)/(\pi - 2)}, \quad \bar{\bar{z}} = 0.$$

(b) $I_{xy}(\Gamma) = \sqrt{2}(\pi - 2), \quad I_{yz}(\Gamma) = \sqrt{2}/3, \quad I_{xz}(\Gamma) = 2\sqrt{2}/3$

$$I_x(\Gamma) = \frac{\sqrt{2}(3\pi - 4)}{3}, \quad I_y(\Gamma) = \frac{\sqrt{2}(3\pi - 5)}{3}, \quad I_z(\Gamma) = \sqrt{2},$$

$$I_0(\Gamma) = \sqrt{2}(\pi - 1),$$

$$\bar{\bar{x}} = \frac{\sqrt{3}}{3}, \quad \bar{\bar{y}} = \frac{\sqrt{6}}{3}, \quad \bar{\bar{z}} = \sqrt{\pi - 2}.$$

(c) $I_{xy}(\Gamma) = \dfrac{72\sqrt{198} + 27\sqrt{55}}{242} - \dfrac{234}{121}\log\left(\dfrac{\sqrt{198} + 12}{\sqrt{55} + 1}\right),$

$$I_{yz}(\Gamma) = \frac{47\sqrt{55} - 36\sqrt{198}}{242} + \frac{117}{121}\log\left(\frac{\sqrt{198} + 12}{\sqrt{55} + 1}\right),$$

$$I_{xz}(\Gamma) = \frac{88\sqrt{198} - 85\sqrt{55}}{242} + \frac{414}{121}\log\left(\frac{\sqrt{198} + 12}{\sqrt{55} + 1}\right).$$

5-2. Surfaces

Surfaces are commonly thought of as certain sets of points in E^3. These sets of points are customarily represented in one of the following three ways: implicitly, explicitly, or parametrically. For example,

$$\{(x, y, z)\colon x^2 + y^2 + z^2 - 1 = 0\}$$

is an implicit representation of the spherical surface;

$$\{(x, y, z): z = \sqrt{1 - x^2 - y^2}\}$$

is an explicit representation of a hemispherical surface; and

$$\{(x, y, z): x = \cos u \sin v, y = \sin u \sin v, z = \cos v,$$
$$0 \leq u \leq \pi/2, \pi/4 \leq v \leq \pi/2\} \quad (5\text{–}2.1)$$

is a parametric representation of "part" of a spherical surface (cf. Figure 5–2.1(b)).

Before trying to define a surface, we first define a "piece" of a surface, a surface element.

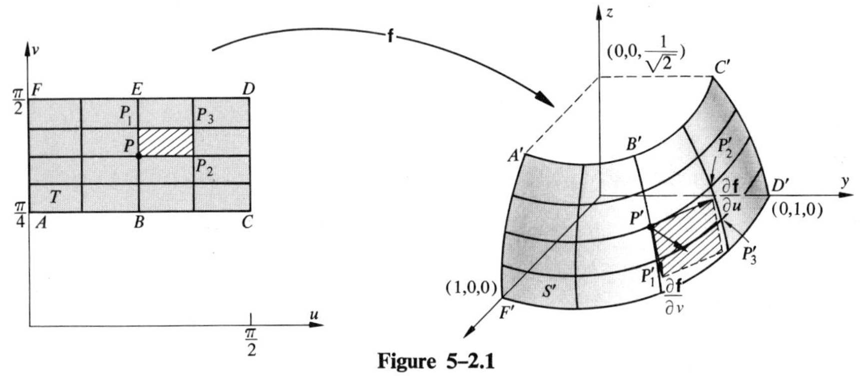

Figure 5–2.1

Definition 5–2.1. *A set of points S in (x, y, z)-space, E^3, is a* ***surface element*** *if there exists a plane regular region T in (u, v)-space, an E^2 space, such that there exists a function* **f** *that maps T onto S such that* **f** *is continuous on T. If it is possible to find T and* **f** *such that* **f** *is 1–1 on T, then S is a simple surface element. If it is possible to find T and* **f** *such that* **f** *is continuously differentiable on T,*

$$\left\| \frac{\partial}{\partial u} \mathbf{f}(u, v) \times \frac{\partial}{\partial v} \mathbf{f}(u, v) \right\| \neq 0$$

for all $(u, v) \in T^{\circ}$, and

$$\lim_{(u,v)\to(u_0,v_0)} \left\| \frac{\partial}{\partial u} \mathbf{f}(u, v) \times \frac{\partial}{\partial v} \mathbf{f}(u, v) \right\| \neq 0$$

for $(u_0, v_0) \in$ bdy *(T), then S is a smooth surface element.*

For each of the above circumstances, the function **f** with the plane regular region T is a parametric representation of S. Given that S is a surface element and $\mathbf{g}$ is a function that maps a plane regular region, T_1, onto S, then $\mathbf{g}$ with T_1 is a *parametric representation* of S only if $\mathbf{g}$ is continuous on T_1. Similarly, if S is a simple surface element or a smooth surface element, then $\mathbf{g}$ with T_1 would be a parametric representation only if $\mathbf{g}$ had the required properties indicated in the above definition.

By a simple calculation it can be shown that

$$\frac{\partial \mathbf{f}}{\partial u} \times \frac{\partial \mathbf{f}}{\partial v} = \left(\frac{\partial(f_2, f_3)}{\partial(u, v)}, \frac{\partial(f_3, f_1)}{\partial(u, v)}, \frac{\partial(f_1, f_2)}{\partial(u, v)}\right). \tag{5–2.2}$$

In Figure 5–2.1(a) a plane regular region T is shown. By the mapping $\mathbf{f}$ defined in (5–2.1), the region T is mapped onto the set S, a part of a spherical surface (cf. Figure 5–2.1(b)). Since on the set T the function $\mathbf{f}$ is 1–1, this surface element is simple. By $\mathbf{f}$, the line segments in T where u and v are constant become the longitude and latitude curves on S. At the point $(u, v) = P$, the vectors $\partial\mathbf{f}/\partial u$ and $\partial\mathbf{f}/\partial v$ become the tangent vectors at P' of these two curves on S (cf. Figure 5–2.1(b)). The quantity

$$\left\| \frac{\partial}{\partial u}\mathbf{f}(u, v) \times \frac{\partial}{\partial v}\mathbf{f}(u, v) \right\| \tag{5–2.3}$$

at $(u, v) = P$ is the area of the shaded parallelogram whose sides are line segments given by the vectors $(\partial/\partial u)\mathbf{f}(u, v)$ and $(\partial/\partial v)\mathbf{f}(u, v)$ evaluated at P. Clearly, the quantity of (5–2.3) will be zero only if

$$\frac{\partial}{\partial u}\mathbf{f}(u, v) = \mathbf{0} \quad \text{or} \quad \frac{\partial}{\partial v}\mathbf{f}(u, v) = \mathbf{0}$$

or if at P' the curves in S corresponding to the curves of T where u and v are constant are tangent. In this example it is easily seen that none of these things happens in T°; nor do they have limiting values of zero on the boundary of T. Hence S is a smooth surface element.

The following example illustrates some of the problems encountered in deciding whether or not certain sets in E^n are surface elements.

Example 5–2.1. Is the set

$$S = \{(x, y, z) \colon x^2 + y^2 + z^2 = 1, x \geq 0, y \geq 0, z \geq 0\}$$

a simple smooth surface element? The answer is yes, but it is instructive to look at two parametric representations of S that fail to indicate completely these properties of S.

The function $\mathbf{f}$ defined by

$$\mathbf{f}(u, v) = (\cos u \sin v, \sin u \sin v, \cos v)$$

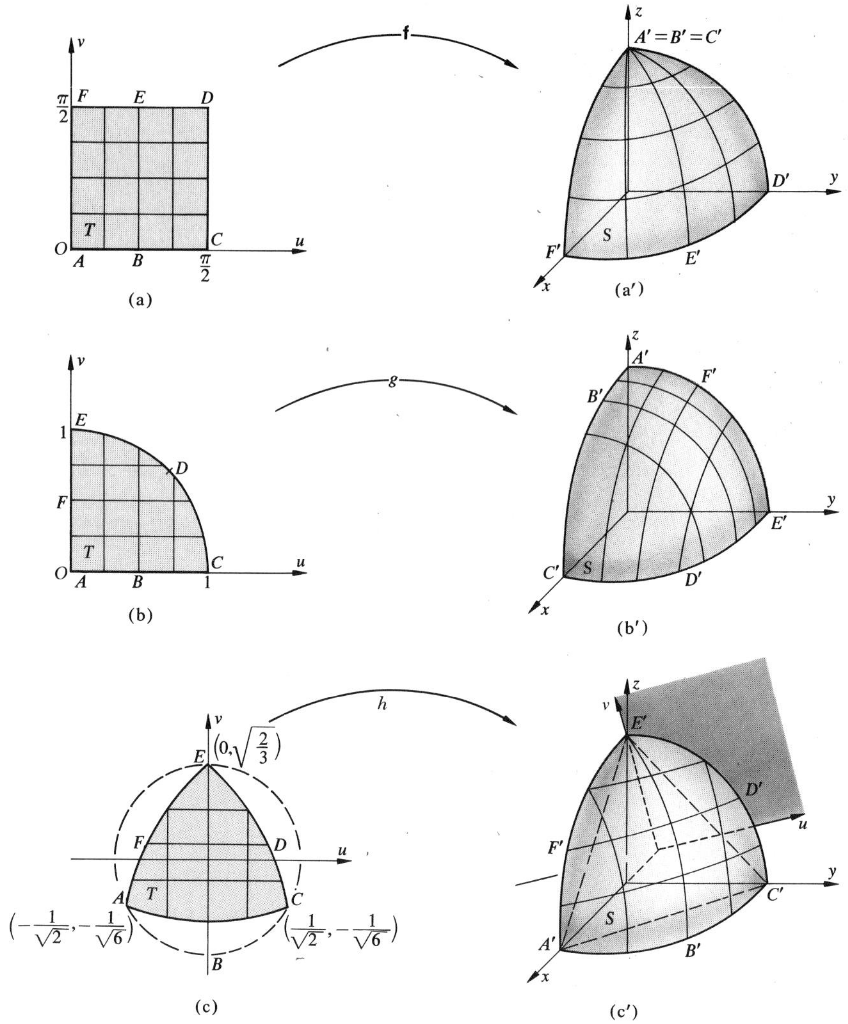

Figure 5–2.2

on $T = \{(u, v): 0 \le u \le \pi/2,\ 0 \le v \le \pi/2\}$ maps T onto S, but **f** is not 1–1 in that all of the points on $\overline{AC}$ map into A' (cf. Figure 5–2.2(a) and (a′)). The function **g** defined by

$$\mathbf{g}(u, v) = (u, v, \sqrt{1 - u^2 - v^2})$$

on $T = \{(u, v): u^2 + v^2 \le 1,\ u \ge 0,\ v \ge 0\}$ maps T in a 1–1 continuous manner onto S (cf. Figure 5–2.2(b) and (b′)). However, the first partials of **g** on T^o cannot be extended to continuous functions on T, since both first partials of the z component become infinite along the boundary curve,

$$\{(u, v): u^2 + v^2 = 1, u \geq 0, v \geq 0\}.$$

Thus $\mathbf{g}$ is not continuously differentiable on T. The function $\mathbf{h}$ defined by

$$\mathbf{h}(u, v) = \left(\frac{1}{\sqrt{2}}u - \frac{1}{\sqrt{6}}v + \frac{1}{\sqrt{3}}\sqrt{1 - u^2 - v^2},\right.$$
$$-\frac{1}{\sqrt{2}}u - \frac{1}{\sqrt{6}}v + \frac{1}{\sqrt{3}}\sqrt{1 - u^2 - v^2},$$
$$\left.\frac{2}{\sqrt{6}}v + \frac{1}{\sqrt{3}}\sqrt{1 - u^2 - v^2}\right)$$

on the set T in Figure 5–2.2(c) where the points (u, v) on the boundary curves from C to E, E to A, and A to C satisfy the following three equations, respectively:

$$-\frac{1}{\sqrt{2}}u + \frac{1}{\sqrt{6}}v = \frac{1}{\sqrt{3}}\sqrt{1 - u^2 - v^2},$$

$$\frac{1}{\sqrt{2}}u + \frac{1}{\sqrt{6}}v = \frac{1}{\sqrt{3}}\sqrt{1 - u^2 - v^2},$$

$$-\frac{2}{\sqrt{6}}v = \frac{1}{\sqrt{3}}\sqrt{1 - u^2 - v^2}.$$

Since all of T lies within the disk of radius $\sqrt{2/3}$ centered at the origin, it is clear that $\mathbf{h}$ is continuously differentiable on T. The mapping $\mathbf{h}$ is best understood by noting that it is obtained by projecting in E^3 parallel to the vector $(-1, -1, -1)$ the set S onto the plane determined by the three points A', C', and E', which is the plane whose equation is $x + y + z = 1$. The (u, v)-plane becomes this plane, with $(1/3, 1/3, 1/3)$ taken as the origin and with the u and v axes taken as shown in Figure 5–2.2(c′). The set T is the image of S by this projection, and thus $\mathbf{h}$ is the inverse mapping of this same projection. Thus it is clear that $\mathbf{h}$ is 1–1, and it is at least intuitively clear from Figure 5–2.2(c′) that

$$\frac{\partial}{\partial u}\mathbf{h}(u, v) \times \frac{\partial}{\partial v}\mathbf{h}(u, v) \neq \mathbf{0}$$

on T^0; nor does it have a limiting value of zero at a boundary point of T. Thus S is a simple smooth surface element.

This example shows that if S is a set in (x, y, z)-space, then there exists a 2-dimensional subspace of E^3 orthonormal basis $\{\mathbf{b}_1, \mathbf{b}_2\}$, with $\{\mathbf{b}_1, \mathbf{b}_2, \mathbf{b}_3\}$

an orthonormal basis of E^3 such that for $(x_0, y_0, z_0) \in S$ there exists a unique (u_0, v_0, t_0) such that

$$(x_0, y_0, z_0) + t_0\mathbf{b}_3 = u_0\mathbf{b}_1 + v_0\mathbf{b}_2. \tag{5-2.4}$$

If the set T of (u_0, v_0) so obtained forms a plane regular region of the (u, v)-plane, the plane whose basis is $\{\mathbf{b}_1, \mathbf{b}_2\}$, and if the function φ defined by $t = \varphi(u, v)$ on T where (u, v, t) satisfies (5-2.4) for some $(x, y, z) \in S$ is continuously differentiable on T, then S is a simple smooth surface element. Such a representation of a surface element S is an *explicit representation.*

In general, if φ is a continuously differentiable function on a plane regular region T in (x, y)-space, then

$$S = \{(x, y, z): z = \varphi(x, y), (x, y) \in T\}$$

is a surface element in (x, y, z)-space where x and y are the parameters, like u and v of Definition 5–2.1, and where

$$\mathbf{f}(x, y) = (x, y, \varphi(x, y)).$$

Clearly, $\mathbf{f}$ is 1–1 on T, and hence S is simple. The surface element S is smooth since

$$\frac{\partial}{\partial x}\mathbf{f}(x, y) \times \frac{\partial}{\partial y}\mathbf{f}(x, y) = (-\varphi_x(x, y), -\varphi_y(x, y), 1) \tag{5–2.5}$$

is not zero for any $(x, y) \in T$.

Definition 5–2.2. *If S is a smooth surface element, the tangent plane to S at P', the image of $P = (u_0, v_0)$ of T° under* $\mathbf{f}$, *is*

$$\left\{(x, y, z): ((x, y, z) - \mathbf{f}(u_0, v_0)) \cdot \left(\frac{\partial}{\partial u}\mathbf{f}(u_0, v_0) \times \frac{\partial}{\partial v}\mathbf{f}(u_0, v_0)\right) = 0\right\}. \tag{5–2.6}$$

If $(u_0, v_0) \in$ bdy (T), *then the cross product is replaced by*

$$\lim_{(u,v)\to(u_0,v_0)} \frac{\partial}{\partial u}\mathbf{f}(u, v) \times \frac{\partial}{\partial v}\mathbf{f}(u, v).$$

There are two unit vectors $\mathbf{n}_1$ *and* $\mathbf{n}_2$ *that are normal to S at P', namely,*

$$\mathbf{n}_1(u_0, v_0) = \frac{\dfrac{\partial}{\partial u}\mathbf{f}(u_0, v_0) \times \dfrac{\partial}{\partial v}\mathbf{f}(u_0, v_0)}{\left\|\dfrac{\partial}{\partial u}\mathbf{f}(u_0, v_0) \times \dfrac{\partial}{\partial v}\mathbf{f}(u_0, v_0)\right\|} \tag{5–2.7}$$

and $\mathbf{n}_2 = -\mathbf{n}_1$, *where the same interpretation as above is made for the case that* $(u_0, v_0) \in$ bdy (T).

If Γ is any smooth curve in T so that $\Gamma = \{(u(t), v(t)): a \le t \le b\} \subset T$ and if

$$\mathbf{g}(t) = (u(t), v(t))$$

is such that $\mathbf{g}$ is continuously differentiable on $[a, b]$, $\mathbf{g}'(t) \neq \mathbf{0}$ for $t \in (a, b)$, $\lim_{t\to a^+} \mathbf{g}'(t) \neq \mathbf{0}$, $\lim_{t\to b^-} \mathbf{g}'(t) \neq \mathbf{0}$, and if

$$\mathbf{g}(b) = (u_0, v_0),$$

then the tangent line to $f(\Gamma)$ at P' lies in the tangent plane given by (5–2.6). The tangent line to $\mathbf{f}(\Gamma)$ at P' is given by

$$\{\mathbf{f}(u_0, v_0) + \lambda(\mathbf{f}(\mathbf{g}(b)))': \lambda, \text{real}\}$$

(cf. Problem 9 of Section 1-5).
This tangent line is in the required tangent plane if

$$(\mathbf{f}(\mathbf{g}(b)))' \cdot \left(\frac{\partial}{\partial u}\mathbf{f}(u_0, v_0) \times \frac{\partial}{\partial v}\mathbf{f}(u_0, v_0)\right) = 0.$$

However, the left side becomes

$$\left(\frac{\partial}{\partial u}\mathbf{f}(u_0, v_0)\frac{d}{dt}u(b) + \frac{\partial}{\partial v}\mathbf{f}(u_0, v_0)\frac{d}{dt}v(b)\right) \cdot \left(\frac{\partial}{\partial u}\mathbf{f}(u_0, v_0) \times \frac{\partial}{\partial v}\mathbf{f}(u_0, v_0)\right)$$

$$= \frac{d}{dt}u(b)\left[\frac{\partial}{\partial u}\mathbf{f}(u_0, v_0) \cdot \left(\frac{\partial}{\partial u}\mathbf{f}(u_0, v_0) \times \frac{\partial}{\partial v}\mathbf{f}(u_0, v_0)\right)\right]$$

$$+ \frac{d}{dt}v(b)\left[\frac{\partial}{\partial u}\mathbf{f}(u_0, v_0) \cdot \left(\frac{\partial}{\partial u}\mathbf{f}(u_0, v_0) \times \frac{\partial}{\partial v}\mathbf{f}(u_0, v_0)\right)\right],$$

which is zero since each bracket is zero. Thus, if there are three curves in S that have P' as the terminal end point of each and if the three tangent lines to each curve at P' do not lie in the same plane, then S is not a smooth surface element. The following example indicates two such sets.

Example 5–2.2. In E^3 the sets

$$S_1 = \{(x, y, z): |x| \le 1, |y| \le 1, \text{ and } z = |x|\}$$

and

$$S_2 = \{(x, y, z): z^2 = x^2 + y^2, 0 \le z \le 1\}$$

are not smooth surface elements (cf. Figure 5–2.3(a) and (b)). In both cases at the origin it is possible to get the three required types of curves with non-coplanar tangent lines.

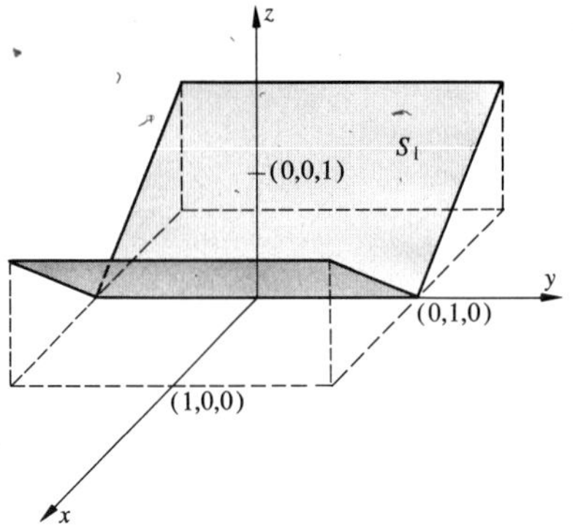

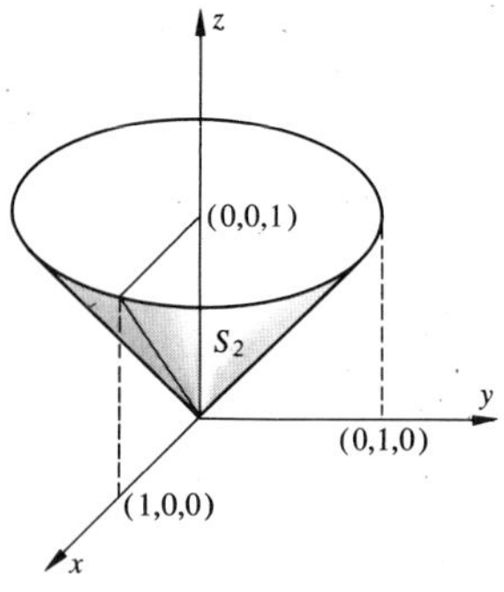

Figure 5–2.3

It is interesting to note that S_2 is still not a smooth surface element if S_2 is modified to include only those points over some sector of the (x, y)-plane, as for example, the set

$$\left\{(x, y, z): z^2 = x^2 + y^2, 0 \le z \le 1, (\tan \alpha)x \le y \le (\tan \beta)x, \right.$$
$$\left. y \ge 0, 0 < \alpha < \beta < \frac{\pi}{2}\right\}.$$

In Figure 5–2.4, we have an example of a surface element that is smooth but not simple. It is

$$S = \{(x, y, z): x = \cos u, y = \sin u, z = v, 0 \le u \le 2\pi, 0 \le v \le 1\}.$$

It is easy to see that $\mathbf{f}(u, v) = (\cos u, \sin u, v)$ is not 1–1 on

$$T = \{(u, v): 0 \le u \le 2\pi, 0 \le v \le 1\}$$

for this parametric representation, but to prove that S is not simple, it would be necessary to show that the one-to-oneness fails for *any* parametric representation of S.

In the case of a surface element S, a plane regular region T that maps onto S by $\mathbf{f}$ has as its boundary a simple piecewise smooth closed curve, Γ. Thus there exists a function $\mathbf{g}$ whose domain is $[a, b] \subset R$ and whose range is Γ; that is, for $t \in [a, b]$, $\mathbf{g}(t) \in \Gamma$. Furthermore, $\mathbf{g}$ is a piecewise differentiable function on $[a, b]$, $\mathbf{g}(a) = \mathbf{g}(b)$, $\mathbf{g}$ is 1–1 on $[a, b)$, and $\lim_{t \to t_0^+} \mathbf{g}'(t)$ and $\lim_{t \to t_0^-} \mathbf{g}'(t) \ne \mathbf{0}$ for any point $t_0 \in [a, b]$. The set of points

$$\{\mathbf{f}(\mathbf{g}(t)): t \in [a, b]\}$$

is the *edge* of the surface element S. Since the first partials of $\mathbf{f}$ may be extended to continuous functions on T, the composite function $\mathbf{fg}$ is

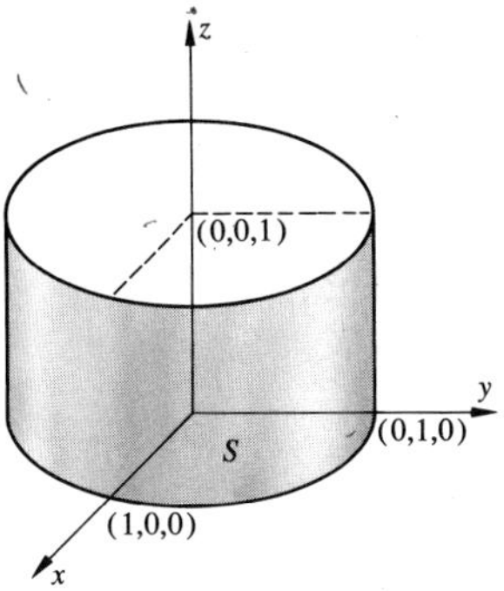

Figure 5–2.4

continuous on $[a, b]$. Furthermore, **fg** is piecewise differentiable on $[a, b]$. By the chain rule the components of $(d/dt)\mathbf{fg}$ are

$$\frac{\partial f_i}{\partial u}\frac{dg_1}{dt} + \frac{\partial f_1}{\partial v}\frac{\partial g_2}{dt} = 0, \quad i = 1, 2, 3.$$

If at $t_0 \in [a, b]$

$$\lim_{t\to t_0^+} \frac{d}{dt}\mathbf{f}(\mathbf{g}(t)) = \mathbf{0},$$

then, since $\lim_{t\to t_0^+} \mathbf{g}'(t) \neq \mathbf{0}$, it follows that the rank of the matrix

$$\begin{pmatrix} \dfrac{\partial f_1}{\partial u} & \dfrac{\partial f_1}{\partial v} \\ \dfrac{\partial f_2}{\partial u} & \dfrac{\partial f_2}{\partial v} \\ \dfrac{\partial f_3}{\partial u} & \dfrac{\partial f_3}{\partial v} \end{pmatrix}$$

at $(u, v) = \mathbf{g}(t_0)$ is less than or equal to one. However, for smooth surface elements, this would contradict the assumption that

$$\left\| \frac{\partial}{\partial u}\mathbf{f}(u, v) \times \frac{\partial}{\partial v}\mathbf{f}(u, v) \right\| \neq 0$$

(cf. (5–2.2)). A similar argument could be given if $\lim_{t\to t_0^-} \frac{d}{dt}\mathbf{f}(\mathbf{g}(t)) = \mathbf{0}$ for $t_0 \in [a, b]$. Therefore, *the edge of a smooth surface element is a simple piecewise smooth closed curve*. Clearly, if S is a simple surface element, then the edge of S is a simple closed curve.

We are now ready to describe how these surface elements may be assembled to form a finite surface.

Definition 5–2.3. *A connected set S in E^3 is a finite surface if there exists a finite collection of simple smooth surface elements $S_1, S_2, \cdots, S_n$ with Γ_i the edge of S_i, $1 \leq i \leq n$, such that:*

(a) $S = \bigcup_{i=1}^{n} S_i$.

(b) *For all $1 \leq i \leq j \leq n$, $S_i \cap S_j$ is empty or consists of a smooth curve or is a single point contained in $\Gamma_i \cap \Gamma_j$.*

(c) *If $S_i \cap S_j$, $i \neq j$, consists of a single point $\mathbf{x}_0$, then there exists a finite sequence $S_{k1}, \cdots, S_{km}$ from our collection of surface elements such that $S_i = S_{k1}$, $S_j = S_{km}$, $S_{k(p-1)} \cap S_{kp}$, $1 < p \leq m$, contains a smooth curve and $\mathbf{x}_0 \in S_{kp}$, $1 \leq p \leq m$.*

(d) *Any three of the surface elements S_i, $1 \leq i \leq n$, meet in, at most, a single point.*

This finite surface, S with these surface elements $S_1, \cdots, S_n$, satisfying (a) through (d) is said to be a sum of these surface elements written $S = S_1 + \cdots + S_n$.

If there is no point $\mathbf{x}$ in $\bigcup_{i=1}^{n} \Gamma_i$ such that $\mathbf{x}$ belongs to only one S_i, $1 \leq i \leq n$, then S is a closed finite surface. If S is not closed, let Γ be the closure of the set of $\mathbf{x}$ that belong to only one set Γ_i, $1 \leq i \leq n$. The set Γ is the edge of S.

This definition renders the "surface" of the regular polyhedrons—that is, tetrahedrons, cubes, octahedrons, etc.—as closed finite surfaces. Furthermore, the "surface" of a sphere, ellipsoid, and a torus are closed finite surfaces. The "surface" of a hemisphere is a finite surface with its edge being a simple closed curve. The lateral "surface" of a finite cylinder is a finite surface whose edge consists of two simple closed curves (cf. Figure 5–2.4).

To see what this last example implies, think of a long garden hose that is ever so badly tangled but has no kinks in it and is separated from touching itself. There would be finite surfaces in E^3 that could be "like" the surface of this garden hose.

Lastly, the Möbius strip—the "surface" realized by taking a strip of paper, giving it a half-twist, and then gluing the ends together—is a finite surface (cf. Figure 5–2.5(d)). The edge of this finite surface is a simple closed curve.

There are some sets in E^3 that may seem to be finite surfaces. First the set S_2, the "surface" of part of a cone containing the vertex, is *not* a finite surface (cf. Figure 5–2.3(b)). Neither are the following sets in (x, y, z)-space

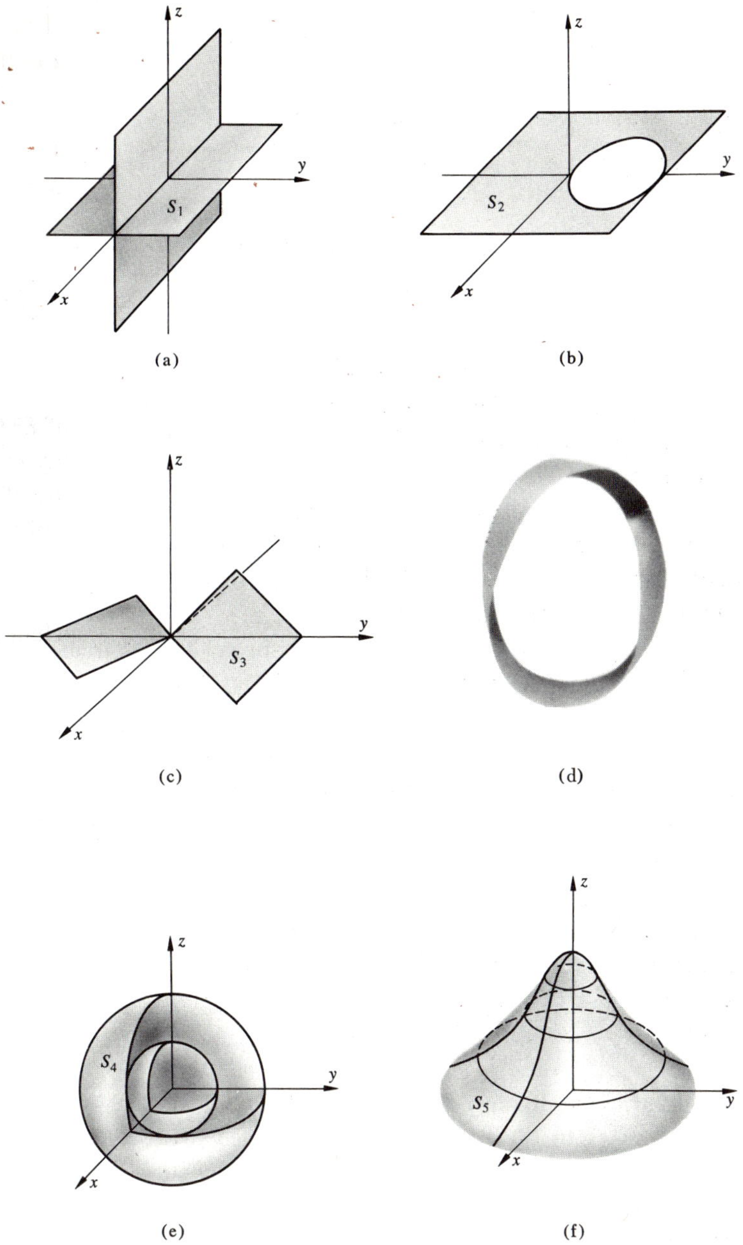

Figure 5–2.5

finite surfaces:

$$S_1 = \{(x, y, z): |x| \leq 1, |z| \leq 1, y = 0\} \cup \{(x, y, z): |y| \leq \tfrac{1}{2}, |x| \leq 1, z = 0\},$$

$$S_2 = \{(x, y, z): |x| \leq 2, |y| \leq 2 \text{ and } x^2 + (y - \tfrac{1}{2})^2 > \tfrac{1}{4}\},$$

$$S_3 = \{(x, y, z): |x| + |y + \tfrac{1}{2}| \leq 1, z = 0\} \cup \{(x, y, z): |y - \tfrac{1}{2}| + |z| \leq 1, x = 0\}$$

(cf. Figure 5–2.5(a), (b), and (c)). Furthermore, the sets

$$S_4 = \{x, y, z): x^2 + y^2 + z^2 = 1 \text{ or } 4\}$$

and

$$S_5 = \{(x, y, z: 1/(1 + x^2 + y^2)\}$$

are not finite surfaces (cf. Figure 5–2.5(e) and (f)).

Thus the idea of a finite surface includes almost all of the sets of E^3 that we may wish to think of as surfaces. However, it fails, most notably in the case of the "part" of a cone, shown in Figure 5–2.3(b), and in the case of the example in Figure 5–2.5(f). The example of the two concentric spherical surfaces of Figure 5–2.5(e) need not be disturbing, for it just serves to remind us that finite surfaces are connected sets.

The following definition of a surface will be broad enough to include the sets shown in Figure 5–2.3(b) and Figure 5–2.5(f).

Definition 5–2.4. *A set S in (x, y, z)-space is a surface*

(a) *if there is a sequence of infinite surfaces, $\{S_n : n = 1, 2, 3, \cdots\}$ such that $S_n \subset S$ for each n, $S_n \subset S_{n+1}$, $n = 1, 2, 3, \cdots$, and such that $\overline{(\bigcup_{n=1}^{\infty} S_n)} = S$ and*

(b) *if for each point $\mathbf{x} \in S$ there is a neighborhood of $\mathbf{x}$, $N(\mathbf{x}; \delta)$ such that $\bar{N}(\mathbf{x}; \delta) \cap S$ is a simple surface element.*

A sequence of finite surfaces related to S like the sequence in (a) is said to generate S. A point $\mathbf{x}$ of S is an exceptional point of S if $\mathbf{x}$ is not contained in any smooth surface element contained in S.

Notice that a smooth surface element and a finite surface are surfaces. Furthermore, notice that a surface is always a closed set, a convention that is suitable for our purposes although it is not universally followed.

In the preceding definition, the sequence of finite surfaces seems to converge monotonically to S. Such a concept will be useful in connection with the theory of integration on surfaces. For the moment, the following definition, which is motivated by the integration theory, will give us what we need.

Definition 5–2.5. *Let S be a surface. Let S^* be a nonempty closed subset of the intersection of a simple piecewise smooth curve and S. If and only if S is unbounded, we shall say ∞ belongs to S^* and allow that S^* need not contain any other points in this case. Let U be any open set containing S^* such that $U \supset \{(x, y, z): \|(x, y, z)\| > r\}$, for some $r > 0$, if $\infty \in S^*$. A sequence of finite surfaces, $\{S_n: n = 1, 2, 3, \cdots\}$, $S_n \subset S - S^*$, $n = 1, 2, 3, \cdots$ that generate S is said to converge to S relative to S^*, written $\lim_{n\to\infty} S_n = S(S^*)$, if there exists a positive integer n_0 such that for $n > n_0$, $S - S_n \subset U$.*

There is no attempt to define the convergence of S_n to S relative to S^* if S^* does not have the required properties.

In Definition 5–2.2 two unit normal vectors, $\mathbf{n}_1(u, v)$ and $\mathbf{n}_2(u, v)$, were defined for a surface element S where

$$S = \{\mathbf{f}(u, v): (u, v) \in T\}.$$

Either of these two functions, either one of which is the negative of the other, is a continuous vector-valued function defining at each point of S, even on the edge, Γ, of S, a unit vector. If now we think of the vector $\mathbf{n}_1(u, v)$ at a point on the edge of S as a person standing on the edge of S in the position of the vector $\mathbf{n}_1(u, v)$ and walking around the edge of S such that S is to the person's left, then the direction of motion is the *direction of positive orientation of Γ with respect to the normal* $\mathbf{n}_1(u, v)$ (cf. Figure 5–2.6). Clearly, the direction

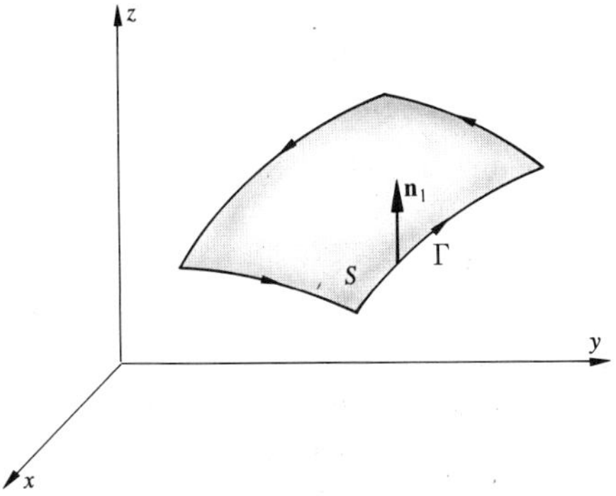

Figure 5–2.6

of positive orientation of Γ with respect to $\mathbf{n}_2(u, v)$ is the direction of negative orientation with respect to $\mathbf{n}_1(u, v)$. Notice that for surface elements in (x, y)-space the positive orientation around the edge of the surface element agrees with the discussion in the foregoing section about line integrals in the positive direction around boundaries of regions bounded by simple closed curves. For in that case the normal to the surface element was the unit vector in the direction of the positive z direction—namely, $(0, 0, 1)$.

A surface element S is said to be *oriented* if it is positively oriented with respect to either of the normals. This is customarily indicated by arrows placed on the edge of the surface element as the surface element is shown in a picture in E^3.

Definition 5–2.6. *A finite surface S is orientable if there exist orientated surface elements $S_1, \cdots, S_n$ such that*

(a) $S = S_1 + \cdots + S_n$, *and such that*

(b) *if $S_i \cap S_j$, $1 \leq i < j \leq n$, in a finite set of arcs, the orientation on each of these arcs induced by the orientation of Γ_i is always opposite to the orientation induced by the orientation of Γ_j.*

All of the examples of finite surfaces that have been mentioned are orientable except the Möbius strip (cf. Figure 5–2.5(d)). It is not obvious

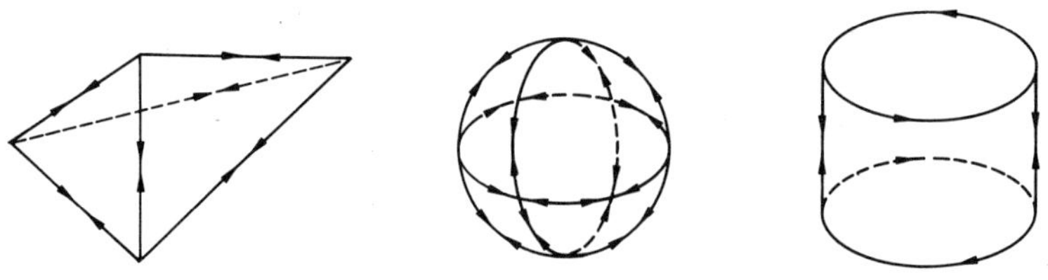

(a) (b) (c)

Figure 5–2.7

that the Möbius strip is not orientable, but it is somewhat convincing to take a set of surface elements whose sum is a Möbius strip and try to orient them successively. In Figure 5–2.7(a), (b), and (c) the "surfaces" of a tetrahedron, a sphere, and a cylinder of finite length with no top or bottom are demonstrated to be orientable.

Let Ω be a region in E^3 whose boundary is a closed finite surface S, as for example the regions illustrated in Figure 5–2.7(a) and (b). Then, since S is a finite surface, there exist surface elements $S_1, S_2, \cdots, S_n$ such that $S = S_1 + \cdots + S_n$. Each of the surface elements S_i is the image by a function $\mathbf{f}_i$ of a plane regular region T_i; that is, $\mathbf{f}_i(u, v) \in S_i$ for each $(u, v) \in T_i$, $1 \leq i \leq n$. At each point of S_i there are two normals to S_i, namely $\mathbf{n}_1(u, v)$ and $\mathbf{n}_2(u, v) = -\mathbf{n}_1(u, v)$ (cf. (5–2.7)). If $(u_0, v_0) \in T_i^{\circ}$, then the *outward normal* to S_i at $\mathbf{f}_i(u_0, v_0)$, $\mathbf{n}_0$ is $\mathbf{n}_1(u_0, v_0)$ if for some $\delta > 0$ $\mathbf{f}_i(u_0, v_0) + t\mathbf{n}_1(u_0, v_0) \notin \Omega$ for $0 < t < \delta$; otherwise $\mathbf{n}_0 = \mathbf{n}_2(u_0, v_0)$. We shall assume that the outward normal is well defined in this way at each point of S_i (that is, not on the edge of S_i); furthermore, we shall assume that it has been shown that $\mathbf{n}_0 = \mathbf{n}_1(u, v)$ for all $(u, v) \in T_i^{\circ}$ or $\mathbf{n}_0 = \mathbf{n}_2(u, v)$ for all $(u, v) \in T_i^{\circ}$. Then take $\mathbf{n}_0 = \mathbf{n}_1(u, v)$ ($\mathbf{n}_0 = \mathbf{n}_2(u, v)$) at points $\mathbf{x} = \mathbf{f}_i(u, v)$ on the edge of S_i if $\mathbf{n}_0 = \mathbf{n}_1(u, v)$ ($\mathbf{n}_0 = \mathbf{n}_2(u, v)$) for $(u, v) \in T_i^{\circ}$, unless this leads to an ambiguity. For example, at a "vertex" of the surface of a tetrahedron that "encloses" a region "within" the surface, we would be led to a circumstance of three outward normals at such points (cf. Figure 5–2.7(a)). This need not be the case, for if S is the boundary of $\Omega = \{\mathbf{x}: \|\mathbf{x}\| \leq r\}$ as illustrated in Figure 5–2.7(c), then Ω has a unique outward normal at each point of S. If $\mathbf{n}_0(\mathbf{x})$ is the outward normal at each point $\mathbf{x} \in S$ where the outward normal of Ω is defined, then $\mathbf{n}_i(x) = -\mathbf{n}_0(x)$ is the *inward normal* of Ω at $x \in S$.

Problems

1. Prove the identity (5–2.2).
2. Draw a sketch similar to Figure 5–2.1(a) and (b) for each of the following sets S in E^3. Classify each set on the basis of whether or not it is a surface element and whether it is simple and smooth with regard to this parametrization.
   (a) $\{(x, y, z): x = u \cos v, y = u \sin v, z = (1/2\pi)v, 1 \leq u \leq 2, 0 \leq v \leq 4\pi\}$.
   (b) $\{(x, y, z): x = (2 + \cos u) \cos v, y = (2 + \cos u) \sin v, z = \sin u, -\pi \leq u \leq \pi, 0 \leq v \leq 2\pi\}$.
   (c) $\{(x, y, z): x = u \cos v, y = u \sin v, z = u^2 \cos v \sin v, 1/2 \leq u \leq 1, 0 \leq v \leq 2\pi\}$.
   (d) $\{(x, y, z): x = \sinh u \sinh v, y = \sinh u \cosh v, z = \cosh u, |u| \leq 1, |v| \leq 1\}$.
3. Demonstrate that all of the following sets are simple smooth surface elements.
   (a) $\{(x, y, z): x^2 + y^2 = z^2, 1 \leq z \leq 2, x \geq 0\}$.
   (b) $\{(x, y, z): xyz = 1, x > 0, y > 0, z > 0, x + y + z \leq 4\}$.
   (c) $\{(x, y, z): x^2 + 2y^2 + 4z^2 = 1, x \geq 0, y \geq 0, z \geq 0\}$.
   (d) $\{(x, y, z): x^2 + y^2 + z^2 = 1, y \geq 0, z \geq 0\}$.
4. Find the equation of the tangent plane of the surface element S, where S is defined in Problem 2, at the point P' that is the image of the point $P = (u, v)$ indicated.
   (a) At $P = (3/2, \pi)$ where S is defined in (a).
   (b) At $P = (0, 0)$ where S is defined in (b).
   (c) At $P = (1/2, 0)$ where S is defined in (c).
   (d) At $P = (-1/2, -1/2)$ where S is defined in (d).

5. Find the vectors $\mathbf{n}_1$ and $\mathbf{n}_2$ of (5–2.7) at each of the points $P = (u, v)$ of the surface elements S defined in Problem 2.

6. Prove that a tangent plane at P', the image of $P \in T^o$, of a surface element S, where S has an explicit representation as given by Definition 5–2.2, is the same plane as the one called a tangent plane at P' by Definition 3–3.1.

7. Demonstrate that the following sets are not smooth surface elements.

   (a) $\{(x, y, z): (x^2 + y^2)^4 = z^6, x \geq 0, y \geq 0, x^2 + y^2 \leq 1\}$,
   (b) $\{(x, y, z): z = |x| + |y|, |x| + |y| \leq 1\}$.

8. Demonstrate that the "surfaces" of the following sets are finite orientable surfaces: (a) a cube, (b) a torus, (c) a hemisphere, and (d) a frustrum.

9. Use intuitive reasoning to demonstrate that on a Möbius strip a unit normal to the surface along a closed curve down the center of the strip would end up being reversed in direction as you returned to the beginning point.

10. Why do the sets of Figure 5–2.5(a), (b), and (c) fail to be finite surfaces?

11. Prove that the edge of a finite surface consists of a finite number of simple piecewise smooth closed curves.

12. Demonstrate that the definition of the tangent plane at a point P' of a surface element S is independent of the parametrization (cf. Section 3-6).

13. It was stated earlier that the surface element S of Figure 5–2.4 was not simple. Show that if $\mathbf{f}$ is a parametric representation of S, that is, $\mathbf{f}$: $T \to S$ where T is a plane regular region and if Γ is a simple closed curve in S, then the pre-image of Γ by $\mathbf{f}$ is a simple closed curve, Γ'. Think of continuously shrinking Γ' to a point in T, and notice the difficulty in the images by $\mathbf{f}$ of each of these curves obtained in the shrinking process.

14. Show that if S is a finite surface contained in the (x, y)-plane of (x, y, z)-space, then S is the sum of a finite set of plane regular regions.

15. Let S be any finite surface such that $S = S_1 + \cdots + S_n$. Show that if $S_i = S_{i1} + \cdots + S_{in_i}$ for $1 \leq i \leq n$, then $S = \sum_{i=1}^{n} \sum_{j=1}^{n} S_{ij}$.

16. In Definition 5–2.4 there is a certain condition on the set S for it to be a surface—namely, that for each $\mathbf{x} \in S$ there exists a neighborhood of $\mathbf{x}$, $N(\mathbf{x}; \delta)$ such that $\bar{N}(\mathbf{x}; \delta) \cap S$ is a simple surface element. Give an example of a set in S in (x, y, z)-space that would be a surface if this condition were not required but would not be a surface as Definition 5–2.4 is stated.

17. Show how the sets shown in Figure 5–2.3(b) and Figure 5–2.5(f) are surfaces.

18. Prove that a simple smooth surface element is a compact set.

19. Prove that a finite surface is arcwise connected. Is a surface arcwise connected?

20. Show that by Definition 5–2.4 a surface need not be orientable. For example, find a sequence of finite surfaces, $\{S_n : n = 1, 2, 3, \cdots\}$, that generate S where S is a Möbius strip. Notice that it is possible to do this with each finite surface S_n being an orientable finite surface.

Answers

2. (a) Simple, smooth. (b) Not simple, smooth.
(c) Not simple, smooth. (d) Not simple, not smooth.

4. (a) $2y + 6\pi z = 3\pi$. (b) $x = 3$. (c) $y - 2z = 0$.
(d) $$(\sinh^2 \tfrac{1}{2})x - \tfrac{1}{2}(\sinh 1)y - (\cosh \tfrac{1}{2})z = \sinh^4 \tfrac{1}{2} + \tfrac{1}{4}\sinh^2 1 - \cosh^2 \tfrac{1}{2}.$$

5. (a) $$\mathbf{n}_1 = \frac{(\sin v, \ -\cos v, \ 2\pi u)}{\sqrt{1 + 4\pi^2 u^2}}, \ \mathbf{n}_2 = -\mathbf{n}_1.$$
(b) $$\mathbf{n}_1 = (\cos v \cos u, \ -\sin v \cos u, \ -\sin u), \ \mathbf{n}_2 = -\mathbf{n}_1.$$
(c) $$\mathbf{n}_1 = \frac{(u \sin v, \ -u \cos v, \ 1)}{\sqrt{1 + u^2}}, \ \mathbf{n}_2 = -\mathbf{n}_1.$$
(d) $$\mathbf{n}_1 = \frac{(\sinh u \sinh v, \ \sinh u, \ -\cosh u)}{\sqrt{\sinh^2 u \cosh 2v + \cosh^2 u}}, \qquad \mathbf{n}_2 = -\mathbf{n}_1.$$

5–3. Surface Integrals

The idea of a surface element, a finite surface, and an orientable surface was purposely discussed in some detail in the previous section in order to give the background for this discussion of surface integrals. However, to establish clearly a beginning point for surface integrals, let us review some of the elementary calculus material that we shall use.

Let S be a bounded set in the (x, y)-plane and let f be a *bounded* real-valued function defined on S. Furthermore, let $I = [(a, b), (a', b')]$ be an interval (rectangle) in the (x, y)-plane containing S (cf. Figure 5–3.1). Define on I a function $\tilde{f}$, an extension of f, as follows:

$$\tilde{f}(x, y) = \begin{cases} f(x, y), & (x, y) \in S, \\ 0, & (x, y) \notin S. \end{cases}$$

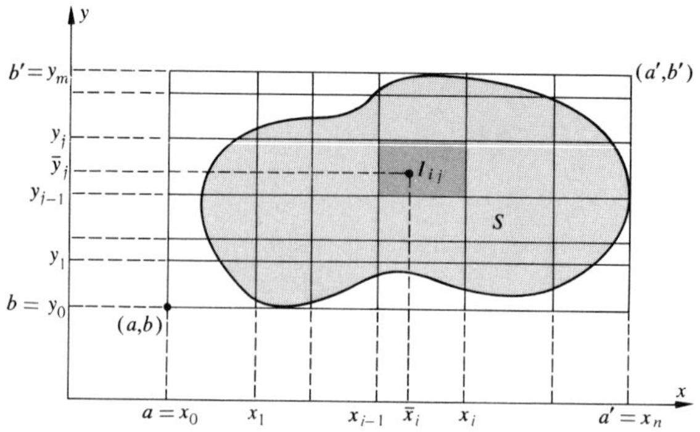

Figure 5–3.1

Then $P = \{I_{ij} : 1 \leq i \leq n,\ 1 \leq j \leq m\}$ is a *partition* of I, if

$$a = x_0 < x_1 < \cdots <_n x = a'$$

and

$$b = y_0 < y_1 < \cdots < y_m = b'$$

and

$$I_{ij} = \{(x, y): x_{i-1} \leq x \leq x_i,\ y_{j-1} \leq y \leq y_j\},\ 1 \leq i \leq n,\ 1 \leq j \leq m.$$

The *mesh* of P, $|P| = \max\{\text{diam}\,(I_{ij}): 1 \leq i \leq n,\ \ 1 \leq j \leq m\}$, where $\text{diam}\,(I_{ij}) = \sup\{\|(x, y) - (x', y')\| : (x, y) \text{ and } (x', y') \in I_{ij}\}$ is the diameter of I_{ij}. The *Riemann integral* of f over S, written $\int_S f(x, y)\, dA_{xy}$, or simply $\int_S f\, dA_{xy}$, is

$$\lim_{|P|\to 0} \textstyle\sum_{i,j=1}^{i=n,j=m} \tilde{f}(\bar{x}_i, \bar{y}_j)\, \Delta_i\, x\, \Delta_j\, y \tag{5–3.1}$$

if this limit exists and is finite where $\Delta_i\, x = x_i - x_{i-1}$, $x_{i-1} \leq \bar{x}_i \leq x_i$ for $1 \leq i \leq n$ and $\Delta_j\, y = y_j - y_{j-1}$, $y_{j-1} \leq \bar{y}_j \leq y_j$ for $1 \leq j \leq m$.

If the limit of (5–3.1) exists, then the function f is *integrable* on S. If $f(x, y, z) = 1$ for points $(x, y, z) \in S$ and f is integrable on S, then

$$A(S) = \int_S dA_{xy}$$

is the *area* of S and S is said to have area. In particular, *if f is continuous on S and S is a plane regular region, then f is integrable on S.*

The limit of (5–3.1) is frequently difficult to evaluate. However, for a continuous function f on certain types of plane regular regions we can readily make the calculation. If $S = S_x$ is a *y-axial region* of the (x, y)-plane and, therefore,

$$S_x = \{(x, y): a \leq x \leq b, g(x) \leq y \leq h(x)\}$$

where g and h are continuous on $[a, b]$ such that $g(x) < h(x)$ for $a < x < b$, then

$$\int_S f(x, y)\, dA_{xy} = \int_a^b \int_{g(x)}^{h(x)} f(x, y)\, dy\, dx, \qquad (5\text{–}3.2)$$

a *double iterated integral.* If the indicated integration of (5–3.2) can be carried out, then we would have the value of the Riemann integral of f over S.

A perfectly analogous remark can be made if the region S is an x-axial region of the (x, y)-plane. In fact, if f is continuous on S, a plane regular region, and $S = S_1 + \cdots + S_n$ where each plane regular region S_i is an *axial region* of the plane (that is, either an x-axial or y-axial region of the plane), then the limit of (5–3.1) can be evaluated by first writing

$$\int_S f\, dA_{xy} = \textstyle\sum_{i=1}^n \displaystyle\int_{S_i} f\, dA_{xy} \qquad (5\text{–}3.3)$$

and then replacing each of the integrals on the right by the appropriate double iterated integral.

It is possible to write (5–3.3) since each of the integrals on the right exists and since the $A(S_i \cap S_j) = 0$, $1 \leq i < j \leq n$ (cf. Proposition 5–3.2). Recall that the intersection of $S_i \cap S_j$ is a point, two points, a simple smooth curve, or the empty set (cf. Definition 2–7.4).

Integrals defined on sets in a plane are frequently referred to as *double integrals* since they are best evaluated by double iterated integrals. The following two propositions summarize the main properties of double integrals.

Proposition 5–3.1. *Let f be a real-valued function defined by $f(x, y)$ for each point $(x, y) \in S$, a bounded set of the (x, y)-plane such that S has area. If f is bounded on S (that is, $f(S)$, is a bounded set) and is continuous on all of S except a subset of S, S^* such that $A(S^*) = 0$, then f is integrable on S.*

This proposition gives sufficient conditions, which are very satisfactory, for determining the existence of the Riemann integral of bounded functions f over S, a bounded set. The following proposition gives us some properties of the double integral that allow us to operate with Riemann integrals that we know exist.

Proposition 5–3.2. *If* $\int_S (s, y)\, dA_{xy}$ *and* $\int_S g(x, y)\, dA_{xy}$ *exist, then* $\int_S \alpha f(x, y)\, dA_{xy}$, *α real, and* $\int_S (f(x, y) + g(x, y))\, dA_{xy}$ *exist and*

$$\int_S \alpha f(x, y)\, dA_{xy} = \alpha \int_S f(x, y)\, dA_{xy}, \tag{5–3.4}$$

$$\int_S (f(x, y) + g(x, y))\, dA_{xy} = \int_S f(x, y)\, dA_{xy} + \int_S g(x, y)\, dA_{xy}. \tag{5–3.5}$$

If, in addition, for all $(x, y) \in S, f(x, y) \leq g(x, y)$, *then*

$$\int_S f(x, y)\, dA_{xy} \leq \int_S g(x, y)\, dA_{xy}. \tag{5–3.6}$$

If $\int_{S_1} f(x, y)\, dA_{xy}$ *and* $\int_{S_2} f(x, y)\, dA_{xy}$ *exist and if* $A(S_1 \cap S_2) = 0$, *then* $\int_{S_1 \cup S_2} f(x, y)\, dA_{xy}$ *exists and*

$$\int_{S_1 \cup S_2} f(x, y)\, dA_{xy} = \int_{S_1} f(x, y)\, dA_{xy} + \int_{S_2} f(x, y)\, dA_{xy}. \tag{5–3.7}$$

We are now ready for a meaningful definition of the area of a simple smooth surface element.

Definition 5–3.1. *Let S be a simple smooth surface element in (x, y, z)-space as defined in Definition 5–2.1. The area of S,*

$$A(S) = \int_T \left\| \frac{\partial}{\partial u} \mathbf{f}(u, v) \times \frac{\partial}{\partial v} \mathbf{f}(u, v) \right\| dA_{uv}. \tag{5–3.8}$$

Clearly, this integral will always exist since the integrand is continuous on T, a plane regular region (cf. Proposition 5–3.1). To demonstrate this definition, let us calculate the area of the surface element shown in Figure 5–2.1 and defined in (5–2.1). Since

$$\mathbf{f}(u, v) = (\cos u \sin v, \sin u \sin v, \cos v),$$

then

$$\left\| \frac{\partial}{\partial u} \mathbf{f}(u, v) \times \frac{\partial}{\partial v} \mathbf{f}(u, v) \right\| = \|(-\cos u \sin^2 v, -\sin u \sin^2 v, -\sin v \cos v)\|$$

$$= \sin v.$$

Therefore,

$$A(S) = \int_T \sin v\, dA_{uv}$$

where

$$T = \left\{(u, v) : 0 \le u \le \frac{\pi}{2}, \frac{\pi}{4} \le v \le \frac{\pi}{2}\right\}$$

(cf. Figure 5–2.1). By (5–3.2),

$$A(S) = \int_0^{\pi/2} \int_{\pi/4}^{\pi/2} \sin v \, dv \, du$$

$$= \frac{\pi}{2\sqrt{2}}.$$

A deeper look at this example helps us to see the motivation for Definition 5–3.1. The subintervals, I_{ij}, of T have dimensions $\Delta u = \pi/8$ and $\Delta v = \pi/16$, and thus their area is $\pi^2/128$. Applying the definition of the integral to (5–3.8), we see that

$$\int_T \left\| \frac{\partial}{\partial u} \mathbf{f}(u, v) \times \frac{\partial}{\partial v} \mathbf{f}(u, v) \right\| dA_{uv}$$

$$= \lim_{|\mathrm{P}| \to 0} \sum_{i,j=1}^{i=n,j=m} \left\| \frac{\partial}{\partial u} \mathbf{f}(u_i, v_j) \times \frac{\partial}{\partial v} \mathbf{f}(u_i, v_j) \right\| \Delta_i u \, \Delta_j v \quad (5\text{–}3.9)$$

where $(u_i, v_j) \in I_{ij}$, $1 \le i \le n$, $1 \le j \le m$. Thus if P is a decomposition of T into the sum of these 16 subintervals such that, for each i, $\Delta_i u = \Delta u = \pi/8$ and, for each j, $\Delta_j v = \Delta v = \pi/16$, then the terms of the summation of (5–3.9) become

$$\left\| \frac{\partial}{\partial u} \mathbf{f}(u_i, v_j) \times \frac{\partial}{\partial v} \mathbf{f}(u_i, v_j) \right\| \Delta u \, \Delta v$$

$$= \left\| \frac{\partial}{\partial u} \mathbf{f}(u_i, v_j) \right\| \cdot \left\| \frac{\partial}{\partial v} \mathbf{f}(u_i, v_j) \right\| (\sin \theta_{ij}) \, \Delta u \, \Delta v$$

$$= \left(\left\| \frac{\partial}{\partial u} \mathbf{f}(u_i, v_j) \right\| \Delta u \right) \left(\left\| \frac{\partial}{\partial v} \mathbf{f}(u_i, v_j) \right\| \Delta v \right) \sin \theta_{ij}. \quad (5\text{–}3.10)$$

We noted in the foregoing section that if $(u, v) = P$, then

$$\left\| \frac{\partial}{\partial u} \mathbf{f}(u, v) \times \frac{\partial}{\partial v} \mathbf{f}(u, v) \right\|$$

is the area of the hashed parallelogram shown in Figure 5–2.1(b). Thus from the last line of (5–3.10) we can see that the terms of the sum (5–3.9) are equal to the area of a parallelogram whose sides are of length

$$\left\| \frac{\partial}{\partial u} \mathbf{f}(u_i, v_j) \right\| \Delta u, \quad \left\| \frac{\partial}{\partial v} \mathbf{f}(u_i, v_j) \right\| \Delta v.$$

Thus

$$\left\| \frac{\partial}{\partial u} \mathbf{f}(u_i, v_j) \right\| \Delta u = \left\| \frac{\partial}{\partial u} f_1(u_i, v_j)\, \Delta u, \frac{\partial}{\partial u} f_2(u_i, v_j)\, \Delta u, \frac{\partial}{\partial u} f_3(u_i, v_j)\, \Delta u \right\|;$$

but

$$\frac{\partial}{\partial u} f_k(u_i, v_j)\, \Delta u \approx \Delta f_k(u_i, v_j\,; \Delta u, 0), \quad k = 1, 2, 3$$

where $\approx$ means approximately equal. Hence

$$\left(\textstyle\sum_{k=1}^{3}[\Delta f_k(u_i, v_j\,; \Delta u, 0)]^2\right)^{1/2} \approx \widehat{P'P_1'}$$

where $\widehat{P'P_1'}$ is the length of the longitudinal arc from P' to P_1'. Similarly,

$$\left\| \frac{\partial}{\partial v} \mathbf{f}(u_i, v_j) \right\| \Delta v \approx \widehat{P'P_2'}.$$

Thus, if $(u_i, v_j) = P$ and I_{ij} is the subinterval with vertices P, P_1, P_2, and P_3 in Figure 5–2.1(a), then

$$\left\| \frac{\partial}{\partial u} \mathbf{f}(u_i, v_j) \times \frac{\partial}{\partial v} \mathbf{f}(u_i, v_j) \right\| \Delta_i u\, \Delta_j v$$

is the area of a parallelogram that appears to be approximately equal to what we "feel" is almost equal to the part of the surface element S that is the image of I_{ij} by $\mathbf{f}$. As the partition of T, which is already a rectangle in this case, is refined, it appears that this approximation would improve. Thus it is natural to accept as the area of S, the limit of the area of these approximating parallelograms.

It is appropriate to wonder if the area of a surface element has been defined in such a way that if S has two parametric representations, then the area of S calculated by either parametric representation is the same. We shall assume for the moment that this is the case and leave the verification for a problem (Problem 15 of this section).

There are several other forms for the integrand of the integral of (5–3.8)—for example,

$$\left\| \frac{\partial \mathbf{f}}{\partial u} \times \frac{\partial \mathbf{f}}{\partial v} \right\| = \left\| \frac{\partial(f_2, f_3)}{\partial(u, v)}, \frac{\partial(f_3, f_1)}{\partial(u, v)}, \frac{\partial(f_1, f_2)}{\partial(u, v)} \right\|, \tag{5–3.11}$$

discussed in Section 5–2, and

$$\left\| \frac{\partial}{\partial u} \mathbf{f}(u, v) \times \frac{\partial}{\partial v} \mathbf{f}(u, v) \right\| = \sqrt{EG - F^2} \tag{5–3.12}$$

where

$$E = \left\| \frac{\partial}{\partial u} \mathbf{f}(u, v) \right\|^2, \qquad G = \left\| \frac{\partial}{\partial v} \mathbf{f}(u, v) \right\|^2$$

and

$$F = \left(\frac{\partial}{\partial u}\mathbf{f}(u, v)\right) \cdot \left(\frac{\partial}{\partial v}\mathbf{f}(u, v)\right).$$

The equation of (1–3.10) is useful in demonstrating this last identity. Furthermore, if S has an explicit representation, so that

$$S = \{(x, y, z): z = \varphi(x, y), (x, y) \in T\},$$

then

$$\left\|\frac{\partial}{\partial u}\mathbf{f}(u, v) \times \frac{\partial}{\partial v}\mathbf{f}(u, v)\right\| = \sqrt{1 + (\varphi_x(x, y))^2 + (\varphi_y(x, y))^2} \qquad (5\text{–}3.13)$$

where $u = x$, $v = y$ and $\mathbf{f}(x, y) = (x, y, \varphi(x, y))$ (cf. (5–2.5)).

Example 5–3.1. Let us calculate the area of the surface element

$$S = \{(x, y, z): x + y + z = 1, x \geq 0, y \geq 0, z \geq 0\}$$

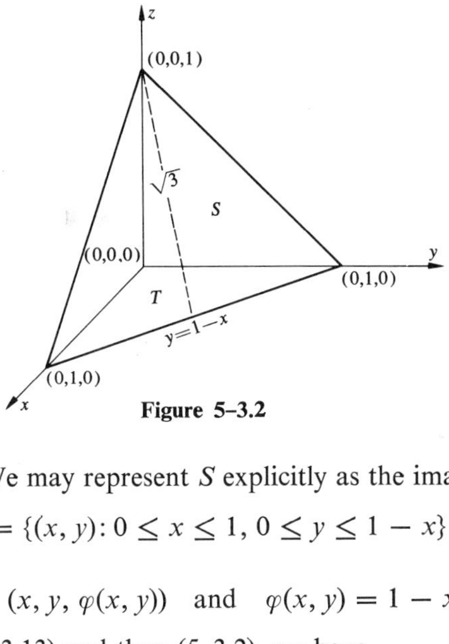

Figure 5–3.2

(cf. Figure 5–3.2). We may represent S explicitly as the image of

$$T = \{(x, y): 0 \leq x \leq 1, 0 \leq y \leq 1 - x\}$$

by $\mathbf{f}$ where

$$\mathbf{f}(x, y) = (x, y, \varphi(x, y)) \quad \text{and} \quad \varphi(x, y) = 1 - x - y.$$

Therefore, using (5–3.13) and then (5–3.2), we have

$$\begin{aligned} A(S) &= \int_T \sqrt{1 + (\varphi_x(x, y))^2 + (\varphi_y(x, y))^2}\, dA_{xy} \\ &= \int_0^1 \int_0^{1-x} \sqrt{1 + (-1)^2 + (-1)^2}\, dy\, dx \\ &= \sqrt{3}\int_0^1 (1 - x)\, dx = \frac{\sqrt{3}}{2}. \end{aligned}$$

Since S is an equilateral triangle of side length $\sqrt{2}$ and altitude $\sqrt{3}/\sqrt{2}$, we can check the area to be indeed $\sqrt{3}/2$.

We are now ready to define an integral on a finite surface.

Definition 5–3.2. *Let S be a simple smooth surface element in (x, y, z)-space and let the function $\mathbf{f}$ with the plane regular region T in the (u, v)-plane be a parametric representation of S. Let g be a bounded function defined by $g(x, y, z)$ for each point (x, y, z) of S. The Riemann integral of g over S, written $\int_S g(x, y, z)\, dA$, or simply $\int_S g\, dA$, is given by*

$$\int_S g(x, y, z)\, dA = \int_T g(\mathbf{f}(u, v)) \left\| \frac{\partial}{\partial u} \mathbf{f}(u, v) \times \frac{\partial}{\partial v} \mathbf{f}(u, v) \right\| dA_{uv}, \quad (5\text{–}3.14)$$

if the integral on the right exists.

If S is a finite surface ($S = S_1 + \cdots + S_n$) and the function g is defined on S, then the Riemann integral of g over S, written $\int_S g(x, y, z)\, dA$, or simply $\int_S g\, dA$, is given by

$$\int_S g(x, y, z)\, dA = \textstyle\sum_{i=1}^n \displaystyle\int_S g(x, y, z)\, dA \quad (5\text{–}3.15)$$

if each of the integrals on the right exists.

In either case, when the required integrals exist, we say that g is integrable over S.

It is natural to wonder if this definition for a simple smooth surface element is independent of the parametric representation, as it should be. For the moment, we shall assume that it is true, and we shall defer until later the verification (Problem 14 of Section 5–5). Also there is a question about the definition of an integral over a finite surface: Is the value of the integral independent of the manner of decomposing S? Consideration of this question will also be deferred, and for the present we shall assume that the definition is completely satisfactory (cf. Problem 21). By Proposition 5–3.1 it is clear that g is integrable over S, a simple smooth surface element or a finite surface, if g is continuous on S. Hence, in particular, if $g(x, y, z) = 1$ for all $(x, y, z) \in S$, then $A(S) = \int_S dA$, the *area* of S, always exists. Thus, if $S = S_1 + \cdots + S_n$, then

$$A(S) = A(S_1) + \cdots + A(S_n). \quad (5\text{–}3.16)$$

The following example illustrates the application of Definition 5–3.2.

Example 5–3.2. Let $S = \text{bdy}\,\{(x, y, z)\colon x^2 + y^2 \leq 1, 0 \leq z \leq 1\}$ in E^3, and let $g(x, y, z) = 1$ for all points $(x, y, z) \in S$ (cf. Figure 5–3.3). Then $S = S_1 + \cdots + S_6$ where $S_1, \cdots, S_6$ are shown in the figure. Thus

$$A(S) = A(S_1) + \cdots + A(S_6).$$

Now S_1 is the image of $T_1 = \{(y, z)\colon |y| \leq 1/\sqrt{2}, 0 \leq z \leq 1\}$ by

$$\mathbf{f}_1(y, z) = (\varphi(y, z), y, z)$$

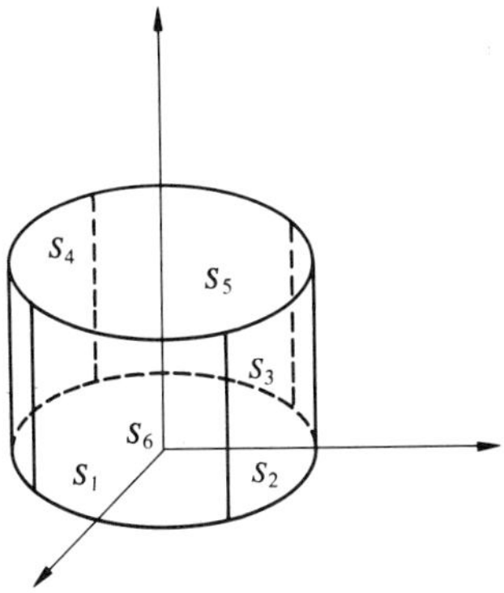

Figure 5–3.3

where $\varphi(y, z) = \sqrt{1 - y^2}$, and hence

$$\begin{aligned} A(S_1) &= \int_{T_1} \sqrt{1 + (\varphi_y(y, z))^2 + (\varphi_z(y, z))^2}\; dA_{yz} \\ &= \int_0^1 \int_{-1/\sqrt{2}}^{1/\sqrt{2}} \sqrt{1 + y^2/(1 - y^2)}\; dy\, dz \\ &= 2 \sin^{-1} y\Big]_0^{1/\sqrt{2}} = \pi/2. \end{aligned}$$

Similarly, $A(S_i) = \pi/2$, $i = 2, 3, 4$. The surface element S_5 is the image of $T_5 = \{(x, y)\colon x^2 + y^2 \leq 1\}$ by $\mathbf{f}_5(x, y) = (x, y, \psi(x, y))$ where $\psi(x, y) = 1$, and hence

$$\begin{aligned} A(S_5) &= \int_{T_5} \sqrt{1 + (\psi_x(x, y))^2 + (\psi_y(x, y))^2}\; dA_{xy} \\ &= \int_{-1}^{1} \int_{\sqrt{1-x^2}}^{\sqrt{1-x^2}} (1)\; dy\, dx = \pi. \end{aligned}$$

Similarly, $A(S_6) = \pi$. Hence $A(S) = 4(\pi/2) + 2\pi = 4\pi$.

Related to the above example is a particular family of finite surfaces that merits special attention—namely, the surfaces of revolution. The next example, which is somewhat general in nature, illustrates a simple way of how the area of this type of finite surface can be found.

Example 5–3.3. Let $f: [a, b] \to R$ where $[a, b] \subset R$. Suppose that f is continuously differentiable on $[a, b]$ and that $f(x) > 0$ for all $x \in [a, b]$. Let

$$S = \{(x, y, z): a \leq x \leq b, y^2 + z^2 = (f(x))^2\}$$

(cf. Figure 5–3.4). Then $S = S_1 + S_2 + S_3 + S_4$ where each S_i, $1 \leq i \leq 4$, is shown in the figure. Thus S_1 is the image by $\mathbf{g}$ of

$$T_1 = \left\{(x, y)\ \ a \leq x \leq b, |y| \leq \frac{1}{\sqrt{2}} f(x)\right\}$$

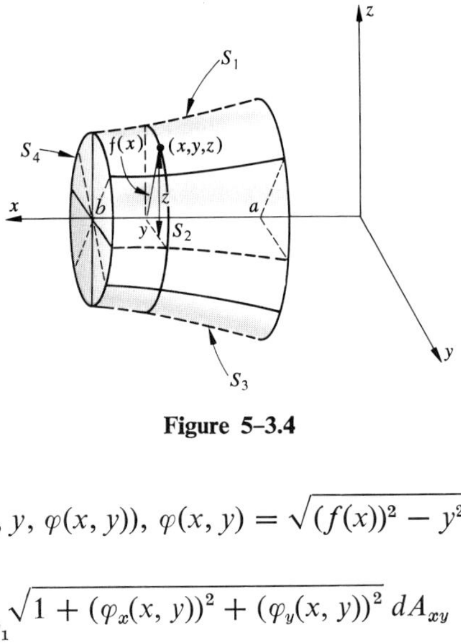

Figure 5–3.4

where $\mathbf{g}(x, y) = (x, y, \varphi(x, y))$, $\varphi(x, y) = \sqrt{(f(x))^2 - y^2}$. Therefore,

$$\begin{aligned} A(S_1) &= \int_{T_1} \sqrt{1 + (\varphi_x(x, y))^2 + (\varphi_y(x, y))^2}\, dA_{xy} \\ &= \int_a^b \int_{-(1/\sqrt{2})f(x)}^{(1/\sqrt{2})f(x)} \left[\frac{f(x)\sqrt{1 + (f'(x))^2}}{\sqrt{(f(x))^2 - y^2}}\right] dy\, dx \\ &= 2\int_a^b \sqrt{1 + (f'(x))^2} \left[f(x) \sin^{-1} \frac{y}{f(x)}\right]_0^{-(1/\sqrt{2})f(x)} dx \\ &= \frac{\pi}{2} \int_a^b f(x)\sqrt{1 + (f'(x))^2}\, dx. \end{aligned}$$

Similarly,

$$A(S_i) = \frac{\pi}{2}\int_a^b f(x)\sqrt{1 + (f'(x))^2}\,dx, \quad i = 2, 3, 4.$$

Hence

$$A(S) = 2\pi\int_a^b f(x)\sqrt{1 + (f'(x))^2}\,dx. \tag{5–3.17}$$

Thus in Example 5–3.2, if $f(z) = 1, 0 \leq z \leq 1, a = 0$, and $b = 1$ in (5–3.17), we have

$$A(S_1 + S_2 + S_3 + S_4) = 2\pi\int_0^1 dx = 2\pi,$$

as we found earlier.

The next example shows how Definition 5–3.2 is used to evaluate a given integral over a finite surface when g is not identically equal to one.

Example 5–3.4. Let S be the boundary of the set,

$$\{(x, y, z): 2 - z = \sqrt{x^2 + y^2}, 0 \leq z \leq 1, x \geq 0, y \geq 0\},$$

a quarter of a frustrum (cf. Figure 5–3.5). Let $g(x, y, z) = z$. Then decompose S into $\{S_1, S_2, S_3, S_4, S_5\}$, as shown in the figure. Then

$$S_1 = \{\mathbf{f}_1(x, y): 0 \leq x^2 + y^2 \leq 4, x \geq 0, y \geq 0\}$$
$$\text{where } \mathbf{f}_1(x, y) = (x, y, 0),$$

$$S_2 = \{\mathbf{f}_2(x, y): 0 \leq x^2 + y^2 \leq 1, x \geq 0, y \geq 0\}$$
$$\text{where } \mathbf{f}_2(x, y) = (x, y, 1),$$

$$S_3 = \{\mathbf{f}_3(x, z): 0 \leq x \leq 2 - z, 0 \leq z \leq 1\}$$
$$\text{where } \mathbf{f}_3(x, z) = (x, 0, z),$$

$$S_4 = \{\mathbf{f}_4(y, z): 0 \leq y \leq 2 - z, 0 \leq z \leq 1\}$$
$$\text{where } \mathbf{f}_4(y, z) = (0, y, z),$$

and

$$S_5 = \{\mathbf{f}_5(u, v): 1 \leq u \leq 2, 0 \leq v \leq \pi/2\}$$
$$\text{where } \mathbf{f}_5(u, v) = (u\cos v, u\sin v, 2 - u).$$

Thus by (5–3.15),

$$\int_S z\,dA = \int_{T_1}(0)\,\|(0, 0, 1)\|\,dA_{xy} + \int_{T_2}(1)\,\|(0, 0, 1)\|\,dA_{xy}$$
$$+ \int_{T_3} z\,\|(0, -1, 0)\|\,dA_{xz} + \int_{T_4} z\,\|(1, 0, 0)\|\,dA_{yz}$$
$$+ \int_{T_5}(2 - u)\,\|(u\cos v, u\sin v, u)\|\,dA_{uv}.$$

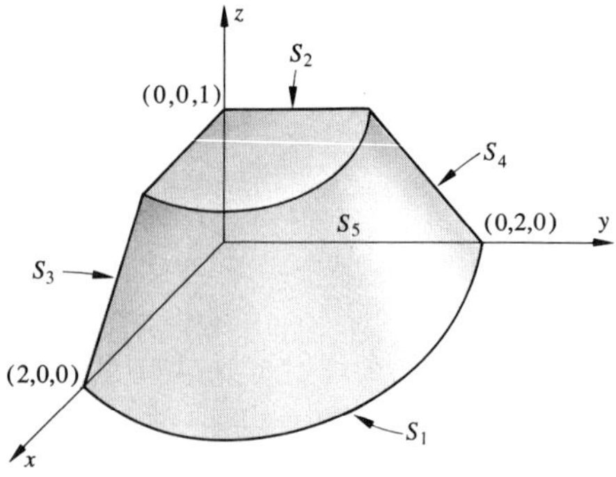

Figure 5–3.5

Then by use of (5–3.2) on each of the last four of these integrals and by the fact that the first integral is zero, we have

$$\begin{aligned}\int_S z\,dA &= \int_0^1\int_0^{\sqrt{1-x^2}} dy\,dx + \int_0^1\int_0^{2-z} dx\,dz + \int_0^1\int_0^{2-z} z\,dy\,dz \\ &\quad + \sqrt{2}\int_1^2\int_0^{\pi/2} (2-u)u\,dv\,du \\ &= \frac{\pi}{4} + \frac{2}{3} + \frac{2}{3} + \frac{\pi\sqrt{2}}{3} \\ &= \left(\frac{\sqrt{2}}{3} + \frac{1}{4}\right)\pi + \frac{4}{3} \approx 3.59.\end{aligned}$$

Three properties of integrals over a finite surface S that follow from Definition 5–3.2 and Proposition 5–3.2 are the following: Suppose that

$$\int_S f(x, y, z)\,dA \quad \text{and} \quad \int_S g(x, y, z)\,dA$$

exist. Then

$$\int_S \alpha f(x, y, z)\,dA \quad \text{and} \quad \int_S (f(x, y, z) + g(x, y, z))\,dA$$

exist and

$$\int_S \alpha f(x, y, z)\,dA = \alpha\int_S f(x, y, z)\,dA \tag{5–3.18}$$

for any real α, and

$$\int_S (f(x, y, z) + g(x, y, z))\,dA = \int_S f(x, y, z)\,dA + \int_S g(x, y, z)\,dA. \tag{5–3.19}$$

If, in addition, for all $(x, y, z) \in S$

$$f(x, y, z) \leq g(x, y, z),$$

then

$$\int_S f(x, y, z)\, dA \leq \int_S g(x, y, z)\, dA. \tag{5–3.20}$$

Lastly, we shall define the Riemann integral of g over a surface S. In a sense these integrals are like improper integrals of the elementary calculus. For example, in the following definition, the function g need not be bounded on all of S. The surface S may have a subset that prevents it from being a finite surface (like the vertex of a part of a cone as in Figure 5–2.3(b)), or S may be unbounded as in the case of a surface like S_5 in Figure 5–2.5(f).

Definition 5–3.3. *Let S be a surface and let g be a non-negative real-valued function defined on a subset of S. Let S^* be a nonempty subset of S that consists of all of the exceptional points of S, which are all of the points of S such that g is unbounded in a neighborhood of the point and ∞ if S is unbounded. Furthermore, suppose that S^* is such that there exists a sequence of finite surfaces $\{S_n : n = 1, 2, 3, \cdots\}$ converging to S relative to S^* (cf. Definition 5–2.5). Let $g \in C(S - S^*)$. Then the Riemann integral of g over S,*

$$\int_S g(x, y, z)\, dA,$$

or simply $\int_S g\, dA$, is given by

$$\lim_{n\to\infty} \int_{S_n} g\, dA$$

if this limit exists and is finite. If g is not necessarily a non-negative function on S, but $\int_S g^+\, dA$ and $\int_S g^-\, dA$ exist where $g^+ = \frac{1}{2}(g + |g|)$ and $g^- = \frac{1}{2}(-g + |g|)$, then the Riemann integral of g over S is

$$\int_S g\, dA = \int_S g^+\, dA - \int_S g^-\, dA. \tag{5–3.21}$$

If g is such that $g(x, y, z) = 1$ for all $(x, y, z) \in S$, then for the area of S, $A(S) = \int_S dA$ if $\int_S dA$ exists.

This definition cannot cover the same circumstance as covered in Definition 5–3.2 since S^* is nonempty. That is, S is not a finite surface, or g is unbounded on S. Notice that since $S - S^*$ contains only points where g is continuous, then $\int_{S_n} g\, dA$ exists for each n.

The initial restriction of g to being non-negative is essential. This is one of the main differences between improper multiple integrals and improper integrals of a single variable. With the restriction of g being non-negative, we have $\left\{\int_{S_n} g\, dA\right\}$ as a monotone increasing sequence of real numbers since $S_{n+1} \supset S_n$ for all n. Furthermore, notice that g^+ and g^- are non-negative functions that are continuous where g is continuous. Since $g = g^+ - g^-$, the Riemann integral of g over S for g not necessarily non-negative is very natural. In fact, all of the usual properties of integrals are preserved under this definition.

Any integral of the kind defined in Definition 5–3.2 or 5–3.3 is a *surface integral.* The notation

$$\int_{-\infty}^{\infty}\int_{-\infty}^{\infty} g(x, y, z)\, dx\, dy, \quad \int_{0}^{\infty}\int_{0}^{\infty} g(x, y, z)\, dx\, dy, \text{ etc.},$$

that is used in other writings could have the meaning

$$\int_S g(x, y, z)\, dA_{xy}, \quad \int_S g(x, y, z)\, dA_{xy}, \text{ etc.},$$

where S is the (x, y)-plane, $S = \{(x, y)\colon x \geq 0, y \geq 0\}$, etc., respectively. However, it could be that these are to be iterated improper integrals. The reader should beware of this point. The following examples indicate how this definition is applied.

Example 5–3.5. Let $S = \{(x, y, z)\colon |x| \leq 1, |y| \leq 1, z = 0\}$ and let $g(x, y, z) = 1/\sqrt{1 - x^2}$. Then S is a finite surface, but g is not continuous on S, and

$$S^* = \{(x, y, z)\colon |x| \leq 1 - (1/(n + 1)), |y| \leq 1, z = 0\}.$$

Then for $n = 1, 2, 3, \cdots$, S_n is a finite surface, $g \in C(S_n)$, and $\lim_{n\to\infty} S_n = S(S^*)$. Now

$$\int_{S_n} g(x, y, z)\, dA = \int_{-1}^{1}\int_{-1+(1/(n+1))}^{1-(1/(n+1))} \frac{1}{\sqrt{1 - x^2}}\, dx\, dy$$

$$= 2\left[\sin^{-1}\left(\frac{n}{n + 1}\right) - \sin^{-1}\left(\frac{-n}{n + 1}\right)\right],$$

and hence

$$\lim_{n\to\infty} \int_{S_n} g(x, y, z)\, dA = 2\pi.$$

Therefore,

$$\int_S \frac{1}{\sqrt{1 - x^2}}\, dA = 2\pi.$$

Example 5–3.6. Let $S = \{(x, y, z): z^2 = x^2 + y^2, 0 \leq z \leq 1\}$ and let g be defined by $g(x, y, z) = 1$ for all $(x, y, z) \in S$. Then S is not a finite surface and $S^* = \{\mathbf{0}\}$. Take

$$S_n = \left\{(x, y, z): z^2 = x^2 + y^2, \quad \frac{1}{n+1} \leq z \leq 1\right\}.$$

then for $n = 1, 2, 3, \cdots, S_n$ is a finite surface, $g \in C(S_n)$, and $\lim_{n\to\infty} S_n = S(S^*)$. Each S_n is a surface of revolution. Hence by Example 5–3.3,

$$\begin{aligned}\int_{S_n} dA &= 2\pi \int_{1/(n+1)}^{1} z\sqrt{1 + (1)^2}\, dz \\ &= 2\pi\sqrt{2}\,\frac{1}{2}\left(1 - \frac{1}{(n+1)^2}\right).\end{aligned}$$

Therefore,

$$\int_S dA = \lim_{n\to\infty} \int_{S_n} dA = \sqrt{2}\,\pi.$$

Example 5–3.7. Let

$$S = \{(x, y, z): z = x + y, x \geq 0, y \geq 0\}.$$

Thus S is not a finite surface; S is unbounded. Let g be defined by $g(x, y, z) = xye^{-x^2-y^2}$ for all $(x, y, z) \in S$. Then $S^* = \{\infty\}$. Take

$$S_n = \{(x, y, z): z = x + y, 0 \leq x \leq n, 0 \leq y \leq n\}.$$

Then for $n = 1, 2, 3, \cdots, S_n$ is a finite surface, $g \in C(S_n)$, and $\lim_{n\to\infty} S_n = S$. Now

$$\int_{S_n} g(x, y, z)\, dA = \int_0^n \int_0^n xye^{-x^2-y^2}\sqrt{1 + (\varphi_x(x, y))^2 + (\varphi_y(x, y))^2}\, dy\, dx$$

where φ is defined by $\varphi(x, y) = x + y$ for all (x, y). Hence

$$\begin{aligned}\int_{S_n} g(x, y, z)\, dA &= \sqrt{3}\left(\int_0^n xe^{-x^2}\, dx\right)\left(\int_0^n ye^{-y^2}\, dy\right) \\ &= \sqrt{3}[\tfrac{1}{2}(1 - e^{-n^2})]^2.\end{aligned}$$

Therefore,

$$\int_S g(x, y, z)\, dA = \lim_{n\to\infty} \int_{S_n} g(x, y, z)\, dA = \frac{\sqrt{3}}{4}.$$

It is natural to wonder whether the definition of $\int_S g(x, y, z)\, dA$ is independent of the choice of the sequence, $\{S_n\}$. This is the case, but a proof of that fact is essential (cf. Problem 20). However, continuing on the assumption of the truth of this fact, we shall see how certain integrals over surface elements may be more easily calculated.

Example 5–3.8. Consider the surface element

$$S = \{(x, y, z): z = \sqrt{1 - x^2 - y^2},\ x^2 + y^2 \leq 1\},$$

a hemispherical surface. The function $\mathbf{f}$ defined by

$$\mathbf{f}(x, y) = (x, y, \sqrt{1 - x^2 - y^2}),$$

which naturally suggests itself, is not continuously differentiable on $T = \{(x, y): x^2 + y^2 \leq 1\}$, but it is continuously differentiable on

$$T_n = \{(x, y): x^2 + y^2 \leq 1 - (1/(n + 1))\},$$

and hence $S_n = \{\mathbf{f}(x, y): (x, y) \in T_n\}$ is a smooth surface element and

$$S^* = \{(x, y, z): x^2 + y^2 = 1,\ z = 0\}.$$

Then

$$A(S) = \int_S dA = \lim_{n\to\infty} \int_{S_n} dA$$

$$= \lim_{n\to\infty} \int_{-r_n}^{r_n} \int_{-\sqrt{r_n^2-x^2}}^{\sqrt{r_n^2-x^2}} \sqrt{1 + (\varphi_x(x, y))^2 + (\varphi_y(x, y))^2}\, dy\, dx$$

where $r_n = 1 - (1/(n + 1))$ and $\varphi(x, y) = \sqrt{1 - x^2 - y^2}$ for $x^2 + y^2 \leq r_n^2$. Hence,

$$A(S) = \int_S dA = \lim_{n\to\infty} (2\pi)[1 - (1 - r_n^2)^{1/2}] = 2\pi.$$

It is very convenient in evaluating certain surface integrals to use the variables of polar coordinates, r and θ. This is particularly the case if the plane regular region T is an r-axial region. In other words, a region T is an *r-axial region* (*plane r-axial region*) if

$$T = \{(r, \theta): \alpha \leq \theta \leq \beta,\ 0 \leq g(\theta) \leq r \leq h(\theta)\},$$

where $0 < \beta - \alpha \leq 2\pi$, $g(\theta) < h(\theta)$ for $\alpha < \theta < \beta$, and g and $h \in C([\alpha, \beta])$. If T is such a region and if $f \in C(T)$, then

$$\int_T f(x, y)\, dA_{xy} = \int_\alpha^\beta \int_{g(\theta)}^{h(\theta)} f(r \cos \theta, r \sin \theta) r\, dr\, d\theta. \qquad (5\text{–}3.22)$$

For instance, in Example 5–3.8 the integral over T_n could have been written by use of (5–3.22) as

$$\int_0^{r_n}\int_0^{2\pi}\frac{1}{\sqrt{1-r^2}}\,r\,dr\,d\theta.$$

Then the integration would have been quite easy.

Justification of (5–3.22) will be given later (cf. Problem 13 of Section 5–5).

Problems

1. Find the area of the following plane regular regions:

   (a) $S = \{(x, y): x^2 \le y \le \sqrt{x}\}$,
   (b) $S = \{(x, y): x^2 + 4y^2 \le 4,\ x^2 + y^2 \ge 1/4,\ x \ge 0\}$,
   (c) $S = \{(x, y): x^2 - y^2 \le 1,\ 2y \ge x,\ y \ge 0\}$,
   (d) $S = \{(x, y): x^2 + y^2 \le 2,\ (x + 1)^2 + y^2 \le 1\}$.

2. Consider a plane lamina occupying the position of the points of a set S in the (x, y)-plane. If the density (mass per unit area) at (x, y) is $\rho(x, y)$, then the mass of the lamina is

$$M(S) = \int_S \rho(x, y)\,dA_{xy}.$$

   The moment of the lamina about the y-axis (x-axis) is

$$M_y(S) = \int_S x\rho(x, y)\,dA_{xy}, \qquad \left(M_x(S) = \int_S y\rho(x, y)\,dA_{xy}\right).$$

   The center of mass, $(\bar{x}, \bar{y})$, is given by

$$M(S)\bar{x} = M_y(S) \quad\text{and}\quad M(S)\bar{y} = M_x(S).$$

   Let $\rho(x, y) = 1$ for all $(x, y) \in S$, and find the center of mass for a plane lamina in the position of S for each case of Problem 1.

3. Use Definition 5–3.1 to calculate the area of the following simple smooth surface elements:

   (a) $S = \{(u, v, uv): u^2 + v^2 \le 1\}$,
   (b) $S = \{(\cosh v, \cos u \sinh v, \sin u \sinh v): 0 \le u \le \pi,$ $\cosh^{-1} 2 \le v \le \cosh^{-1} 3\}$,
   (c) $S = \{(\cos\theta \sin\varphi, \sin\theta\sin\varphi, \cos\varphi), 0 \le \theta \le \pi,$ $\pi/4 \le \varphi \le 3\pi/4\}$,
   (d) $S = \{(y^2 + z^2, y, z): y^2 + z^2 \le 1\}$.

   Make a sketch of each set S.

4. For Problem 3(c), where $\theta = u$ and $\varphi = v$, approximate the value of $A(S)$ by taking the sum of (5–3.9), a Riemann sum, for $n = 4$ and then for $n = 16$. In particular, let the subintervals T_i of T be formed by using the lines $\theta = \pi/2$ and $\varphi = \pi/2$ for the case $n = 4$, and use the additional lines $\theta = \pi/4$, $\theta = 3\pi/4$, $\varphi = 3\pi/8$, and $\varphi = 5\pi/8$ to form the subintervals for $n = 16$. Each time let (θ_i, φ_i) be the midpoint of the T_i subinterval. Draw sketches to illustrate.

5. Find the area of the finite surfaces that are the boundaries of the following sets:

   (a) $\{(x, y, z): x^2 + y^2 + z^2 \leq 1, x \geq 0, y \geq 0\}$,
   (b) $\{(x, y, z): 0 \leq z \leq x + y, x + y \leq 1, x \geq 0, y \geq 0\}$,
   (c) $\{(x, y, z): |x| + |y| + |z| \leq 1\}$.

6. Example 5–3.3 states, near the end, that "Similarly,

$$A(S_i) = \frac{\pi}{2}\int_a^b f(x)\sqrt{1 + (f'(x))^2}\,dx, \quad i = 2, 3, 4."$$

   Demonstrate that this is true for $i = 2$.

7. Combine the techniques of Examples 5–3.3 and 5–3.5 to find the areas of the following surfaces of revolution:

   (a) a sphere of radius a,
   (b) a torus generated by revolving a circle of radius a about a line at a distance b from the center of the circle, $b > a$,
   (c) an arch of a cosine curve, $\{(x, y): y = \cos x, -\pi/2 \leq x \leq \pi/2\}$, about $y = 0$,
   (d) an arch of a cycloid, $\{(x, y): x = a(\theta - \sin\theta), y = a(1 - \cos\theta), 0 \leq \theta \leq 2\pi\}$.

8. Appropriately modify the definitions in Problem 2 for the circumstance of a thin lamina occupying the position of a finite surface, S. Find the center of mass if S is the same as in Problem 5(a) and (b) where $\rho(x, y, z) = 1$ for all $(x, y, z) \in S$.

9. Prove the following theorem of Pappus: Let S be a plane regular region with boundary curve Γ such that S fails to meet the y-axis. If $\bar{x}$ is the x-coordinate of the center of mass of a plane lamina in the position of the set S, then $(2\pi\bar{x})$(length of Γ) is equal to the area of the surface generated by revolving S around the y-axis.

10. Consider a lamina occupying the position of the points of a simple smooth surface element S (cf. Problem 2). Let $\rho(x, y, z)$ be the mass of the lamina per unit area at the point (x, y, z) of S where the function ρ defined by $\rho = \rho(x, y, z)$ is continuous on S.

    Define analogously to Problem 5–1.8, the moment of inertia about the (x, y)-plane, $I_{xy}(S)$ (also about the (x, z)-plane and the (y, z)-plane). Define the moments of inertia of the lamina about each of the three axes and about the origin. Also define $(\bar{\bar{x}}, \bar{\bar{y}}, \bar{\bar{z}})$.

Find each of the quantities mentioned above for a lamina occupying the position of the surface element S of Problem 3(a) if $\rho(x, y, z) = 1$ for all $(x, y, z) \in S$.

11. Evaluate the following integrals:

(a) $$\int_S \frac{1}{(x^2 + y^2)^{3/2}}\, dA_{xy}, \quad S = \{(x, y): x \geq 1, y \geq 1\},$$

(b) $$\int_S \frac{1}{(1 + x^2 + y^2)^2}\, dA_{xy}, \quad S = \{(x, y): x \geq 0, y \geq 0\},$$

(c) $$\int_S e^{-x^2-y^2}\, dA_{xy}, \quad S \text{ is all of } (x, y)\text{-space}.$$

12. At the beginning of this chapter, the area of the set $S = \{(x, y): a \leq x \leq b,\ 0 \leq y \leq f(x)\}$ is defined for f, a continuous function on $[a, b]$. Similarly, at the beginning of this section, the area of a y-axial set $S_x = \{(x, y): a \leq x \leq b,\ g(x) \leq y \leq h(x)\}$ where g and h are continuous on $[a, b]$ and $g(x) < h(x)$ for $a < x < b$ is defined in (5–3.2), with $f(x, y) = 1$ for all $(x, y) \in S$. Show that there is no conflict in these two definitions when the set $S = S_x$. Give an example of a set that has area by the first definition but not by the second, and vice versa.

13. In Definition 5–3.1 let T be a plane regular region in the (x, y)-plane and let $\mathbf{f}(x, y) = (x, y, 0)$. Let $S = T$. Show that by the definition $A(S)$ is the area of the plane regular region T as defined earlier in the section. In other words, show that Definition 5–3.1 is consistent with the earlier definitions.

14. Prove Proposition 5–3.2.

15. Prove that $\int_S f(x, y)\, dA_{xy}$ for S, a bounded set, does not depend upon the choice of the closed interval I that contains S.

16. In connection with Definition 5–3.3, show that g^+ and g^- are non-negative functions that are continuous where g is continuous. Show that $g = g^+ - g^-$.

17. Show that (5–3.18), (5–3.19), and (5–3.20) hold under the conditions stated in the text. State and prove the extensions of these properties to the circumstances of Definition 5–3.3.

18. Is it always possible to decompose a plane regular region T into a sum of plane regular regions T_i, $1 \leq i \leq n$, such that each T_i is axial?

19. Give an example to show that it is essential in Definition 5–3.3 to require initially that g be a non-negative function.

20. Show that in Definition 5–3.3 the value of the integral does not depend upon the choice of the sequence of finite surfaces. Suppose there are two sequences of the required type, $\{S_n\}$ and $\{S_n'\}$, and show that for any n, $S_n \subset S_k'$ for some k. Use the fact that S_n and S^* are disjoint

sets that are a finite distance apart to get an open set U of the required type such that $S_n \cap U = \emptyset$.

21. Show that in the definition of an integral over a finite surface, Definition 5–3.2, the value of the integral is independent of the manner in which S is decomposed. Let $S = S_1 + \cdots + S_n$ and $S = S_1' + \cdots + S_m'$. Consider the integrals over $S_{ij} = S_i \cap S_j'$, $1 \le i \le n$, $1 \le j \le m$ where $A(S_{ij})$ exists. Try to show that the integrals over these surface elements S_{ij} exist and that the integral over S is the sum of the integrals over S_{ij}. Notice that, with this proven, the definition of an integral over a simple smooth surface element would have the same value if it were treated as a finite surface (cf. Definition 5–3.2).

22. Show that a plane regular region always has area.

23. Use the results of Problem 5–2.19 to show that if f is a continuous function on a finite surface S, then there exists a point $(x_0, y_0, z_0) \in S$ such that

$$\int_S f(x, y, z)\, \rho(x, y, z)\, dA = f(x_0, y_0, z_0) \int_S \rho(x, y, z)\, dA$$

where $\rho(x, y, z) \ge 0$ for all $(x, y, z) \in \Omega$ and where ρ is continuous on S.

Answers

1. (a) $\dfrac{1}{3}$, (b) $\dfrac{7\pi}{8}$, (c) $\left(\dfrac{1}{4}\right) \log 3$, (d) $\pi - 1$.

2. (a) $(\bar{x}, \bar{y}) = (9/20, 9/20)$,
   (b) $(\bar{x}, \bar{y}) = (62/(21\pi), 0)$,
   (c) $(\bar{x}, \bar{y}) = \left(\dfrac{4}{3\sqrt{3}\log 3}, \dfrac{8\sqrt{3} - 12}{9 \log 3}\right)$
   (d) $(\bar{x}, \bar{y}) = \left(-\dfrac{8 + 3\pi}{6\pi - 6}, 0\right)$.

3. (a) $2\pi \dfrac{2\sqrt{2} - 1}{3}$,
   (b) $\dfrac{\pi}{2}\left\{3\sqrt{19} - 6 + \dfrac{1}{\sqrt{2}} \log \dfrac{3\sqrt{2} + \sqrt{19}}{2\sqrt{2} + 3}\right\}$,
   (c) $\sqrt{2}\,\pi$,
   (d) $\dfrac{\pi}{6}(5\sqrt{5} - 1)$.

4. For $n = 4$, 4.56 and for $n = 16$, 4.37.

5. (a) 2π, (b) $(3 + 2\sqrt{2} + \sqrt{3})/2$, (c) $4\sqrt{3}$.

7. (a) $4\pi a^2$, (b) $4\pi^2 ab$, (c) $2\pi(\sqrt{2} + \log(1 + \sqrt{2}))$, (d) $32\pi a^2/5$.

8. (a) $\bar{x} = \bar{y} = \dfrac{4 + 3\pi}{18\pi}$, $\bar{z} = 0$,

(b) $\bar{x} = \bar{y} = \dfrac{3 + 3\sqrt{2} + \sqrt{3}}{3(3 + 2\sqrt{2} + \sqrt{3})}$, $\bar{z} = \dfrac{2 + 3\sqrt{2} + 2\sqrt{3}}{3(3 + 2\sqrt{2} + \sqrt{3})}$.

10. (a) $I_{xy}(S) = \dfrac{\pi(11\sqrt{2} - 4)}{420}$, $I_{xz}(S) = I_{yz}(S) = \dfrac{2\pi(1 + \sqrt{2})}{15}$,

$I_x(S) = I_y(S) = \dfrac{\pi(67\sqrt{2} + 52)}{420}$, $I_z(S) = \dfrac{4\pi(1 + \sqrt{2})}{15}$,

$I_0(S) = \dfrac{\pi(123\sqrt{2} + 108)}{420}$, $\bar{\bar{x}} = \bar{\bar{y}} = \sqrt{105\sqrt{2} + 175}/35$,

$z = \dfrac{\sqrt{400 + 30\sqrt{2}}}{140}$

11. (a) $2 - \sqrt{2}$, (b) $\dfrac{\pi}{4}$, (c) π.

5–4. Volume Integrals

The idea of a volume integral can be developed in much the same way as the surface integral.

Let Ω be a bounded set in (x, y, z)-space, and let f be a real-valued bounded function defined on S. Furthermore, let $I = [(a, b, c), (a', b', c')]$ be an interval (rectangular parallelopiped) containing Ω (cf. Figure 5–4.1). Define on I, an extension of f, $\tilde{f}$, that is identical with f on Ω and zero elsewhere. Then

$$\mathrm{P} = \{I_{ijk}: 1 \le i \le n,\ 1 \le j \le m,\ 1 \le k \le p\}$$

is a *partition* of I if

$$a = x_0 < x_1 < \cdots < x_n = a', \qquad b = y_0 < y_1 < \cdots < y_m = b'$$

and

$$c = z_0 < z_1 < \cdots < z_p = c'$$

and

$$I_{ijk} = \{(x, y, z): x_{i-1} \le x \le x_i,\ y_{j-1} \le y \le y_j,\ z_{k-1} \le z \le z_k\},$$

$1 \le i \le n$, $1 \le j \le m$, $1 \le k \le p$. The *mesh* of P,

$$|\mathrm{P}| = \max\{\operatorname{diam}(I_{ijk}): 1 \le i \le n,\ 1 \le j \le m,\ 1 \le k \le p\}$$

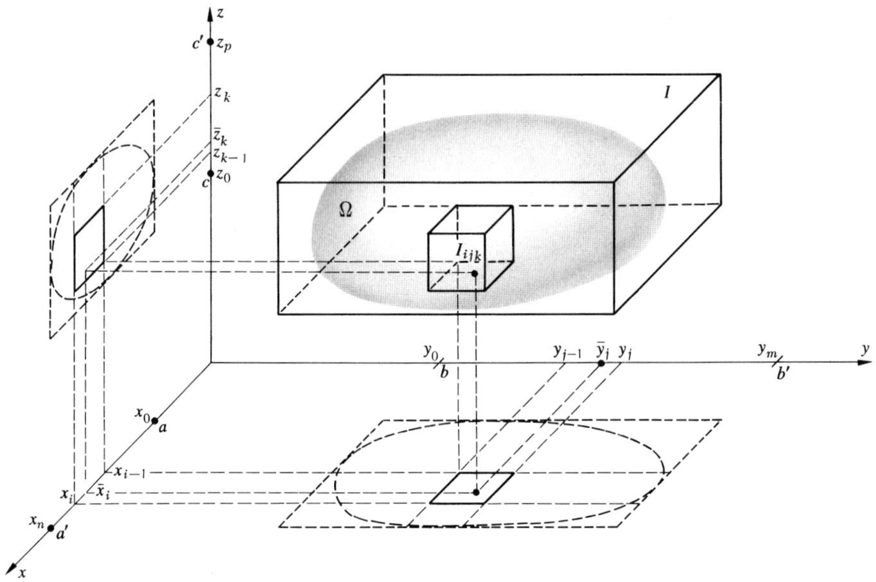

Figure 5–4.1

where $\text{diam}\,(I_{ijk}) = \sup\{\|(x, y, z) - (x', y', z')\| : (x, y, z) \text{ and } (x', y', z') \in I_{ijk}\}$ is the diameter of I_{ijk}. The *Riemann integral* of f over Ω, written

$$\int_\Omega f(x, y, z)\, dV_{xyz}, \qquad \int_\Omega f(x, y, z)\, dV,$$

or simply $\int_\Omega f\, dV$, is

$$\lim_{|P|\to 0} \sum_{i,j,k=1}^{\substack{i=n\\ j=m\\ k=p}} f(\bar{x}_i, \bar{y}_j, \bar{z}_k)\, \Delta_i x\, \Delta_j y\, \Delta_k z, \tag{5–4.1}$$

if this limit exists and is finite where

$$\Delta_i x = x_i - x_{i-1},\ x_{i-1} \leq \bar{x}_i \leq x_i \quad \text{for } 1 \leq i \leq n,$$

$$\Delta_j y = y_j - y_{j-1},\ y_{j-1} \leq \bar{y}_j \leq y_j \quad \text{for } 1 \leq j \leq m,$$

and

$$\Delta_k z = z_k - z_{k-1},\ z_{k-1} \leq \bar{z}_k \leq z_k \quad \text{for } 1 \leq k \leq p.$$

If the limit of (5–4.1) exists, then f is *integrable* on Ω; and if, in particular, $f(x, y, z) = 1$ for all $(x, y, z) \in \Omega$ and f is integrable on Ω, then

$$V(S) = \int_\Omega dV$$

is the *volume* of Ω and Ω is said to have volume.

The limit of (5–4.1) can be calculated if $\Omega = \Omega_{xy}$ where Ω_{xy} is an *axial region* of (x, y, z)-space, or, in other words, an x-, y-, or z-axial region of (x, y, z)-space. The set Ω_{xy} is a *z-axial region* of (x, y, z)-space if

$$\Omega_{xy} = \{(x, y, z): (x, y) \in T,\ g(x, y) \leq z \leq h(x, y)\}$$

where g and h are continuous functions on T, $g(x, y) < h(x, y)$ for $(x, y) \in T^0$ and T is a sum of plane regular regions (cf. Figure 5–4.2). If, for example,

$$T = \{(x, y): a \leq x \leq b,\ \varphi(x) \leq y \leq \psi(x)\}$$

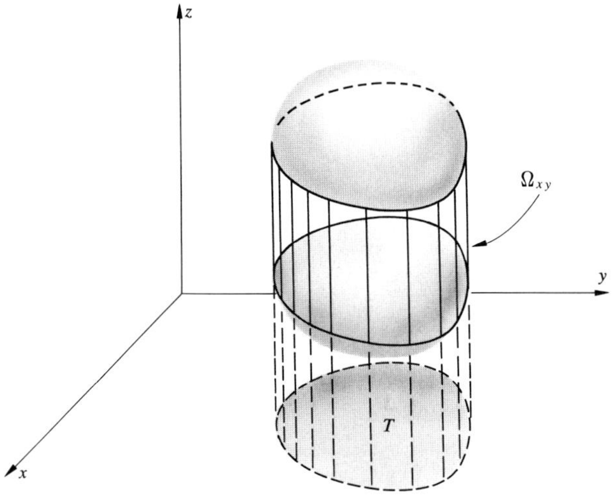

Figure 5–4.2

where φ and ψ are continuous on $[a, b]$ and $\varphi(x) < \psi(x)$ for $a < x < b$ (that is, T is a y-axial region of the (x, y)-plane), then

$$\int_\Omega f(x, y, z)\, dV = \int_a^b \int_{\varphi(x)}^{\psi(x)} \int_{g(x,y)}^{h(x,y)} f(x, y, z)\, dz\, dy\, dx, \qquad (5\text{–}4.2)$$

a *triple iterated integral*. Thus we have the value of the limit of (5–4.1), if the integration on the right side of (5–4.2) can be carried out. The integrals like

$$\int_\Omega f\, dV$$

are frequently called *triple integrals* or *volume integrals*. The properties of triple integrals are exactly analogous to the properties of double integrals. For example, we have the following existence proposition that is analogous to Proposition 5–3.1.

Proposition 5–4.1. *Let f be a real-valued function defined by $f(x, y, z)$ for each point $(x, y, z) \in \Omega$, a bounded set in (x, y, z)-space, such that Ω has volume. If f is bounded on Ω so that $f(\Omega)$ is a bounded set and f is continuous on all of Ω except a subset of Ω, Ω^* such that $V(\Omega^*) = 0$, then f is integrable on Ω.*

A proposition analogous to Proposition 5–3.2 gives us the properties of operating with triple integrals.

Proposition 5–4.2. *If* $\int_\Omega f(x, y, z)\,dV$ and $\int_\Omega g(x, y, z)\,dV$ *exist, then for α real,* $\int_\Omega \alpha f(x, y, z)\,dV$ *exists,* $\int_\Omega (f(x, y, z) + g(x, y, z))\,dV$ *exists, and*

$$\int_\Omega \alpha f(x, y, z)\,dV = \alpha \int_\Omega f(x, y, z)\,dV, \tag{5–4.3}$$

$$\int_\Omega (f(x, y, z) + g(x, y, z))\,dV = \int_\Omega f(x, y, z)\,dV + \int_\Omega g(x, y, z)\,dV. \tag{5–4.4}$$

If, in addition, for all $(x, y, z) \in \Omega$,

$$f(x, y, z) \leq g(x, y, z),$$

then

$$\int_\Omega f(x, y, z)\,dV \leq \int_\Omega g(x, y, z)\,dV. \tag{5–4.5}$$

If $\int_{\Omega_1} f(x, y, z)\,dV$ *and* $\int_{\Omega_2} f(x, y, z)\,dV$ *exist, and if* $V(\Omega_1 \cap \Omega_2) = 0$, *then* $\int_{\Omega_1 \cup \Omega_2} f(x, y, z)\,dV$ *exists and*

$$\int_{\Omega_1 \cup \Omega_2} f(x, y, z)\,dV = \int_{\Omega_1} f(x, y, z)\,dV + \int_{\Omega_2} f(x, y, z)\,dV. \tag{5–4.6}$$

The regions of (x, y, z)-space that play a fundamental role in the development of volume integrals are space region elements. These regions play a role analogous to that of a plane regular region in the development of surface integrals. These regions are defined explicitly in the following definition.

Definition 5–4.1. *A compact region Ω of (x, y, z)-space such that bdy (Ω) is a closed finite surface is a space region element.*

It is clear that the region Ω_{xy} of Figure 5–4.2 is a z-axial region that is also a space region element. However, the region

$$\Omega = \{(x, y, z): 0 \leq x \leq 1, 0 \leq y \leq 1, g(x, y) \leq z \leq h(x, y)\}$$

where $g(x, y) = 0$ for all (x, y) and

$$h(x, y) = \begin{cases} \frac{1}{6}\left[x^3\left(\sin\left(\frac{2\pi}{x}\right) + 1\right) + y\right], & x \neq 0, \\ 0, \quad x = 0, \end{cases}$$

is a z-axial region that is not a space region element since bdy (Ω) is not a finite surface (cf. Figure 5–4.3).

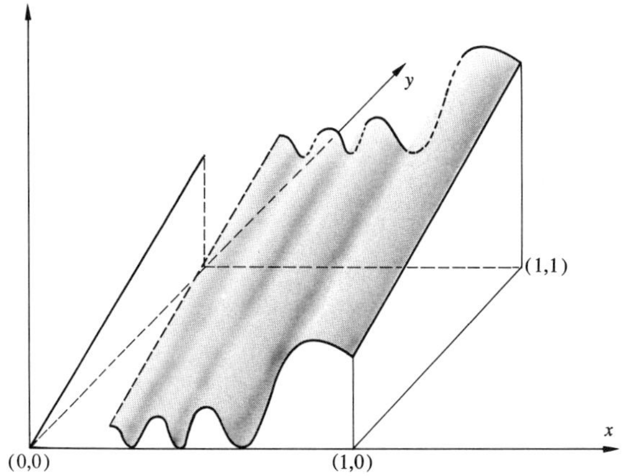

Figure 5–4.3

Since finite surfaces are decomposed into sums of surface elements, we will do something analogous with the finite space regions defined next.

Definition 5–4.2. *A region Ω in (x, y, z)-space is a finite space region if there exist space region elements $\Omega_1, \cdots, \Omega_n$, such that each of the following hold:*

(a) *The set $\Omega = \bigcup_{i=1}^{n} \Omega_i$.*

(b) *For $1 \leq i < j \leq n$, $\Omega_i \cap \Omega_j$ is a simple smooth surface element, simple smooth curve, or a single point contained in (bdy (Ω_i)) $\cap$ bdy (Ω_j).*

(c) *If $\Omega_i \cap \Omega_j$, $i \neq j$, consists of a simple smooth curve Γ, then there exists a sequence $\Omega_{k1}, \cdots, \Omega_{km}$ from our collection of space region elements such that $\Omega_i = \Omega_{k1}$, $\Omega_j = \Omega_{km}$, and $\Omega_{k(p-1)} \cap \Omega_{kp}$, $1 < p \leq m$, contains a simple smooth surface element and $\Gamma \in \Omega_{kp}$, $1 \leq p \leq m$.*

(d) *If $\Omega_i \cap \Omega_j$, $i \neq j$, consists of a single point $\mathbf{x}_0$, then there exists a sequence $\Omega_{k1}, \cdots, \Omega_{km}$ from our collection of space region elements* *such that* $\Omega_1 = \Omega_{k1}$, $\Omega_j = \Omega_{km}$, *and* $\Omega_{k(p-1)} \cap \Omega_{kp}$, $1 < p \leq m$, *contains a simple smooth surface element and* $\mathbf{x}_0 \in \Omega_{kp}$, $1 \leq p \leq m$.

(e) *Any three of the space region elements meet in at most a simple smooth curve, and at most two of these space region elements meet in a simple smooth surface element.*

In such a case Ω *is said to be the sum of the space region elements* $\Omega_1, \cdots, \Omega_n$, *written* $\Omega = \Omega_1 + \cdots + \Omega_n$.

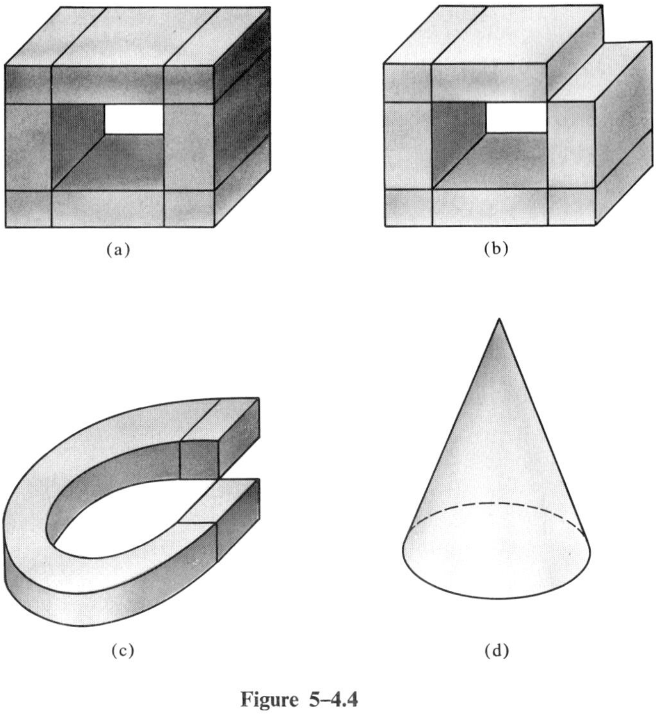

Figure 5–4.4

In Figure 5–4.4 the region illustrated by (a) is a finite space region, and those regions illustrated in (b), (c), and (d) are not finite space regions. Since any finite space region is the union of compact sets, it is clear that a finite space region must be closed and bounded.

If f is a continuous function on a finite space region, the Riemann integral of f over Ω will always exist. This follows from Proposition 5–4.1, if we know that Ω has volume, but Ω will always have volume since its boundary is a finite union of smooth surface elements (cf. Problem 12 at the end of this section). In order to evaluate an integral of f over a finite space region Ω, we try to decompose Ω into space region elements Ω_i,

$\Omega = \Omega_1 + \cdots + \Omega_n$ such that each Ω_i is axial. Then we can write

$$\int_\Omega f\, dV = \sum_{i=1}^{n} \int_{\Omega_i} f\, dV \tag{5–4.7}$$

by use of (5–4.6), and each integral on the right can be evaluated by (5–4.2) (cf. Problem 10). The following example illustrates this procedure.

Example 5–4.1. Let the function f be defined by $f(x, y, z) = x^2$ and let

$$\Omega = \{(x, y, z): x^2 + z^2 \leq 4,\ y^2 + z^2 \leq 4,\ z \geq 0\}$$

(cf. Figure 5–4.5). Then $\Omega = \Omega_1 + \cdots + \Omega_4$, as shown in the figure. Each

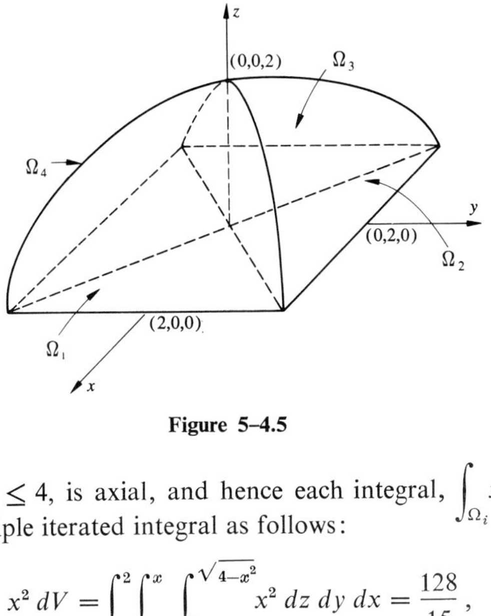

Figure 5–4.5

region Ω_i, $1 \leq i \leq 4$, is axial, and hence each integral, $\int_{\Omega_i} x^2\, dV$, can be evaluated by a triple iterated integral as follows:

$$\int_{\Omega_1} x^2\, dV = \int_0^2 \int_{-x}^{x} \int_0^{\sqrt{4-x^2}} x^2\, dz\, dy\, dx = \frac{128}{15},$$

$$\int_{\Omega_2} x^2\, dV = \int_0^2 \int_{-y}^{y} \int_0^{\sqrt{4-y^2}} x^2\, dz\, dx\, dy = \frac{128}{45},$$

$$\int_{\Omega_3} x^2\, dV = \int_{-2}^0 \int_{x}^{-x} \int_0^{\sqrt{4-x^2}} x^2\, dz\, dy\, dx = \frac{128}{15},$$

$$\int_{\Omega_4} x^2\, dV = \int_{-2}^0 \int_{y}^{-y} \int_0^{\sqrt{4-y^2}} x^2\, dz\, dx\, dy = \frac{128}{45}.$$

Therefore,

$$\int_\Omega x^2\, dV = \frac{1024}{45}.$$

We shall now consider improper volume integrals. That is, we shall discuss integrals of functions that become infinite on certain bounded sets, and we shall consider integrals over unbounded sets. Naturally, this discussion will parallel very closely the presentation of improper integrals in the preceding section. First, we shall give a definition analogous to Definition 5–2.4.

Definition 5–4.3. *A closed region Ω is said to be generated by a sequence of finite space regions, $\{\Omega_n : n = 1, 2, 3, \cdots\}$ if*

(a) $\Omega_n \subset \Omega$ *for each n,*
(b) $\Omega_n \subset \Omega_{n+1}$, $n = 1, 2, 3, \cdots$, *and*
(c) $\Omega = \overline{(\bigcup_{n=1}^{\infty} \Omega_n)}$.

A point $\mathbf{x} \in \Omega$ is an exceptional point of Ω if $\mathbf{x}$ is not contained in a space region element contained in Ω.

Clearly any finite space region is a closed region generated by a sequence of finite space regions.

In order to be able to evaluate improper volume integrals, it will be necessary to define the idea of convergence of a sequence of finite space regions. The following definition, which is analogous to Definition 5–2.5, will make precise this meaning of convergence.

Definition 5–4.4. *Let Ω be a closed region generated by a sequence of finite space regions. Let Ω^* be a nonempty closed subset of the intersection of a finite surface and Ω. If and only if Ω is unbounded, we shall say ∞ belongs to Ω^* and allow that Ω^* can be otherwise empty. Let U be any open set containing Ω^* such that U contains $\{(x, y, z) : \|(x, y, z)\| > r\}$, for some $r > 0$, if $\infty \in \Omega^*$. A sequence of finite space regions, $\{\Omega_n : n = 1, 2, 3, \cdots\}$, $\Omega_n \subset \Omega - \Omega^*$, $n = 1, 2, 3, \cdots$, that generates Ω is said to converge to Ω relative to Ω^*, written $\lim_{n\to\infty} \Omega_n = \Omega(\Omega^*)$, if there exists an integer $n_0 > 0$ such that for $n > n_0$, $\Omega - \Omega_n \subset U$.*

Again in the case of these regions as in the case of surfaces, we do not consider convergence of a sequence $\{\Omega_n\}$ relative to Ω^* unless Ω^* has the above required properties. We are now ready to consider integrals over these regions.

Definition 5–4.5. *Let Ω be a region generated by a sequence of finite space regions and let f be a non-negative real-valued function defined on a subset of S. Let Ω^* be a nonempty subset of Ω that consists of all of the points of Ω such that f is unbounded in a neighborhood of the point and ∞ if Ω is* *unbounded. Furthermore, suppose that* Ω^* *is such that there exists a sequence of finite space regions,* $\{\Omega_n : n = 1, 2, 3, \cdots\}$, *converging to* Ω *relative to* Ω^*. *Let* $f \in C(\Omega - \Omega^*)$. *Then the Riemann integral of* f *over* Ω, $\int_\Omega f(x, y, z)\, dV_{xyz}$, $\int_\Omega f(x, y, z)\, dV$, *or simply* $\int_\Omega f\, dV$, *is given by* $\lim_{n\to\infty} \int_{\Omega_n} dV$ *if this limit exists and is finite.*

If f *is not necessarily a non-negative function on* Ω *but* $\int_\Omega f^+\, dV$ *and* $\int_\Omega f^-\, dV$ *exist where* $f^+ = \tfrac{1}{2}(f + |f|)$ *and* $f^- = \tfrac{1}{2}(-f + |f|)$, *then the Riemann integral of* f *over* Ω *is*

$$\int_\Omega f\, dV = \int_\Omega f^+\, dV - \int_\Omega f^-\, dV. \tag{5–4.8}$$

This definition cannot cover the same circumstance as that covered at the beginning of this section since, if Ω is bounded, Ω^* being nonempty implies that f is not bounded on Ω. Notice that since $f \in C(\Omega - \Omega^*)$, the function $f \in C(\Omega_n)$ for each n and hence by Proposition 5–4.1, $\int_{\Omega_n} f\, dV$ always exists. Recall that Ω_n, being a finite space region, always has volume.

Any integral defined at the beginning of this section or defined in Definition 5–4.5 is a *volume integral.* Three examples of the evaluation of integrals using Definition 5–4.5 follow.

Example 5–4.2. (a) Let f be defined by $f(x, y, z) = (xyz)^{-2}$ and let $\Omega = \{(x, y, z): x \geq 1, y \geq 1, z \geq 1\}$. Thus Ω is not a finite space region, and $\Omega^* = \{\infty\}$. Taking $\Omega_n = \{(x, y, z): 1 \leq x \leq n, 1 \leq y \leq n, 1 \leq z \leq n\}$ for $n = 2, 3, 4, \cdots$ makes Ω_n converge to Ω relative to Ω^*. Then

$$\begin{aligned}\int_\Omega f(x, y, z)\, dV &= \lim_{n\to\infty} \int_1^n \int_1^n \int_1^n (xyz)^{-2}\, dz\, dy\, dx \\ &= \lim_{n\to\infty} (1 - n^{-1})^3 \\ &= 1.\end{aligned}$$

(b) Let f be defined by $f(x, y, z) = (xyz)^{-1/2}$ and let

$$\Omega = \{(x, y, z): 0 \leq x \leq 1, 0 \leq y \leq 1, 0 \leq z \leq 1\}.$$

Thus f is not continuous on Ω, a finite space region, and $\Omega^* = \{\mathbf{0}\}$. Then take $\Omega_n = \{(x, y, z): 1/n \leq x \leq 1, 1/n \leq y \leq 1, 1/n \leq z \leq 1\}$ for $n = 2, 3, 4, \cdots$ and thus $f \in C(\Omega_n)$, $n = 2, 3, 4, \cdots$. Furthermore, Ω_n converges to Ω relative to Ω^*.

$$\int_\Omega f(x, y, z)\, dV = \lim_{n\to\infty} \int_{1/n}^{1} \int_{1/n}^{1} \int_{1/n}^{1} (xyz)^{-1/2}\, dz\, dy\, dx$$
$$= \lim_{n\to\infty} 2^3(1 - n^{-1/2})^3$$
$$= 8.$$

(c) Let f be defined by $f(x, y, z) = 1$ and let

$$\Omega = \{(x, y, z): x^2 + y^2 \leq z^2 \leq 1\}.$$

The set Ω is not a finite space region since $\mathbf{0}$ is an exceptional point of Ω. In this case let

$$\Omega_n = \{(x, y, z): 1/n^2 \leq x^2 + y^2 \leq z^2 \leq 1\}, \quad n = 2, 3, 4, \cdots.$$

Then

$$V(\Omega) = \int_\Omega dV = \lim_{n\to\infty} \int_{\Omega_n} dV$$

where

$$\int_{\Omega_n} dV = \pi \int_{1/n}^{1} z^2\, dz = \frac{\pi}{3}\left(1 - \left(\frac{1}{n}\right)^3\right).$$

Therefore,

$$V(\Omega) = \frac{\pi}{3}.$$

In other writings on volume integrals, an integral of the type of Example 5–4.2(a) is denoted by

$$\int_1^\infty \int_1^\infty \int_1^\infty (xyz)^{-2}\, dz\, dy\, dx.$$

However, this could, and most likely would, mean an improper iterated integral. The reader should always be careful to understand the author's interpretation.

As in the case of certain surface integrals it is also convenient in evaluating certain volume integrals to use variables of other coordinate systems—for example, cylindrical coordinates, r, θ, and z. In particular, let Ω be given by

$$\Omega = \{(r, \theta, z): (r, \theta) \in T,\ \varphi(r, \theta) \leq z \leq \psi(r, \theta)\}$$

where T is an r-axial region of (r, θ)-space defined by

$$T = \{(r, \theta): \alpha \leq \theta \leq \beta, 0 \leq g(\theta) \leq r \leq h(\theta)\},$$

where $0 < \beta - \alpha \leq 2\pi$, $g(\theta) < h(\theta)$ for $\alpha < \theta < \beta$ and for both g and $h \in C([\alpha, \beta])$. Suppose g and $h \in C(T)$ and $\varphi(r, \theta) < \psi(r, \theta)$ for $(r, \theta) \in T^\circ$.

If the function $f \in C(\Omega)$, then

$$\int_{\Omega} f\,dV = \int_{\alpha}^{\beta} \int_{g(\theta)}^{h(\theta)} \int_{\varphi(r,\theta)}^{\psi(r,\theta)} f(r\cos\theta, r\sin\theta, z) r\,dz\,dr\,d\theta \qquad (5\text{–}4.9)$$

(cf. Problem 13(b) page 282). For example, in Problem 1(a), which follows,

$$V(\Omega) = \int_0^{2\pi} \int_0^1 \int_0^{\sqrt{4-r^2}} r\,dz\,dr\,d\theta.$$

Problems

1. Find the volume of the following axial regions in (x, y, z)-space. In each case make a sketch of the region.
   (a) $\{(x, y, z): x^2 + y^2 \leq 1, 0 \leq z \leq \sqrt{4 - x^2 - y^2}\}$,
   (b) $\{(x, y, z): 1 \leq x^2 + y^2 \leq 4, 0 \leq z \leq 2 + xy\}$,
   (c) $\{(x, y, z): z \geq 0, 0 \leq y \leq 2 - 2z, 0 \leq x \leq y^2 + z^2\}$,
   (d) $\{(x, y, z): 1 \leq y \leq 2, x \geq 0, 4y^2 \geq x^2 + z^2\}$.
2. Sketch the following regions in (x, y, z)-space. Give reasons why you think that the regions are or are not space region elements.
   (a) $\{(x, y, z): |x| + |y| + |z| \leq 1\}$.
   (b) $\{(x, y, z): x^2 + y^2 \leq 1\}$.
   (c) $\{(x, y, z): 1 \leq x^2 + y^2 + z^2 \leq 4\}$.
   (d) $$\left\{(x, y, z): \left(x - \frac{1}{t}\cos(t-1)\right)^2 + \left(y - \frac{1}{t}\sin(t-1)\right)^2 + z^2 \leq t^{-2}(t + 2\pi)^{-2}, t \geq 1\right\}.$$

   Is the region in (d) like the house of a snail?
3. In Figure 5–4.4 why are the regions illustrated by (b), (c), and (d) not finite space regions?
4. Consider the function f defined by $f(x, y, z) = x$ on the region

   $$\Omega = \left\{(x, y, z): 0 \leq x, 0 \leq y, 0 \leq z \leq \left(\frac{1}{6}\right)(6 - 2x - 3y)\right\}.$$

   Let $I = [(0, 0, 0), (3, 2, 1)]$ and let P be a partition of I consisting of eight subintervals made by dividing the x, y, and z side of I into two equal pieces. Find the Riemann sum,

   $$\sum_{i,j,k=1}^{i,j,k=2} \tilde{f}(\bar{x}_i, \bar{y}_j, \bar{z}_k)\,\Delta_i x\,\Delta_j y\,\Delta_k z,$$

   where $(\bar{x}_i, \bar{y}_j, \bar{z}_k)$ is the center of I_{ijk}. Work the same problem again with the condition that this time each side of I is divided into three equal pieces.
5. Consider a region Ω enclosed by a surface of revolution so that

   $$\Omega = \{(x, y, z): a \leq x \leq b, y^2 + z^2 \leq (f(x))^2\}$$

where f is a continuously differentiable function on $[a, b]$, and $f(x) > 0$ for all $x \in [a, b]$. Show that Ω is a finite space region and that

$$\int_{\Omega} dV = \pi \int_a^b (f(x))^2 \, dx.$$

Show the equality still holds even if the restriction that $f(x) > 0$ for $x \in [a, b]$ is removed.

6. If $\rho(x, y, z)$ is the volume density of some "solid material" or simply "solid" occupying a finite space region Ω, the mass of the solid,

$$M(\Omega) = \int_{\Omega} \rho(x, y, z) \, dV,$$

and the moment (first moment) of the solid about the (x, y)-plane, $M_{xy}(\Omega)$, is given by

$$M_{xy}(\Omega) = \int_{\Omega} z\rho(x, y, z) \, dV.$$

Similarly, $M_{xz}(\Omega)$ and $M_{yz}(\Omega)$ are defined. The point $(\bar{x}, \bar{y}, \bar{z})$, where $\bar{z}\, M(\Omega) = M_{xy}(\Omega)$, $\bar{y}\, M(\Omega) = M_{xz}(\Omega)$, and $\bar{x}\, M(\Omega) = M_{yz}(\Omega)$, is the center of mass of the solid. For the solids occupying the following regions Ω with the given density functions ρ, find $M(\Omega)$, $M_{xy}(\Omega)$, $M_{yz}(\Omega)$, $M_{xz}(\Omega)$, and $(\bar{x}, \bar{y}, \bar{z})$.

(a) $\Omega = \{(x, y, z): 1 \le z \le \sqrt{4 - x^2 - y^2}\}$, $\quad \rho(x, y, z) = 1$.
(b) $\Omega = \{(x, y, z): x^2 + (y^2/4) \le 1, x \ge 0, y \ge 0, 0 \le z \le 1\}$, $\quad \rho(x, y, z) = 1$.
(c) $\Omega = \{(x, y, z): 0 \le z \le x^2 + y^2 \le 1, y \ge 0\}$, $\quad \rho(x, y, z) = y$.
(d) $\Omega = \{(x, y, z): 0 \le z \le \min(1 - y, 1 - x), x^2 + y^2 \le 1\}$, $\quad \rho(x, y, z) = z$.

7. Following the setting of Problem 6, the moment of inertia (2nd moment) of the solid about the (x, y)-plane, $I_{xy}(\Omega)$, is given by

$$I_{xy}(\Omega) = \int_{\Omega} z^2 \rho(x, y, z) \, dV.$$

Similarly, $I_{yz}(\Omega)$ and $I_{xz}(\Omega)$ are defined. The moment of inertia of the "solid" about the x-axis is

$$I_x(\Omega) = \int_{\Omega} (y^2 + z^2) \rho(x, y, z) \, dV.$$

Thus $I_x(\Omega) = I_{xy}(\Omega) + I_{xz}(\Omega)$. Similarly, $I_y(\Omega)$ and $I_z(\Omega)$ are defined. The moment of inertia of the solid about the origin is

$$I_0(\Omega) = \int_{\Omega} (x^2 + y^2 + z^2)\rho(x, y, z) \, dV,$$

and $I_0(\Omega) = I_x(\Omega) + I_y(\Omega) + I_z(\Omega)$. Find each of these quantities for each of the solids given in Problem 6(a) and (b). Also in each case find

$(\bar{\bar{x}}, \bar{\bar{y}}, \bar{\bar{z}})$, where $\bar{\bar{x}}$, the radius of gyration about the (y, z)-plane, is defined by $\bar{\bar{x}}^2 M(\Omega) = I_{yz}(\Omega)$ and $\bar{\bar{y}}$ and $\bar{\bar{z}}$ are defined similarly.

8. Following the ideas of Problem 7, define the moment of inertia, I_L, of a solid occupying a region Ω about an arbitrary line L. Let L_C be a line through the center of mass of the solid parallel to L at a distance s from the line L. If I_{L_C} is the moment of inertia of the solid about L_C, prove that $I_L = I_{L_C} + Ms^2$ where M is the mass of the solid.

9. Notice that in Example 5–4.1 the required integral is the moment of inertia $I_{yz}(\Omega)$ of the solid of density $\rho(x, y, z) = 1$ occupying Ω about the (y, z)-plane. Check the answer of this example by first finding the moment of inertia about the (y, z)-plane of a solid occupying the region
$$\Omega_{x_0} = \{(x, y, z) \colon |x| \le x_0, |y| \le x_0, 0 \le z_0 \le z \le z_0 + \Delta z \le 2\}$$
where $x_0^2 + z_0^2 = 4$, namely, $I_{yz}(\Omega_{x_0}) = (4/3)x_0^4(\Delta z)$. Then show that
$$I_{yz}(\Omega) = \left(\frac{4}{3}\right)\int_0^2 x^4\, dz$$
and give reasons why this is so.

10. Is it always possible to decompose a space region element into a sum of axial space region elements?

11. Show that in Definition 5–4.5 the value of the integral does not depend upon the choice of the sequence of finite space regions (cf. Problem 20, page 225).

12. Show that a space region element always has volume.

13. Show that a finite space region is arcwise connected. If Ω is a closed region generated by a sequence of finite space regions, is Ω arcwise connected?

14. Prove that the properties (5–4.3), (5–4.4), and (5–4.5) hold for the integrals of Definition 5–4.5.

15. Let Ω be a bounded closed region with Ω^* a nonempty set of exceptional points of Ω. Suppose that $\{\Omega_n \colon n = 1, 2, 3, \cdots\}$ is a sequence of finite space regions such that $\lim \Omega_n = \Omega(\Omega^*)$. Let f be a bounded real-valued function on Ω that is continuous on $\Omega - \Omega^*$. Then, by Proposition 5–4.1, f is integrable on Ω since Ω^* has $V(\Omega^*) = 0$. Show that the value of the limit of the Riemann sums, (5–4.1), is equal to the limit of Definition 5–4.5; that is,
$$\lim_{n\to\infty} \int_{\Omega_n} f\, dV.$$

16. Use the results of Problem 13 to show that if f is a continuous function on a finite space region Ω, then there exists a point $(x_0, y_0, z_0) \in \Omega$ such that
$$\int_\Omega f(x, y, z)\rho(x, y, z)\, dV = f(x_0, y_0, z_0)\int_\Omega \rho(x, y, z)\, dV$$
where $\rho(x, y, z) \ge 0$ for all $(x, y, z) \in \Omega$ and where ρ is continuous on Ω.

Answers

1. (a) $2\pi\dfrac{8 - 3\sqrt{3}}{3}$, (b) 6π, (c) $\dfrac{5}{6}$ (d) $\dfrac{7\pi}{3}$.

2. (a) Yes. (b) No. (c) No. (d) No.

4. The value of the Riemann sum for eight subintervals is $9/16$, and for 27 it is $2/3$. The actual value of the integral is $3/4$.

6. (a) $M(\Omega) = \dfrac{5\pi}{3}$, $M_{xy}(\Omega) = \dfrac{9\pi}{4}$, $M_{xz}(\Omega) = M_{yz}(\Omega) = 0$,

$\bar{x} = \bar{y} = 0$, $\bar{z} = 27/20$.

(b) $M(\Omega) = \dfrac{\pi}{2}$, $M_{xy}(\Omega) = \pi$, $M_{xz}(\Omega) = \dfrac{4}{3}$, $M_{yz}(\Omega) = \dfrac{2}{3}$,

$\bar{x} = \dfrac{4}{3\pi}$, $\bar{y} = \dfrac{8}{3\pi}$, $z = \dfrac{1}{2}$.

(c) $M(\Omega) = \dfrac{2}{5}$, $M_{xy}(\Omega) = \dfrac{1}{7}$, $M_{xz}(\Omega) = \dfrac{\pi}{12}$, $M_{yz}(\Omega) = 0$,

$\bar{x} = 0$, $\bar{y} = \dfrac{5\pi}{24}$, $\bar{z} = \dfrac{5}{14}$.

(d) $M(\Omega) = \dfrac{3\pi - 2\sqrt{2}}{3}$, $M_{xy}(\Omega) = \dfrac{105\pi - 142}{60}$,

$M_{xz}(\Omega) = M_{yz}(\Omega) = \dfrac{52\sqrt{2} - 15\pi}{120}$,

$\bar{x} = \bar{y} = \dfrac{52\sqrt{2} - 15\pi}{120\pi - 80\sqrt{2}}$

$\bar{z} = \dfrac{105\pi - 142\sqrt{2}}{60\pi - 40\sqrt{2}}$.

7. (a) $I_{xy}(\Omega) = \dfrac{47\pi}{15}$, $I_{xz}(\Omega) = I_{yz}(\Omega) = \dfrac{53\pi}{60}$,

$I_x(\Omega) = I_y(\Omega) = \dfrac{241\pi}{60}$, $I_z(\Omega) = \dfrac{53\pi}{30}$, $I_0(\Omega) = \dfrac{49\pi}{10}$,

$x = \bar{\bar{y}} = \dfrac{\sqrt{53}}{10}$, $\bar{\bar{z}} = \dfrac{\sqrt{47}}{5}$

(b) $I_{xy}(\Omega) = \dfrac{\pi}{6}$, $I_{xz}(\Omega) = \dfrac{\pi}{2}$, $I_{yz}(\Omega) = \dfrac{\pi}{8}$,

$I_x(\Omega) = \dfrac{2\pi}{3}$, $I_y(\Omega) = \dfrac{7\pi}{24}$, $I_z(\Omega) = \dfrac{5\pi}{8}$,

$I_0(\Omega) = \dfrac{19\pi}{24}$, $\bar{\bar{x}} = \dfrac{1}{2}$, $y = 1$, $z = \dfrac{\sqrt{3}}{3}$.

5–5. Change of Variables of Integrals

It is frequently desirable to be able to change the variables of integration of a multiple integral. In order to prepare for this, let us review the details of changing the variable of an integral involving a single variable.

Let h be a function such that h: $[a, b] \to R$ where $[a, b]$ is a closed interval in E^1 and suppose that $h \in C([a, b])$. By a part of the Fundamental Theorem of the Calculus, the function H defined by

$$H(t) = \int_a^t h(u)\, du$$

for $a \leq t \leq b$ is such that $H \in C([a, b])$ and such that $H'(t) = h(t)$ for $a < t < b$. Hence H is continuously differentiable on $[a, b]$, and we may write $H'(t) = h(t)$ for $t \in [a, b]$. Let f be a function such that f: $[c, d] \to [a, b]$ where $[c, d]$ is a closed interval in E^1. Suppose that f is continuously differentiable on $[c, d]$, $f'(x) > 0$ for $c < x < d$ and $f(c) = a$, $f(d) = b$. Then it is easily possible, with the function f having these properties, to extend the definitions of f and H to open sets containing $[c, d]$ and $[a, b]$, respectively, such that the functions f and H satisfy Proposition 3–4.1, the chain rule. Hence

$$\frac{d}{dx} H(f(x)) = H'(f(x))f'(x) = h(f(x))f'(x). \tag{5–5.1}$$

Thus, by another part of the Fundamental Theorem of Calculus, we have

$$\begin{aligned}\int_c^d h(f(x))f'(x)\, dx &= H(f(d)) - H(f(c)) \\ &= H(b) - H(a).\end{aligned}$$

However, by this same theorem,

$$\int_a^b h(u)\, du = H(b) - H(a).$$

Therefore, we obtain the *change of variable* identity of integrals of a single variable—namely,

$$\int_a^b h(u)\, du = \int_c^d h(f(x))f'(x)\, dx. \tag{5–5.2}$$

If in the above setting we require, instead, that $f'(x) < 0$ at all interior points of its domain so that f: $[d, c] \to [a, b]$, then again we get (5–5.2)

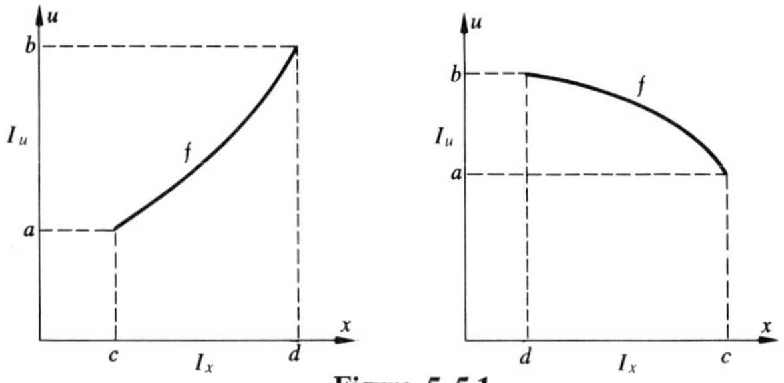

Figure 5–5.1

(cf. Figure 5–5.1). However, if the symbol $\int_{[a,b]}$ is used in place of the symbol $\int_a^b$ when $a < b$, then (5–5.2) becomes

$$\int_{[a,b]} h(u)\, du = \int_{[c,d]} h(f(x))f'(x)\, dx, \tag{5–5.3}$$

and

$$\int_{[a,b]} h(u)\, du = -\int_{[d,c]} h(f(x))f'(x)\, dx, \tag{5–5.4}$$

respectively, for the cases of the derivative of f positive and negative on the interior of its domain. Notice that (5–5.3) and (5–5.4) can be summarized into one case as indicated by the following proposition. Here the symbol

$$\int_I g(t)\, dt = \int_{t_1}^{t_2} g(t)\, dt$$

where $I = [t_1, t_2]$.

Proposition 5–5.1. *Let f be a function such that $f\colon I_x \to R$ where I_x is a closed interval in E^1 and suppose f, which is defined by $u = f(x)$, is continuously differentiable on I_x and that $f'(x) \neq 0$ on I_x where $f'(x_0) = \lim_{x\to x_0} f'(x)$ for x_0 a boundary point of I_x. Then $I_u = f(I_x)$ is a closed interval, and f is* 1–1 *on I_x. Furthermore, if h is a function such that $h\colon I_u \to R$ and h is continuous on I_x, then*

$$\int_{I_u} h(u)\, du = \int_{I_x} h(f(x))\, |f'(x)|\, dx. \tag{5–5.5}$$

The proof that $I_u = f(I_x)$ is a closed interval and that f is 1–1 on I_x is not difficult. The remainder of the proposition is just a more concise statement of the earlier conclusions.

Now we shall consider the two-dimensional analogy to the foregoing proposition.

Proposition 5–5.2. *Let* $\mathbf{f} = (f_1, f_2)$ *be a function such that* $\mathbf{f}\colon S_{xy} \to E^2$, *the* (u, v)*-plane, where* S_{xy} *is a plane regular region in the* (x, y)*-plane and suppose that* $\mathbf{f}$, *which is defined by* $u = f_1(x, y)$, $v = f_2(x, y)$, *is* 1–1 *and continuously differentiable on* S_{xy}. *Suppose, furthermore, that*

$$\frac{\partial(u, v)}{\partial(x, y)} \neq 0$$

on the interior of S_{xy}, *nor does it have a limiting value of zero on the boundary of* S_{xy}. *Then* $S_{uv} = \mathbf{f}(S_{xy})$ *is a plane regular region of the* (u, v)*-plane. Furthermore, if* h *is a function such that* $h\colon S_{uv} \to R$ *and if* h *is continuous on* S_{uv}, *then*

$$\int_{S_{uv}} h(u, v)\, dA_{uv} = \int_{S_{xy}} h(\mathbf{f}(x, y)) \left| \frac{\partial(u, v)}{\partial(x, y)} \right| dA_{xy}. \tag{5–5.6}$$

Proof. Let $\mathbf{F} = (F_1, F_2, F_3)$ be a function defined by $F_1(x, y) = f_1(x, y)$, $F_2(x, y) = f_2(x, y)$ and $F_3(x, y) = 0$. Then $\mathbf{F}$ is continuously differentiable on S_{xy}, and, by Definition 5–2.1, S_{uv} is a simple surface element. Then

$$\frac{\partial}{\partial x}\mathbf{F}(x, y) \times \frac{\partial}{\partial y}\mathbf{F}(x, y) = \left(0, 0, \frac{\partial(f_1, f_2)}{\partial(x, y)}\right) \neq \mathbf{0},$$

and hence S_{uv} is a simple smooth surface element that lies in the (u, v)-plane. As was indicated in 5–2, the edge of the surface element S_{uv} is a simple piecewise smooth closed curve. By Problem 18, page 237, S_{uv} is a compact set. If we assume that the boundary of S_{uv} is the edge of S_{uv}, then it follows that S_{uv} is a plane regular region.

By Definition 5–3.2 where S is a simple smooth surface element, it follows from (5–3.14) that

$$\int_{S_{uv}} h(u, v)\, dA_{uv} = \int_{S_{xy}} h(f_1(x, y), f_2(x, y)) \left| \frac{\partial(f_1, f_2)}{\partial(x, y)} \right| dA_{xy}. \quad \blacksquare$$

Notice that in this proposition it is not possible to obtain the fact that $\mathbf{f}$ is 1–1 on S_{xy} from the fact that the Jacobian is not zero on the interior of S_{xy}. Nor does it have the value zero in the limit on the boundary of S_{xy}, for consider:

$$f_1(x, y) = x \cos y, \qquad f_2(x, y) = x \sin y \tag{5–5.7}$$

where $S_{xy} = \{(x, y)\colon 1 \leq x \leq 2, 0 \leq y \leq 2\pi\}$. Here the Jacobian has the required properties, but $\mathbf{f}$ is not 1–1 on S_{xy}. In contrast, in Proposition 5–5.1, f is 1–1 on I_x because $f'(x) \neq 0$ on I_x.

In order to check the one-to-oneness of the function $\mathbf{f}$, it may be simple enough to show directly that if $\mathbf{f}(x_1, y_1) = \mathbf{f}(x_2, y_2)$, then $(x_1, y_1) = (x_2, y_2)$.

However, it may be simpler to plot the level curves

$$\{(x, y): u_0 = f_1(x, y)\} \quad \text{and} \quad \{(x, y): v_0 = f_2(x, y)\}$$

where u_0 and v_0 are constants and then notice that there is a unique pair of these curves through each point in S_{xy}.

The following example illustrates the application of the foregoing proposition.

Example 5–5.1. Let $\mathbf{f} = (f_1, f_2)$ be the transformation given by (5–5.7) and let

$$S_{xy} = \{(x, y): 1 \leq x \leq 2, 0 \leq y \leq \pi/2\}.$$

Then S_{uv} becomes a region that is one-fourth of an annulus (cf. Figure 5–5.2). Then with $h(u, v) = 1$ we have, by (5–5.6),

$$\int_{S_{uv}} dA_{uv} = \int_{S_{xy}} \left| \frac{\partial(u, v)}{\partial(x, y)} \right| dA_{xy} = \int_1^2 \int_0^{\pi/2} x \, dy \, dx = \frac{3\pi}{4}.$$

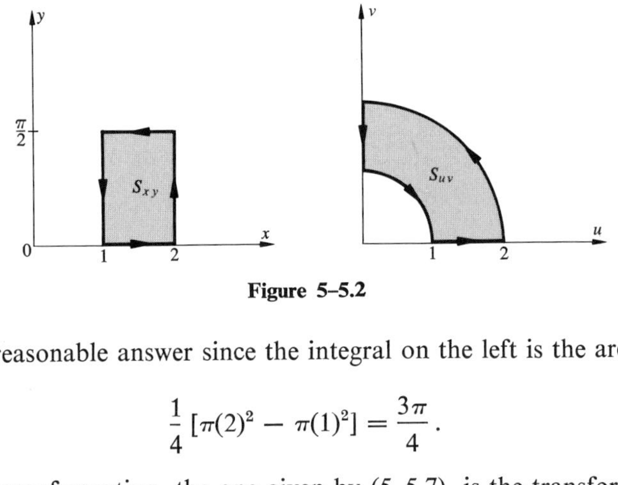

Figure 5–5.2

This is a reasonable answer since the integral on the left is the area of S_{uv}. That is,

$$\frac{1}{4}[\pi(2)^2 - \pi(1)^2] = \frac{3\pi}{4}.$$

This transformation, the one given by (5–5.7), is the transformation of polar coordinates to rectangular coordinates. Notice that $|x| \, dx \, dy$, which is usually written $r \, dr \, d\theta$, $r \geq 0$, is the commonly referred to "element of area of polar coordinates."

Example 5–5.2. Find the moment about the u-axis of the region S_{uv} that is the image of $S_{xy} = \{(x, y): 0 \leq x \leq 1, 0 \leq y \leq 1\}$ by the transformation

$$u = x + y^2, \qquad v = 2x - y \tag{5–5.8}$$

where the density of S_{uv} at (u, v), $\rho(u, v) = \rho_0$, a constant.

In order to get a sketch of S_{uv}, eliminate x and then y between the two equations of (5–5.8), obtaining

$$v = 2u - (2y^2 + y),$$
$$u - x = (v - 2x)^2.$$

Plot in the (u, v)-plane the set of points where $y = 0, \frac{1}{4}, \frac{1}{2}, \frac{3}{4}, 1$ and similarly for x (cf. Figure 5–5.3). Notice that the image of the curve joining

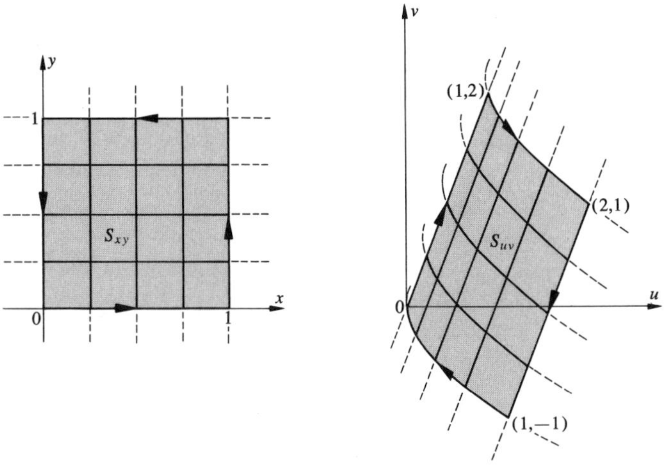

Figure 5–5.3

(0, 0) to (1, 0) in the (x, y)-plane is given by

$$\{(u, v): u = x,\ v = 2x,\ 0 \leq x \leq 1\},$$

or, in other words, by the line joining (0, 0) to (1, 2) in the (u, v)-plane. Similarly, the images of the other boundary curves of S_{xy} can be found until we are sure that S_{uv} is the region indicated in Figure 5–5.3.

Then, by Proposition 5–5.2,

$$M_u(S_{uv}) = \rho_0 \int_{S_{uv}} v\, dA_{uv}$$
$$= \rho_0 \int_{S_{xy}} (2x - y)\, |-1 - 4y|\, dA_{xy}.$$

In S_{xy}, $|-1 - 4y| = 4y + 1$; hence

$$M_u(S_{uv}) = \rho_0 \int_0^1 \int_0^1 (2x - y)(4y + 1)\, dy\, dx = \frac{7}{6}\, \rho_0.$$

The next proposition, which is given without proof, is an extension of Propositions 5–5.1 and 5–5.2 to the case of transformations of three variables. In general, extensions can be made in an analogous manner for transformations of n variables.

Proposition 5–5.3. *Let* $\mathbf{f}=(f_1, f_2, f_3)$ *be a function such that* $\mathbf{f}\colon \Omega_{xyz} \to E^3$, *$(u, v, w)$-space, where* Ω_{xyz} *is a space region element in (x, y, z)-space. Suppose that* $\mathbf{f}$, *which is defined by* $u = f_1(x, y, z)$, $v = f_2(x, y, z)$, *and* $w = f_3(x, y, z)$, *is* 1–1 *and continuously differentiable on* Ω_{xyz}. *Suppose, furthermore, that*

$$\frac{\partial(u, v, w)}{\partial(x, y, z)} \neq 0$$

on the interior of Ω_{xyz}, *nor does it have a limiting value of zero on the boundary of* Ω_{xyz}. *If* $\Omega_{uvw} = \mathbf{f}(\Omega_{xyz})$ *is a space region element and if* $h\colon \Omega_{uvw} \to R$ *and* h *is continuous on* Ω_{uvw}, *then*

$$\int_{\Omega_{uvw}} h(u, v, w)\, dV_{uvw} = \int_{\Omega_{xyz}} h(\mathbf{f}(x, y, z)) \left| \frac{\partial(u, v, w)}{\partial(x, y, z)} \right| dV_{xyz}. \quad (5\text{–}5.9)$$

The following example will illustrate the use of this proposition.

Example 5–5.3. Let us consider the transformation $\mathbf{f}$ of spherical coordinates:

$$u = f_1(x, y, z) = x \sin y \cos z,$$
$$v = f_2(x, y, z) = x \sin y \sin z,$$
$$w = f_3(x, y, z) = x \cos y.$$

Let $\Omega_{xyz} = \{(x, y, z)\colon 1 \le x \le 2,\ 0 \le y \le \pi/4,\ \pi/2 \le z \le \pi\}$. Then find the volume of Ω_{uvw}, the image of Ω_{xyz} by $\mathbf{f}$ (cf. Figure 5–5.4).

By Section 5–4, the volume of Ω_{uvw} is

$$V(\Omega_{uvw}) = \int_{\Omega_{uvw}} dV_{uvw};$$

but by Proposition 5–5.3

$$\int_{\Omega_{uvw}} dV_{uvw} = \int_{\Omega_{xyz}} \left| \frac{\partial(u, v, w)}{\partial(x, y, z)} \right| dV_{xyz}.$$

Therefore, by (5–4.2)

$$V(\Omega_{uvw}) = \int_1^2 \int_0^{\pi/4} \int_{\pi/2}^{\pi} x^2 \sin y\, dz\, dy\, dx$$
$$= 7(2 - \sqrt{2})\pi/12.$$

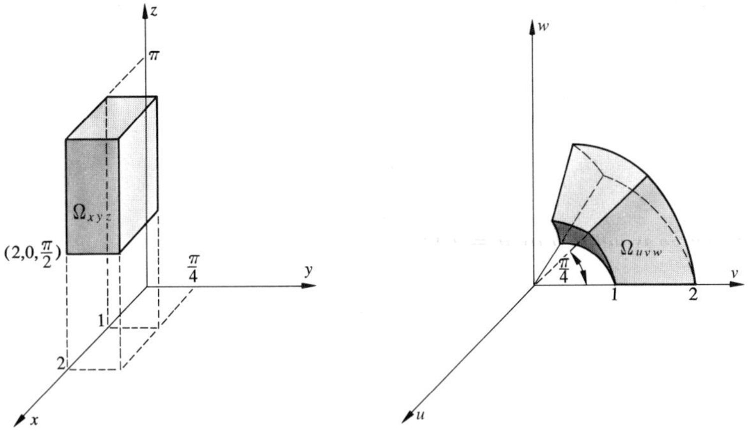

Figure 5–5.4

Example 5–5.4. The foregoing transformation is 1–1 as long as $x \leq 0$, $0 < y < \pi$ and $0 \leq z < 2\pi$. Thus if

$$\Omega_{xyz}^n = \left\{(x, y, z) \colon \frac{1}{n} \leq x \leq r, \frac{1}{n} \leq y \leq \pi - \frac{1}{n}, 0 \leq z \leq 2\pi - \frac{1}{n}\right\}$$

where $n > \max\,(1/r, 2/\pi)$, then **f** is 1–1 on Ω_{xyz}^n for each n, and Ω_{uvw}^n, the image of Ω_{xyz}^n by **f**, converges to Ω_r, the spherical region of radius r with center at the origin, relative to

$$\Omega_r^* = \{(x, y, z) \colon 0 \leq x \leq 1, y = 0 \text{ or } \pi\}.$$

Thus, by Definition 5–4.5,

$$\begin{aligned} V(\Omega_r) &= \lim_{n \to \infty} \int_{\Omega_{uvw}^n} dV_{uvw} \\ &= \lim_{n \to \infty} \int_{\Omega_{xyz}^n} |x^2 \sin y| \, dV_{xyz} \\ &= \lim_{n \to \infty} \int_{1/n}^{r} \int_{1/n}^{\pi - (1/n)} \int_{0}^{2\pi - (1/n)} x^2 \sin y \, dz \, dy \, dx \\ &= \lim_{n \to \infty} \left(\frac{1}{3}\right)(r^3 - n^{-3})\left(2\pi - \frac{1}{n}\right)\left(\cos\frac{1}{n} - \cos\left(\pi - \frac{1}{n}\right)\right) \\ &= \left(\frac{r^3}{3}\right)(2\pi)(2) = \frac{4}{3}\,\pi r^3. \end{aligned}$$

This is, of course, the expected value of the volume of a spherical region of radius r. Notice that this shows that **f** need not be 1–1 for (5–5.9) to hold.

If in Proposition 5–5.3 the function h maps each point (u, v, w) into 1, then (5–5.9) becomes

$$\int_{\Omega_{uvw}} dV_{uvw} = \int_{\Omega_{xyz}} \left| \frac{\partial(u, v, w)}{\partial(x, y, z)} \right| dV_{xyz} .$$

By Problem 16, page 269, there exists a point $(x_0, y_0, z_0) \in \Omega_{xyz}$ such that

$$\int_{\Omega_{uvw}} dV_{uvw} = \left| \frac{\partial(u, v, w)}{\partial(x, y, z)} \right| \Bigg|_{(x_0, y_0, z_0)} \int_{\Omega_{xyz}} dV_{xyz} .$$

Hence

$$V(\Omega_{uvw}) = \left| \frac{\partial(u, v, w)}{\partial(x, y, z)} \right| \Bigg|_{(x_0, y_0, z_0)} \cdot V(\Omega_{xyz}). \tag{5–5.10}$$

If Ω^k_{xyz}, $k = 1, 2, 3, \cdots$, is a sequence of space region elements and

$$\Omega^k_{uvw} = \mathbf{f}(\Omega_{xyz}), \quad k = 1, 2, 3, \cdots,$$

such that Proposition 5–5.3 is satisfied for each k, if for $k = 1$,

$$\Omega^1_{xyz} = \Omega_{xyz}, \ \Omega^{k+1}_{xyz} \subset \Omega^k_{xyz}, \quad k = 1, 2, 3, \cdots,$$

and if

$$\textstyle\bigcap_{k=1}^{\infty} \Omega^k_{xyz} = (x_0, y_0, z_0),$$

then

$$\lim_{k\to\infty} \frac{V(\Omega^k_{uvw})}{V(\Omega^k_{xyz})} = \left| \frac{\partial(u, v, w)}{\partial(x, y, z)} \right| \Bigg|_{(x_0; y_0, z_0)} .$$

Thus, the value of the Jacobian of the transformation at a point gives the limit of the ratios of the volumes of corresponding regions as they shrink to the point. Thus the absolute value of the Jacobian of a transformation is called the *local magnification factor* of the transformation. The results are analogous for $n = 2$ and $n > 3$.

In Example 5–5.2 the local magnification factor of the transformation is $4y + 1$. Thus the local magnification factor increases as we move away from the x-axis. Parallel to the x-axis this factor is constant. Notice that the images of the four horizontal rows of square regions of S_{xy} are correspondingly larger as we consider successively rows of square regions farther and farther from the x-axis.

Problems

1. Express the following integrals in terms of repeated integrals of x and y, and then evaluate. Check your answers by evaluating an appropriate repeated integral in u and v.

(a) $\int_{S_{uv}} (u + 2uv - v^2)\, dA_{uv}$ where $x = u - 2v,\ y = u + v,$ and $S_{uv} = \{(u, v): 0 \le u - 2v \le 1, 1 \le u + v \le 2\}.$

(b) $\int_{S_{uv}} (u^2 + v^2)\, dA_{uv}$ where $x = u^2 - v^2,\ y = 2uv,$ and $S_{uv} = \{(u, v): 1 \le u^2 + v^2 \le 4, u \ge 0, v \ge 0\}.$

2. Express the following repeated integrals in terms of repeated integrals of x and y, and then evaluate. Check your answer by evaluating the given repeated integral directly.

(a) $\int_{1/\sqrt{2}}^{1} \int_{\sqrt{1-u^2}}^{u} v\, dv\, du$ where $u = \dfrac{x}{x^2 + y^2},\quad v = \dfrac{-y}{x^2 + y^2}.$

(b) $\int_{-1}^{1} \int_{\sqrt{1-u^2}}^{\sqrt{e^2-u^2}} (u^2 + v^2)\, dv\, du$ where $u = e^x \cos y, v = e^x \sin y.$

3. Express the following integrals in terms of repeated integrals of the indicated new variables, and then evaluate. As a check, evaluate each of the integrals by first expressing them as repeated integrals involving x, y, and z.

(a) $\int_{\Omega_{xyz}} \left(\dfrac{1}{x^2 + y^2}\right) dV_{xyz}$

where $x = \rho \sin \varphi \cos \theta, y = \rho \sin \varphi \sin \theta, z = \rho \cos \varphi$, and where

$$\Omega_{xyz} = \left\{(x, y, z): \left(\frac{1}{4}\right) \le x^2 + y^2 + z^2 \le 1, 0 \le z \le \sqrt{x^2 + y^2}, x \ge 0\right\}.$$

(b) $\int_{\Omega_{xyz}} \sqrt{x^2 + y^2}\, dV_{xyz}$

where $x = r \cos \theta,\ y = r \sin \theta,\ z = z'$, the cylindrical coordinate transformation, and where

$$\Omega_{xyz} = \left\{(x, y, z): x^2 + y^2 \ge \frac{1}{4}, 4x^2 + 4y^2 + z^2 \le 4\right\}.$$

(c) $\int_{\Omega_{xyz}} (x + yz)\, dV_{xyz}$

where $x = u + v + w, y = u + v - 2w, z = 2v + w$, and where

$$\Omega_{xyz} = \{(x, y, z): 0 \le x \le 1, |y| \le 1, |z| \le 1\}.$$

4. (a) Draw a sketch and give an argument that makes it appear reasonable that a parallelopiped in (r, φ, θ)-space with faces determined by the planes $r = r_0, r = r_0 + \Delta r, r_0 > 0, \theta = \theta_0, \theta = \theta_0 + \Delta\theta,$

and $\varphi_0 = \varphi_0$, $\varphi = \varphi_0 + \Delta\varphi$, corresponds by the spherical coordinate transformation to a region whose volume is approximately $r_0^2 \sin \varphi_0$, if Δr, $\Delta\theta$, and $\Delta\varphi$ are "small."

(b) Do the analogous problem for cylindrical coordinates.

5. Use the methods of Example 5–5.3 to find the volume of the following regions in (x, y, z)-space:

(a) $\Omega_a = \{(x, y, z)\colon x^2 + y^2 \le a^2, 0 \le z \le 1\}$ using the transformation:

$$x = a \cos \theta, \qquad y = a \sin \theta, \qquad z = z',$$

(b) $\Omega_a = \{(x, y, z)\colon 1 \le x^2 + y^2 \le 4, 0 \le z \le 1\}$ using the transformation:

$$x = u^2 - v^2, \qquad y = 2uv, \qquad z = w.$$

6. The transformation of (5–5.7), the transformation of polar coordinates, could be written

$$x_1 = r \sin \theta_1,$$
$$x_2 = r \cos \theta_2.$$

Then the Jacobian of this transformation is

$$\frac{\partial(x_1, x_2)}{\partial(r, \theta_1)} = -r.$$

The transformation of Example 5–5.3, the transformation of spherical coordinates (3-dimensional), could be written

$$x_1 = r \sin \theta_1 \sin \theta_2,$$
$$x_2 = r \sin \theta_1 \cos \theta_2.$$
$$x_3 = r \cos \theta_1.$$

(a) Make a sketch showing the geometrical meaning of each of these variables and demonstrate why they are related as indicated. Show that

$$\frac{\partial(x_1, x_2, x_3)}{\partial(r, \theta_1, \theta_2)} = -r^2 \sin \theta_1.$$

(b) In a similar way we can define spherical coordinates for four dimensions, namely,

$$x_1 = r \sin \theta_1 \sin \theta_2 \sin \theta_3,$$
$$x_2 = r \sin \theta_1 \sin \theta_2 \cos \theta_3,$$
$$x_3 = r \sin \theta_1 \cos \theta_2,$$
$$x_4 = r \cos \theta_1.$$

Modify the reasoning of (a) to demonstrate this transformation. Show that

$$\frac{\partial(x_1, x_2, x_3, x_4)}{\partial(r, \theta_1, \theta_2, \theta_3)} = r^3 \sin^2 \theta_1 \sin \theta_2.$$

(c) Write out the n-dimensional spherical coordinate transformation and show that the Jacobian is

$$(-1)^{[n/2]} r^{n-1} \sin^{n-2}\theta_1 \sin^{n-3}\theta_2 \cdots \sin\theta_{n-2}$$

where $[n/2]$ is the largest integer less than or equal to $n/2$.

7. If in E^n, $n \geq 2$, $S_n = \{\mathbf{x}\colon \|\mathbf{x}\| \leq 1\}$, then for $n = 2$, $V(S_2)$, the area of S_2, is π and $A(S_2)$, the length of the boundary curve of S_2, is 2π. For $n = 3$, $V(S_3)$, the volume of S_3, and $A(S_3)$, the area of S_3, are

$$V(S_3) = \int_{S_3} dV_{x_1 x_2 x_3} = \int_0^1 \int_0^{2\pi} \int_0^{\pi} r^2 \sin\theta_1 \, d\theta_1 \, d\theta_2 \, dr = \frac{4\pi}{3}$$

and

$$A(S_3) = \int_{\mathrm{bdy}(S_3)} dA_{\theta_1\theta_2} = \int_0^{2\pi} \int_0^{\pi} \sin\theta_1 \, d\theta_1 \, d\theta_2 = 4\pi,$$

respectively (cf. Examples 5–5.4 and 5–3.7). By using the results of Problem 6 show in an analogous manner that

$$V(S_n) = \frac{\pi^{(n/2)}}{\Gamma\left(\dfrac{n}{2} + 1\right)}$$

and that $A(S_n) = n\, V(S_n)$ where $\Gamma\left(\dfrac{n}{2} + 1\right) = \left(\dfrac{n}{2}\right)!$ if n is even and

$$\Gamma\left(\frac{n}{2} + 1\right) = \left(\frac{n}{2}\right)\left(\frac{n}{2} - 1\right) \cdots \left(\frac{1}{2}\right) \sqrt{\pi}$$

if n is odd.

8. Calculate the local magnification factor for the following transformations:

(a) $x = \dfrac{a \sinh u}{\cosh u + \cos v}$, $\quad y = \dfrac{a \sin v}{\cosh u + \cos v}$,

the bi-polar coordinate transformation,

(b) $x = \frac{1}{2} a \cosh u \cos v$, $\quad y = \frac{1}{2} a \sinh u \sin v$,

the elliptic coordinate transformation.

9. Early in the discussion of this section the functions f and H were found to be continuously differentiable on $[c, d]$ and $[a, b]$, respectively. Furthermore, in one case $f'(x) > 0$, $c < x < d$, and $f(c) = a$, $f(d) = b$. Show that the domains of f and H can be extended to open sets such that f and H are differentiable on their respective domains and such that the composite Hf exists—that is, H and f satisfy the hypothesis of Proposition 3–4.1.

10. Let f be a continuously differentiable function on a closed interval I_x as in Proposition 5–5.1. Show that if $f'(x) \neq 0$ in the interior of I_x, that $I_u = f(I_x)$ is a closed interval and that f is 1–1 on I_x. Notice that the part of the hypothesis of Proposition 5–5.1 that requires that

$f'(x_0) \neq 0$ at a boundary point of I_x is not an essential part of the hypothesis, but is present only for the sake of analogy with Proposition 5–5.2 where this part of the hypothesis is essential in order to be sure that $\mathbf{f}(S_{xy})$ is a smooth simple surface element.

11. In the proof of Proposition 5–5.2 it was assumed that the boundary of S_{uv} is the edge of S_{uv}. Prove that this is indeed the case under these circumstances. First use Theorem 4–2.1 to show that a boundary point of S_{uv} is the image by $\mathbf{f}$ of a point of the edge of S_{xy}. Then show that $\mathbf{f}^{-1}$ is continuous on S_{uv} and that its Jacobian is not zero at an interior point of S_{uv}. Then use Theorem 4–2.1 on $\mathbf{f}^{-1}$.

12. Use Theorem 4–2.1 to show that if $\mathbf{f}$ of Proposition 5–5.2 is 1–1 from the edge of S_{xy} onto the edge of S_{uv} and if the Jacobian of the transformation is not zero on the interior of S_{xy}, then $\mathbf{f}$ is 1–1 on S_{xy}.

13. (a) Justify (5–3.22). Notice that if $g(\theta) = 0$ for $\alpha \leq \theta < \beta$, then the Jacobian of the transformation has the limiting value of zero on the boundary of the r-axial region and hence Proposition 5–5.2 is not directly applicable. Use the technique of Examples 5–3.5, 5–3.6, etc.
    (b) State and prove analogies for the cylindrical coordinate transformation.
    (c) Do as in (b) for the spherical coordinate transformation (cf. Example 5–5.4).

14. Prove that in the definition of a Riemann integral over a simple smooth surface element the value of the integral is independent of the parametrization (cf. Definition 5–3.2).

Answers

1. (a) $\frac{61}{128}$, (b) $\frac{\pi}{8}$.

2. (a) $\frac{\sqrt{2} - 1}{6}$,

   (b) $\frac{4(e^2 - 1)^{3/2} + 6\,e^2(e^2 - 1)^{1/2} + 6e^4 \sin^{-1}(e^{-1}) - 3\pi}{24}$.

3. (a) $\left(\frac{\pi}{2}\right) \log(1 + \sqrt{2})$,

   (b) $\pi\left[\left(\frac{\pi}{6}\right) + \left(\frac{\sqrt{3}}{16}\right)\right]$,

   (c) $\frac{1}{2}$.

5. (a) πa^2, (b) $\frac{3\pi}{4}$.

8. (a) $a^2[(\cosh u \cos v + 1)^2 + \sinh^2 u \sin^2 v](\cosh u + \cos v)^{-4}$,

(b) $a^2 \left(\dfrac{\sinh^2 u \cos^2 v + \cosh^2 u \sin^2 v}{4} \right)$.

5-6. Differentiation of Integrals

There are often circumstances in mathematics where functions are defined by expressions that involve integrals. As with other functions, there is the need for being able to take derivatives of these functions. In this section we shall consider the problem of taking these derivatives. We shall start with the simple case given in the following proposition which is considered as part of the Fundamental Theorem of Calculus.

Proposition 5-6.1. *Let f be a function such that $f\colon [a, b] \to R$ where $[a, b]$ is a closed interval in E^1. Suppose that $f \in C([a, b])$. If the function F is defined by*

$$F(x) = \int_a^x f(t)\, dt$$

for $a \leq x \leq b$, then $F \in C([a, b])$ and $F'(x) = f(x)$ for all $a < x < b$.

Proof. Let x_0 and $x \in [a, b]$, $x \neq x_0$; then

$$\begin{aligned} F(x) - F(x_0) &= \int_a^x f(t)\, dt - \int_a^{x_0} f(t)\, dt \\ &= \int_{x_0}^x f(t)\, dt. \end{aligned} \tag{5-6.1}$$

Then by the mean-value theorem for integrals, (5-6.1) becomes

$$F(x) - F(x_0) = f(x_0 + \theta(x - x_0)) \int_{x_0}^x dt \tag{5-6.2}$$

for some $0 < \theta < 1$. Thus

$$F(x) - F(x_0) = (x - x_0) f(x_0 + \theta(x - x_0)).$$

Then for $a < x_0 < b$,

$$F'(x) = \lim_{x \to x_0} \frac{F(x) - F(x_0)}{x - x_0} = f(x_0). \quad \blacksquare$$

In the above case the variable of the function f appeared as an upper limit of the integral. In other cases the variable could be a parameter of the integrand of the integral, so that

$$F(x) = \int_a^b f(x, t)\, dt$$

where f: $[(a, c), (b, d)] \to R$. The set $[(a, c), (b, d)]$ is an interval in (x, t)-space, and so

$$[(a, c), (b, d)] = \{(x, t) : c \leq x \leq d, a \leq t \leq b\}$$

(cf. Definition 2–1.2). The following proposition will give us a sufficient condition that the function F is continuous on the interval $[c, d]$.

Proposition 5–6.2. *Let f be a function such that f: $[(a, c), (b, d)] \to R$. If f is continuous on $[(a, c), (b, d)]$, then the function F defined by*

$$F(x) = \int_a^b f(x, t)\, dt \tag{5–6.3}$$

is continuous on $[c, d]$.

Proof. Since f is continuous on $[(a, c), (b, d)]$, then $f(x, \ )$ is continuous on $[a, b]$ for each $x \in [c, d]$ (cf. Problem 14 of Section 3–1). Hence for each $x \in [c, d]$ the Riemann integral of (5–6.3) exists, and thus F is defined at each point of the interval $[c, d]$.

Let x and $x_0 \in [c, d]$; then

$$\begin{aligned} F(x) - F(x_0) &= \int_a^b f(x, t)\, dt - \int_a^b f(x_0, t)\, dt \\ &= \int_a^b (f(x, t) - f(x_0, t))\, dt. \end{aligned}$$

Taking the absolute value of both sides and using the fact that the absolute value of an integral is less than or equal to the integral of the absolute value of the integrand, we have

$$|F(x) - F(x_0)| \leq \int_a^b |f(x, t) - f(x_0, t)|\, dt. \tag{5–6.4}$$

Since f is continuous on $[(a, c), (b, d)]$, a compact set, f is uniformly continuous on this same set (cf. Theorem 3–1.4). Hence, given $\varepsilon > 0$, there exists $\delta > 0$ such that if

$$(x_i, t_i) \in [(a, c), (b, d)], \quad i = 1, 2 \quad \text{and} \quad \|(x_1, t_1) - (x_2, t_2)\| < \delta,$$

then

$$|f(x_1, t_1) - f(x_2, t_2)| < \frac{\varepsilon}{b - a}.$$

Therefore, if $|x - x_0| < \delta$, $a \le t \le b$, then $\|(x, t) - (x_0, t)\| < \delta$ and thus from (5–6.4)

$$|F(x) - F(x_0)| < \int_a^b \left(\frac{\varepsilon}{b - a}\right) dt = \varepsilon.$$

Therefore, F is continuous at x_0, an arbitrary point of $[c, d]$. ▮

It is now possible to consider the most general possible circumstance of the derivative of a proper integral. The following proposition, which summarizes this procedure for calculating the derivative, is known as Leibnitz's rule.

Proposition 5–6.3. *Let f be a function such that $f\colon [(a, c), (b, d)] \to R$ and let g_1 and g_2 be two functions such that $g_i : [c, d] \to R$ and $g_i(x) \in (a, b)$ for $c \le x \le d$, $i = 1, 2$. If the functions f, g_1, and g_2 are continuously differentiable on their respective domains, then the function F defined by*

$$F(x) = \int_{g_1(x)}^{g_2(x)} f(x, t)\, dt \tag{5–6.5}$$

is continuously differentiable on $[c, d]$ and

$$F'(x) = f(x, g_2(x))g_2'(x) - f(x, g_1(x))g_1'(x) + \int_{g_1(x)}^{g_2(x)} f_x(x, t)\, dt \tag{5–6.6}$$

for $x \in (c, d)$.

Proof. The function F is clearly defined for each $x \in [c, d]$ since f is continuous on $[(a, c), (b, d)]$. Let $u = g_1(x)$, $v = g_2(x)$, then, by (5–6.5),

$$F(x) = G(x, u, v) = G(x, g_1(x), g_2(x))$$

and, by the chain rule,

$$F'(x) = \frac{\partial}{\partial x} G(x, u, v)\bigg)_{u,v} + \frac{\partial}{\partial u} G(x, u, v)\bigg)_{x,v} g_1'(x) + \frac{\partial}{\partial v} G(x, u, v)\bigg)_{x,u} g_2'(x) \tag{5–6.7}$$

for $c < x < d$ (cf. Proposition 3–4.1). By Proposition 5–6.1,

$$\frac{\partial}{\partial v} G(x, u, v)\bigg)_{x,u} = \frac{\partial}{\partial v}\left(\int_u^v f(x, t)\, dt\right) = f(x, g_2(x)),$$

and

$$\frac{\partial}{\partial u} G(x, u, v)\bigg)_{x,v} = \frac{\partial}{\partial u}\left(-\int_u^v f(x, t)\, dt\right) = -f(x, g_1(x)).$$

Lastly, by the definition of a partial derivative, we have

$$\frac{\partial}{\partial x} G(x, u, v)\Big)_{u,v} = \lim_{\Delta x \to 0} \frac{G(x + \Delta x, u, v) - G(x, u, v)}{\Delta x}$$

$$= \lim_{\Delta x \to 0} \left(\frac{1}{\Delta x}\right) \int_u^v (f(x + \Delta x, t) - f(x, t))\, dt \quad (5\text{–}6.8)$$

(cf. Section 3–2). By the mean-value theorem for x and $x + \Delta x \in (c, d)$, we have

$$f(x + \Delta x, t) - f(x, t) = (\Delta x) \cdot f_x(x + \theta\, \Delta x, t)$$

for some θ such that $0 < \theta < 1$. Thus, substituting in (5–6.8), we have

$$\frac{\partial}{\partial x} G(x, u, v)\Big)_{u,v} = \lim_{\Delta x \to 0} \int_u^v f_x(x + \theta\, \Delta x, t)\, dt. \quad (5\text{–}6.9)$$

By the hypothesis of this proposition, f_x is continuous on $[(a, c), (b, d)]$, and hence by Proposition 5–6.2, the limit of (5–6.9) is $\int_u^v f_x(x, t)\, dt$. Thus for $x \in (c, d)$, (5–6.7) becomes (5–6.6).

From (5–6.6) it follows from the hypothesis of this proposition and Theorem 3–1.3 that F' can be extended to a continuous function on $[c, d]$. ∎

The following example demonstrates the use of the foregoing proposition.

Example 5–6.1. Calculate the indicated derivative:

$$\frac{d}{dx} \int_{(x^2-6x)/9}^{6/x} \frac{1}{x + t}\, dt. \quad (5\text{–}6.10)$$

By (5–6.6), we can see that the derivative is

$$\frac{1}{x + (6/x)}\left(-\frac{6}{x^2}\right) - \frac{1}{x + (x^2 - 6x)/9}\,\frac{2x - 6}{9} - \int_{(x^2-6x)/9}^{6/x} \frac{1}{(x + t)^2}\, dt$$

$$= -\frac{x^3 + 18x + 36}{x(x^2 + 6)(x + 3)}.$$

As a check let us first integrate (5–6.10) and then differentiate, so that (5–6.10) becomes

$$\frac{d}{dx}[\log (x + t)]_{(x^2-6x)/9}^{6/x} = \frac{d}{dx}\left[\log\left(x + \frac{6}{x}\right) - \log\left(x + \frac{x^2 - 6x}{9}\right)\right]$$

$$= -\frac{x^3 + 18x + 36}{x(x^2 + 6)(x + 3)}.$$

In Figure 5–6.1 the area of the shaded plane region, $F(x)$, is the value of the integral of (5–6.10), and hence the derivative of this integral is the instantaneous rate of change of this area with respect to x.

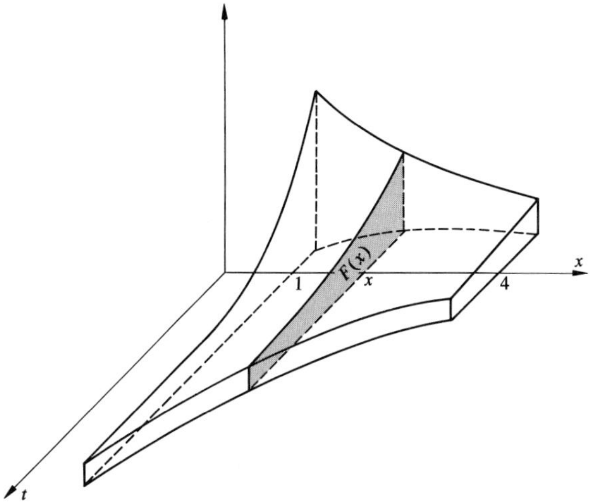

Figure 5–6.1

Problems

1. Calculate the derivative with respect to x of the function F where $F(x)$ is given by the following integrals:

(a) $\int_0^x e^{-t^2}\, dt$, (b) $\int_\pi^x \sin \frac{1}{t}\, dt$,

(c) $\int_x^{\pi/4} \frac{\tan t}{t}\, dt$, (d) $\int_x^1 \log^2 t\, dt$.

Give the subset of the set of points of continuity of F where Proposition 5–6.1 asserts that F exists.

2. Using Leibnitz's rule, calculate the derivative with respect to x of the function F where $F(x)$ is given by the following integrals:

(a) $\int_1^{x^2} \cos(x + t)\, dt$, (b) $\int_{\sqrt{x}}^{x} e^{xt}\, dt$,

(c) $\int_{x+2}^{\log x} (1 + x^2 t^2)\, dt$, (d) $\int_{1/(x+1)}^{x^3} \sin 2t\, dt$.

Upon the basis of Proposition 5–6.3, on what intervals $[c, d]$ is F continuously differentiable? Check each of the answers found above by first integrating and then differentiating.

3. (a) Let the function F be defined by

$$F(x) = \int_0^1 \frac{t^x - 1}{\log t}\, dt.$$

Show that the hypothesis of Proposition 5–6.3 is not satisfied for any set $[(0, c), (1, d)]$.

(b) Show, however, that $F(x)$ is defined for $x > 0$; that is, show that the integral is convergent by showing $\lim_{t \to 1} (t - 1)^p f(x, t)$ exists and is finite for some p, $0 < p < 1$, and for $x > 0$. Assume that $F'(x)$ can be calculated by (5–6.6) for $x > 0$, and show

$$F(x) = (x + 1)^{-1}$$

for $x > 0$. In such a case would F be continuously differentiable on $[0, d]$, $d > 0$?

(c) Show that $F(x) = \log (x + 1)$, $x \geq 0$.

4. Let the function f which is defined by $f(x, t)$ on $((a, c), (b, d))$ be such that for each $x \in (c, d)$, $f(x, \quad)$ is extendable to a continuous function on $[a, b]$. Then the function f defined by

$$F(x) = \int_a^b f(x, t)\, dt$$

exists for each $x \in (c, d)$. Prove that if f_x is extendable to a continuous function on $[(a, c), (b, d)]$, then $F'(x)$ exists on (c, d) and

$$F'(x) = \int_a^b f_x(x, t)\, dt.$$

5. Let the function F be defined by

$$F(x) = \int_0^1 t^x\, dt = \frac{1}{x + 1}, \quad x \geq 0.$$

By successive differentiation, each time using Problem 4, show that

$$\int_0^1 t^x (\log t)^n\, dt = (-1)^n \frac{n!}{(1 + x)^{n+1}}, \quad x > 0.$$

Is F continuous at $x = 0$? Explain.

6. Let the function F be defined by

$$F(x) = \int_0^\pi \log (1 + x \cos t)\, dt, \quad |x| < 1.$$

Prove by Theorem 3–1.3 that f defined by $f(x, t) = \log (1 + x \cos t)$ satisfies the hypothesis of Problem 4 on $((0, -1), (\pi, -1))$, and hence

$$F'(x) = \pi \left(\frac{1}{x} - \frac{1}{x\sqrt{1 - x^2}} \right), \quad 0 < |x| < 1,\ F'(0) = 0.$$

Find $F(x)$ by integrating and then using the fact that $F(0) = 0$ to evaluate the arbitrary constant.

7. Let the function F be defined by

$$F(x) = \int_0^1 \sin xt\, dt.$$

Find $F^{(n)}(0)$ without first integrating. Write the Maclaurin series for F and find its interval of convergence.

8. Let f be a function defined by $y = f(x)$ on the interval (a, b) such that $0 \in (a, b)$, and suppose that $f \in C^{n+1}((a, b))$ such that the hypothesis of Taylor's theorem is satisfied. Consider the function R defined by

$$R(x) = \int_0^x \left[\frac{(x - t)^n}{n}\right] f^{(n+1)}(t)\, dt.$$

By the use of Proposition 5–6.3, calculate $R'(x)$, $R''(x)$, $\cdots$, $R^{(n)}(x)$. Show that each of these derivatives is zero at $x = 0$ and show that

$$R^{(n+1)}(x) = f^{(n+1)}(x).$$

By successively integrating both sides of this last equality, establish Taylor's formula about $x = 0$. By using the mean-value theorem for integrals for a fixed x change $R(x)$ to the usual Lagrange remainder.

Answers

1. (a) e^{-x^2}, (b) $\sin \frac{1}{x}$, (c) $-\frac{\tan x}{x}$, (d) $-\log^2 x$.

2. (a) $(2x + 1) \cos (x + x^2) - \cos (x + 1)$,

(b) $$\frac{(2x^2 - 1)e^{x^2} + \left(1 - \left(\frac{3}{2}\right)x^{(3/2)}\right)e^{(3/2)}}{x^2},$$

(c) $$\frac{1}{x} - 1 + x \log^2 x + \frac{2}{3} x \log^3 x - x^2(x + 2)^2 - \frac{2}{3} x(x + 2)^3,$$

(d) $$3x^2 \sin 2x^3 + \frac{1}{(x + 1)^2} \sin \frac{2}{x + 1}.$$

6. $$F(x) = \log \frac{1 + \sqrt{1 - x^2}}{2}.$$

7. $$F(x) = \sum_{n=0}^{\infty} \frac{x^{2n+1}}{(2n + 2)!},$$

interval of convergence: $(-\infty, \infty)$.

6

Differential and Integral Calculus of Vector Fields

6–1. Vector and Scalar Fields

There are numerous physical concepts that motivate the consideration of both vector and scalar fields. For example, force, velocity, and acceleration lead to the consideration of vector fields; and quantities like temperature, density, and pressure lead us to consider scalar fields. In other words, quantities that are assumed to have non-negative values (magnitudes) and direction are described by a vector field, and those that just have (real) values, without any associated direction, are described by a scalar field. Although the concepts of vector and scalar fields can be described in E^n space for any n, it is usually the case that $n \leq 3$ and most often $n = 3$.

The following definition states precisely what is meant by a vector field and a scalar field.

Definition 6–1.1. *Let* $\mathbf{f} = (f_1, f_2, f_3)$ *be a vector-valued function defined on a region* $\Omega \subset E^3$*; then* $f_i(\mathbf{x}) = f_i(x_1, x_2, x_3)$, $i = 1, 2, 3$, *is real for each*

$\mathbf{x} = (x_1, x_2, x_3) \in \Omega$, *and thus* $\mathbf{f}: \Omega \to E^3$. *The set of vectors* $\mathbf{f}(\mathbf{x})$ *for each* $\mathbf{x} \in \Omega$ *is the (static) vector field on* Ω *defined by* $\mathbf{f}$. *If the function* $\mathbf{f} = (f_1, f_2, f_3)$ *is defined by* $\mathbf{f}(\mathbf{x}; t) = \mathbf{f}(x_1, x_2, x_3, t)$ *for all* $\mathbf{x} \in \Omega$, $t \in I$ *where* I *is an interval of the real line (that is,* $\mathbf{f}: \Omega \times I \to E^3$*) then the set of vectors* $\mathbf{f}(\mathbf{x}; t)$ *for each* $(\mathbf{x}; t) \in \Omega \times I$ *is the (dynamic) vector field on* Ω *defined by* $\mathbf{f}$ *for all* $t \in I$. *Similarly, if* ϕ *is a real-valued function defined on* Ω *such that* $\phi: \Omega \to R$, *then the set of reals,* $\phi(\mathbf{x})$, *is the (static) scalar field on* Ω *defined by* ϕ. *Furthermore, if* $\phi: \Omega \times I \to R$, *then the set of reals,* $\phi(\mathbf{x}; t)$, *is the (dynamic) scalar field on* Ω *defined by* ϕ *for all* $t \in I$.

Vector fields and scalar fields are usually mentioned without reference to the words static and dynamic. It is usually clear from the context which

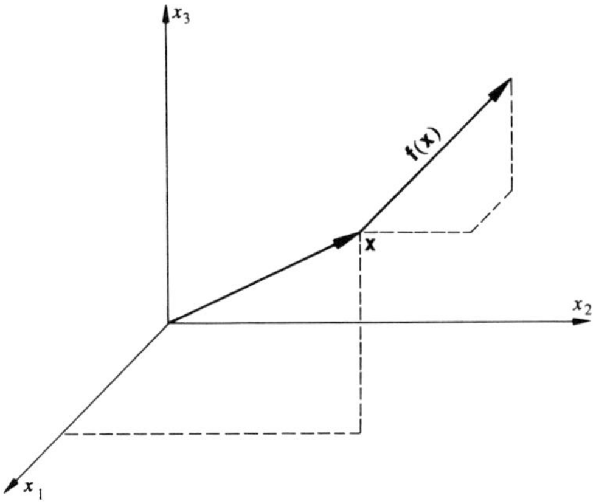

Figure 6–1.1

of the two types is being discussed. If t is a fixed value $t_0 \in I$, then a dynamic vector field becomes a static vector field.

A static vector field can be visualized by taking $\{\mathbf{x}_1, \cdots, \mathbf{x}_m\}$, a set of points uniformly distributed throughout Ω, and plotting at each $\mathbf{x}_i$, $\mathbf{f}(\mathbf{x}_i)$, a directed line segment with initial point $\mathbf{x}_i$ (cf. Figure 6–1.1). This collection of directed line segments tends to give one a "feel" for the vector field given by $\mathbf{f}$. The following example illustrates this point.

Example 6–1.1. If $\mathbf{f}$ is the force function of the gravitational field "in the small"—such as when we are close to the surface of the earth—and if we consider only a small portion of the points near the earth, then $\mathbf{f}(\mathbf{x}) = (0, 0, -k)$, $k > 0$ and constant, and $\mathbf{x} \in \Omega = \{\mathbf{x}: x_3 \geq 0\}$ (cf. Figure 6–1.2(a)).

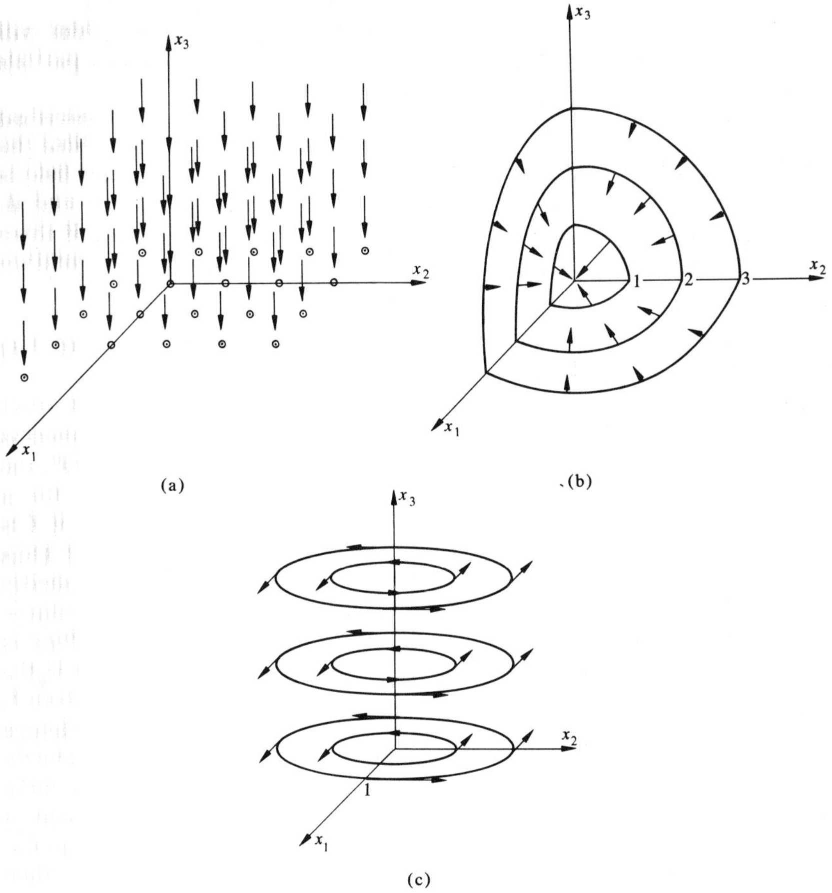

Figure 6–1.2

If $\mathbf{g}$ is the electrostatic field on a test particle that is positively charged because of a positive unit charge at the origin in E^3, then by Coulomb's law

$$\mathbf{g}(\mathbf{x}) = k\frac{\mathbf{x}}{\|\mathbf{x}\|^3}$$

for all $\mathbf{x} \in \Omega = \{\mathbf{x}: \mathbf{x} \neq 0\}$ (cf. Figure 6–1.2(b)).

Let $\mathbf{h}$ be a vector field defined by

$$\mathbf{h}(\mathbf{x}) = (-x_2\sqrt{x_1^2 + x_2^2}, x_1\sqrt{x_1^2 + x_2^2}, 0)$$

for all $\mathbf{x} \in \Omega = \{\mathbf{x}: x_1^2 + x_2^2 \leq 1\}$. Then the field given by $\mathbf{h}$ resembles the velocity field of a fluid placed in a cylinder and stirred rapidly in a circular manner (cf. Figure 6–1.2(c)).

A vector field on Ω defined by $\mathbf{f}$ is k-dimensional if the subspace spanned by the set $\{\mathbf{f}(\mathbf{x}) : \mathbf{x} \in \Omega\}$ is k-dimensional. Thus in the preceding example the fields given by $\mathbf{f}$, $\mathbf{g}$, and $\mathbf{h}$ are 1-, 3-, and 2-dimensional respectively.

The functions $\mathbf{f}$ and ϕ of Definition 6–1.1 that we shall consider will usually be functions that are continuous and have continuous first partials on the interior of Ω (or $\Omega \times I$).

In the above discussion the quantities $\mathbf{f}(\mathbf{x}; t)$ and $\phi(\mathbf{x}; t)$ are described in terms of the point in space, $\mathbf{x}$, and t, time. These variables are called the *spatial coordinates*, and this description of a vector field and a scalar field is commonly referred to as the *Eulerian* (or *spatial*) *description* of $\mathbf{f}$ and ϕ. Suppose that $\mathbf{f}$ defined by $\mathbf{f}(\mathbf{x}; t)$ is known and that we wish to see if there is a family of curves such that if $\mathbf{x}(t)$, $t \in I$, is a parametric representation of a member of the family; then at each point,

$$\frac{d\mathbf{x}}{dt} = \mathbf{f}(\mathbf{x}; t). \tag{6–1.1}$$

Thus we wish to know if it is possible to solve this system of three, first-order ordinary differential equations. From the elementary existence and uniqueness theorem of differential equations, we know that given any point $\mathbf{c} \in \Omega^{o}$, the interior of Ω, there exists a unique function $\mathbf{x}(\mathbf{c};\ \ )$ such that, for a neighborhood of $t = 0$, I_0, $\mathbf{x}(\mathbf{c}; t)$ satisfies (6–1.1) and $\mathbf{x}(\mathbf{c}; 0) = \mathbf{c}$ if $\mathbf{f}$ is continuous on Ω and has continuous first partials on the interior of Ω.† Thus through each point $\mathbf{c} \in \Omega$ there is a unique curve $\Gamma_{\mathbf{c}}$ whose parametric representation is given by $\mathbf{x}(\mathbf{c}; t)$, and this curve passes through the point $\mathbf{c}$. These curves are the *path lines* of the field given by $\mathbf{f}$. This terminology is motivated by the study of fluid dynamics. In such a setting, $\mathbf{f}(\mathbf{x}; t)$ is the velocity of the fluid at $\mathbf{x}$ at time t. Clearly, if t is missing from the given $\mathbf{f}$, then the direction of the path line through a given point does not change with time. Such a flow (field) is said to be *steady*. Given a continuous-velocity vector field defined by $\mathbf{f}(\mathbf{x})$ or by $\mathbf{f}(\mathbf{x}; t)$ for a fixed value of t, then a curve Γ is a *streamline* of this velocity field if the curve always has its tangent at any point in the same (or in the opposite direction) of $\mathbf{f}(\mathbf{x})$ at this same point. Thus if $\mathbf{x}(s)$, $s \in I$, is a parametric representation of Γ, a streamline, then

$$\frac{d}{ds}\mathbf{x}(s) \times \mathbf{f}(\mathbf{x}(s)) = \mathbf{0}$$

for each $s \in I$. Clearly, if the velocity field defined by $\mathbf{f}$ is steady, the path lines and the streamlines are identical.

If the field is a force field, so that $\mathbf{f}(\mathbf{x}; t)$ is a force at $\mathbf{x}$ at time t, then the curves Γ are called *lines of force* rather than streamlines. In general, if

† Cf. Walter Leighton, *Ordinary Differential Equations*, pp. 164 ff.

$\mathbf{f}(\mathbf{x}; t)$ gives a vector field without any physical interpretation, then the curves Γ are called *vector lines.*

The next examples indicate methods of obtaining the functions $\mathbf{x}(\mathbf{c};\ )$ if the functions $\mathbf{f}(\mathbf{x}; t)$ are given.

Example 6–1.2. Let $\mathbf{f}(\mathbf{x}; t)$ be defined by

$$\mathbf{f}(\mathbf{x}; t) = A\mathbf{x} + \mathbf{g}(t)$$

where

$$A = \begin{pmatrix} 1 & 2 & 0 \\ 0 & -1 & 1 \\ 0 & 0 & 2 \end{pmatrix}$$

and $\mathbf{g}(t) = (t, t^2, -1)$. To solve the equation of (6–1.1) subject to the initial conditions $\mathbf{x}(\mathbf{c}; 0) = \mathbf{c}$, first solve the homogeneous system

$$\frac{d\mathbf{x}}{dt} - A\mathbf{x} = \mathbf{0} \tag{6–1.2}$$

by substituting $\mathbf{x}(t) = e^{\lambda t}(a, b, c)$, where a, b, and c are constants, into (6–1.2), obtaining

$$e^{\lambda t}(\lambda I - A) = (0)$$

where I is the identity matrix. Thus we have a solution of the homogeneous system if λ is a *characteristic root* (*eigenvalue*) of the matrix A. In our case these values are $\lambda_1 = 1$, $\lambda_2 = -1$, and $\lambda_3 = 2$. For each characteristic value we find the corresponding *characteristic vectors* (*eigenvectors*). That is, for $\lambda_1 = 1$, we have $e^t(a_1, 0, 0)$; for $\lambda_2 = -1$, $e^{-t}(a_2, -a_2, 0)$; and for $\lambda_3 = 2$, $e^{2t}(a_3, (1/2)a_3, (3/2)a_3)$. Thus

$$\mathbf{x}_h(t) = e^t(a_1, 0, 0) + e^{-t}(a_2, -a_2, 0) + e^{2t}\left(a_3, \frac{1}{2}a_3, \frac{3}{2}a_3\right) \tag{6–1.3}$$

is a *general solution* of (6–1.2).

To find a particular solution of the nonhomogeneous equation

$$\frac{d\mathbf{x}}{dt} - A\mathbf{x} = \mathbf{g}(t), \tag{6–1.4}$$

use the method of undetermined coefficients; that is, substitute $\mathbf{x}_p(t)$ defined by $x_{pi}(t) = a_i' + b_i't + c_i't^2$, $i = 1, 2, 3$, into (6–1.4), since the components of $\mathbf{g}(t)$ are polynomials of second degree or less. Then determine the coefficients a_i', b_i', c_i' such that we obtain a solution to the nonhomogeneous equation. Thus in this problem

$$x_{p1}(t) = -6 - t - 2t^2,$$

$$x_{p2}(t) = \frac{5}{2} - 2t + t^2,$$

$$x_{p3}(t) = \frac{1}{2},$$

and hence the general solution of the nonhomogeneous system is $\mathbf{x}_h(t) + \mathbf{x}_p(t)$. In order that the solution have the correct initial values, we need to determine a_1, a_2, a_3 such that $\mathbf{x}_h(0) + \mathbf{x}_p(0) = \mathbf{c}$, that solve

$$a_1 + a_2 + a_3 = 6 + c_1$$

$$-a_2 + \frac{1}{2}a_3 = -\frac{5}{2} + c_2$$

$$\frac{3}{2}a_3 = -\frac{1}{2} + c_3$$

for a_1, a_2, a_3 in terms of c_1, c_2, and c_3. The resulting function obtained by substituting into (6–1.3) is $\mathbf{x}(\mathbf{c};\ )$. Thus,

$$\mathbf{x}(\mathbf{c}; t) = (3 + c_1 + c_2 + c_3)e^t(1, 0, 0) + \left(\frac{8}{3} - c_2 - \frac{1}{3}c_3\right)e^{-t}(1, -1, 0) + \left(\frac{1}{3} - \frac{2}{3}c_3\right)e^{2t}\left(1, \frac{1}{2}, \frac{3}{2}\right). \quad (6\text{–}1.5)$$

The foregoing example demonstrated the technique of finding $\mathbf{x}(\mathbf{c}; t)$ if $\mathbf{f}$ is such that the resulting system of differential equations is *linear*. The next example illustrates a technique that is possible if the resulting system of differential equations is nonlinear.

Example 6–1.3. Suppose that $\mathbf{f}$ is such that the vector field determined by $\mathbf{f}$ is steady; for example,

$$\mathbf{f}(\mathbf{x}; t) = (x_1 x_2 - x_1 x_3, x_2 x_3 - x_2 x_1, x_3 x_1 - x_3 x_2).$$

Then (6–1.1) becomes

$$\frac{dx_1}{dt} = x_1 x_2 - x_1 x_3,$$

$$\frac{dx_2}{dt} = x_2 x_3 - x_2 x_1,$$

$$\frac{dx_3}{dt} = x_3 x_1 - x_3 x_2.$$

Thus, if each of these equations is multiplied by dt, we see that $\mathbf{dx}$ is proportional to $\mathbf{f}(\mathbf{x}; t)$. Hence we can write

$$\frac{dx_1}{x_1 x_2 - x_1 x_3} = \frac{dx_2}{x_2 x_3 - x_2 x_1} = \frac{dx_3}{x_3 x_1 - x_3 x_2} = \lambda. \quad (6\text{–}1.6)$$

Thus

$$dx_1 + dx_2 + dx_3 = \lambda(x_1 x_2 - x_1 x_3 + x_2 x_3 - x_2 x_1 + x_3 x_1 - x_3 x_2),$$

or

$$dx_1 + dx_2 + dx_3 = 0.$$

Integrating, we find that

$$u_1(\mathbf{x}) = x_1 + x_2 + x_3 = k_1, \tag{6–1.7}$$

a *first integral* of (6–1.6). Similarly, if we write (6–1.6) in the following equivalent way,

$$\frac{\frac{1}{x_1}\,dx_1}{x_2 - x_3} = \frac{\frac{1}{x_2}\,dx_2}{x_3 - x_1} = \frac{\frac{1}{x_3}\,dx_3}{x_1 - x_2},$$

and then add the three numerators and equate them to λ times the sum of the three denominators, we get

$$\frac{dx_1}{x_1} + \frac{dx_2}{x_2} + \frac{dx_3}{x_3} = 0.$$

Integrating this time, we obtain another first integral of (6–1.6)—namely,

$$u_2(\mathbf{x}) = x_1 x_2 x_3 = k_2. \tag{6–1.8}$$

Each of the equations, (6–1.7) and (6–1.8), can be thought of as determining a one-parameter family of surfaces, the intersection of any two of which is one of the curves Γ. In Figure 6–1.3, S_1 and S_2 are surfaces determined by (6–1.7) and (6–1.8), respectively, for $k_1 = 6$ and $k_2 = 1$.

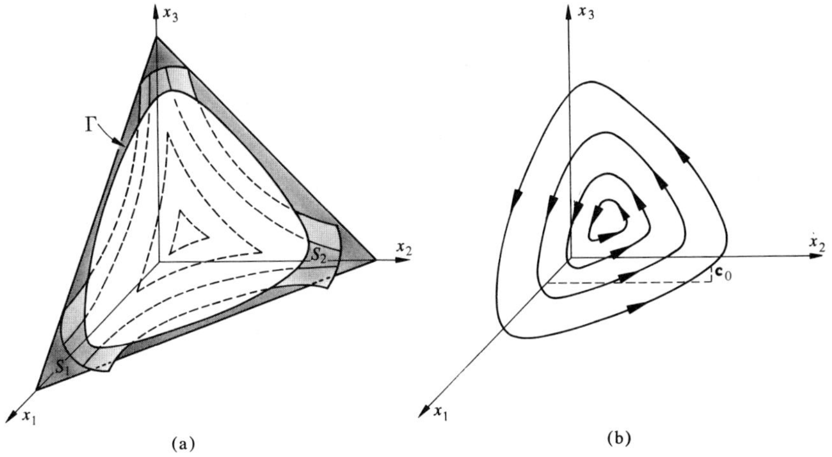

Figure 6–1.3

The collection of these curves Γ forms the solution curves to the given system of differential equations. They are the streamlines of a flow field determined by this $\mathbf{f}$. If the curve Γ which passes through a particular point $\mathbf{c}$ is written parametrically, with t the parameter, then we again obtain the form of $\mathbf{x}(\mathbf{c}; t)$. For example, if

$$\mathbf{c}_0 = \left(1, \frac{5 + \sqrt{21}}{2}, \frac{2}{5 + \sqrt{21}}\right),$$

then the curve Γ of Figure 6–1.3 passes through this point and

$$\mathbf{x}(\mathbf{c}_0; t) = \left(t + 1, \frac{1}{2}\left(5 - t \pm \sqrt{(t - 5)^2 - \frac{4}{t + 1}}\right), \frac{2}{\left(5 - t \pm \sqrt{(t - 5)^2 - \frac{4}{t + 1}}\right)(t + 1)}\right).$$

These two examples demonstrate how the function $\mathbf{x}(\mathbf{c};\ )$ defined by $\mathbf{x}(\mathbf{c}; t)$ can be found if $\mathbf{f}(\mathbf{x}; t)$ is given. If we continue to use the model of a fluid flowing, then $\mathbf{f}(\mathbf{x}; t)$ (or perhaps more appropriately $\mathbf{v}(\mathbf{x}; t)$) represents the velocity of the particle at the point $\mathbf{x}$ at time t and $\mathbf{x}(\mathbf{c}; t)$ represents the position (vector) at time t of the particle that was at $\mathbf{c}$ at time $t = 0$. Various other physical quantities, both scalar-valued and vector-valued quantities, can be expressed in terms of $\mathbf{c}$ and t, the *material coordinates*. For example, $\phi(\mathbf{c}; t)$ may be the temperature or the density of a particle at time t that was at $\mathbf{c}$ at time $t = 0$, and $\mathbf{f}(\mathbf{c}; t)$ may be the velocity or acceleration at time t of a particle that was at position $\mathbf{c}$ at time $t = 0$. Such a description of these physical quantities is frequently referred to as the *Lagrangian* (or *material*) *description* of the quantity.

Thus in the Eulerian description we have the viewpoint of an observer stationed at $\mathbf{x}$ and observing what happens at $\mathbf{x}$ at time t. In the Lagrangian viewpoint the observer "rides" with a particle that was at the point $\mathbf{c}$ at time $t = 0$ and observes what takes place at time t (cf. Figure 6–1.4). If we have found $\mathbf{x}(\mathbf{c}; t)$, then we have the equation

$$\mathbf{x} = \mathbf{x}(\mathbf{c}; t). \tag{6–1.9}$$

In a sense for t fixed, this is a transformation of the point $\mathbf{c}$ into $\mathbf{x}$. In many circumstances of interest the inverse of this transformation exists, so that we can find

$$\mathbf{c} = \mathbf{c}(\mathbf{x}; t) \tag{6–1.10}$$

from (6–1.9). In such circumstances it is possible to change from either one of the descriptions of a quantity to the other.

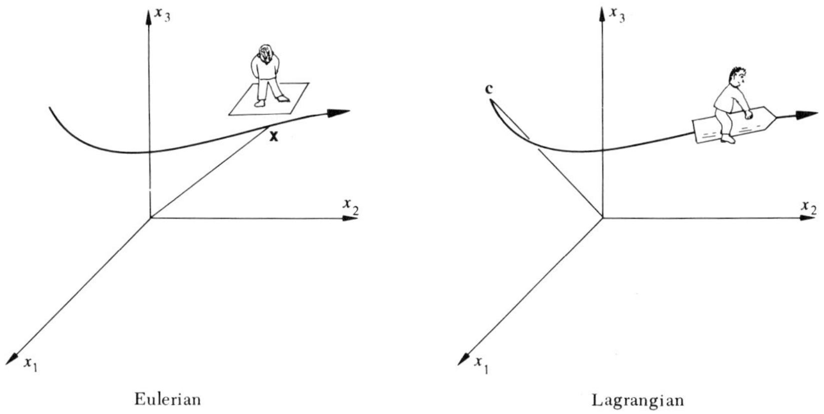

Figure 6–1.4

We shall now consider time derivatives of both scalar-valued and vector-valued functions that are described either with the Eulerian or Lagrangian description.

Consider the scalar-valued function ϕ with the Eulerian description, $\phi(\mathbf{x}; t)$; then $(\partial/\partial t)\phi(\mathbf{x}; t)$ is the *instantaneous local rate of change* of $\phi(\mathbf{x}; t)$ at $\mathbf{x}$ with respect to t, time. If we consider instead a vector-valued function $\mathbf{f}$ with Eulerian description ($\mathbf{f}$ is defined by $\mathbf{f}(\mathbf{x}; t)$), then each component considered separately is a scalar-valued function like the previous function ϕ. Hence,

$$\frac{\partial}{\partial t}\mathbf{f}(\mathbf{x}; t) = \left(\frac{\partial}{\partial t}f_1(\mathbf{x}; t), \frac{\partial}{\partial t}f_2(\mathbf{x}; t), \frac{\partial}{\partial t}f_3(\mathbf{x}; t)\right)$$

is the *instantaneous local rate of change* of $\mathbf{f}(\mathbf{x}; t)$ at $\mathbf{x}$ with respect to t.

Now consider ϕ defined by $\phi(\mathbf{c}; t)$, a scalar-valued function with the Lagrangian description. A derivative of ϕ with respect to time depends on knowing the position at time t of the particle that was at $\mathbf{c}$ at time $t = 0$. Therefore, we need to know the equation of (6–1.9), $\mathbf{x} = \mathbf{x}(\mathbf{c}; t)$, or more appropriately the equation of (6–1.10), $\mathbf{c} = \mathbf{c}(\mathbf{x}; t)$. Thus, by substitution from (6–1.10),

$$\phi(\mathbf{c}(\mathbf{x}; t); t) = \Phi(\mathbf{x}; t),$$

and by the chain rule, Proposition 3–4.1,

$$\begin{aligned}\frac{d}{dt}\phi(\mathbf{c}(\mathbf{x}; t); t) &= \frac{d}{dt}\Phi(\mathbf{x}; t) = \frac{\partial}{\partial x_1}\Phi(\mathbf{x}; t)\frac{d}{dt}x_1(\mathbf{c}; t) + \frac{\partial}{\partial x_2}\Phi(\mathbf{x}; t)\frac{d}{dt}x_2(\mathbf{c}; t) \\ &\quad + \frac{\partial}{\partial x_3}\Phi(\mathbf{x}; t)\frac{d}{dt}x_3(\mathbf{c}; t) + \frac{\partial}{\partial t}\Phi(\mathbf{x}; t). \quad (6\text{–}1.11)\end{aligned}$$

This derivative is usually written as $D\Phi/Dt$, and it is referred to as the *Stokes' total time derivative*. Notice that this derivative of ϕ is made up of two parts:

$$\frac{\partial \Phi}{\partial x_1}\frac{dx_1}{dt} + \frac{\partial \Phi}{\partial x_2}\frac{dx_2}{dt} + \frac{\partial \Phi}{\partial x_3}\frac{dx_3}{dt} = \left(\frac{\partial \Phi}{\partial x_1}, \frac{\partial \Phi}{\partial x_2}, \frac{\partial \Phi}{\partial x_3}\right) \cdot \left(\frac{dx_1}{dt}, \frac{dx_2}{dt}, \frac{dx_3}{dt}\right)$$

$$= \mathbf{grad}\,\Phi \cdot \frac{d\mathbf{x}}{dt}, \tag{6–1.12}$$

the *instantaneous convective rate of change* of $\Phi(\mathbf{x}; t)$ at $\mathbf{x}(\mathbf{c}; t)$ at time t, and $\partial\Phi/\partial t$, which, as already indicated, is the instantaneous local rate of change of $\Phi(\mathbf{x}; t)$ at $\mathbf{x}(\mathbf{c}; t)$ at time t.

Similarly, if $\mathbf{f}(\mathbf{c}; t)$ is the description of a vector-valued function, or the Lagrangian description, then, using (6–1.10) and substituting, we have

$$\mathbf{f}(\mathbf{c}(\mathbf{x}; t); t) = \mathbf{F}(\mathbf{x}; t),$$

and thus

$$\frac{D}{Dt}\mathbf{F}(\mathbf{x}; t) = \left(\frac{D}{Dt}F_1(\mathbf{x}; t), \frac{D}{Dt}F_2(\mathbf{x}; t), \frac{D}{Dt}F_3(\mathbf{x}; t)\right).$$

Using (6–1.11) and (6–1.12) on the ith coordinate, we have

$$\frac{D}{Dt}F_i(\mathbf{x}; t) = \mathbf{grad}\,F_i(\mathbf{x}; t) \cdot \frac{d}{dt}\mathbf{x}(\mathbf{c}; t) + \frac{\partial}{\partial t}F_i(\mathbf{x}; t) \tag{6–1.13}$$

for $i = 1, 2, 3$. This could be written in the following vector form:

$$\frac{D}{Dt}\mathbf{F}(\mathbf{x}; t) = \mathbf{grad}\,\mathbf{F}(\mathbf{x}; t) \cdot \frac{d}{dt}\mathbf{x}(\mathbf{c}; t) + \frac{\partial}{\partial t}\mathbf{F}(\mathbf{x}; t). \tag{6–1.14}$$

Here the understanding is that (6–1.14) stands for the three equations of (6–1.13).

If, for example, $\mathbf{x} = \mathbf{x}(\mathbf{c}; t)$ is the position vector at time t of the particle that was at $\mathbf{c}$ at time $t = 0$, then

$$\mathbf{x}(\mathbf{c}(\mathbf{x}; t); t) = \mathbf{X}(\mathbf{x}; t),$$

and

$$\mathbf{v}(\mathbf{c}; t) = \frac{D}{Dt}\mathbf{X}(\mathbf{x}; t)$$

is the velocity at time t of this same particle. Similarly,

$$\mathbf{v}(\mathbf{c}(\mathbf{x}; t); t) = \mathbf{V}(\mathbf{x}; t),$$

and

$$\mathbf{a}(\mathbf{c}; t) = \frac{D}{Dt}\mathbf{V}(\mathbf{x}; t) = \frac{D^2}{Dt^2}\mathbf{X}(\mathbf{x}; t)$$

is the acceleration at time t of this same particle.

Problems

1. Draw figures like Figure 6–1.2 of each of the following velocity vector fields and determine the dimension of the vector field. Also sketch in several of the streamlines in each case.

   (a) $\mathbf{f}(\mathbf{x}) = \mathbf{x}, \mathbf{x} \in E^3$, (b) $\mathbf{f}(\mathbf{x}) = (1, 1, 0)$,

   (c) $\mathbf{f}(\mathbf{x}) = (-x_2, x_1, 0)$, (d) $\mathbf{f}(\mathbf{x}) = \left(\dfrac{1}{1 + x_1^2}, 0, \dfrac{1}{1 + x_3^2}\right)$,

   (e) $\mathbf{f}(\mathbf{x}) = (1, x_1, x_3^2)$, (f) $\mathbf{f}(\mathbf{x}) = -(\log \|\mathbf{x}\|)\dfrac{\mathbf{x}}{\|\mathbf{x}\|}$, $\mathbf{x} \in E^2$.

2. Draw figures like Figure 6–1.2 for each of the following velocity vector fields for each of these values of time: $t = 0$, 1, and 2. Draw in several of the streamlines for each value of t.

   (a) $\mathbf{f}(\mathbf{x}; t) = \left(\cos\dfrac{\pi t}{4}, \sin\dfrac{\pi t}{4}, 0\right)$,

   (b) $\mathbf{f}(\mathbf{x}; t) = (x_1 + t, x_2 + t, 0)$.

3. Find approximates of the path lines that pass through $\mathbf{x}_0$ of each of the velocity vector fields defined in Problem 2 by drawing successively the following vectors as directed line segments:

   $$\mathbf{x}_1 = \mathbf{f}(\mathbf{x}_0; 0)\, dt, \qquad \mathbf{x}_2 = \mathbf{f}(\mathbf{x}_1; dt)\, dt, \cdots,$$
   $$\mathbf{x}_n = \mathbf{f}(\mathbf{x}_{n-1}; (n-1)\, dt)\, dt.$$

   with $\mathbf{x}_i$ having its initial point at $\mathbf{x}_{i-1}$ for $1 \leq i \leq n$. In particular, let $\mathbf{x}_0 = (1, 0, 0)$, and then let $dt = 1$ and $n = 4$. Next let $dt = \frac{1}{2}$ and $n = 8$, and then let $dt = \frac{1}{4}$ and let $n = 16$.

4. Use the technique of Example 6–1.1 to find the functions $\mathbf{x}(\mathbf{c};\ )$ for each of the cases of $\mathbf{f}$ given in Problem 2. For $0 \leq t \leq 4$, plot $\mathbf{x}(\mathbf{c}; t)$ for $\mathbf{c} = (1, 0, 0)$. Compare this path line with the approximate path lines obtained in Problem 3.

5. Use the technique of Example 6–1.1 to find the function $\mathbf{x}(\mathbf{c};\ )$ for each of the cases of $\mathbf{f}$ defined below:

   (a) $\mathbf{f}(\mathbf{x}) = (3x_1 + 2x_3, 2x_2, 3x_2 - 2x_3)$,
   (b) $\mathbf{f}(\mathbf{x}) = (2x_2, 2x_3, 2x_1)$,
   (c) $\mathbf{f}(\mathbf{x}; t) = (x_1 - 2x_2 + 1, -x_2 + t, -x_1 + x_2 + 2x_3)$,
   (d) $\mathbf{f}(\mathbf{x}; t) = (-5x_1 - 7x_2 - 5x_3 + t, 2x_1 + 4x_2 + x_3 - t,$ $2x_1 + 2x_2 + 3x_3 - 1)$.

6. Given the function $\mathbf{x}(\mathbf{c};\ )$ defined by (6–1.5), find the related function $\mathbf{c}(\mathbf{x};\ )$ as defined in (6–1.10).

7. For the function $\mathbf{x}(\mathbf{c};\ )$ defined in (6–1.5) plot $\mathbf{x}(\mathbf{0}; t)$ for $t \in [0, 3]$.

8. Use the technique of Example 6–1.2 to find two one-parameter families of surfaces given by equations like (6–1.7) and (6–1.8), whose curves of

intersection are the solution curves of the system of (6–1.1), for the following functions **f**. In each case plot several of the solution curves as shown in Figure 6–1.3.

(a) $\mathbf{f}(\mathbf{x}) = \left(-x_2 x_3, x_3 x_1, \left(-\frac{1}{3}\right)x_1 x_2\right)$.

(b) $\mathbf{f}(\mathbf{x}) = (x_1 x_3 - x_2, x_2 x_3 - x_1, 1 - x_3^2)$.
(c) $\mathbf{f}(\mathbf{x}) = (x_1^2(x_2^3 - x_3^3), x_2^2(x_3^3 - x_1^3), x_3^2(x_1^3 - x_2^3))$.
(d) $\mathbf{f}(\mathbf{x}) = (x_1^2 + 4, x_1 x_2 - 2x_3, x_1 x_3 + 2x_2)$.

9. Given two first integrals, $u_1(\mathbf{x}) = k_1$, $u_2(\mathbf{x}) = k_2$, like those of (6–1.7) and (6–1.8). The vector ∇u_i, is normal to the surface given by $u_i(\mathbf{x}) = k_i$, $i = 1, 2$ for a particular constant k_i. Show that $\nabla u_i(\mathbf{x})$ is orthogonal to $\mathbf{f}(\mathbf{x})$ for $i = 1, 2$ and show $\nabla u_i \times \nabla u_2$ is a vector in the direction of the curve of intersection of the surfaces determined by these two first integrals. Hence these curves of intersection are solution curves of the system of (6–1.1) (cf. Problem 6).

10. Find the instantaneous local rate of change of the following quantities:
(a) $\phi(\mathbf{x}; t) = (x_1 + x_2 + x_3)t$,
(b) $\phi(\mathbf{x}; t) = x_1 + x_2 t + x_3 t^2 + t^3$,
(c) $\mathbf{f}(\mathbf{x}; t) = (x_1 + t, x_2 t, x_2 x_3 t^2)$,
(d) $\mathbf{f}(\mathbf{x}; t) = ((x_2 + x_3)t, (x_1 + x_2) \sin t, x_3 \cos t)$.

11. For each of the functions ϕ and **f** defined next, find the Stokes' total time derivatives

$$\frac{D}{Dt}\phi \text{ and } \frac{D}{Dt}\mathbf{f},$$

respectively. In each case, use the function $\mathbf{x}(\mathbf{c};\ )$ indicated. Also calculate the instantaneous convective and the instantaneous local rates of change.
(a) $\phi(\mathbf{c}; t) = (c_1 + c_2 + c_3)t, \quad \mathbf{x}(\mathbf{c}; t) = \mathbf{c}e^t$.
(b) $\phi(\mathbf{c}; t) = c_1 \cos t - c_2 \sin t + c_3$,
$\mathbf{x}(\mathbf{c}; t) = (c_1 \cos t, c_2 \sin t, c_3(t + 1))$.
(c) $\mathbf{f}(\mathbf{c}; t) = (c_1 + c_2 t, c_2 + c_3 t^2, c_3 - c_1 e^t)$,
$\mathbf{x}(\mathbf{c}; t) = (c_1 e^{-t} + c_2 e^t, c_3 e^t - c_1 e^{2t}, c_1 e^{2t})$.
(d) $\mathbf{f}(\mathbf{c}; t) = (c_2 t - c_3, c_3 + c_2 e^{-t}, c_1 c_2)$,
$\mathbf{x}(\mathbf{c}; t) = (c_1 \cos t + c_3 \sin t, -c_1 \sin t + c_3 \cos t, c_2 e^t)$.

Answers

1. The dimension of the field is three for (a) and (e), two for (c), (d), and (f), and one for (b).

4. (a) $\mathbf{x}(\mathbf{c}; t) = \left(1 + \frac{4}{\pi} \sin \frac{\pi t}{4}, \frac{4}{\pi}\left(1 - \cos \frac{\pi t}{4}\right), 0\right)$,

(b) $\mathbf{x}(\mathbf{c}; t) = (2e^t - t - 1, e^t - t - 1, 0)$.

5. (a) $$\mathbf{x}(\mathbf{c};t) = \left[c_1 + \frac{6}{5}c_2 + \frac{2}{5}c_3\right]e^{3t}(1,0,0) + \frac{1}{4}c_2 e^{2t}(-6,4,3) + \left[\frac{-3}{20}c_2 + \frac{1}{5}c_3 e^{-2t}\right](-2,0,5),$$

(b) $$\mathbf{x}(\mathbf{c};t) = \left[\frac{1}{3}(c_1+c_2+c_3)\right]e^{2t}(1,1,1) + \left(\frac{1}{6}c_1 + \frac{1}{12}\bar{\nu}c_2 + \frac{1}{12}\nu c_3\right)e^{\nu t}(2,\nu,\bar{\nu}) + \left(\frac{1}{6}c_3 + \frac{1}{12}\nu c_3 + \frac{1}{12}\bar{\nu}c_3\right)e^{\bar{\nu}t}(2,\bar{\nu},\nu),$$

where $\nu = -1 + \sqrt{3}\,i$ and $\bar{\nu} = -1 - \sqrt{3}\,i$,

(c) $$\mathbf{x}(\mathbf{c};t) = (c_1 - c_2)e^t(1,0,1) + (c_2+1)e^{-t}(1,1,0) + \left(c_3 - \frac{1}{4}\right)e^{2t}(0,0,1) + \left(-1+2t,\ -1+t,\ \frac{1}{4}+\frac{1}{2}t\right),$$

(d) $$\mathbf{x}(\mathbf{c};t) = \left[\frac{1}{5}(c_1+c_2+2c_3-2)\right]e^t(2,1,1) + \left[\frac{1}{5}(-c_1-c_2+3c_3-13)\right]e^{-t}(-3,1,1) + \left[-c_2+c_3-\frac{1}{4}\right]e^{2t}(1,-1,0) + \left(\frac{-27}{4} - \tfrac{1}{2}t,\ \frac{11}{4} + \tfrac{1}{2}t,\ 3\right).$$

8. (a) $x_1^2 + 3x_3^2 = k_1, \quad x_1^2 + x_2^2 = k_2,$
   (b) $(x_1^2 - x_2^2)(1 - x_3^2) = k_1, \quad (x_1 + x_2 x_3)(x_2 + x_1 x_3)^{-1} = k_2,$
   (c) $x_1^{-1} + x_2^{-1} + x_3^{-1} = k_1, \quad x_1^2 + x_2^2 + x_3^2 = k_2,$
   (d) $(x_1 x_3 + x_2)(x_3 - x_1 x_2)^{-1} = k_1, \quad (x_2^2 + x_3^2)(x_1^2 + 4)^{-1} = k_2.$

10. (a) $x_1 + x_2 + x_3$, (b) $x_2 + 2x_3 t + 3t^2$,
   (c) $(x_1, x_2, 2x_2 x_3 t)$, (d) $(x_2 + x_3, (x_1 + x_2)\cos t, -x_3 \sin t)$.

11. (a) $\dfrac{D}{Dt}\phi(\mathbf{c};t) = c_1 + c_2 + c_3$, instantaneous convective rate of change of $\phi = (x_1 + x_2 + x_3)te^{-t}$; instantaneous local rate of change of $\phi = (x_1 + x_2 + x_3)(1 - t)e^{-t}$.
   (b) $-c_1 \cos t - c_2 \sin t$; $-x_1 \tan t - x_2 \cot t + (t+1)^{-2}(x_3)$, $-x_3(t+1)^{-2}$.
   (c) $\dfrac{D}{Dt}\mathbf{f}(\mathbf{c};t) = (c_2, 2c_3, -c_1 e^t)$; convective, $(x_1 te^{-t} + x_3(2e^{-2t}) \times (1 - 2te^{-2t}),\ x_1 e^{-t} + x_2 t^2 e^{-t} + x_3(t^2 e^{-t} - 4e^{-4t}),\ x_2 e^{-t} - x_3 e^{-t})$; local, $(x_1 e^{-t}(1-t) + x_3 e^{-2t}(4te^{-2t} - e^{-2t} - 2),\ -x_1 e^{-t} + x_2 te^{-t}(2-t) + x_3 e^{-t}(4e^{-3t} + 2t - t^2),\ -x_2 e^{-t})$.

(d) $\dfrac{D}{Dt}\mathbf{f}(\mathbf{c}; t) = (c_2, -c_2 e^{-t}, 0)$; convective,
$(x_3(e^{-t} + \cos t - \sin t), x_3(e^{-2t} + \cos t + \sin t),$
$x_1 x_3 e^{-t} \cos t + x_3^2(e^{-t} \cos t - 2e^{-t} \sin t))$;
local, $(-x_1 \cos t - x_3(e^{-t} + \sin t), x_1 \cos t + x_3(-2e^{-2t} - \sin t),$
$x_1 x_3(-e^{-t} \cos t - e^{-t} \sin t) + x_3^2(-e^{-t} \cos t + e^{-t} \sin t))$.

6–2. Differential Operators on E^3

Three differential operators that are used quite extensively are the gradient, the divergence, and the curl. Although it is convenient to define the gradient and the divergence on E^n for any integer n, it is always true that the curl is used only in 3-space. For our interests now we shall discuss all three operators on E^3. In the section that follows, it will be shown how these definitions can be extended to other spaces.

In this section we shall be taking derivatives of both scalar and vector functions that are either static or dynamic. Thus these functions are defined on a region Ω or a region $\Omega \times I$ where I is an interval. *It will be assumed in every case that any of these functions being discussed is continuous* on Ω (or $\Omega \times I$), *and the first partials are continuous on the interior of* Ω (or $\Omega \times I$). Furthermore, if second partials of these functions are involved in the discussion, then it is assumed that the first partials also have the above properties.

Recall that in Definition 3–6.2 if $\phi: S \to R$ where S is an open subset of E^3 and if the first partials of ϕ exist at a point $\mathbf{x}$ of S, then

$$\mathbf{grad}\, \phi(\mathbf{x}) = (\phi_{x_1}(\mathbf{x}), \phi_{x_2}(\mathbf{x}), \phi_{x_3}(\mathbf{x})), \tag{6–2.1}$$

or we can write $\nabla\phi(\mathbf{x}) = \mathbf{grad}\, \phi(\mathbf{x})$. That is, we say that "del" of ϕ at $\mathbf{x}$ equals the gradient of ϕ at $\mathbf{x}$.

A word of caution is necessary if ϕ is a dynamic scalar function. That is, ϕ is defined on $\Omega \times I$ by $\phi(\mathbf{x}; t)$ where Ω is an open region in E^3 and I is an interval of the real line; then

$$\mathbf{grad}\, \phi(\mathbf{x}; t) = (\phi_{x_1}(\mathbf{x}; t), \phi_{x_2}(\mathbf{x}; t), \phi_{x_3}(\mathbf{x}; t)). \tag{6–2.2}$$

Thus only the partials with respect to the space variables go into making up the components of the gradient vector for any scalar function. Furthermore, notice that if ϕ is a scalar function, then the domain of the space variable, $\mathbf{x}$, is Ω, a region which may not be an open set. However, by the earlier assumptions about ϕ, it is clear that $\mathbf{grad}\, \phi$ always exists on the interior of Ω (or $\Omega \times I$).

If $\mathbf{i} = (1, 0, 0)$, $\mathbf{j} = (0, 1, 0)$, $\mathbf{k} = (0, 0, 1)$ then $\{\mathbf{i}, \mathbf{j}, \mathbf{k}\}$ is an orthonormal basis of E^3, and we could write

$$\nabla\phi(\ ) = \frac{\partial}{\partial x_1}\phi(\ )\mathbf{i} + \frac{\partial}{\partial x_2}\phi(\ )\mathbf{j} + \frac{\partial}{\partial x_3}\phi(\ )\mathbf{k}. \tag{6–2.3}$$

On the right side of this equation it appears as though $\phi(\ )$, meaning either $\phi(\mathbf{x})$ or $\phi(\mathbf{x}; t)$, is a common factor. This fact motivates the following notational convention, which is

$$\left(\frac{\partial}{\partial x_1}\mathbf{i} + \frac{\partial}{\partial x_2}\mathbf{j} + \frac{\partial}{\partial x_3}\mathbf{k}\right)\phi(\ ) = \frac{\partial}{\partial x_1}\phi(\ )\mathbf{i} + \frac{\partial}{\partial x_2}\phi(\ )\mathbf{j} + \frac{\partial}{\partial x_3}\phi(\ )\mathbf{k}. \tag{6–2.4}$$

Comparing (6–2.3) and (6–2.4) we can see that it is natural to write

$$\nabla = \frac{\partial}{\partial x_1}\mathbf{i} + \frac{\partial}{\partial x_2}\mathbf{j} + \frac{\partial}{\partial x_3}\mathbf{k}, \tag{6–2.5}$$

the *del operator*. Notice that the del operator operates or maps a scalar function into a vector function, or, in another sense, it is correct to say that it maps a scalar field into a vector field.

In fact, the del operator, ∇, is a linear operator in that

$$\nabla(\alpha\phi + \beta\psi) = \alpha\,\nabla\,\phi + \beta\,\nabla\,\psi, \tag{6–2.6}$$

where α and β are real or complex constants and where ϕ and ψ are scalar functions. Furthermore,

$$\nabla(\phi\psi) = \phi\,\nabla\,\psi + \psi\,\nabla\,\phi \tag{6–2.7}$$

under the same assumptions on ϕ and ψ.

Recall also from Section 3–6 that

$$\|\mathbf{grad}\ \phi(\mathbf{x})\| = \max\left\{\frac{d}{ds}\phi(\mathbf{x}; \mathbf{u})\right\},$$

the maximum value of the directional derivative of ϕ at $\mathbf{x}$ for all unit vectors, $\mathbf{u}$. Furthermore, $\mathbf{grad}\ \phi(\mathbf{x})$ is normal (perpendicular) to the level surface of ϕ that passes through $\mathbf{x}$, and the direction of $\mathbf{grad}\ \phi(\mathbf{x})$ is in the direction that $\phi(\mathbf{x})$ is increasing the most rapidly.

The precise definition of the other two basic differential operators with which we shall concern ourselves is given in the following definition.

Definition 6–2.1. *Let* $\mathbf{f} = (f_1, f_2, f_3)$ *be a continuously differentiable vector-valued function defined on an open region* $\Omega \subset E^3$; *that is,* $\mathbf{f}\colon \Omega \to R^3$. *Then for* $\mathbf{x} \in \Omega$,

$$\operatorname{div} \mathbf{f}(\mathbf{x}) = \frac{\partial}{\partial x_1}f_1(\mathbf{x}) + \frac{\partial}{\partial x_2}f_2(\mathbf{x}) + \frac{\partial}{\partial x_3}f_3(\mathbf{x}) \tag{6–2.8}$$

is the divergence of **f** *at* **x** *and*

$$\mathbf{curl}\,\mathbf{f}(\mathbf{x}) = \left(\frac{\partial}{\partial x_2} f_3(\mathbf{x}) - \frac{\partial}{\partial x_3} f_2(\mathbf{x})\right)\mathbf{i} + \left(\frac{\partial}{\partial x_3} f_1(\mathbf{x}) - \frac{\partial}{\partial x_1} f_3(\mathbf{x})\right)\mathbf{j} + \left(\frac{\partial}{\partial x_1} f_2(\mathbf{x}) - \frac{\partial}{\partial x_2} f_1(\mathbf{x})\right)\mathbf{k} \quad (6\text{–}2.9)$$

is the curl of **f** *at* **x**.

The following example will illustrate the definition.

Example 6–2.1. Let the vector field **f** be given by

$$\mathbf{f}(x, y, z) = 2xy\mathbf{i} + 10yz^2\mathbf{j} - 5xz\mathbf{k}.$$

Then

$$\text{div}\,\mathbf{f}(x, y, z) = 2y + 10z^2 - 5x$$

and

$$\mathbf{curl}\,\mathbf{f}(x, y, z) = -20yz\mathbf{i} + 5z\mathbf{j} - 2x\mathbf{k}.$$

Notice that div **f**(**x**) is a scalar quantity and that **curl f**(**x**) is a vector-valued quantity. Notice also that

$$\nabla \cdot \mathbf{f}(\mathbf{x}) = \text{div}\,\mathbf{f}(\mathbf{x}) \quad (6\text{–}2.10)$$

and that

$$\nabla \times \mathbf{f}(\mathbf{x}) = \mathbf{curl}\,\mathbf{f}(\mathbf{x}) \quad (6\text{–}2.11)$$

(cf. (1–2.4) and (1–3.1)).

Again, if **f** is a function that defines a dynamic vector field, then the argument **x** in (6–2.8) and (6–2.9) is replaced by $(\mathbf{x}; t)$ just as was analogously mentioned in the discussion of the gradient.

In the case of the *divergence operator*, $\nabla\,\cdot$, we have an operator that maps a vector function into a scalar function, but in the case of the *curl operator*, $\nabla\,\times$, we have an operator that maps a vector function into a vector function. In fact, these two operators are linear transformations, as was the del operator, ∇—that is,

$$\nabla \cdot (\alpha\mathbf{f} + \beta\mathbf{g}) = \alpha\,\nabla \cdot \mathbf{f} + \beta\,\nabla \cdot \mathbf{g}, \quad (6\text{–}2.12)$$

$$\nabla \times (\alpha\mathbf{f} + \beta\mathbf{g}) = \alpha\,\nabla \times \mathbf{f} + \beta\,\nabla \times \mathbf{g} \quad (6\text{–}2.13)$$

where α and β are real or complex constants.

Strict analogies to (6–2.7) cannot be considered relative to these two operators, since the product of two scalar functions is a scalar function. However, other products would be appropriate to consider, namely, the product of a scalar function and a vector function and the cross product of two vector functions. In the first case we can establish the following two identities:

$$\nabla \cdot (\phi \mathbf{f}) = \phi \, \nabla \cdot \mathbf{f} + (\nabla \, \phi) \cdot \mathbf{f}, \tag{6–2.14}$$

$$\nabla \times (\phi \mathbf{f}) = \phi \, \nabla \times \mathbf{f} + \nabla \, \phi \times \mathbf{f}. \tag{6–2.15}$$

In the case of the cross product of two functions it can be established that

$$\nabla \cdot (\mathbf{f} \times \mathbf{g}) = \mathbf{g} \cdot \nabla \times \mathbf{f} - \mathbf{f} \cdot \nabla \times \mathbf{g}. \tag{6–2.16}$$

Before considering $\nabla \times (\mathbf{f} \times \mathbf{g})$, it is best to introduce the following notation so that this identity and some others can be written more concisely. By $\mathbf{f} \cdot \nabla$ we shall mean the differential operator

$$\mathbf{f} \cdot \nabla = f_1 \frac{\partial}{\partial x_1} + f_2 \frac{\partial}{\partial x_2} + f_3 \frac{\partial}{\partial x_3}. \tag{6–2.17}$$

That is, if ϕ is a scalar function and $\mathbf{g}$ is a vector function,

$$(\mathbf{f} \cdot \nabla)\phi = f_1 \frac{\partial \phi}{\partial x_1} + f_2 \frac{\partial \phi}{\partial x_2} + f_3 \frac{\partial \phi}{\partial x_3} = \mathbf{f} \cdot \nabla \phi \tag{6–2.18}$$

and

$$(\mathbf{f} \cdot \nabla)\mathbf{g} = f_1 \frac{\partial \mathbf{g}}{\partial x_1} + f_2 \frac{\partial \mathbf{g}}{\partial x_2} + f_3 \frac{\partial \mathbf{g}}{\partial x_3} \tag{6–2.19}$$

where

$$\frac{\partial \mathbf{g}}{\partial x_i} = \frac{\partial g_1}{\partial x_i} \mathbf{i} + \frac{\partial g_2}{\partial x_i} \mathbf{j} + \frac{\partial g_3}{\partial x_i} \mathbf{k}$$

for $i = 1, 2, 3$. Notice that the symbols ∇ and $\mathbf{f}$ cannot be commuted as in the case of the ordinary dot product of two vectors, since $(\nabla \cdot \mathbf{f})\phi$ and $(\nabla \cdot \mathbf{f})\mathbf{g}$ would not be the same as (6–2.18) and (6–2.19), respectively.

With this new notation the identity evaluating the curl of a cross product can be written

$$\nabla \times (\mathbf{f} \times \mathbf{g}) = \mathbf{f} \, \nabla \cdot \mathbf{g} - \mathbf{g} \, \nabla \cdot \mathbf{f} + (\mathbf{g} \cdot \nabla)\mathbf{f} - (\mathbf{f} \cdot \nabla)\mathbf{g}. \tag{6–2.20}$$

Certain composite transformations can be formed among these three differential operators. For example, it is meaningful to consider both $\nabla \cdot (\nabla \, \phi)$ and $\nabla \times (\nabla \, \phi)$ since $\nabla \, \phi$ is a vector function, but it is only appropriate to consider $\nabla(\nabla \cdot \mathbf{f})$ since $\nabla \cdot \mathbf{f}$ is a scalar function. Lastly, it is meaningful to consider $\nabla \cdot (\nabla \times \mathbf{f})$ and $\nabla \times (\nabla \times \mathbf{f})$ since $\nabla \times \mathbf{f}$ is a vector function. Below are two easy-to-establish identities that give the result of considering two of these five composite operators:

$$\mathbf{curl\ grad}\ \phi = \nabla \times \nabla \phi = \mathbf{0}, \tag{6–2.21}$$

$$\text{div}\ \mathbf{curl\ f} = \nabla \cdot \nabla \times \mathbf{f} = 0. \tag{6–2.22}$$

The divergence operator of the del operator is a well-known operator called the *Laplacian operator*, Δ; that is,

$$\begin{aligned} \operatorname{div} \mathbf{grad}\, \phi &= \nabla \cdot \nabla \phi \\ &= \frac{\partial^2 \phi}{\partial {x_1}^2} + \frac{\partial^2 \phi}{\partial {x_2}^2} + \frac{\partial^2 \phi}{\partial {x_3}^2} \\ &= \Delta \phi. \end{aligned} \qquad (6\text{–}2.23)$$

Occasionally some authors use the notation ∇^2 in place of Δ for the Laplacian operator.

The following example proves an identity that relates the remaining two composites mentioned—namely, $\nabla \times \nabla \times \mathbf{f}$ and $\nabla(\nabla \cdot \mathbf{f})$.

Example 6–2.2. Prove the following vector identity:

$$\nabla \times \nabla \times \mathbf{f} = \nabla(\nabla \cdot \mathbf{f}) - \nabla^2 \mathbf{f} \qquad (6\text{–}2.24)$$

where $\nabla^2 \mathbf{f} = \nabla^2 f_1\, \mathbf{i} + \nabla^2 f_2\, \mathbf{j} + \nabla^2 f_3\, \mathbf{k}$ and where $\mathbf{f} = (f_1, f_2, f_3)$ defines a vector field.

Begin by recalling the definition of $\nabla \times \mathbf{f}$ in (6–2.9),

$$\nabla \times \mathbf{f} = \left(\frac{\partial f_3}{\partial x_2} - \frac{\partial f_2}{\partial x_3}\right)\mathbf{i} + \left(\frac{\partial f_1}{\partial x_3} - \frac{\partial f_3}{\partial x_1}\right)\mathbf{j} + \left(\frac{\partial f_2}{\partial x_1} - \frac{\partial f_1}{\partial x_2}\right)\mathbf{k}.$$

Then, using the definition again, we get

$$\nabla \times (\nabla \times \mathbf{f}) = \left[\frac{\partial}{\partial x_2}\left(\frac{\partial f_2}{\partial x_1} - \frac{\partial f_1}{\partial x_2}\right) - \frac{\partial}{\partial x_3}\left(\frac{\partial f_1}{\partial x_3} - \frac{\partial f_3}{\partial x_1}\right)\right]\mathbf{i} + \cdots,$$

Then add to the right side the following zero vector,

$$\mathbf{0} = \left(\frac{\partial^2 f_1}{\partial {x_1}^2} - \frac{\partial^2 f_1}{\partial {x_1}^2}\right)\mathbf{i} + \left(\frac{\partial^2 f_2}{\partial {x_2}^2} - \frac{\partial^2 f_2}{\partial {x_2}^2}\right)\mathbf{j} + \left(\frac{\partial^2 f_3}{\partial {x_3}^2} - \frac{\partial^2 f_3}{\partial {x_3}^2}\right)\mathbf{k}.$$

Collecting all of the positive terms into one term and all of the negative terms into another gives

$$\begin{aligned} \nabla \times \nabla \times \mathbf{f} = &\left[\left(\frac{\partial^2 f_1}{\partial {x_1}^2} + \frac{\partial^2 f_2}{\partial x_2\, \partial x_1} + \frac{\partial^2 f_3}{\partial x_3\, \partial x_1}\right)\mathbf{i} + \cdots\right] \\ &- \left[\left(\frac{\partial^2 f_1}{\partial {x_1}^2} + \frac{\partial^2 f_1}{\partial {x_2}^2} + \frac{\partial^2 f_1}{\partial {x_3}^2}\right)\mathbf{i} + \cdots\right]. \end{aligned}$$

Hence

$$\begin{aligned} \nabla \times \nabla \times \mathbf{f} &= \left[\frac{\partial}{\partial x_1}(\nabla \cdot \mathbf{f})\mathbf{i} + \frac{\partial}{\partial x_2}(\nabla \cdot \mathbf{f})\mathbf{j} + \frac{\partial}{\partial x_3}(\nabla \cdot \mathbf{f})\mathbf{k}\right] \\ &\quad - [\nabla^2 f_1\, \mathbf{i} + \nabla^2 f_2\, \mathbf{j} + \nabla^2 f_3\, \mathbf{k}] \\ &= \nabla(\nabla \cdot \mathbf{f}) - \nabla^2 \mathbf{f}. \end{aligned}$$

Problems

1. Prove the two identities (6–2.6) and (6–2.7).

2. Suppose that the chain rule is applicable at the point $\mathbf{x}$ to the real-valued composite function $\phi\mathbf{f}$ where ϕ is defined by $z = \phi(u_1, \cdots, u_k)$ and the function f_i is defined by $u_i = f_i(\mathbf{x})$ for $1 \leq i \leq k$. Then show that

$$\nabla\phi(\mathbf{f}(\mathbf{x})) = \frac{\partial}{\partial x_1}\phi(\mathbf{f}(\mathbf{x}))\,\nabla f_1(\mathbf{x}) + \cdots + \frac{\partial}{\partial x_k}\phi(\mathbf{f}(\mathbf{x}))\,\nabla f_k(\mathbf{x}).$$

3. Given that ϕ_1, ϕ_2, and ϕ_3 are scalar functions for $1 \leq i \leq 3$, show that

$$\nabla\phi_1 \cdot (\nabla\phi_2 \times \nabla\phi_3) = \frac{\partial(\phi_1, \phi_2, \phi_3)}{\partial(x_1, x_2, x_3)}.$$

4. If ϕ is a scalar function defined by $z = \phi(\mathbf{x}; t)$, then show

$$d\phi(\mathbf{x}_0, t; \mathbf{dx}, dt) = \mathbf{dx} \cdot \nabla\phi(\mathbf{x}_0; t) + \frac{\partial}{\partial t}\phi(\mathbf{x}_0; t)\, dt.$$

5. Prove that ϕ is a scalar function defined by $z = \phi(\mathbf{x})$ for all $\mathbf{x} \in \Omega$, an *open* region in E^3, and that if $\nabla\phi(\mathbf{x}) = \mathbf{0}$ for all $\mathbf{x} \in \Omega$, then $\phi(\mathbf{x}) = z_0$, a constant. Show that this result is still true for Ω being a closed region if ϕ is continuously differentiable on Ω (cf. Section 3–3).

6. Let ϕ be a scalar function defined on E^n by

$$z = \phi(\mathbf{x}) = F(\|\mathbf{x}\|)$$

where F is a real-valued function such that $F \in C^1((0, \infty))$. Calculate $\nabla\phi(\mathbf{x})$ for $\mathbf{x} \neq \mathbf{0}$. What is the direction of $\nabla\phi(\mathbf{x})$ if $F(\|\mathbf{x}\|)$ is decreasing on $(0, \infty)$?

7. Let ϕ, ψ, $\mathbf{f}$, and $\mathbf{g}$ be functions defined as follows:

$$\phi(x, y, z) = x^2 - yz, \qquad \psi(x, y, z, t) = y^2 t + xz^2,$$
$$\mathbf{f}(x, y, z) = xy^2\,\mathbf{i} + 2yz\mathbf{j} - xy\mathbf{k},$$
$$\mathbf{g}(x, y, z; t) = (zy + t)\mathbf{i} - (yt^2 + x)\mathbf{j} + (xyzt)\mathbf{k}.$$

Find

(a) $\mathbf{curl\,f}$, (b) $\nabla \cdot \mathbf{g}$,
(c) $\nabla^2\phi$, (d) $\operatorname{div}\mathbf{f} - \nabla \cdot \mathbf{g}$,
(e) $\nabla \times \nabla\psi$, (f) $\nabla \cdot (\psi\mathbf{g})$,
(g) $\mathbf{curl\,curl\,g}$, (h) $\nabla \cdot (\mathbf{f} \times \mathbf{g})$,
(i) $\nabla \times (\mathbf{f} \times \nabla\phi)$, (j) $\Delta\psi$,
(k) $\nabla(\mathbf{f} \cdot \mathbf{g})$, (l) $2\,\mathbf{grad}\,\psi - 3\,\mathbf{curl\,g}$.

8. Prove the identities (6–2.12) and (6–2.13).

9. Prove the identities (6–2.14), (6–2.15), and (6–2.16).

10. Prove that the differential operator $\mathbf{f} \cdot \nabla$ is a linear operator; that is, prove an identity analogous to (6–2.6).

11. Prove the following two identities:

(a) $(\mathbf{f} \cdot \nabla)\mathbf{x} = \mathbf{f}$,
(b) $(\mathbf{f} \cdot \nabla)\mathbf{f} = \frac{1}{2} \nabla(\mathbf{f} \cdot \mathbf{f}) - \mathbf{f} \times (\nabla \times \mathbf{f})$.

12. Prove the identity (6–2.20).

13. Prove the following identity:

$$\nabla(\mathbf{f} \cdot \mathbf{g}) = (\mathbf{f} \cdot \nabla)\mathbf{g} + (\mathbf{g} \cdot \nabla)\mathbf{f} + (\mathbf{f} \times \nabla \times \mathbf{g}) + (\mathbf{g} \times \nabla \times \mathbf{f}).$$

Notice that this is the only product of vector functions to which the del operator can be applied. Calculate $\nabla(\mathbf{f} \cdot \mathbf{f})$ by the use of this identity. If $\mathbf{f}$ is a unit vector, is it true that

$$(\mathbf{f} \cdot \nabla)\mathbf{f} = -\mathbf{f} \times \nabla \times \mathbf{f}?$$

14. Demonstrate that

$$(\nabla \times \mathbf{f}) \cdot (\mathbf{g} \times \mathbf{h}) = ((\mathbf{g} \cdot \nabla)\mathbf{f}) \cdot \mathbf{h} - ((\mathbf{h} \cdot \nabla)\mathbf{f}) \cdot \mathbf{g}.$$

15. Prove the two identities (6–2.21) and (6–2.22).

16. Prove that

(a) $\nabla \cdot \mathbf{x} = 3$, (b) $\nabla \cdot \left(\dfrac{\mathbf{x}}{\|\mathbf{x}\|^3}\right) = 0, \quad \mathbf{x} \neq \mathbf{0}$,

(c) $\nabla \times \mathbf{x} = \mathbf{0}$, (d) $\Delta \left(\dfrac{1}{\|\mathbf{x}\|}\right) = 0, \quad \mathbf{x} \neq \mathbf{0}$.

17. Use some of the above identities to calculate the following:

(a) $\nabla \cdot (\mathbf{f} \times (\mathbf{g} \times \mathbf{h}))$, (b) $\nabla \times (\mathbf{f} \times (\mathbf{g} \times \mathbf{h}))$,
(c) $\nabla(\mathbf{f} \cdot (\mathbf{g} \times \mathbf{h}))$, (d) $\nabla \times (\mathbf{f} \times \nabla \times \mathbf{g})$.

Answers

6. $\nabla\phi(\mathbf{x}) = \left\{\dfrac{F'(\|\mathbf{x}\|)}{\|\mathbf{x}\|}\right\} \mathbf{x}$, toward the origin, $\mathbf{0}$.

7. (a) $(-x - 2y, y, -2xy)$,
(b) $-t^2 + xyt$, (c) 2,
(d) $y^2 + 2z + t^2 - xyt$, (e) 0,
(f) $xy^3t + 3x^2yz^2t + yz^3 + z^2t - y^2t^2 - xz^2t^2 - 2y^2t^3 - 2xyt$,
(g) $(yt, xt + x, 0)$,
(h) $2y^2z^2t - y^2t^2 - 2xy - 2xyz - xt - 3x^2y^2zt - 4y^2z - 2yt$,
(i) $(-2xyz - 4xz, y^2z + 4yz - 2y^2 - xy, y^3 + 4yz + xz - 4xy)$,
(j) $2(x + t)$,
(k) $(y^3z + y^2t - 2yz - 2xy^2zt,$
$3xy^2z + 2xyt - 4yzt^2 - 2xz - 2x^2yzt,$
$xy^3 - 2y^2t^2 - 2xy - x^2y^2t)$,
(l) $(2z^2 - 3xzt, 4yt - 3y + 3yzt, 4xz + 3 + 3z)$.

17. (a) $\mathbf{f} \cdot \mathbf{h}\nabla \cdot \mathbf{g} - (\mathbf{f} \cdot \mathbf{g})\nabla \cdot \mathbf{h} + \nabla(\mathbf{f} \cdot \mathbf{h}) \cdot \mathbf{g} - \nabla(\mathbf{f} \cdot \mathbf{g}) \cdot \mathbf{h}$.

6–3. Differential Operators on Other Spaces

In this section we shall extend to other spaces than E^3 the definitions of the main differential operators discussed in the previous section, the del operator, the divergence operator, and the curl operator.

There are other spaces than E^3 that arise quite naturally in the consideration of physical problems. It is frequently more convenient to define scalar functions and vector functions in terms of the coordinates of some space other than the coordinates of E^3. In such cases it will be desirable to know how to operate with the differential operators.

A good example is spherical coordinate space, or (ρ, φ, θ)-space (cf. Problem 6, page 161). The transformation of coordinates that gives the relations between the coordinates of (x, y, z)-space (that is, E^3 space and (ρ, φ, θ)-space) is

$$x = \rho \cos \theta \sin \varphi,$$
$$y = \rho \sin \theta \sin \varphi, \tag{6–3.1}$$
$$z = \rho \cos \varphi$$

(cf. Figure 6–3.1).

Given a scalar function ϕ defined by $\phi(x, y, z)$ at (x, y, z), the problem is to define the del operator on ψ where ψ is the composite function defined by

$$\psi(\rho, \varphi, \theta) = \phi(\rho \cos \theta \sin \varphi, \rho \sin \theta \sin \varphi, \rho \cos \phi)$$

at (ρ, φ, θ) such that

$$\nabla\psi(\rho_0, \varphi_0, \theta_0) = \nabla\phi(x_0, y_0, z_0) \tag{6–3.2}$$

where $(\rho_0, \varphi_0, \theta_0)$ is the image of (x_0, y_0, z_0) by the given transformation. By so defining $\nabla\psi(\rho, \varphi, \theta)$, the gradient vector will be the same or invariant, at the point P_0, which is labeled (x_0, y_0, z_0) in (x, y, z)-space and $(\rho_0, \varphi_0, \theta_0)$ in (ρ, φ, θ)-space. We also shall want to make analogous definitions for the divergence operator and the curl operator when we are given a vector function.

In order to know how to define $\nabla\psi$ properly, let us first look closely at Figure 6–3.1. The curve given parametrically by taking $\theta = \theta_0$, $\varphi = \varphi_0$, and $\rho \geq \rho_0$ in the transformation of (6–3.1) is called the ρ-curve, and it is labeled as such in Figure 6–3.1. Similarly, the φ-curve and θ-curve are defined. The vectors labeled $\mathbf{u}_\rho$, $\mathbf{u}_\varphi$, and $\mathbf{u}_\theta$ are the unit tangent vectors to the ρ-, φ-, and θ-curves, respectively, in the direction of increasing parameter. Thus

$$\mathbf{u}_\rho = \frac{1}{h_\rho}\mathbf{r}_\rho, \quad \mathbf{u}_\varphi = \frac{1}{h_\rho}\mathbf{r}_\varphi, \quad \mathbf{u}_\theta = \frac{1}{h_\theta}\mathbf{r}_\theta \tag{6–3.3}$$

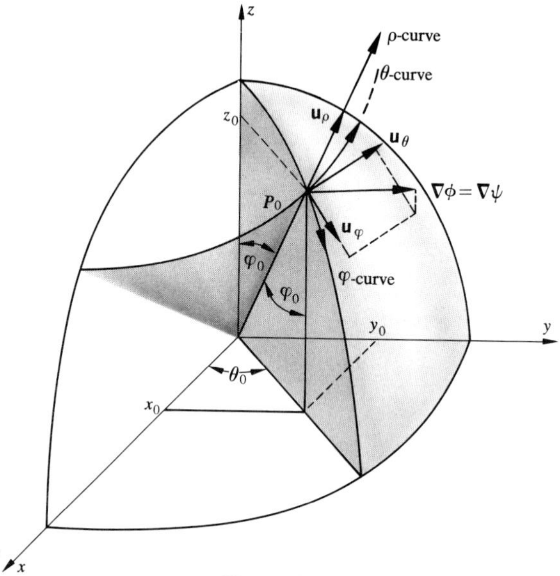

Figure 6–3.1

where

$$\mathbf{r}_\rho = \left(\frac{\partial x}{\partial \rho}, \frac{\partial y}{\partial \rho}, \frac{\partial z}{\partial \rho}\right) = (\cos\theta \sin\varphi, \sin\theta \sin\varphi, \cos\varphi),$$

$$\mathbf{r}_\varphi = \left(\frac{\partial x}{\partial \varphi}, \frac{\partial y}{\partial \varphi}, \frac{\partial z}{\partial \varphi}\right) = (\rho\cos\theta \cos\varphi, \rho\sin\varphi \cos\varphi, -\rho\sin\varphi), \quad (6\text{–}3.4)$$

$$\mathbf{r}_\theta = \left(\frac{\partial x}{\partial \theta}, \frac{\partial y}{\partial \theta}, \frac{\partial z}{\partial \theta}\right) = (-\rho\sin\theta \sin\varphi, \rho\cos\theta \sin\varphi, 0)$$

and where

$$h_\rho = \|\mathbf{r}_\rho\|, \; h_\varphi = \|\mathbf{r}_\varphi\|, \; h_\theta = \|\mathbf{r}_\theta\|.$$

That is,

$$h_\rho^{\,2} = \left(\frac{\partial x}{\partial \rho}\right)^2 + \left(\frac{\partial y}{\partial \rho}\right)^2 + \left(\frac{\partial z}{\partial \rho}\right)^2 = 1,$$

$$h_\varphi^{\,2} = \left(\frac{\partial x}{\partial \varphi}\right)^2 + \left(\frac{\partial y}{\partial \varphi}\right)^2 + \left(\frac{\partial z}{\partial \varphi}\right)^2 = \rho^2, \quad (6\text{–}3.5)$$

$$h_\theta^{\,2} = \left(\frac{\partial x}{\partial \theta}\right)^2 + \left(\frac{\partial y}{\partial \theta}\right)^2 + \left(\frac{\partial z}{\partial \theta}\right)^2 = \rho^2 \sin^2\varphi.$$

Notice that $\{\mathbf{u}_\rho, \mathbf{u}_\varphi, \mathbf{u}_\theta\}$ is an orthonormal set of vectors when $\rho \sin\varphi \neq 0$, as for example, at point P_0 in Figure 6–3.1. Thus $\{\mathbf{u}_\rho, \mathbf{u}_\varphi, \mathbf{u}_\theta\}$ is an orthonormal basis for E^3. For this reason and the fact that $(\mathbf{u}_\rho, \mathbf{u}_\varphi, \mathbf{u}_\theta) > 0$, the condition of a positive triple, the spherical coordinates $\{\rho, \varphi, \theta\}$ are called *orthogonal coordinates*. In our case

$$\mathbf{u}_\rho \cdot (\mathbf{u}_\varphi \times \mathbf{u}_\theta) = (\mathbf{u}_\rho, \mathbf{u}_\varphi, \mathbf{u}_\theta) = \frac{1}{h_\rho h_\varphi h_\theta} \cdot J = 1$$

where

$$J = \frac{\partial(x, y, z)}{\partial(\rho, \varphi, \theta)}$$

is the Jacobian of the transformation (cf. (1–3.9)).

Notice, furthermore, that

$$\begin{aligned} &\mathbf{r}_\rho \cdot \nabla\rho = 1, \quad &&\mathbf{r}_\rho \cdot \nabla\varphi = 0, \quad &&\mathbf{r}_\rho \cdot \nabla\theta = 0, \\ &\mathbf{r}_\varphi \cdot \nabla\rho = 0, &&\mathbf{r}_\varphi \cdot \nabla\varphi = 1, &&\mathbf{r}_\varphi \cdot \nabla\theta = 0, \\ &\mathbf{r}_\theta \cdot \nabla\rho = 0, &&\mathbf{r}_\theta \cdot \nabla\varphi = 0, &&\mathbf{r}_\theta \cdot \nabla\theta = 1, \end{aligned} \tag{6–3.6}$$

since, by the chain rule,

$$\mathbf{r}_\rho \cdot \nabla\rho = \frac{\partial x}{\partial\rho}\frac{\partial\rho}{\partial x} + \frac{\partial y}{\partial\rho}\frac{\partial\rho}{\partial y} + \frac{\partial z}{\partial\rho}\frac{\partial\rho}{\partial z} = \left.\frac{\partial\rho}{\partial\rho}\right)_{\varphi,\theta} = 1,$$

$$\mathbf{r}_\rho \cdot \nabla\varphi = \frac{\partial x}{\partial\rho}\frac{\partial\varphi}{\partial x} + \frac{\partial y}{\partial\rho}\frac{\partial\varphi}{\partial y} + \frac{\partial z}{\partial\rho}\frac{\partial\varphi}{\partial z} = \left.\frac{\partial\varphi}{\partial\rho}\right)_{\varphi,\theta} = 0,$$

and so forth.

Thus $\nabla\rho$ is orthogonal to $\mathbf{r}_\varphi$, and $\mathbf{r}_\theta$ and has magnitude $1/h_\rho$, the reciprocal of the magnitude of $\mathbf{r}_\rho$. The situation is analogous for $\nabla\varphi$ and $\nabla\theta$. Therefore,

$$\begin{aligned} \mathbf{u}_\rho &= \frac{1}{h_\rho}\mathbf{r}_\rho = h_\rho \nabla\rho, \\ \mathbf{u}_\varphi &= \frac{1}{h_\varphi}\mathbf{r}_\varphi = h_\varphi \nabla\varphi, \\ \mathbf{u}_\theta &= \frac{1}{h_\theta}\mathbf{r}_\theta = h_\theta \nabla\theta. \end{aligned} \tag{6–3.7}$$

The basis $\{\nabla\rho, \nabla\varphi, \nabla\theta\}$ is called the *reciprocal basis* of $\{\mathbf{r}_\rho, \mathbf{r}_\varphi, \mathbf{r}_\theta\}$.

The vector $\nabla\phi$ at P_0 is like the vector labeled $\nabla\phi$ in Figure 6–3.1. Since we, of course, want $\nabla\psi = \nabla\phi$, let us find the coordinates of $\nabla\phi$ with respect to the basis $\{\mathbf{u}_\rho, \mathbf{u}_\varphi, \mathbf{u}_\theta\}$; that is, we shall find α_ρ, α_φ, and α_θ such that

$$(\phi_x, \phi_y, \phi_z) = \alpha_\rho \mathbf{u}_\rho + \alpha_\varphi \mathbf{u}_\varphi + \alpha_\theta \mathbf{u}_\theta.$$

It follows that

$$\alpha_\rho = (\phi_x, \phi_y, \phi_z) \cdot \mathbf{u}_\rho = \frac{1}{h_\rho}\left(\phi_x \frac{\partial x}{\partial\rho} + \phi_y \frac{\partial y}{\partial\rho} + \phi_z \frac{\partial z}{\partial\rho}\right),$$

and, by the chain rule,

$$\alpha_\rho = \frac{1}{h_\rho}\frac{\partial\psi}{\partial\rho}.$$

Similarly,

$$\alpha_\varphi = \frac{1}{h_\varphi}\frac{\partial \psi}{\partial \varphi}$$

and

$$\alpha_\theta = \frac{1}{h_\theta}\frac{\partial \psi}{\partial \theta}.$$

Therefore,

$$\nabla \psi = \frac{1}{h_\rho}\frac{\partial \psi}{\partial \rho}\mathbf{u}_\rho + \frac{1}{h_\varphi}\frac{\partial \psi}{\partial \varphi}\mathbf{u}_\varphi + \frac{1}{h_\theta}\frac{\partial \psi}{\partial \theta}\mathbf{u}_\theta \qquad (6\text{–}3.8)$$

is the ∇ operator in (ρ, φ, θ)-space.

The following example verifies the correctness of (6–3.8) at a particular point.

Example 6–3.1. Suppose that $\phi(x, y, z) = x^2 + yz$. Then at the point, $(\rho, \varphi, \theta) = (2, \pi/2, \pi/4)$, we have $(x, y, z) = (\sqrt{2}, \sqrt{2}, 0)$ by (6–3.1). Thus

$$\nabla\phi(\sqrt{2}, \sqrt{2}, 0) = 2\sqrt{2}\,\mathbf{u}_1 + \sqrt{2}\,\mathbf{u}_3 \qquad (6\text{–}3.9)$$

where $\mathbf{u}_1 = \mathbf{i}$ and $\mathbf{u}_3 = \mathbf{k}$. Now

$$\psi(\rho, \varphi, \theta) = \rho^2(\cos^2\theta \sin^2\varphi + (1/2)\sin\theta \sin 2\varphi)$$

and

$$\nabla\psi\left(2, \frac{\pi}{2}, \frac{\pi}{4}\right) = 2\mathbf{u}_\rho - \sqrt{2}\,\mathbf{u}_\varphi - 2\mathbf{u}_\theta.$$

Hence, by (6–3.3) and (6–3.4),

$$\begin{aligned}\nabla\psi\left(2, \frac{\pi}{2}, \frac{\pi}{4}\right) &= 2\left(\frac{1}{\sqrt{2}}, \frac{1}{\sqrt{2}}, 0\right) - \sqrt{2}(0, 0, -1) - 2\left(-\frac{1}{\sqrt{2}}, \frac{1}{\sqrt{2}}, 0\right)\\ &= 2\sqrt{2}\,\mathbf{u}_1 + \sqrt{2}\,\mathbf{u}_3,\end{aligned}$$

the same as in (6–3.9).

Let us now continue with this same transformation (transformation (6–3.1)) and see how the divergence operator and the curl operator should be defined for the vector function $\mathbf{F}$ in (ρ, φ, θ)-space where

$$\mathbf{F}(\rho, \varphi, \theta) = \mathbf{f}(\rho\cos\theta\sin\varphi, \rho\sin\theta\sin\varphi, \rho\cos\varphi). \qquad (6\text{–}3.10)$$

First let us write the vector $\mathbf{F}$ in terms of the orthonormal basis $\{\mathbf{u}_\rho, \mathbf{u}_\varphi, \mathbf{u}_\theta\}$:

$$\mathbf{F} = F_\rho\,\mathbf{u}_\rho + F_\varphi\,\mathbf{u}_\varphi + F_\theta\,\mathbf{u}_\theta \qquad (6\text{–}3.11)$$

where F_ρ, F_φ, and F_θ are the coordinates of **F** relative to this basis. Then, by (6–3.7), we can write

$$\mathbf{F} = F_\rho(h_\rho \nabla \rho) + F_\varphi(h_\varphi \nabla \varphi) + F_\theta(h_\theta \nabla \theta), \tag{6–3.12}$$

or, since by (6–3.7)

$$\begin{aligned} h_\rho \nabla \rho &= h_\varphi h_\theta \nabla \varphi \times \nabla \theta, \\ h_\varphi \nabla \varphi &= h_\theta h_\rho \nabla \theta \times \nabla \rho, \\ h_\theta \nabla \theta &= h_\rho h_\varphi \nabla \rho \times \nabla \varphi, \end{aligned} \tag{6–3.13}$$

we could write instead

$$\mathbf{F} = (F_\rho h_\varphi h_\theta)\nabla \varphi \times \nabla \theta + (F_\varphi h_\theta h_\rho)\nabla \theta \times \nabla \rho + (F_\theta h_\rho h_\varphi)\nabla \rho \times \nabla \varphi. \tag{6–3.14}$$

However, by (6–2.15),

$$\begin{aligned} \nabla \times (\varphi \nabla \theta) &= \varphi \nabla \times \nabla \theta + \nabla \varphi \times \nabla \theta \\ &= \nabla \varphi \times \nabla \theta, \end{aligned} \tag{6–3.15}$$

since $\nabla \times \nabla \theta = \mathbf{0}$ (cf. (6–2.21)). Thus we can substitute for $\nabla \varphi \times \nabla \theta$ in the first term of (6–3.14), and, after making similar substitutions in the second and third terms, we can have

$$\begin{aligned} \mathbf{F} = (F_\rho h_\varphi h_\theta)\nabla \times (\varphi \nabla \theta) + (F_\varphi h_\theta h_\rho)\nabla \times (\theta \nabla \rho) \\ + (F_\theta h_\rho h_\varphi) \nabla \times (\rho \nabla \varphi). \end{aligned} \tag{6–3.16}$$

Now by taking the divergence operator on both sides of (6–3.16), we have

$$\nabla \cdot \mathbf{F} = \nabla \cdot ((F_\rho h_\varphi h_\theta) \nabla \times (\varphi \nabla \theta)) + \cdots$$

(cf. (6–2.12)). Using (6–2.14) on the first term on the right side gives

$$\nabla \cdot \mathbf{F} = [(F_\rho h_\varphi h_\theta)\nabla \cdot \nabla \times (\varphi \nabla \theta) + \nabla(F_\rho h_\varphi h_\theta) \cdot \nabla \times (\varphi \nabla \theta)] + \cdots.$$

However, $\nabla \cdot \nabla \times (\varphi \nabla \theta) = 0$ by (6–2.22) and thus

$$\nabla \cdot \mathbf{F} = \nabla (F_\rho h_\varphi h_\theta) \cdot \nabla \times (\varphi \nabla \theta) + \cdots,$$

or, by using (6–3.15) again, we get

$$\nabla \cdot \mathbf{F} = \nabla (F_\rho h_\varphi h_\theta) \cdot (\nabla \varphi \times \nabla \theta) + \cdots.$$

Then from (6–3.13) it follows that

$$\nabla \cdot \mathbf{F} = \nabla (F_\rho h_\varphi h_\theta) \cdot \left(\frac{1}{h_\varphi h_\theta} h_\rho \nabla \rho\right) + \cdots,$$

and finally, by using (6–3.7), we have

$$\nabla \cdot \mathbf{F} = \frac{1}{h_\varphi h_\theta} \nabla (F_\rho h_\varphi h_\theta) \cdot \mathbf{u}_\rho + \cdots.$$

From (6–3.8) we know that

$$\nabla\psi\cdot\mathbf{u}_\rho=\frac{1}{h_\rho}\frac{\partial\psi}{\partial\rho};$$

hence

$$\nabla\cdot\mathbf{F}=\frac{1}{h_\rho\,h_\varphi\,h_\theta}\left[\frac{\partial}{\partial\rho}(h_\varphi\,h_\theta\,F_\rho)+\frac{\partial}{\partial\varphi}(h_\theta\,h_\rho\,F_\varphi)+\frac{\partial}{\partial\theta}(h_\rho\,h_\varphi\,F_\theta)\right].\qquad(6\text{–}3.17)$$

The functions F_ρ, F_φ, and F_θ can be calculated by noting that

$$F_\rho=\mathbf{F}\cdot\mathbf{u}_\rho,\qquad F_\varphi=\mathbf{F}\cdot\mathbf{u}_\varphi,\quad\text{and}\quad F_\theta=\mathbf{F}\cdot\mathbf{u}_\theta.$$

For example,

$$F_\rho=\mathbf{F}\cdot\mathbf{u}_\rho=\frac{1}{h_\rho}(f_1,f_2,f_3)\cdot\left(\frac{\partial x}{\partial\rho},\frac{\partial y}{\partial\rho},\frac{\partial z}{\partial\rho}\right)\qquad(6\text{–}3.18)$$

where $\mathbf{f}=(f_1,f_2,f_3)$, and

$$f_i(x,y,z)=f_i(\rho\cos\theta\sin\varphi,\ \rho\sin\theta\sin\varphi,\ \rho\cos\varphi),$$

$i=1,2,3$.

The curl operator can be handled in a similar manner. Again writing (6–3.12), we have

$$\mathbf{F}=(h_\rho\,F_\rho)\nabla\rho+(h_\varphi\,F_\varphi)\nabla\varphi+(h_\theta\,F_\theta)\nabla\theta.$$

Using (6–2.13), we have

$$\nabla\times\mathbf{F}=\nabla\times(h_\rho\,F_\rho)\nabla\rho+\cdots.\qquad(6\text{–}3.19)$$

By (6–2.15),

$$\nabla\times(h_\rho\,F_\rho)\nabla\rho=(h_\rho\,F_\rho)\nabla\times\nabla\rho+\nabla(h_\rho\,F_\rho)\times\nabla\rho,$$

but $\nabla\times\nabla\rho=\mathbf{0}$ by (6–2.21), and hence (6–3.19) can be written

$$\nabla\times\mathbf{F}=\nabla(h_\rho\,F_\rho)\times\nabla\rho+\nabla(h_\varphi\,F_\varphi)\times\nabla\varphi+\nabla(h_\theta\,F_\theta)\times\nabla\theta.\qquad(6\text{–}3.20)$$

In order to get the coordinates of $\nabla\times\mathbf{F}$ relative to $\{\mathbf{u}_\rho,\mathbf{u}_\varphi,\mathbf{u}_\theta\}$, we shall use (6–3.8) to write

$$\nabla(h_\rho\,F_\rho)=\frac{1}{h_\rho}\frac{\partial}{\partial\rho}(h_\rho\,F_\rho)\mathbf{u}_\rho+\frac{1}{h_\varphi}\frac{\partial}{\partial\varphi}(h_\rho\,F_\rho)\mathbf{u}_\varphi+\frac{1}{h_\theta}\frac{\partial}{\partial\theta}(h_\rho\,F_\rho)\mathbf{u}_\theta,$$

or

$$\nabla(h_\rho\,F_\rho)=\frac{\partial}{\partial\rho}(h_\rho\,F_\rho)\nabla\rho+\frac{\partial}{\partial\varphi}(h_\rho\,F_\rho)\nabla\varphi+\frac{\partial}{\partial\theta}(h_\rho\,F_\rho)\nabla\theta$$

if we use (6–3.7). Thus the first term of the right side of (6–3.20) becomes

$$\frac{\partial}{\partial\varphi}(h_\rho\,F_\rho)\nabla\varphi\times\nabla\rho+\frac{\partial}{\partial\theta}(h_\rho\,F_\rho)\nabla\theta\times\nabla\rho,\qquad(6\text{–}3.21)$$

since $\nabla\rho \times \nabla\rho = \mathbf{0}$. However, using (6–3.13) to substitute for $\nabla\varphi \times \nabla\rho$ and $\nabla\theta \times \nabla\rho$ in (6–3.21) gives

$$\frac{1}{h_\theta h_\rho}\frac{\partial}{\partial\theta}(h_\rho F_\rho)(h_\varphi \nabla\varphi) - \frac{1}{h_\varphi h_\rho}\frac{\partial}{\partial\varphi}(h_\rho F_\rho)h_\theta \nabla\theta, \tag{6–3.22}$$

since $\nabla\varphi \times \nabla\rho = -\nabla\rho \times \nabla\varphi$. In a similar way the other terms of (6–3.20) can be revaluated, and we can write

$$\begin{aligned}\nabla \times \mathbf{F} = {} & \frac{1}{h_\theta h_\rho}\frac{\partial}{\partial\theta}(h_\rho F_\rho)(h_\varphi \nabla\varphi) - \frac{1}{h_\varphi h_\rho}\frac{\partial}{\partial\varphi}(h_\rho F_\rho)(h_\theta \nabla\theta) \\ & + \frac{1}{h_\rho h_\varphi}\frac{\partial}{\partial\rho}(h_\varphi F_\varphi)(h_\theta \nabla\theta) - \frac{1}{h_\theta h_\varphi}\frac{\partial}{\partial\theta}(h_\varphi F_\varphi)(h_\rho \nabla\rho) \\ & + \frac{1}{h_\varphi h_\theta}\frac{\partial}{\partial\varphi}(h_\theta F_\theta)(h_\rho \nabla\rho) - \frac{1}{h_\rho h_\theta}\frac{\partial}{\partial\rho}(h_\theta F_\theta)(h_\varphi \nabla\varphi).\end{aligned} \tag{6–3.23}$$

By regrouping and using (6–3.7), we have

$$\begin{aligned}\nabla \times \mathbf{F} = {} & \frac{1}{h_\varphi h_\theta}\left[\frac{\partial}{\partial\varphi}(h_\theta F_\theta) - \frac{\partial}{\partial\theta}(h_\varphi F_\varphi)\right]\mathbf{u}_\rho \\ & + \frac{1}{h_\theta h_\rho}\left[\frac{\partial}{\partial\theta}(h_\rho F_\rho) - \frac{\partial}{\partial\rho}(h_\theta F_\theta)\right]\mathbf{u}_\varphi \\ & + \frac{1}{h_\rho h_\varphi}\left[\frac{\partial}{\partial\rho}(h_\varphi F_\varphi) - \frac{\partial}{\partial\varphi}(h_\rho F_\rho)\right]\mathbf{u}_\theta .\end{aligned} \tag{6–3.24}$$

Again F_ρ, F_φ, and F_θ can be found as mentioned earlier in (6–3.18).

The following example illustrates the use of the divergence operator as defined by (6–3.17) and the curl operator as defined by (6–3.24) for (ρ, φ, θ)-space.

Example 6–3.2. Suppose that $\mathbf{f} = (f_1, f_2, f_3)$ is a vector function defined by

$$f_1(x, y, z) = x^2, \qquad f_2(x, y, z) = yz, \qquad f_3(x, y, z) = z - 2x.$$

Then $\mathbf{F}$ of (6–3.10) is given by

$$\mathbf{F}(\rho, \varphi, \theta) = \mathbf{f}(\rho\cos\theta\sin\varphi, \rho\sin\theta\sin\varphi, \rho\cos\varphi),$$

where

$$f_1(\rho\cos\theta\sin\varphi, \rho\sin\theta\sin\varphi, \rho\cos\varphi) = \rho^2\cos^2\theta\sin^2\varphi,$$

$$f_2(\rho\cos\theta\sin\varphi, \rho\sin\theta\sin\varphi, \rho\cos\varphi) = \rho^2\sin\theta\sin\varphi\cos\varphi,$$

$$f_3(\rho\cos\theta\sin\varphi, \rho\sin\theta\sin\varphi, \rho\cos\varphi) = \rho\cos\varphi - 2\rho\cos\theta\sin\varphi.$$

Thus by (6–3.18)

$$F_\rho = \rho^2 \cos^3\theta \sin^3\varphi + \rho^2 \sin^2\theta \sin^2\varphi \cos\varphi + \rho \cos^2\varphi - 2\rho \cos\theta \sin\varphi \cos\varphi,$$

$$F_\varphi = \rho^2 \cos^3\theta \sin^2\varphi \cos\varphi + \rho^2 \sin^2\theta \sin\varphi \cos^2\varphi - \rho \sin\varphi \cos\varphi + 2\rho \cos\theta \sin^2\varphi,$$

$$F_\theta = -\rho^2 \sin\theta \cos^2\theta \sin^2\varphi + \rho^2 \sin\theta \cos\theta \sin\varphi \cos\varphi.$$

If we consider the point $(\rho, \varphi, \theta) = (2, \pi/2, \pi/4)$, the image point in (x, y, z)-space is $(x, y, z) = (\sqrt{2}, \sqrt{2}, 0)$. Using (6–3.17) to calculate the divergence of $\mathbf{F}$ at $(\rho, \varphi, \theta) = (2, \pi/2, \pi/4)$, we get

$$\nabla \cdot \mathbf{F}\left(2, \frac{\pi}{2}, \frac{\pi}{4}\right)$$

$$= \frac{1}{\rho^2 \sin\varphi}\left[\frac{\partial}{\partial\rho}(\rho^2 \sin\varphi)(F_\rho) + \frac{\partial}{\partial\varphi}(\rho \sin\varphi)(F_\varphi) + \frac{\partial}{\partial\theta}(\rho)(F_\theta)\right]_{\rho=2,\varphi=\pi/2,\theta=\pi/4}$$

$$= 2\sqrt{2} + 1.$$

Now using (6–2.8) to evaluate the divergence of $\mathbf{f}$ at the image point $(x, y, z) = (\sqrt{2}, \sqrt{2}, 0)$ we have

$$\nabla \cdot \mathbf{f}(\sqrt{2}, \sqrt{2}, 0) = 2x + z + 1\Big|_{x=\sqrt{2},y=\sqrt{2},z=0}$$
$$= 2\sqrt{2} + 1,$$

which agrees with the value obtained previously.

Now let us make the analogous comparison for the curl operator. Using (6–3.24) to calculate the curl of $\mathbf{F}$ at $(\rho, \varphi, \theta) = (2, \pi/2, \pi/4)$, gives

$$\nabla \times \mathbf{F}\left(2, \frac{\pi}{2}, \frac{\pi}{4}\right) = \frac{1}{\rho^2 \sin\varphi}\left[\frac{\partial}{\partial\varphi}(\rho \sin\varphi F_\theta) - \frac{\partial}{\partial\theta}(\rho F_\varphi)\right]\mathbf{u}_\rho$$
$$+ \frac{1}{\rho \sin\varphi}\left[\frac{\partial}{\partial\theta}(F_\rho) - \frac{\partial}{\partial\rho}(\rho \sin\varphi F_\theta)\right]\mathbf{u}_\varphi$$
$$+ \frac{1}{\rho}\left[\frac{\partial}{\partial\rho}(\rho F_\varphi) - \frac{\partial}{\partial\varphi}(F_\rho)\right]\mathbf{u}_\theta\Bigg|_{\rho=2,\varphi=\pi/2,\theta=\pi/4}$$
$$= (\sqrt{2} - 1)\mathbf{u}_\rho + (\sqrt{2} + 1)\mathbf{u}_\theta. \qquad (6\text{–}3.25)$$

Using (6–3.3) and (6–3.4), $\mathbf{u}_\rho = (1/\sqrt{2}, 1/\sqrt{2}, 0)$ and $\mathbf{u}_\theta = (-1/\sqrt{2}, 1/\sqrt{2}, 0)$, and hence (6–3.25) becomes

$$\nabla \times \mathbf{F}\left(2, \frac{\pi}{2}, \frac{\pi}{4}\right) = (-\sqrt{2}, 2, 0).$$

Using (6–2.9) to evaluate the curl of **f** at the image point, $(x, y, z) = (\sqrt{2}, \sqrt{2}, 0)$, we have

$$\nabla \times \mathbf{f}(\sqrt{2}, \sqrt{2}, 0) = (0 - y, 0 + 2, 0 - 0)_{x=\sqrt{2}, y=\sqrt{2}, z=0}$$
$$= (-\sqrt{2}, 2, 0),$$

the same as the above.

Throughout this section we have used as our example the transformation of (6–3.1), the transformation giving the relation between the coordinates of (x, y, z)-space and (ρ, φ, θ)-space. This is a very good example since the discussion has not been as specialized as it may appear. *The essential property possessed by this transformation that facilitated our discussion is that* $\{\rho, \varphi, \theta\}$ *are orthogonal coordinates. In other words,* $\{\mathbf{u}_\rho, \mathbf{u}_\varphi, \mathbf{u}_\theta\}$ *forms an orthonormal basis for* (x, y, z)*-space and is a positive triple. Thus, given an arbitrary transformation from* (ρ, φ, θ)*-space to* (x, y, z)*-space such that* $\{\rho, \varphi, \theta\}$ *are orthogonal coordinates, and if the del, divergence, and curl operators in* (ρ, φ, θ)*-space are defined as in* (6–3.8), (6–3.17), *and* (6–3.24), *they will enjoy the invariance properties that were demonstrated in Examples* 6–3.1 *and* 6–3.2.

In the foregoing summary statement, it is, of course, necessary that the transformation have an inverse transformation and that both of these transformations need the existence of all of the partial derivatives encountered. For example, in the transformation (6–3.1), we must beware of the points along the z-axis, where $\rho \sin \varphi = 0$. Notice that each of these three operators is undefined at these points.

One important combination of the above operators that merits special attention in (ρ, φ, θ)-space is the Laplacian operator (cf. 6–3.23). This is now easily calculated by letting **F** of (6–3.17) be $\nabla\psi$ of (6–3.8), so that

$$F_u = \frac{1}{h_\rho}\frac{\partial \psi}{\partial \rho}, \qquad F_\varphi = \frac{1}{h_\varphi}\frac{\partial \psi}{\partial \varphi}, \qquad F_\theta = \frac{1}{h_\theta}\frac{\partial \psi}{\partial \theta}. \tag{6–3.26}$$

Then

$$\Delta\psi = \nabla \cdot \nabla\psi = \frac{1}{h_\rho h_\varphi h_\theta}\left[\frac{\partial}{\partial \rho}\left(\frac{h_\varphi h_\theta}{h_\rho}\frac{\partial \psi}{\partial \rho}\right) + \frac{\partial}{\partial \varphi}\left(\frac{h_\theta h_\rho}{h_\varphi}\frac{\partial \psi}{\partial \varphi}\right) + \frac{\partial}{\partial \theta}\left(\frac{h_\rho h_\varphi}{h_\theta}\frac{\partial \psi}{\partial \theta}\right)\right]. \tag{6–3.27}$$

The following example calculates the Laplacian operator in (ρ, φ, θ)-space where (ρ, φ, θ) are spherical coordinates.

Example 6–3.3. In the transformation given by (6–3.1), $h_\rho = 1$, $h_\varphi = \rho$, and $h_\theta = \rho \sin \varphi$. Thus

$$\Delta\psi = \frac{\partial^2 \psi}{\partial \rho^2} + \frac{2}{\rho}\frac{\partial \psi}{\partial \rho} + \frac{1}{\rho^2}\frac{\partial^2 \psi}{\partial \varphi^2} + \frac{\cot \varphi}{\rho^2}\frac{\partial \psi}{\partial \varphi} + \frac{1}{\rho^2 \sin^2 \varphi}\frac{\partial^2 \psi}{\partial \theta^2}. \tag{6–3.28}$$

Notice that if the transformation (6–3.1) is the identity transformation—that is, if

$$x = \rho, \qquad y = \varphi, \qquad z = \theta$$

then $h_\rho = h_\varphi = h_\theta = 1$, $\psi(x, y, z) = \phi(x, y, z)$, $F(x, y, z) = \mathbf{f}(x, y, z)$, and the gradient of ψ as given by (6–3.8), the divergence of $\mathbf{F}$ given by (6–3.17), and the curl of $\mathbf{F}$ given by (6–3.24) agree with the definitions of gradient of ϕ, divergence of $\mathbf{f}$, and curl of $\mathbf{f}$, respectively, as given in Section 6–2. So in this sense the definitions of this section are an extension of those in the previous section.

Problems

1. Consider the following transformation of (x, y, z)-space to (r, θ, z')-space, the space of cylindrical coordinates: $x = r \cos \theta$, $y = r \sin \theta$, $z = z'$.

   (a) Make a sketch like Figure 6–3.1 for this transformation.
   (b) Define analogously the vectors $\mathbf{u}_r$, $\mathbf{u}_\theta$, and $\mathbf{u}_{z'}$ and the quantities h_r, h_θ, and $h_{z'}$ at a point, P_0, where the Jacobian of the transformation is not zero. Find each of these quantities.
   (c) Show that at point P_0 $\{\mathbf{u}_r, \mathbf{u}_\theta, \mathbf{u}_{z'}\}$ is an orthonormal basis and that $(\mathbf{u}_r, \mathbf{u}_\theta, \mathbf{u}_{z'})$ is a positive triple.
   (d) Since by (c) $\{r, \theta, z'\}$ are orthogonal coordinates and since (6–3.8), (6–3.17), (6–3.24), and (6–3.27) hold for any orthogonal coordinates, we can use these equations to calculate $\nabla\psi$, $\nabla \cdot \mathbf{F}$, $\nabla \times \mathbf{F}$ and $\Delta\psi$, respectively, where ψ is a scalar function of r, θ, and z'. Calculate each of these functions.

2. Prove each of the identities of (6–3.6) that are not verified in the text.

3. For each of the following transformations of coordinates, follow analogously the instructions of Problem 1.

   (a) $x = a \cosh u \cos v, \qquad y = a \sinh u \sin v, \qquad z = z'.$
   (b) $x = a \cosh \alpha \cos \theta \sin \varphi, \qquad y = a \cosh \alpha \sin \theta \sin \varphi,$
   $z = a \sinh \alpha \cos \varphi.$
   (c) $x = a \sinh \alpha \cos \theta \sin \varphi, \qquad y = a \sinh \alpha \sin \theta \sin \varphi,$
   $z = a \cosh \alpha \cos \varphi.$
   (d) $x = a^2 \xi \rho^{-2}, \qquad y = a^2 \eta \rho^{-2}, \qquad z = a^2 \zeta \rho^{-2},$ where
   $\rho^2 = \xi^2 + \eta^2 + \zeta^2.$

   The transformations of (a), (b), and (c) are, respectively, the transformations of elliptic cylindrical, oblate spheroidal, and prolate spheroidal coordinates. The transformation of (d) is known as inversion in a sphere of radius a.

4. Consider the transformation of polar coordinates, $x = r\cos\theta$, $y = r\sin\theta$.

(a) At a point (r_0, θ_0), $r_0 > 0$, draw a figure like Figure 6–3.1 showing the r-curve and the θ-curve.

(b) Define analogously the vectors $\mathbf{u}_r$ and $\mathbf{u}_\theta$ and the quantities h_r and h_θ at the point (r_0, θ_0) and show that $\{\mathbf{u}_r, \mathbf{u}_\theta\}$ is an orthonormal basis of E^2. Show that $h_r = 1$ and $h_\theta = r$.

(c) Follow the method of this section to show that if

$$\psi(r, \theta) = \phi(r\cos\theta, r\sin\theta),$$

then

$$\nabla\psi = h_r^{-1}\psi_r\,\mathbf{u}_r + h_\theta^{-1}\psi_\theta\,\mathbf{u}_\theta = \psi_r\,\mathbf{u}_r + \left(\frac{1}{r}\right)\psi_\theta\,\mathbf{u}_\theta$$

if we are to have $\nabla\psi(r_0, \theta_0) = \nabla\phi(x_0, y_0)$ where $x_0 = r_0\cos\theta_0$, $y_0 = r_0\sin\theta_0$.

(d) Assume that if $\mathbf{F}(r, \theta) = \mathbf{f}(r\cos\theta, r\sin\theta)$, then

$$\nabla\cdot\mathbf{F} = (h_r h_\theta)^{-1}\left(\frac{\partial}{\partial r}(h_\theta F_r) + \frac{\partial}{\partial\theta}(h_r F_\theta)\right)$$

if we are to have

$$\nabla\cdot\mathbf{F}(r_0, \theta_0) = \nabla\cdot\mathbf{f}(x_0, y_0).$$

Then show that

$$\nabla\cdot\mathbf{F} = r^{-1}\left(\frac{\partial}{\partial r}(rF_r) + \frac{\partial}{\partial\theta}(F_\theta)\right).$$

(e) Use the results of (c) to demonstrate that if $\phi(x, y) = y^2 - x$ and $r_0 = 2$, $\theta_0 = \pi/6$,

$$\nabla\psi(r_0, \theta_0) = \nabla\phi(x_0, y_0)$$

(cf. Example 6–3.1).

(f) Given $\mathbf{f}(x, y) = (xy, y^2)$ and (r_0, θ_0) as in (e), use the result of (d) to show that

$$\nabla\cdot\mathbf{F}(r_0, \theta_0) = \nabla\cdot\mathbf{f}(x_0, y_0)$$

(cf. Example 6–3.2).

(g) Use the results of (c) and (d) and the fact that $\Delta\psi = \nabla\cdot\nabla\psi$ to show that

$$\Delta\psi = (h_r h_\theta)^{-1}\left[\frac{\partial}{\partial r}(h_\theta h_r^{-1}\psi_r) + \frac{\partial}{\partial\theta}(h_r h_\theta^{-1}\psi_\theta)\right]$$

$$= \psi_{rr} + r^{-1}\psi_r + r^{-2}\psi_{\theta\theta}.$$

5. Use the results of Problem 4(b) and (c) to show that

$$(\phi_x)^2 + (\phi_y)^2 = (\psi_r)^2 + r^{-2}\psi_{\theta\theta}\,.$$

Prove this identity directly by use of the chain rule.

6. From the results that $\{\rho, \varphi, \theta\}$ are orthogonal coordinates and (6–3.8), show that

$$(\phi_x)^2 + (\phi_y)^2 + (\phi_z)^2 = (\psi_\rho)^2 + \rho^{-2}(\psi_\varphi)^2 + (\rho \sin \varphi)^{-2}(\psi_\theta)^2.$$

Prove this identity directly by using the chain rule.

7. Consider each of the following transformations of coordinates:

(a) $x = \dfrac{u}{u^2 + v^2}, \qquad y = \dfrac{-v}{u^2 + v^2},$

(b) $x = u^2 + 2v, \qquad y = u - v,$

(c) $x = e^u \cos v, \qquad y = e^u \sin v,$

(d) $x = \dfrac{u}{1 + u + v}, \qquad y = \dfrac{v}{1 + u + v},$

(e) $x = u^2 - v^2, \qquad y = 2uv.$

At a point (u_0, v_0) where the Jacobian of the transformation is finite and not zero, draw a figure like Figure 6–3.1 showing the u-curve and v-curve. Given that $\{u, v\}$ are orthogonal coordinates, use the results of Problem 6–3.4(c), (d), and (g) to find $\nabla\psi$, $\nabla \cdot \mathbf{F}$, and $\nabla\psi$, respectively.

8. In the text the details of the development of (6–3.17) and (6–3.24) were not completely written out but were limited to the consideration of only the first term on the right sides of (6–3.14) and (6–3.20), respectively. Write out these developments completely to give the missing details.

9. Prove the following identities:

(a) $\mathbf{r}_\rho = J(\nabla\varphi \times \nabla\theta), \qquad \mathbf{r}_\varphi = J(\nabla\theta \times \nabla\rho), \qquad \mathbf{r}_0 = J(\nabla\rho \times \nabla\varphi),$
(b) $\nabla\rho = J^{-1}(\mathbf{r}_\varphi \times \mathbf{r}_\theta), \qquad \nabla\varphi = J^{-1}(\mathbf{r}_\theta \times \mathbf{r}_\rho), \qquad \nabla\theta = J^{-1}(\mathbf{r}_\rho \times \mathbf{r}_\varphi).$

10. Use (6–2.15) to prove the following identities:

$$\mathbf{r}_\rho = J\nabla(\varphi\nabla\theta), \qquad \mathbf{r}_\varphi = J\nabla(\theta\nabla\rho), \qquad \mathbf{r}_\theta = J\nabla(\rho\nabla\varphi).$$

Then use (6–2.22) to prove that

$$\nabla \cdot (J^{-1}\mathbf{r}_\rho) = 0, \qquad \nabla \cdot (J^{-1}\mathbf{r}_\varphi) = 0, \qquad \nabla \cdot (J^{-1}\mathbf{r}_\theta) = 0.$$

11. Use (6–3.7) and (6–2.21) to prove the identities:

$$\nabla \times (h_\rho^{-2}\mathbf{r}_\rho) = \mathbf{0}, \qquad \nabla \times (h_\varphi^{-2}\mathbf{r}_\varphi) = \mathbf{0}, \qquad \nabla \times (h_\theta^{-2}\mathbf{r}_\theta) = \mathbf{0}.$$

12. Derive (6–3.17) by writing (6–3.11) in the form

$$\mathbf{F} = (h_\varphi\, h_\theta\, F_\rho)(J^{-1}\mathbf{r}_\rho) + (h_\theta\, h_\rho\, F_\varphi)(J^{-1}\mathbf{r}_\varphi) + (h_\rho\, h_\varphi\, F_\theta)(J^{-1}\mathbf{r}_\theta),$$

and then operate with the divergence operator on both sides. Use (6–2.14) and the last set of identities of Problem 10.

13. Derive the identity that was assumed in Problem 4(d).

14. In the text, the gradient, divergence, and Laplacian operators are given for 3-dimensional transformations of coordinates (cf. (6–3.8), (6–3.17), and (6–3.24)). In Problem 4 these same operators were developed for 2-dimensional transformation of coordinates. In both cases the new coordinates are orthogonal. Show that each of these operators can be rewritten such that if the transformation of coordinates is written $x_j = f_j(y_1, \cdots, y_n)$, $1 \le j \le n$, and if $\{y_1, \cdots, y_n\}$ are orthogonal coordinates, then

(a) $\nabla\psi = \sum_{i=1}^{n} h_{y_i}{}^{-1} \dfrac{\partial\psi}{\partial y_i} \mathbf{u}_{y_i}$,

(b) $\nabla \cdot \mathbf{F} = J^{-1} \sum_{i=1}^{n} \dfrac{\partial}{\partial y_i} (J h_{y_i}{}^{-1} F_{y_i})$,

(c) $\Delta\psi = J^{-1} \sum_{i=1}^{n} \dfrac{\partial\psi}{\partial y_i}\left((J h_{y_i}{}^{-2} \dfrac{\partial\psi}{\partial y_i}\right)$

Answers

1. (b)

$$h_r = 1,\ h_\theta = r,\ h_{z'} = 1; \qquad \mathbf{u}_r = (\cos\theta, \sin\theta, 0),$$
$$\mathbf{u}_\theta = (-\sin\theta, \cos\theta, 0), \qquad \mathbf{u}_{z'} = (0, 0, 1).$$

(d)

$$\nabla\psi = \frac{\partial\psi}{\partial r}\mathbf{u}_r + \frac{1}{r}\frac{\partial\psi}{\partial\theta}\mathbf{u}_\theta + \frac{\partial\psi}{\partial z'}\mathbf{u}_{z'},$$

$$\nabla \cdot \mathbf{F} = \frac{1}{r}\left[\frac{\partial}{\partial r}(rF_r) + \frac{\partial}{\partial r}(F_\theta) + \frac{\partial}{\partial z'}(rF_{z'})\right],$$

$$\nabla \times \mathbf{F} = \frac{1}{r}\left[\frac{\partial}{\partial\theta}(F_{z'}) - r\frac{\partial}{\partial z'}(F_\theta)\right]\mathbf{u}_r + \left[\frac{\partial}{\partial z'}(F_r) - \frac{\partial}{\partial r}(F_{z'})\right]\mathbf{u}_\theta$$
$$+ \frac{1}{r}\left[\frac{\partial}{\partial r}(rF_\theta) - \frac{\partial}{\partial\theta}(F_r)\right]\mathbf{u}_r,$$

$$\Delta\psi = \frac{\partial^2\psi}{\partial r^2} + \frac{1}{r}\frac{\partial\psi}{\partial r} + \frac{1}{r^2}\frac{\partial^2\psi}{\partial\theta^2} + \frac{\partial^2\psi}{\partial z'^2}.$$

3. (a)
$$\begin{aligned} h_u &= h_v = a\alpha, \qquad h_{z'} = 1, \\ \mathbf{u}_u &= \alpha^{-1}(\sinh u \cos v, \cosh u \sin v, 0), \\ \mathbf{u}_v &= \alpha^{-1}(-\cosh u \sin v, \sinh u \cos v, 0), \\ \mathbf{u}_{z'} &= (0, 0, 1), \end{aligned}$$

$$\nabla\psi = (a\alpha)^{-1}\left[\frac{\partial\psi}{\partial u}\mathbf{u}_u + \frac{\partial\psi}{\partial v}\mathbf{u}_v\right] + \frac{\partial\psi}{\partial z'}\mathbf{u}_{z'},$$

$$\nabla\cdot\mathbf{F} = a^{-1}\alpha^{-2}\left[\frac{\partial}{\partial u}(\alpha F_u) + \frac{\partial}{\partial v}(\alpha F_v) + \frac{\partial}{\partial z'}(a\alpha^2 F_{z'})\right],$$

$$\begin{aligned} \nabla\times\mathbf{F} &= (a\alpha)^{-1}\left[\frac{\partial}{\partial v}(F_{z'}) - \frac{\partial}{\partial z'}(a\alpha F_v)\right]\mathbf{u}_v \\ &\quad + (a\alpha)^{-1}\left[\frac{\partial}{\partial z'}(a\alpha F_u) - \frac{\partial}{\partial u}(F_{z'})\right]\mathbf{u}_v \\ &\quad + (a\alpha)^{-2}\left[\frac{\partial}{\partial u}(a\alpha F_v) - \frac{\partial}{\partial v}(a\alpha F_u)\right]\mathbf{u}_{z'}, \end{aligned}$$

$$\Delta\psi = (a\alpha)^{-2}\left[\frac{\partial^2\psi}{\partial u^2} + \frac{\partial^2\psi}{\partial v^2} + a^2\frac{\partial}{\partial z'}\left(\alpha^2\frac{\partial\psi}{\partial z'}\right)\right],$$

where $\alpha = (\sinh^2 u + \sin^2 v)^{1/2}$.

(b)
$$\begin{aligned} h_\alpha &= h_\varphi = a\lambda, \qquad h_\theta = a\cosh\alpha\sin\varphi, \\ \mathbf{u}_\alpha &= \lambda^{-1}(\sinh\alpha\cos\theta\sin\varphi, \sinh\alpha\sin\theta\sin\varphi, \cosh\alpha\cos\varphi), \\ \mathbf{u}_\varphi &= \lambda^{-1}(\cosh\alpha\cos\theta\cos\varphi, \cosh\alpha\sin\theta\cos\varphi, -\sinh\alpha\sin\varphi), \\ \mathbf{u}_\theta &= (-\sin\theta, \cos\theta, 0), \end{aligned}$$

$$\nabla\psi = (a\lambda)^{-1}\left[\frac{\partial\psi}{\partial\alpha}\mathbf{u}_\alpha + \frac{\partial\psi}{\partial\varphi}\mathbf{u}_\varphi\right] + h_\theta^{-1}\frac{\partial\psi}{\partial\theta}\mathbf{u}_\theta,$$

$$\nabla\cdot\mathbf{F} = a^{-1}\lambda^{-2}h_\theta^{-1}\left[\frac{\partial}{\partial\alpha}(\lambda h_\theta F_\alpha) + \frac{\partial}{\partial\varphi}(\lambda h_\theta F_\varphi) + \frac{\partial}{\partial\theta}(a\lambda^2 F_\theta)\right],$$

$$\begin{aligned} \nabla\times\mathbf{F} &= (a\lambda h_\theta)^{-1}\left[\frac{\partial}{\partial\varphi}(h_\theta F_\theta) - \frac{\partial}{\partial\theta}(a\lambda F_\varphi)\right]\mathbf{u}_\alpha \\ &\quad + (a\lambda h_\theta)^{-1}\left[\frac{\partial}{\partial\theta}(a\lambda F_\alpha) - \frac{\partial}{\partial\alpha}(h_\theta F_\theta)\right]\mathbf{u}_\varphi \\ &\quad + (a\lambda)^{-2}\left[\frac{\partial}{\partial\alpha}(a\lambda F_\varphi) - \frac{\partial}{\partial\varphi}(a\lambda F_\alpha)\right]\mathbf{u}_\theta, \end{aligned}$$

$$\Delta\psi = (a\lambda)^{-2}[\psi_{\alpha\alpha} + (\tanh\alpha)\psi_\alpha + \psi_{\varphi\varphi} + (\cot\varphi)\psi_\varphi] + h_\theta^{-2}\psi_{\theta\theta},$$

where $\lambda = (\cosh^2\alpha - \sin^2\varphi)^{1/2}$.

(c)
$$
\begin{aligned}
h_\alpha &= h_\varphi = a\nu, \qquad h_\theta = a \sinh \alpha \sin \varphi,\\
\mathbf{u}_\alpha &= \nu^{-1}(\cosh \alpha \cos \theta \sin \varphi,\ \cosh \alpha \sin \theta \sin \varphi,\ \sinh \alpha \cos \varphi),\\
\mathbf{u}_\varphi &= \nu^{-1}(\sinh \alpha \cos \theta \cos \varphi,\ \sinh \alpha \sin \theta \cos \varphi,\ -\cosh \alpha \sin \varphi),\\
\mathbf{u}_\theta &= (-\sin \theta, \cos \theta, 0),
\end{aligned}
$$

$$\nabla \psi = (a\nu)^{-1} \left[\frac{\partial \psi}{\partial \alpha} \mathbf{u}_\alpha + \frac{\partial \psi}{\partial \varphi} \mathbf{u}_\varphi\right] + h_\theta^{-1} \frac{\partial \psi}{\partial \theta} \mathbf{u}_\theta,$$

$$\nabla \cdot \mathbf{F} = a^{-1}\nu^{-2} h_\theta^{-1} \left[\frac{\partial}{\partial \alpha} (\nu h_\theta F_\alpha) + \frac{\partial}{\partial \varphi} (\nu h_\theta F_\varphi) + \frac{\partial}{\partial \theta} (a\nu^2 F_\theta)\right],$$

$$
\begin{aligned}
\nabla \times \mathbf{F} = {} & (a\nu h_\theta)^{-1}\left[\frac{\partial}{\partial \varphi} (h_\theta F_\theta) - \frac{\partial}{\partial \theta} (a\nu F_\varphi)\right]\mathbf{u}_\alpha\\
& + (a\nu h_\theta)^{-1}\left[\frac{\partial}{\partial \theta} (a\nu F_\alpha) - \frac{\partial}{\partial \alpha} (h_\theta F_\theta)\right]\mathbf{u}_\varphi\\
& + (a\nu)^{-2}\left[\frac{\partial}{\partial \alpha} (a\nu F_\varphi) - \frac{\partial}{\partial \varphi} (a\nu F_\alpha)\right]\mathbf{u}_\theta,
\end{aligned}
$$

$$\Delta \psi = (a\nu)^{-2}[\psi_{\alpha\alpha} + (\coth \alpha)\psi_\alpha + \psi_{\varphi\varphi} + (\cot \varphi)\psi_\varphi] + h_\theta^{-2}\, \psi_{\theta\theta},$$

where $\nu = (\cosh^2 \alpha - \cos^2 \varphi)^{1/2}$.

(d)
$$
\begin{aligned}
h_\xi &= h_\eta = h_\zeta = a^2 \rho^{-2},\\
\mathbf{u}_\xi &= \rho^{-2}(\eta^2 + \zeta^2 - \xi^2,\ -2\xi\eta,\ -2\xi\zeta),\\
\mathbf{u}_\eta &= \rho^{-2}(-2\xi\eta,\ \xi^2 + \zeta^2 - \eta^2,\ -2\xi\eta),\\
\mathbf{u}_\zeta &= \rho^{-2}(-2\xi\zeta,\ -2\xi\eta,\ \xi^2 + \eta^2 - \zeta^2),\\
\nabla \psi &= \rho^2 a^{-2}(\psi_\xi \mathbf{u}_\xi + \psi_\eta \mathbf{u}_\eta + \psi_\zeta \mathbf{u}_\zeta),
\end{aligned}
$$

$$\nabla \cdot \mathbf{F} = \rho^6 a^{-2}\left[\frac{\partial}{\partial \xi} (\rho^{-4}F_\xi) + \frac{\partial}{\partial \eta} (\rho^{-4}F_\eta) + \frac{\partial}{\partial \zeta} (\rho^{-4}F_\zeta)\right],$$

$$
\begin{aligned}
\nabla \times \mathbf{F} = \rho^4 a^{-2}\Bigg\{ & \left[\frac{\partial}{\partial \eta} (\rho^{-2}F_\zeta) - \frac{\partial}{\partial \zeta} (\rho^{-2}F_\eta)\right]\mathbf{u}_\xi\\
& + \left[\frac{\partial}{\partial \zeta} (\rho^{-2}F_\xi) - \frac{\partial}{\partial \xi} (\rho^{-2}F_\zeta)\right]\mathbf{u}\\
& + \left[\frac{\partial}{\partial \xi} (\rho^{-2}F_\eta) + \frac{\partial}{\partial \eta} (\rho^{-2}F_\xi)\right]\mathbf{u}_\zeta\Bigg\},
\end{aligned}
$$

$$\Delta \psi = \rho^6 a^{-4}\left[\frac{\partial}{\partial \xi}\left(\rho^{-2}\frac{\partial \psi}{\partial \xi}\right) + \frac{\partial}{\partial \eta}\left(\rho^{-2}\frac{\partial \psi}{\partial \eta}\right) + \frac{\partial}{\partial \zeta}\left(\rho^{-2}\frac{\partial \psi}{\partial \zeta}\right)\right].$$

7. (a) $\nabla \psi = (u^2 + v^2)^2[\psi_u \mathbf{u}_u + \psi_v \mathbf{u}_v]$,

$$\nabla \cdot \mathbf{F} = (u^2 + v^2)^4 \left[\frac{\partial}{\partial u} ((u^2 + v^2)^{-2} F_u) + \frac{\partial}{\partial v} ((u^2 + v^2)^{-2} F_v)\right],$$

$$\Delta \psi = (u^2 + v^2)\, [\psi_{uu} + \psi_{vv}].$$

(b) $\{u, v\}$ are not orthogonal coordinates.
(c) $\nabla\psi = e^{-u}[\psi_u \mathbf{u}_u + \psi_v \mathbf{u}_v]$,

$$\nabla \cdot \mathbf{F} = e^{-2u}\left[\frac{\partial}{\partial u}(e^u F_u) + \frac{\partial}{\partial v}(e^u F_v)\right],$$

$$\Delta\psi = e^{-2u}[\psi_{uu} + \psi_{vv}].$$

6–4. Divergence Theorem

The divergence theorem gives sufficient conditions on a vector-valued function $\mathbf{f}$ and on a region Ω such that the following equality of integrals holds:

$$\int_\Omega \nabla \cdot \mathbf{f}(x, y, z)\, dV = \int_S \mathbf{f}(x, y, z) \cdot \mathbf{n}(x, y, z)\, dA, \qquad (6\text{–}4.1)$$

where S is the surface that is the boundary of Ω and where $\mathbf{n}(x, y, z)$ is the (unit) outward normal of the region Ω at the point (x, y, z) belonging to S (cf. 5–2). Notice that on the left side, the integral is a volume integral and on the right side, it is a surface integral. Since the divergence operator is present in the volume integral, the theorem is called the divergence theorem.

Before trying to prove this very basic theorem of mathematical analysis, we shall first look at an example in which the equality of (6–4.1) is true.

Example 6–4.1. Consider the region

$$\Omega = \{(x, y, z): 0 \le z \le 4 - x^2 - y^2, x \ge 0, y \ge 0\}$$

and the vector field, $\mathbf{f}(x, y, z) = (z, xy, xz)$ (cf. Figure 6–4.1). Then

$$\nabla \cdot \mathbf{f}(x, y, z) = 2x$$

and hence

$$\begin{aligned}\int_\Omega \nabla \cdot \mathbf{f}(x, y, z)\, dV &= \int_0^2 \int_0^{\sqrt{4-x^2}} \int_0^{4-x^2-y^2} 2x\, dz\, dy\, dx \\ &= 2\int_0^2 \int_0^{\sqrt{4-x^2}} x(4 - x^2 - y^2)\, dy\, dx \\ &= 2\int_0^2 \int_0^{\pi/2} (r\cos\theta)(4 - r^2) r\, d\theta\, dr \\ &= \frac{128}{15}.\end{aligned}$$

Then, by making use of Proposition 5–3.3, we can write

$$\int_S \mathbf{f}(x, y, z) \cdot \mathbf{n}(x, y, z)\, dA = \sum_{i=1}^n \int_{S_i} \mathbf{f}(x, y, z) \cdot \mathbf{n}(x, y, z)\, dA \quad (6\text{–}4.2)$$

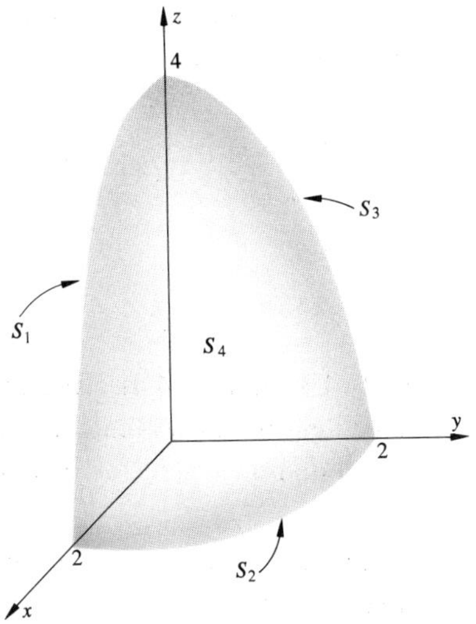

Figure 6–4.1

where S_1, S_2, S_3, and S_4 are the smooth surface elements shown in Figure 6–4.1. On S_1, $\mathbf{n}(x, y, z) = (0, -1, 0)$, and therefore,

$$\mathbf{f}(x, y, z) \cdot \mathbf{n}(x, y, z) = -\ xy = 0,$$

since $y = 0$ on S_1. Similarly, on S_2, $\mathbf{f} \cdot \mathbf{n} = 0$ since $z = 0$ on S_2. Thus the first two integrals of the summation of (6–4.2) are zero. On S_3,

$$\mathbf{f}(x, y, z) \cdot \mathbf{n}(x, y, z) = (z, xy, xz) \cdot (-1, 0, 0) = -z,$$

and hence

$$\int_{S_3} \mathbf{f}(x, y, z) \cdot \mathbf{n}(x, y, z)\, dA = \int_{S_3} (-z)\, dA$$
$$= -\int_0^2 \int_0^{4-y^2} z\, dz\, dy$$
$$= -\frac{128}{15}. \qquad (6\text{–}4.3)$$

Lastly, on S_4 we have

$$\mathbf{n}(x, y, z) = (1 + 4x^2 + 4y^2)^{-(1/2)}(2x, 2y, 1),$$

since $\mathbf{n} = -\ \nabla h/\|\nabla h\|$ where $h(x, y) = 4 - x^2 - y^2$. Thus on S_4,

$$\mathbf{f}(x, y, z) \cdot \mathbf{n}(x, y, z) = (2xz + 2xy^2 + xz)(1 + 4x^2 + 4y^2)^{-1/2},$$

and hence

$$\int_{S_4} \mathbf{f}(x, y, z) \cdot \mathbf{n}(x, y, z)\, dA = \int_{T_{xy}} (\mathbf{f} \cdot \mathbf{n})\sqrt{1 + h_x{}^2 + h_y{}^2}\, dA_{xy}$$

where $T_{xy} = S_2$ and where $(x, y, z) = (x, y, h(x, y))$ in the integrand on the right. Canceling the radical quantity gives

$$\int_{S_4} \mathbf{f}(x, y, z) \cdot \mathbf{n}(x, y, z)\, dA = \int_0^2 \int_0^{\sqrt{4-x^2}} (3x(4 - x^2 - y^2) + 2x^2y^2)\, dy\, dx$$

$$= \frac{256}{15}. \qquad (6\text{–}4.4)$$

Thus, adding the results of (6–4.3) and (6–4.4) gives us the surface integral of (6–4.2) equal to the volume integral that was calculated initially. Therefore, the equality of (6–4.1) has been demonstrated for this example.

The function of this example behaves very nicely on this region Ω. If Ω is a space region element and if $\mathbf{f} \in C(\Omega)$, $\mathbf{f} \in C^1(\Omega^0)$ and if each first partial of $\mathbf{f}$ on Ω^0 is extendable to a continuous function on Ω, then $\mathbf{f}$ is *continuously differentiable* on Ω. A first partial of $\mathbf{f}$, for instance $\mathbf{f}_x$, on Ω^0 is extendable to a continuous function on Ω, if it is possible to define the functions f_{1x}, f_{2x}, and f_{3x} on the bdy (Ω) such that each function becomes continuous on Ω. If Ω can be written $\Omega = \Omega_1 + \cdots + \Omega_n$ where Ω_i, $1 \le i \le n$, is a space region element and if $\mathbf{f}$ is continuous on Ω_i^0 for each i and if $\mathbf{f}$ on Ω^0 is extendable to a continuous function on Ω_i, $1 \le i \le n$, then $\mathbf{f}$ is *piecewise continuous* on Ω. If Ω can be written $\Omega = \Omega_1 + \cdots + \Omega_n$ and if $\mathbf{f}$ is continuously differentiable on each Ω_i, $1 \le i \le n$, then $\mathbf{f}$ is *piecewise differentiable* on Ω. Notice that these definitions are analogous to the definitions in Section 3–3 where f was a real-valued function defined on a plane regular region or on a finite sum of plane regular regions. Since in Example 6–4.1 the components of $\mathbf{f}$ were polynomials, then $\mathbf{f}$ and its first partials are continuous everywhere in (x, y, z)-space, and hence $\mathbf{f}$ is continuously differentiable in Ω.

The region, Ω, of Example 6–4.1 also has some very nice properties. It is an x-, y-, and z-axial region, and it is even a space region element, but notice that S_4 does not have an explicit representation in (y, z)-space since the first partial derivatives of φ defined by $\varphi(y, z) = \sqrt{4 - z - y^2}$ approach ∞ as (y, z) approaches a point (y, z) where $y^2 = 4 - z$. There are similar difficulties in finding an explicit representation in (x, z)-space, but there is no difficulty in (x, y)-space.

The following definition will give us a region for which we can establish (6–4.1). From this region we can "build" more complicated regions for which (6–4.1) holds.

Definition 6–4.1. *Let* Ω *be a space region element that is a z-axial region in* (x, y, z)*-space. Then*

$$\Omega = \{(x, y, z): (x, y) \in T, g(x, y) \leq z \leq h(x, y)\} \qquad (6\text{–}4.5)$$

where T is a sum of plane regular regions. Let S, the finite surface that is the boundary of Ω*, and T be such that* $S = (S_1' + \cdots + S_n') + (S_1'' + \cdots + S_n'') + (S_1 + \cdots + S_r)$ and $T = T_1 + \cdots + T_n$ where

$$\{(x, y, g(x, y)): (x, y) \in T_i\} \quad \textit{and} \quad \{(x, y, h(x, y)): (x, y) \in T_i\}$$

are explicit parametric representations of S_i' *and* S_i''*, respectively,* $1 \leq i \leq n$*. If* Ω *also has these same properties with respect to the* (x, z)*- and* (y, z)*-plane, then* Ω *is a regular space region element.*

Notice that a "solid tetrahedron" and a "solid cube," as well as all of the "regular solids" are regular space region elements. However, Ω of Example 6–4.1 is not a regular space region element since it does not have the required properties with respect to the (x, z)- and (y, z)-plane.

It is not sufficient to require only that Ω be a space region element that is an x-, y-, and z-axial region, even though g and h each provides explicit parametric representations on a decomposition of T. It may be that one single decomposition of T is not simultaneously satisfactory for both g and h (cf. Problem 8 of Section 2–7.). One should also notice that an explicit representation is not assured even though Ω is a space region element that is an x-, y-, and z-axial region and there is a common decomposition each time of the plane regular region T for both the f and the g functions. As an example, consider the following space region element:

$$\{(x, y, z): |x|^{(1/3)} + |y|^{(1/3)} \leq 1, 0 \leq z \leq 1\}.$$

The following proposition will give us a sufficient condition on the function **f** such that (6–4.1) holds for a regular space region element.

Proposition 6–4.1. *Let* Ω *be a regular space region element with S the boundary of* Ω*. If the function* **f** *is continuously differentiable on* Ω*, then*

$$\int_\Omega \nabla \cdot \mathbf{f}\, dV = \int_S \mathbf{f} \cdot \mathbf{n}\, dA, \qquad (6\text{–}4.6)$$

where **n** *is the outward normal of* Ω*.*

Proof. The desired equality will be demonstrated if we can show that

$$\int_\Omega \frac{\partial f_1}{\partial x}\, dV = \int_S f_1\, n_1\, dA, \tag{6–4.7(a)}$$

$$\int_\Omega \frac{\partial f_2}{\partial y}\, dV = \int_S f_2\, n_2\, dA, \tag{6–4.7(b)}$$

and

$$\int_\Omega \frac{\partial f_3}{\partial z}\, dV = \int_S f_3\, n_3\, dA, \tag{6–4.7(c)}$$

since by adding these three equalities we can get (6–4.6).

We shall demonstrate (6–4.7(c)) using the fact that Ω can be represented a certain way with respect to the (x, y)-plane. The other two equalities would follow by an analogous argument since Ω has exactly the corresponding properties with respect to the (x, z)- and (y, z)-plane

In order to establish (6–4.7(c)), it will be desirable to decompose Ω. That is, since Ω is a regular space region element, it is possible to write Ω as in (6–4.5), where $T = T_1 + \cdots + T_n$. Let

$$\Omega_i = \{(x, y, z)\colon (x, y) \in T_i, g(x, y,) \leq z \leq h(x, y)\}, \tag{6–4.8}$$

$1 \leq i \leq n$. Recall that $g(x, y) < h(x, y)$ if $(x, y) \in T^0$. It is easy to demonstrate that Ω_i, $1 \leq i \leq n$, is a space region element, and $\Omega = \Omega_1 + \cdots + \Omega_n$. Furthermore, g and f give explicit parametric representations, respectively, of $S_i' = \{(x, y, g(x, y))\colon (x, y) \in T_i\}$, the "bottom" of Ω_i, and $S_i'' = \{(x, y, h(x, y))\colon (x, y) \in T_i\}$, the "top" of Ω_i. Let $S_{i1}, S_{i2}, \cdots, S_{ir_i}$ be the surface elements that make up the "sides" of Ω_i (cf. Figure 6–4.2). In fact, S_i, the finite surface that is the boundary of Ω_i, can be written

$$S_i = S_i' + S_i'' + S_{i1} + \cdots + S_{ir_i}.$$

The outward normal $\mathbf{n}$ can be written as

$$\frac{1}{\sqrt{1 + h_x^2 + h_y^2}}(-h_x, -h_y, 1), \tag{6–4.9}$$

$$\frac{1}{\sqrt{1 + g_x^2 + g_y^2}}(g_x, g_y, -1) \tag{6–4.10}$$

and

$$(n_1, n_2, 0) \tag{6–4.11}$$

for the point belonging to the "top," "bottom," and the "sides," respectively, of Ω_i.

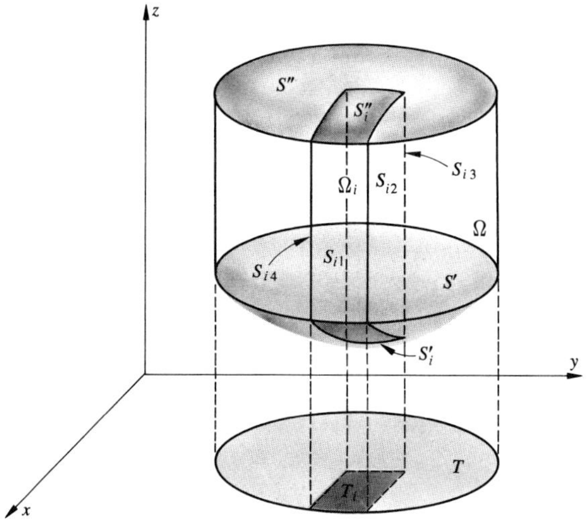

Figure 6–4.2

Since $\partial f_3/\partial z$ is continuous on Ω_i, we can write

$$\int_{\Omega_i} \frac{\partial f_3}{\partial z}\, dV = \int_{T_i}\left[\int_{g(x,y)}^{h(x,y)} \frac{\partial f_3}{\partial z}\, dz\right] dA_{xy}$$

$$= \int_{T_i} [f_3(x, y, h(x, y)) - f_3(x, y, g(x, y))]\, dA_{xy}. \quad (6\text{–}4.12)$$

Thus we have an expression for the left side of (6–4.7(c)). We shall now consider the right side of (6–4.7(c)) and show that it also is equal to the right side of (6–4.12).

By use of (5–3.7) of Proposition 5–3.2, we can write

$$\int_{S_i} f_3\, n_3\, dA = \int_{S_i'} f_3\, n_3\, dA + \int_{S_i''} f_3\, n_3\, dA + \textstyle\sum_{k=1}^{r_i} \displaystyle\int_{S_{ik}} f_3\, n_3\, dA. \quad (6\text{–}4.13)$$

However, since $n_3 = 0$ on S_{iz}, $1 \le k \le r_i$, the integral of $f_3\, n_3$ over S_i is just the sum of the integrals of $f_3\, n_3$ over the "top" and "bottom" of Ω_i. Furthermore, by (5–3.14) of Definition 5–3.2, we can write

$$\int_{S_i'} f_3\, n_3\, dA = \int_{T_i} f_3(x, y, h(x, y))\, dA_{xy}, \quad (6\text{–}4.14)$$

since $n_3 = 1/\sqrt{1 + h_x^2 + h_y^2}$ and the norm quantity in the integrand of (5–3.14) in this case is $\sqrt{1 + h_x^2 + h_y^2}$. That is, the product of these two quantities is one. In an exactly analogous fashion, we can write

$$\int_{S''} f_3\, n_3\, dA = -\int_{T_i} f_3(x, y, g(x, y))\, dA_{xy}. \quad (6\text{–}4.15)$$

Therefore,

$$\int_{S_i} f_3 n_3 \, dA = \int_{T_i} [f_3(x, y, h(x, y)) - f_3(x, y, g(x, y))] \, dA_{xy}, \quad (6\text{–}4.16)$$

and we have established (6–4.7(c)) for Ω_i and hence we have (6–4.6) for Ω_i; that is,

$$\int_{\Omega_i} \nabla \cdot \mathbf{f} \, dV = \int_{S_i} \mathbf{f} \cdot \mathbf{n} \, dA, \quad (6\text{–}4.17)$$

$1 \leq i \leq n$. By adding these n equations we get (6–4.5) for Ω, since

$$\int_{\Omega} \nabla \cdot \mathbf{f} \, dV = \sum_{i=1}^{n} \int_{\Omega_i} \nabla \cdot \mathbf{f} \, dV \quad (6\text{–}4.18)$$

by (5–4.6) of Proposition 5–4.2 and since

$$\int_{S} \mathbf{f} \cdot \mathbf{n} \, dA = \sum_{i=1}^{n} \int_{S_i} \mathbf{f} \quad \mathbf{n} \, dA \quad (6\text{–}4.19)$$

by (5–3.7) of Proposition 5–3.2. In the case of the surface integrals, the use of (5–3.7) is not exactly straightforward. If T_i and T_j have a smooth curve in their intersection, then the "sides" Ω_i and Ω_j will abut. However, recall that the integrals over the "sides" of Ω_i are always zero. ▌

It is, of course, natural to try to extend the results of the above proposition to more general circumstances. First let us make an extension that will allow us to include the region Ω of Example 6–4.1. Clearly, for such a region with **f** continuously differentiable on Ω, we could establish (6–4.7(c)) without difficulty, since we have the hypothesis of Proposition 6–4.1 satisfied with respect to the (x, y)-plane. However, in order to establish (6–4.7(b)) we could do the following: Let

$$\Omega_n = \left\{(x, y, z) : 0 \leq z \leq \min\left(\left(2 - \frac{1}{n}\right)^2 - x^2, 4 - x^2 - y^2\right), \right.$$
$$\left. x \geq 0, y \geq 0\right\},$$

$n = 1, 2, 3, \cdots$, and let S_n be the boundary of Ω_n (cf. Figure 6–4.3). Furthermore, let

$$T_n = \left\{(x, z) : 0 \leq z \leq \left(2 - \frac{1}{n}\right)^2 - x^2, x \geq 0\right\}.$$

If $h(x, z) = \sqrt{4 - z - x^2}$ and $g(x, z) = 0$ for $(x, z) \in T_n$, then

$$\{(x, h(x, z), z) : (x, z) \in T_n\} \quad \text{and} \quad \{(x, g(x, z), z) : (x, z) \in T_n\}$$

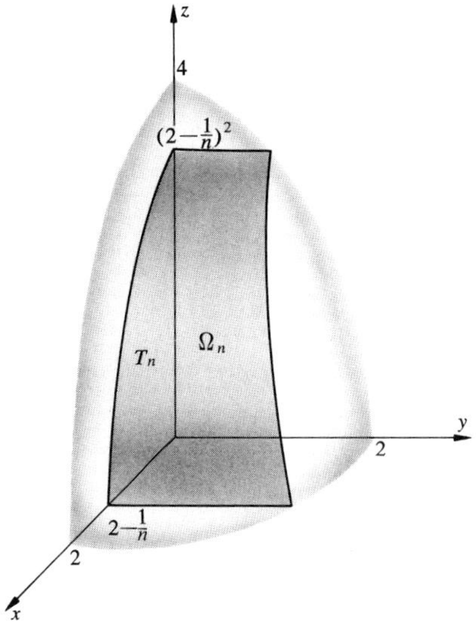

Figure 6–4.3

give an explicit representation for the "right side" (instead of "top") and "left side" (instead of "bottom"), respectively. Thus for Ω_n we can write

$$\int_{\Omega_n} \frac{\partial f_2}{\partial y}\, dV = \int_{S_n} f_2\, n_2\, dA, \tag{6–4.20}$$

or

$$\int_{\Omega_n} \frac{\partial f_2}{\partial y}\, dV = \int_{T_n} (f_2(x, h(x, z), z) - f_2(x, g(x, z), z))\, dA_{xy}. \tag{6–4.21}$$

It can be shown that as $n \to \infty$,

$$\int_{\Omega_n} \frac{\partial f_2}{\partial y}\, dV \to \int_{\Omega} \frac{\partial f_2}{\partial y}\, dV,$$

and the integral on the right of (6–4.21) approaches

$$\int_T (f_2(x, h(x, z), z) - f_2(x, g(x, z), z))\, dA_x = \int_S f_2\, n_2\, dA. \tag{6–4.22}$$

Therefore, we have established (6–4.7(b)) for Ω and, clearly, we could establish (6–4.7(a)) in a similar manner. Hence, (6–4.6) holds for Ω of Example 6–4.1 if **f** is continuously differentiable on Ω.

Clearly, the technique of demonstrating that (6–4.6) holds for Ω could be used any time Ω is a z-axial region in (x, y, z)-space, as in Definition 6–4.1, except for the requirement that the functions g and h allow us to give explicit representations of S_i' and S_i'', respectively. That is, it is possible that on one of the simple smooth curves Γ_{ik} of the boundary T_i the first partial derivatives of h or g may become infinite. This is what happens to the first partials of

$$h(x, z) = \sqrt{4 - z - x^2}$$

on the curve $\{(x, z): z = 4 - x^2, x \geq 0, z \geq 0\}$ in the case of Ω of Example 6–4.1. In fact, it is possible to use this technique even if the first partials become infinite at all points of the boundary of T_i. This possible extension of Proposition 6–4.1 motivates the following definition.

Definition 6–4.2. *A space region element that has all of the properties of a regular space region element except that on each T_i the functions f and g need not be such that their first partials are continuous at the boundary points of T_i, is a quasi-regular space region element* (*cf. Definition* 6–4.1).

This type of region will allow us to accommodate many of the regions frequently encountered, such as spheres, ellipsoids, frustrums, and so forth.

To understand another type of extension of Proposition 6–4.1, consider a finite space region Ω that is a sum of quasi-regular space region elements, Ω_i, $1 \leq i \leq n$, or a *quasi-regular finite space region.* If S is the boundary of Ω and S_i is the boundary of Ω_i, $1 \leq i \leq n$, then for $\mathbf{f}$ a continuously differentiable function on each Ω_i, we have

$$\int_{\Omega_i} \nabla \cdot \mathbf{f}\, dV = \int_{S_i} \mathbf{f} \cdot \mathbf{n}\, dA, \tag{6–4.23}$$

$1 \leq i \leq n$. Adding these n equations, we obtain (6–4.5) by (5–3.7) and (5–4.6) and by the fact that if Ω_i and Ω_j have in common a simple smooth surface element, then the integrals over this common surface element cancel one another. This is due to the fact that the outward normals to Ω_i and Ω_j have exactly opposite directions on this common surface element.

Next is a statement of the *divergence theorem* as we have now established it.

Theorem 6–4.1. *Let Ω be a quasi-regular finite space region such that $\Omega = \Omega_1 + \cdots + \Omega_n$. If $\mathbf{f}$ is a continuously differentiable function on each Ω_i, $1 \leq i \leq n$, then*

$$\int_{\Omega} \nabla \cdot \mathbf{f}\, dV = \int_{S} \mathbf{f} \cdot \mathbf{n}\, dA \tag{6–4.24}$$

where $\mathbf{n}$ is the outward normal of Ω.

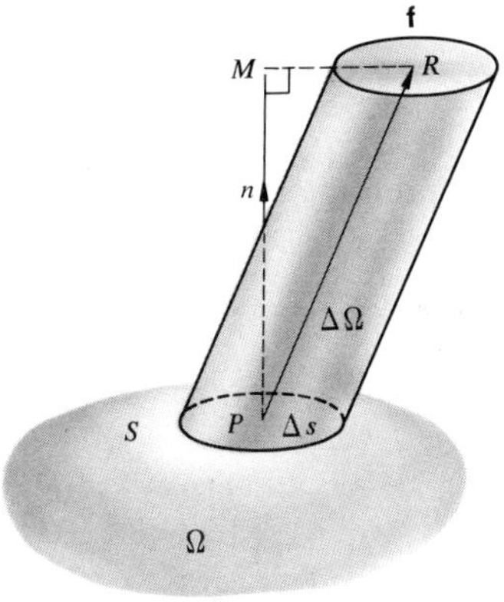

Figure 6–4.4

Further generalization could be made, but this will allow us to accommodate most of the usual circumstances.

One of the main applications of the divergence theorem can be seen in the setting of fluid flow. Let **f**, Ω, and S be as indicated in the divergence theorem, where **f** represents a velocity field. Let ΔS be a "small part" of S with area ΔA (cf. Figure 6–4.4). The fluid passing through ΔS in one unit of time would occupy a region as illustrated by $\Delta\Omega$. The volume of this region $\Delta\Omega$ would be approximately $\mathbf{f} \cdot \mathbf{n}\, \Delta A = \overline{OM}\, \Delta A$, or the altitude times the area of the base. On "summing up" these regions over S, we would have a Riemann sum that would approximate the surface integral of (6–4.24). Hence it is reasonable to call

$$\int_S \mathbf{f} \cdot \mathbf{n}\, dA$$

the net outward flow (flux) from Ω *per unit of time due to the field* **f**. For example, in Example 6–4.1 the net outward flow from Ω per unit of time due to the field **f** is 128/15 (units)^3/unit of time. The divergence theorem allows the calculation of this quantity by using a volume integral, which is frequently easier than evaluating several surface integrals over surface elements that go into making up the boundary of the region. However, notice in Example 6–4.1 the evaluation of the surface integral over S_i, $1 \le i \le 4$, gives some additional information. For example,

$$\int_{S_i} \mathbf{f} \cdot \mathbf{n}\, dA,$$

gives the net outward flow from Ω across S_i per unit of time due to the field **f**. Thus, from (6–4.3) we can conclude that the net flow across S_3 is inward at 128/15 (units)^3/unit of time.

Problems

1. Demonstrate the correctness of the equality of (6–4.1) by evaluating separately each of the integrals for each of the vector functions **f** and the regions Ω given below.

   (a) $\mathbf{f}(x, y, z) = (x^2, xy - z, xz^2)$,
   $\Omega = \{(x, y, z): x^2 + y^2 + z^2 \le 1, z \ge 0\}$,

   (b) $\mathbf{f}(x, y, z) = (x - y, xy^2, x^2z)$,
   $\Omega = \{(x, y, z): 0 \le z \le x^2 + y^2 \le 1\}$,

   (c) $\mathbf{f}(x, y, z) = (x^2 + y^2, xz, zy)$,
   $\Omega = \{(x, y, z): x \ge 0, x^2 + y^2 \le z \le 1\}$.

2. Given

$$\Omega = \{(x, y, z): \|(x, y, z)\| \le r\}, \qquad r > 0,$$

and

$$\mathbf{F}(x, y, z) = (2x, 3y, 4z),$$

show that

$$\int_\Omega \nabla \cdot \mathbf{F}\, dV = 12\pi r^3.$$

3. (a) Given

$$\Omega = \{(x, y, z): |x| \le 1, |y| \le 1, 0 \le z \le 1\}$$

and

$$\mathbf{F}(x, y, z) = (x^2 - y^2, 2xy + z, z^3 - 3xy),$$

find

$$\int_S \mathbf{F} \cdot \mathbf{n}\, dA$$

where **n** is the outward normal and S is the boundary of Ω.

   (b) Find the same quantity as called for in (a) given

$$\Omega = \{(x, y, z): x^2 + z^2 \le 1, 0 \le y \le 1\}.$$

4. Suppose Ω is a quasi-regular finite space region with boundary S. Then let $\mathbf{f}(x, y, z) = (0, 0, z)$ and show by use of the divergence theorem that

$$V(\Omega) = \int_S z\, n_3(x, y, z)\, dA.$$

Similarly, show that

$$V(\Omega) = \int_S y\, n_2(x, y, z)\, dA = \int_S x\, n_1(x, y, z)\, dA$$

where $\mathbf{n} = (n_1, n_2, n_3)$ is the outward normal of Ω, and consequently,

$$V(\Omega) = \frac{1}{3}\int_S (x, y, z) \cdot \mathbf{n}\, dA.$$

Show how these equations for the volume of Ω are consistent with the equation of (5–4.1).

5. Let Ω be a quasi-regular finite space region with boundary S. Are the following equations identities? Explain your answer.

   (a) $I_0(\Omega) = (\tfrac{1}{3}) \int_S (x^3, y^3, z^3) \cdot \mathbf{n}\, dA$, where $I_0(\Omega)$ is the moment of inertia of Ω about the origin.

   (b) $\int_S \mathbf{curl}\, \mathbf{F}(x, y, z) \cdot \mathbf{n}(x, y, z)\, dA = 0$, where the first partials of F_i, $1 = 1, 2, 3$, are continuously differentiable on Ω.

6. Let Ω be a quasi-regular finite space region with boundary S such that $\mathbf{0} \notin \Omega$. Let $\mathbf{F}(x, y, z) = (x, y, z)$.

   (a) Show that

$$\int_S \frac{\mathbf{F} \cdot \mathbf{n}}{\|\mathbf{F}\|^\alpha}\, dA = (3 - \alpha) \int_\Omega \frac{1}{\|\mathbf{F}\|^\alpha}\, dV$$

   (cf. Problem 16(b), page 310).

   (b) Let $\Omega_\varepsilon = \{(x, y, z) \colon \varepsilon^2 \le x^2 + y^2 + z^2 \le r^2\}$ where $0 < \varepsilon < r$, and let $S_\varepsilon \cup S_\varepsilon$ be the boundary of Ω_ε, where

$$S_\varepsilon = \{(x, y, z) \colon x^2 + y^2 + z^2 = \varepsilon^2\},$$

   and $S_r = \{(x, y, z) \colon x^2 + y^2 + z^2 = r^2\}$. Show that

$$\int_{S_r \cup S_\varepsilon} \frac{\mathbf{F} \cdot \mathbf{n}}{\|\mathbf{F}\|}\, dA \to 8\pi r$$

   as $\varepsilon \to 0$.

   (c) Does (6–4.24) hold for $\Omega = \{(x, y, z) \colon x^2 + y^2 + z^2 \le r^2\}$ if

$$\mathbf{f}(x, y, z) = \frac{1}{\|\mathbf{F}(x, y, z)\|}\, \mathbf{F}(x, y, z)?$$

7. Let $\Omega_r = \{(x, y, z) \colon \|(x, y, z) - (\xi, \eta, \zeta)\|\}$ and let $\mathbf{f}$ be the force field created due to a particle of mass m at (ξ, η, ζ).

   (a) Show that the net outward flow from Ω_r due to $\mathbf{f}$ is $-4\pi m$, and thus the net outward flow is independent of r, the radius of Ω_r.

   (b) Show that the result is still the same if Ω is a quasi-regular finite space region such that (ξ, η, ζ) belongs to the interior of Ω. Assume that if r is taken small enough such that Ω_r is contained in the interior of Ω, then $\Omega - \Omega_r^{\,0}$ is a quasi-regular finite space region.

   (c) Show that the result is zero if $(\xi, \eta, \zeta) \notin \Omega$.

   (d) Prove *Gauss' theorem*: The net outward flow across Ω, a quasi-regular finite space region, is equal to -4π times the total mass of the region Ω where the function ρ, giving the mass density at (x, y, z), $\rho(x, y, z)$, is continuous on Ω.

8. Give an example of a quasi-regular space region element Ω with boundary surface $S = S_1 + \cdots + S_n$ and a field $\mathbf{f}$ such that the net outward flow across one of the simple smooth surface elements S_i is positive but $\mathbf{f}(\mathbf{x}_0)$ is inward at some point $\mathbf{x}_0$ of S_i.

9. Let $\mathbf{f}$ be continuously differentiable on a closed region Ω_0 and let $(\xi, \eta, \zeta) \in \Omega_0{}^0$. Show that

$$\nabla \cdot \mathbf{f}(\xi, \eta, \zeta) = \lim_{\text{diam}(\Omega) \to 0} \frac{\int_S \mathbf{f} \cdot \mathbf{n}\, dA}{V(\Omega)}$$

where Ω is a quasi-regular finite space region that contains (ξ, η, ζ) in its interior and $V(\Omega)$ is the volume of Ω. That is, show that *the divergence of a field at a point is the limit of the net outward flow of a region Ω per unit of time per unit of volume as the diameter of the region approaches zero.*

10. The following two identities are known as Green's identities:

I. $$\int_S u \frac{\partial v}{\partial n}\, dA = \int_\Omega (u\, \nabla^2 v + \nabla u \cdot \nabla v)\, dV,$$

II. $$\int_S \left(u \frac{\partial v}{\partial n} - v \frac{\partial u}{\partial n}\right) dA = \int_\Omega (u \nabla^2 v - v \nabla^2 u)\, dV,$$

where $\partial/\partial n$ means the directional derivative on S in the direction of the outward normal. Give some sufficient conditions on the functions u and v in order to establish these two identities when Ω is a quasi-regular finite space region.

11. Give an example of a space region element Ω that is an x-, y-, and z-axial region but is not a regular space region element because the f and g functions of Definition 6–4.1 are such that for no decomposition of T into plane regular regions T_i do both f and g have the properties required in Definition 6–4.1, although each of the functions has these properties for different decompositions of T.

12. Give supplementary arguments to support the following remarks that were made in the proof of Proposition 6–4.1:

    (a) It is easy to demonstrate that Ω_i, $1 \le i \le n$, is a space region element and $\Omega = \Omega_1 + \cdots + \Omega_n$.

    (b) In fact, S_i, the finite surface that is the boundary of Ω_i, can be written

    $$S_i = S_i' + S_i'' + S_{i1} + \cdots + S_{ir_i}.$$

    (c) The sentence of which (6–4.12) is a part.

13. In the discussion of extending Proposition 6–4.1 to the region Ω of Example 6–4.1, it was asserted that it could be shown that the volume integral of (6–4.20) and the surface integral of (6–4.21) converge, respectively, to the volume integral of $\partial f_2/\partial y$ over Ω and to the surface

integral of $f_2(x, h(x, z), z) - f_2(x, g(x, z), z)$ over T. Supply the missing details of this demonstration. Notice that both of the integrands are continuous functions. Assume the fact that the range of a continuous function on a compact set is a bounded set (cf. Theorem 4–20, p. 73, in Apostol's *Mathematical Analysis*.). Notice that $A(T_n) \to A(T)$. Show that $V(\Omega_n) \to V(\Omega)$ by showing that $V(\Omega \backslash \Omega_n{}^0)$ is contained in a region of arbitrarily small volume for n sufficiently large.

Answers

1. (a) 0, (b) $\dfrac{2\pi}{3}$, (c) $\dfrac{8}{15}$.

3. (a) 4, (b) $\dfrac{3\pi}{4}$.

6. (c) No.

6–5. Green's Theorem and Stokes' Theorem

In this section we shall consider two more integral identities and sufficient conditions under which they hold. The first identity, which is a special case of the second identity, is simply the following:

$$\int_T \left(\frac{\partial Q}{\partial x} - \frac{\partial P}{\partial y}\right) dA_{xy} = \oint_\Gamma P\,dx + Q\,dy \tag{6–5.1}$$

where P and Q are functions defined on Ω, a region in the (x, y)-plane, and where Γ is a curve that is the boundary of T. This identity, which is the conclusion of Green's theorem, relates a surface integral to a line integral. Both the curve and the surface lie in the (x, y)-plane.

Let us first look at an example in which (6–5.1) holds.

Example 6–5.1. Let $P(x, y) = x^2$, $Q(x, y) = xy$, and $T = \{(x, y)\colon x^2 + y^2 \le 1\}$. Evaluating the left side of (6–5.1), we get

$$\int_T \left(\frac{\partial Q}{\partial x} - \frac{\partial P}{\partial y}\right) dA_{xy} = \int_{-1}^{1} \int_{-\sqrt{1-x^2}}^{\sqrt{1-x^2}} y\,dy\,dx = 0. \tag{6–5.2}$$

The right side of (6–5.1) would be

$$\oint_\Gamma P\,dx + Q\,dy = \int_\Gamma x^2\,dx + xy\,dy.$$

However, $\Gamma = \{(\cos t, \sin t): 0 \le t \le 2\pi\}$, and hence

$$\int_\Gamma P\,dx + Q\,dy = \int_0^{2\pi} (-\cos^2 t \sin t + \cos^2 t \sin t)\,dt = 0. \qquad (6\text{–}5.3)$$

Thus, in this case the equality of (6–5.1) is demonstrated.

The following example shows that (6–5.1) does not hold without some restrictions.

Example 6–5.2. Let

$$P(x, y) = -y(x^2 + y^2)^{-1}, \qquad Q(x, y) = x(x^2 + y^2)^{-1}$$

and let

$$T = \{(x, y): x^2 + y^2 \le 1\}.$$

The boundary of T, Γ, has the parametric representation $\{(\cos t, \sin t): 0 \le t \le 2\pi\}$. Thus the integral on the right of (6–5.1) becomes

$$\int_0^{2\pi} (\sin^2 t + \cos^2 t)\,dt = 2\pi. \qquad (6\text{–}5.4)$$

However, the integrand of the surface integral of (6–5.1) becomes

$$\frac{\partial}{\partial x}\left(\frac{x}{x^2 + y^2}\right) - \frac{\partial}{\partial y}\left(\frac{-y}{x^2 + y^2}\right) = g(x, y),$$

$g(x, y) = 0$ for $(x, y) \ne (0, 0)$ and $g(0, 0)$ is undefined. Using Definition 5–3.3 to evaluate the surface integral

$$\int^T g\,dA_{xy},$$

which is the left side of (6–5.1), gives zero, which is clearly not equal to the right side of (6–5.1). Thus, the identity of (6–5.1) fails in this case.

Proposition 6–5.1 gives us a sufficient criterion for (6–5.1) to hold.

Proposition 6–5.1. *Let T be a plane regular region of the (x, y)-plane that is both x- and y-axial. Let P and Q be continuously differentiable functions on T. Then*

$$\int_T \left(\frac{\partial Q}{\partial x} - \frac{\partial P}{\partial y}\right) dA_{xy} = \oint_\Gamma P\,dx + Q\,dy. \qquad (6\text{–}5.5)$$

Proof. Recall that Γ is a simple piecewise smooth closed curve. Since T is x-axial, there exist real-valued functions g and h that belong to $C([a, b])$ such that

$$T = \{(x, y): g(x) \le y \le h(x), a \le x \le b\}. \qquad (6\text{–}5.6)$$

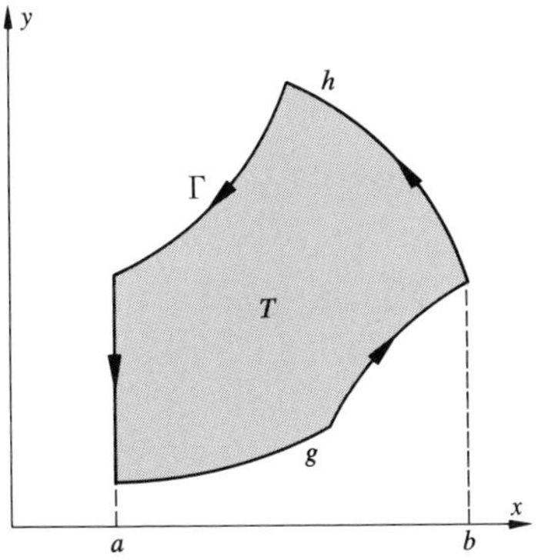

Figure 6–5.1

Then by (5–3.2)

$$-\int_T P_y(x, y)\, dA_{xy} = -\int_a^b \int_{g(x)}^{h(x)} P_y(x, y)\, dy\, dx$$

$$= -\int_a^b (P(x, h(x)) - P(x, g(x)))\, dx$$

$$= \oint_\Gamma P(x, y)\, dx \tag{6–5.7}$$

(cf. Figure 6–5.1).

Since T is also y-axial, it follows in an analogous manner that

$$\int_T Q_x(x, y)\, dA_{xy} = \oint_\Gamma Q(x, y)\, dy. \tag{6–5.8}$$

Adding the results of (6–5.7) and (6–5.8) gives the desired result. ∎

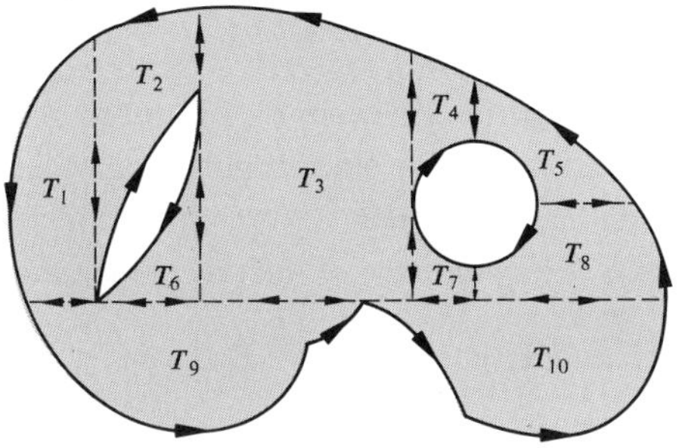

Figure 6–5.2

Figure 6–5.2 illustrates how Proposition 6–5.1 can be extended to a more general region. The region T shown is the sum of the regions T_1, $T_2, \cdots, T_{10}$ where each region T_i, $1 \leq i \leq 10$, is x- and y-axial. If P and Q

are both continuously differentiable on T, then Proposition 6–5.1 holds for each T_i and

$$\int_{T_i} \left(\frac{\partial Q}{\partial x} - \frac{\partial P}{\partial y} \right) dA_{xy} = \oint_{\Gamma_i} P\,dx + Q\,dy, \tag{6–5.9}$$

$1 \leq i \leq 10$, where Γ_i is the boundary curve of T_i. Adding each of these ten equations gives

$$\int_{T} \left(\frac{\partial Q}{\partial x} - \frac{\partial P}{\partial y} \right) dA_{xy} = \oint_{\Gamma} P\,dx + Q\,dy \tag{6–5.10}$$

where Γ is the boundary curve of T. The curve Γ includes the boundary curves of the two "holes." Notice that the line integrals along all of the dotted curves cancel out, since the required integration goes over each of these curves twice, once in each direction.

The foregoing explanation clearly illustrates the proof of the following theorem, *Green's theorem*, which gives us Proposition 6–5.1 extended to this more general type of plane region.

Theorem 6–5.1. *Let T be a finite plane region of the (x, y)-plane, so that $T = T_1 + \cdots + T_n$, where T_i, $1 \leq i \leq n$, is a plane regular region. Suppose, furthermore, that each region T_i, $1 \leq i \leq n$, is both x- and y-axial, and that the functions P and Q are piecewise continuously differentiable with respect to our decomposition of T. Then*

$$\int_{T} \left(\frac{\partial Q}{\partial x} - \frac{\partial P}{\partial y} \right) dA_{xy} = \oint_{\Gamma} P\,dx + Q\,dy \tag{6–5.11}$$

where Γ is the boundary curve of T.

It is possible to extend this theorem to still more general plane regions. In fact, Proposition 6–5.1 is true for plane regular regions, and hence *Theorem* 6–5.1 *holds for a finite plane region.*†

Green's theorem is a special case of the divergence theorem. For let

$$\Omega = \{(x, y, z)\colon (x, y) \in T, 0 \leq z \leq 1\}$$

where T is a plane regular region of the (x, y)-plane that is x- and y-axial, and T is written as in (6–5.6). Suppose, moreover that g and h are piecewise continuously differentiable on $[a, b]$. Let $\mathbf{f} = (Q, -P, 0)$ be a continuously differentiable function on T. The hypothesis of Theorem 6–4.1 is satisfied,

† Cf. T. M. Apostol, *Mathematical Analysis*, Theorem 10–43, p. 289.

and hence from (6–4.24) we have

$$\int_\Omega \left(\frac{\partial Q}{\partial x} - \frac{\partial P}{\partial y}\right) dV = \int_S (Q, -P, 0) \cdot \mathbf{n}\, dA. \tag{6–5.12}$$

Then by (5–4.2) the volume integral becomes

$$\int_0^1 \left(\int_T \left(\frac{\partial Q}{\partial x} - \frac{\partial P}{\partial y}\right) dA_{xy}\right) dz = \int_T \left(\frac{\partial Q}{\partial x} - \frac{\partial P}{\partial y}\right) dA_{xy}. \tag{6–5.13}$$

The finite surface S can be written as

$$S = S' + S'' + S_1 + \cdots + S_n$$

where S' is the "bottom" of Ω, S'' is the "top" of Ω, and $S_1, \cdots, S_n$ are the "sides" of Ω. Figure 6–5.3 illustrates these surface elements in the case that T

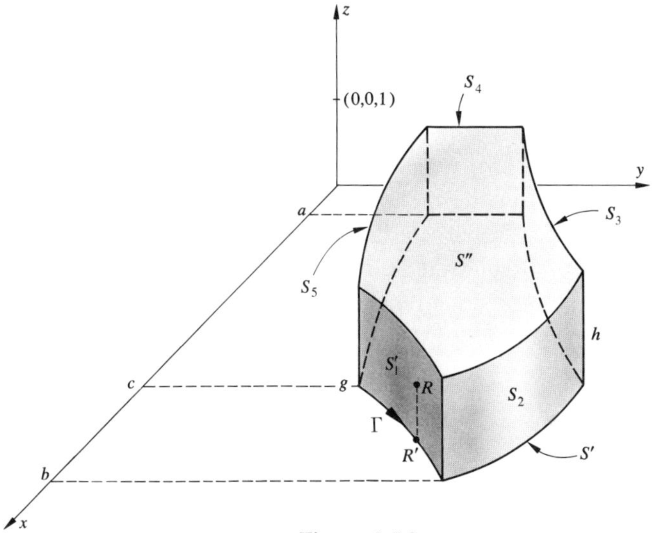

Figure 6–5.3

is as shown in Figure 6–5.1. The outward normal vector $\mathbf{n}$ will be (0, 0, 1) on S'' and $(0, 0, -1)$ on S'. Hence the integral of $(Q, -P, 0) \cdot \mathbf{n}$ over S' and S'' will be zero.

If T is as in (6–5.6), then at a point like R on S_1 in Figure 6–5.3

$$\mathbf{n} = (y'(s), -x'(s), 0) \tag{6–5.14}$$

where $R' = (x(s), y(s), 0)$, since the unit tangent to Γ is $(x'(s), y'(s))$ (cf. (1–5.6)). In this case, Γ is written with a parametric representation where the parameter s is arc length. Of course, Γ is understood to have the positive direction with respect to T (cf. Figure 6–5.3).

The surface element S_1 has a parametric representation,

$$S_1 = \{(x, g(x), z): c \le x \le b, 0 \le z \le 1\}$$

(cf. Figure 6–3.3). This representation cannot fail to be an explicit representation because of the restrictions that have been placed on g. Then by (5–3.14)

$$\begin{aligned}\int_{S_1} (Q, -P, 0) \cdot \mathbf{n}\, dA &= \int_0^1 \int_c^b (Qy' + Px')\sqrt{1 + (g')^2}\, dx\, dz \\ &= \int_c^b (Qy' + Px')\, ds \\ &= \int_\Gamma P\, dx + Q\, dy \end{aligned} \tag{6–5.15}$$

where $\Gamma_1 = \Gamma \cap S_1$ in the direction of Γ. Analogous results can be obtained for each S_i, $1 \le i \le n$. Hence, adding these n integrals gives

$$\int_S (Q, -P, 0) \cdot \mathbf{n}\, dA = \int_\Gamma P\, dx + Q\, dy. \tag{6–5.16}$$

From (6–5.12) and (6–5.13), we can obtain (6–5.11).

The requirements on the region T could be relaxed a great deal, but this presentation is sufficient to illustrate the relationship between two important theorems—Green's theorem and the divergence theorem. Problem 6–5.7 will show that Green's theorem is also a special case of Stokes' theorem, the other important theorem of the integral calculus.

Stokes' theorem, like the other two, also has as its conclusion an integral identity, which is

$$\int_S (R_y - Q_z, P_z - R_x, Q_x - P_y) \cdot \mathbf{n}\, dA = \int_\Gamma P\, dx + Q\, dy + R\, dz. \tag{6–5.17}$$

In vector notation (6–5.17) would be

$$\int_S (\nabla \times \mathbf{f}) \cdot \mathbf{n}\, dA = \int_S (\mathbf{curl\ f}) \cdot \mathbf{n}\, dA = \int_\Gamma \mathbf{f} \cdot \mathbf{t}\, ds \tag{6–5.18}$$

where $\mathbf{f} = (P, Q, R)$ (cf. Problem 11 of Section 5–1). The surface S has Γ as its "boundary curve," a closed curve. There is a direction that is understood on Γ, and the unit tangent $\mathbf{t}$ to Γ is in this direction. The unit normal vector, $\mathbf{n}$, is related to Γ in such a way that Γ has positive orientation with respect to $\mathbf{n}$ (cf. Figure 5–2.6).

The following example will illustrate (6–5.18), the integral identity of Stokes' theorem.

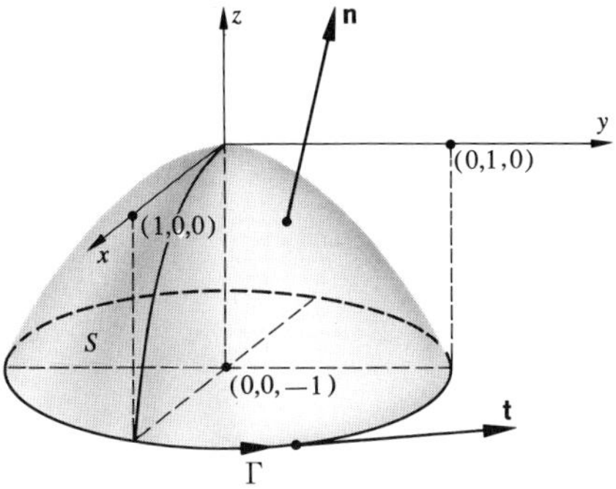

Figure 6–5.4

Example 6–5.3. Consider the surface S given by

$$S = \{(x, y, -(x^2 + y^2)) : x^2 + y^2 \leq 1\}$$

(cf. Figure 6–5.4). Then

$$\Gamma = \{(x, y, z) : x^2 + y^2 = 1, z = -1\}$$

is the "boundary curve," or in this case the edge of S. Writing Γ parametrically we have

$$\Gamma = \{(\cos \lambda, \sin \lambda, -1) : 0 \leq \lambda \leq 2\pi\}.$$

This parametric representation imparts a "direction" to Γ, the "direction" shown in the figure. On S,

$$F(x, y, z) = z + x^2 + y^2 = 0 \quad \text{and} \quad \nabla F(x, y, z) = (2x, 2y, 1)$$

is normal to S. Thus, the correct unit normal is $\pm(\nabla F/\|\nabla F\|)$. From the figure it is clear that if Γ is to have a positive orientation with respect to $\mathbf{n}$, the z component of $\mathbf{n}$ must be positive. Hence

$$\mathbf{n}(x, y, z) = \frac{(2x, 2y, 1)}{\sqrt{1 + 4x^2 + 4y^2}} .$$

Let $\mathbf{f}(x, y, z) = (x^2, xz, z - y)$ be the vector field.

Evaluating the line integral of (6–5.18) gives

$$\begin{aligned} \oint_\Gamma \mathbf{f} \cdot \mathbf{t}\, ds &= \int_0^{2\pi} (\cos^2 \lambda\, (-\sin \lambda) - \cos^2 \lambda)\, d\lambda \\ &= \left[\frac{1}{3} \cos^3 \lambda - \frac{\lambda}{2} - \frac{1}{4} \sin 2\lambda\right]_0^{2\pi} \\ &= -\pi, \end{aligned}$$

since $\mathbf{t} = (-\sin \lambda, \cos \lambda, 0)$ and $s = \lambda$.

For the integrand of the surface integral

$$\nabla \times \mathbf{f}(x, y, z) = (-(1 + x), 0, z)$$

and hence

$$(\nabla \times \mathbf{f}(x, y, z)) \cdot \mathbf{n}(x, y, z) = \frac{-2x(x + 1) + z}{\sqrt{1 + 4x^2 + 4y^2}}.$$

By (5–3.13) and (5–3.14),

$$\int_S (\nabla \times \mathbf{f}) \cdot \mathbf{n}\, dA =$$

$$-\int_{-1}^{1} \int_{-\sqrt{1-y^2}}^{\sqrt{1-y^2}} \frac{2x^2 + 2x + x^2 + y^2}{\sqrt{1 + 4x^2 + 4y^2}} \sqrt{1 + \varphi_x^{\,2}(x, y) + \varphi_y^{\,2}(x, y)}\, dx\, dy,$$

where $\varphi(x, y) = -x^2 - y^2$. Clearly, the radical quantities of the integrand cancel. Changing to polar coordinates,

$$\begin{aligned} \int_S (\nabla \times \mathbf{f}) \cdot \mathbf{n}\, dA &= -\int_0^{2\pi} \int_0^1 (2r^2 \cos^2 \theta + 2r \cos \theta + r^2) r\, dr\, d\theta \\ &= -\int_0^{2\pi} \left(\frac{1}{2} \cos^2 \theta + \frac{2}{3} \cos \theta + \frac{1}{4} \right) d\theta \\ &= -\pi. \end{aligned}$$

Therefore, the identity (6–5.18) has been verified for this special circumstance.

As in the case of the integral identity of Green's theorem, (6–5.11), the integral identity of Stokes' theorem, (6–5.18), does not hold for all circumstances (cf. Example 6–5.2). The following proposition gives us sufficient conditions for (6–5.18) to hold.

Proposition 6–5.2. *Let S be a simple smooth surface element in (x, y, z)-space such that it is possible to give an explicit parametric representation for the surface element as follows:*

$$S = \{(x, y, \psi(x, y)) : (x, y) \in T\}$$

where T is a plane regular region of the (x, y)-plane. Let **n** *be a unit normal defined at each point of S (cf. (5–2.7)). Let Γ have positive orientation with respect to* **n**. *If* **f** *is a function that is continuously differentiable in an open region that contains S, then*

$$\int_S (\nabla \times \mathbf{f}) \cdot \mathbf{n}\, dA = \int_\Gamma \mathbf{f} \cdot \mathbf{t}\, ds \tag{6–5.19}$$

where **t** *is the unit tangent to Γ.*

Proof. Let us try to establish (6–5.19) by demonstrating the equivalent identity, (6–5.17). First, we shall try to show the parts on the left of (6–5.17) that involve the function P equal the part of the right side of (6–5.17) that involves P—that is,

$$\int_S (P_z n_2 - P_y n_3)\,dA = \int_\Gamma P\,dx, \qquad (6\text{–}5.20)$$

where $\mathbf{n} = (n_1, n_2, n_3)$. Starting with the surface integral, we have

$$\int_S (P_z n_2 - P_3 n_3)\,dA = \int_S \left(P_z(x, y, \psi(x, y)) \frac{-\psi_y(x, y)}{\sqrt{1 + \psi_x^2(x, y) + \psi_y^2(x, y)}} - P_y(x, y, \psi(x, y)) \frac{1}{\sqrt{1 + \psi_x^2(x, y) + \psi_y^2(x, y)}} \right) dA$$

since $\mathbf{n} = (-\psi, -\psi_y, 1)/\sqrt{1 + \psi_x^2 + \psi_y^2}$ (cf. Figure 6–5.5). Then by (5–3.14), the integral on the right becomes

$$-\int_T (P_z(x, y, \psi(x, y))\psi_y(x, y) + P_y(x, y, \psi(x, y))\,dA_{xy},$$

which is equal to

$$\int_T -\frac{\partial}{\partial y} P(x, y, \psi(x, y))\bigg)_x \,dA_{xy}.$$

Notice that P_y means the partial derivative of P holding x and z constant. Similarly for P_z. By Green's theorem this becomes

$$\oint_C P(x, y, \psi(x, y))\,dx$$

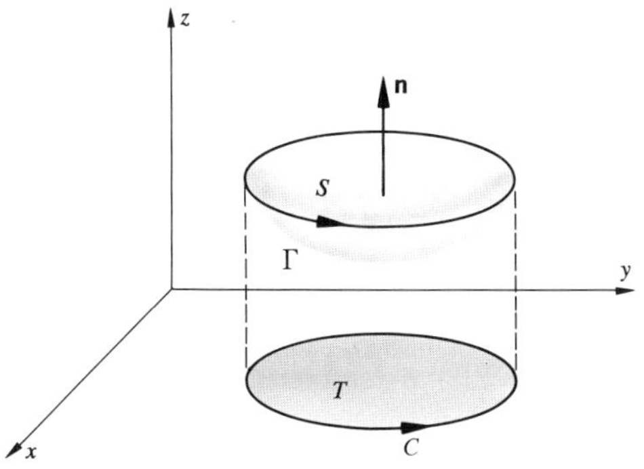

Figure 6–5.5

where C is the boundary curve of T (cf. Figure 6–5.5). However, this line integral is exactly $\int_\Gamma P(x, y, z)\,dx$. Hence we have established (6–5.20). In exactly an analogous manner we could show

$$\int_S (Q_x n_3 - Q_z n_1)\, dA = \int_\Gamma Q\, dy. \tag{6–5.21}$$

Lastly, we shall show that

$$\int_S (R_y n_1 - R_x n_2)\, dA = \int_\Gamma R\, dz, \tag{6–5.22}$$

and then this with the equations (6–5.20) and (6–5.21) easily allows us, by adding, to establish (6–5.19). As before, the surface integral becomes

$$\int_S \Bigg(R_y(x, y, \psi(x, y)) \frac{-\psi_x(x, y)}{\sqrt{1 + \psi_x^2(x, y) + \psi_y^2(x, y)}} + R_x(x, y, \psi(x, y)) \frac{\psi_y(x, y)}{\sqrt{1 + \psi_x^2(x, y) + \psi_y^2(x, y)}} \Bigg)\, dA,$$

which is equal to

$$\int_T [-R_y(x, y, \psi(x, y))\psi_x(x, y) + R_x(x, y, \psi(x, y))\psi_y(x, y)]\, dA_{xy}. \tag{6–5.23}$$

Turning now to the line integral, we have

$$\int_\Gamma R(x, y, z)\, dz = \oint_C R(x, y, \psi(x, y))\, dz,$$

which can be shown to be equal to

$$\oint_C R(x, y, \psi(x, y))\psi_x(x, y)\, dx + R(x, y, \psi(x, y))\psi_y(x, y)\, dy,$$

since $dz = d\psi = \psi_x\, dx + \psi_y\, dy$ (cf. Problem 12 of Section 5–1). By Green's theorem this line integral around the boundary of T becomes

$$\int_T \left[\frac{\partial}{\partial x}(R(x, y, \psi(x, y))\psi_y(x, y))\bigg)_y - \frac{\partial}{\partial y}(R(x, y, \psi(x, y))\psi_x(x, y))\bigg)_x \right] dA_{xy}. \tag{6–5.24}$$

It is easy to see that by expanding the integrand of (6–5.24) we have the integral of (6–5.23), and thus we have demonstrated (6–5.22). ▮

Notice that **n** could have been chosen in the opposite direction, but by so doing, the direction of integrating along Γ would need to be reversed. The result would be a change in sign on both sides of (6–5.19), and hence the equality would still hold. Since Green's theorem holds for finite plane regions, the requirement in the hypothesis that T be a plane regular region is not essential. In fact, the restriction that S have an explicit parametric representation need not be unduly restrictive. For if S is an orientable simple finite surface—that is, the sum of simple smooth surface elements $S_1, \cdots, S_n$ such that for each i, $1 \leq i \leq n$, S_i has an explicit representation—it is possible to establish (6–5.18) where Γ is the edge and where $\mathbf{f}$ is continuously differentiable in an open region containing S. This is true because Proposition 6–5.2 gives us

$$\int_{S_i} (\nabla \times \mathbf{f}) \cdot \mathbf{n}\, dA = \int_{\Gamma_i} \mathbf{f} \cdot \mathbf{t}\, ds, \quad 1 \leq i \leq n, \tag{6–5.25}$$

where Γ_i is the edge of S_i. By adding these n equalities, we obtain (6–5.18), since the line integrals along the surface elements that abut cancel out (cf. Definition 5–2.6). This is like the discussion that extends Proposition 6–5.1 to Theorem 6–5.1 (cf. Figure 6–5.2). *Stokes' theorem*, which follows, summarizes this result.

Theorem 6–5.2. *Let S be an orientable finite surface. Suppose that S can be decomposed into a sum of simple smooth surface elements S_i, $1 \leq i \leq n$, such that each S_i has an explicit representation. Let $\mathbf{n}$ be the unit normal to S such that on S_i, the edge of S_i, Γ_i, has positive orientation with respect to $\mathbf{n}$. If $\mathbf{f}$ is a continuously differentiable function on an open region containing S, then*

$$\int_{S} (\nabla \times \mathbf{f}) \cdot \mathbf{n}\, dA = \int_{\Gamma} (\mathbf{f} \cdot \mathbf{t})\, ds \tag{6–5.26}$$

where Γ is the edge of S and $\mathbf{t}$ is the unit tangent to Γ in the direction of Γ.

The line integral of Stokes' theorem has a particular interpretation in connection with problems in fluid flow. Let $\mathbf{f} = \mathbf{v}$ be the velocity vector field (a static field) in (x, y, z)-space due to a fluid that is flowing. Then $\mathbf{v} \cdot \mathbf{t}$ is the component of the velocity along Γ, and $\int_{\Gamma} \mathbf{v} \cdot \mathbf{t}\, ds$ is a limit of Riemann sums of the form

$$\textstyle\sum_{i=1}^{n} \mathbf{v} \cdot \mathbf{t}\big|_{P_i} \Delta_i s \tag{6–5.27}$$

where $P_i = \mathbf{r}(\bar{s}_i)$ and $\mathbf{r}$ is a parametric representation of Γ. The quantity

$$\Delta_i s = s_i - s_{i-1},\ \bar{s}_i \in [s_{i-1}, s_i],$$

$1 \leq i \leq n$, and $\{s_0, s_1, \cdots, s_n\}$ is a partition of the interval of the parameter (cf. Figure 6–5.6). Since $\mathbf{v} \cdot \mathbf{t}\big|_{P_i} \Delta_i s$ is intuitively a measure of the fluid flowing along Γ near P_i, then the sum of (6–5.27) would be an approximation of the net flow along Γ. Thus

$$C(\Gamma; \mathbf{v}) = \int_{\Gamma} \mathbf{v} \cdot \mathbf{t}\, ds \tag{6–5.28}$$

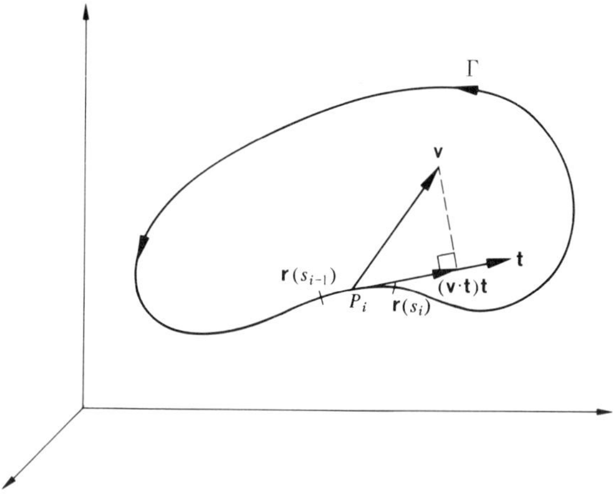

Figure 6–5.6

is called the *circulation along* Γ *due to the field* **f**. Should $C(\Gamma; f)$ be a positive number, we know that on Γ the overall picture is that the fluid is tending to flow in the direction of Γ although at some points of Γ the flow may be opposite to the direction of Γ. Thus in Example 6–5.3, if $\mathbf{v} = \mathbf{f}$, then the flow is tending to flow opposite to the direction of Γ since

$$\int_\Gamma \mathbf{f} \cdot \mathbf{t}\, ds = -\pi.$$

Usually, circulation is of interest only when Γ is a closed curve, but it is not intended that our definition bear this restriction.

In the setting of fluid flow, one can use Stokes' theorem to give a meaning to **curl v** similar to the meaning given to div **f** in Problem 9 of Section 6–4. Let P be a point in an open region Ω of (x, y, z)-space in which **v** is continuously differentiable. Let **n** be a unit vector in the direction of **curl v** and let S_1 be a surface element in Ω. In fact, let S be a disk of radius ε that is centered at P. Suppose furthermore, that **n** is a unit normal to S_1 at P. Let S_k, $k = 2, 3, 4, \cdots$, be closed disks concentric with S_1 of radius ε/k and let Γ_k be the edge of S_k. By Stokes' theorem,

$$C(\Gamma_k; \mathbf{v}) = \int_{S_k} (\mathbf{curl\, v} \cdot \mathbf{n})\, dA, \tag{6–5.29}$$

$k = 1, 2, 3, \cdots$. Applying the results of Problem 23 in Section 5–3 to the surface integral, gives

$$C(\Gamma_k; \mathbf{v}) = (\mathbf{curl\, v} \cdot \mathbf{n})|_{P_k} \cdot A(S_k)$$

where $P_k \in S_k$. Then $k \to \infty$, $P_k \to P$, and we have

$$\|\mathbf{curl\ v}\|\big|_P = \mathbf{curl\ v} \cdot \mathbf{n}\big|_P = \lim_{k\to\infty} \frac{C(\Gamma_k\,;\,\mathbf{v})}{A(S_k)},$$

the limit of the ratio of the circulation around closed circular curves to the areas of the disks determined by these curves as these curves shrink down on P in a plane through P perpendicular to **curl v**. Thus, if $\mathbf{curl\ v} \neq \mathbf{0}$, then $\|\mathbf{curl\ v}\|_P$ is a measure of the tendency of the fluid near P to rotate around the line L through P that is in the direction of **curl v**. For k sufficiently large, it follows from the continuity of **v** that $\mathbf{curl\ v} \cdot \mathbf{n} > 0$ on S_k and hence it follows from (6–5.29) that $C(\Gamma_k\,;\,\mathbf{v})$ is positive. Thus the direction of rotation about L is the direction that a right-hand screw in the direction of **curl v** would be rotated to drive the screw forward. This result helps explain the choice of the word *curl* for this differential operator.

Problems

1. Demonstrate that the identity (6–5.1) holds under each of the following circumstances.
   (a) $P(x, y) = xy$, $Q(x, y) = x + y$ and T is the triangle with vertices $(0, 0)$, $(2, 0)$, and $(2, 1)$.
   (b) $P(x, y) = x^2 + y^3$, $Q(x, y) = x + 2xy$, and $T = \{(x, y): 4x^2 + y^2 \leq 4, y \geq 0\}$.
   (c) $P(x, y) = (y + 1)^2$, $Q(x, y) = (x + 1)^{-1}$, and $T = \{(x, y): |x - \tfrac{1}{2}| \leq \tfrac{1}{2}, |y - \tfrac{1}{2}| \leq \tfrac{1}{2}\}$.
   (d) $P(x, y) = x^3y$, $Q(x, y) = x^2 y^2$, and $T = \{(x, y): x^2 + y^2 \leq 1\}$.

2. Change the following surface integrals to line integrals of equal value by use of Green's theorem and then evaluate both integrals.
   (a) $\int_T (x + y)\, dA_{xy}$ where $T = \{(x, y): |x| \leq 1, 0 \leq y \leq 1\}$.
   (b) $\int_T (xy + y^2)\, dA_{xy}$ where $T = \{(x, y): x^2 + y^2 \leq 4\}$.
   (c) $\int_T x^3y\, dA_{xy}$ where T is the triangle with vertices $(1, 0)$, $(2, 0)$, and $(0, 1)$.
   (d) $\int_T \left(\frac{y}{x^2 + y^2} - \frac{x}{x^2 + y^2}\right) dA_{xy}$ where $T = \{(x, y): 1 \leq x^2 + y^2 \leq 9\}$.

3. Use Green's theorem to change the following line integrals to a surface integral of equal value and then evaluate both integrals.

   (a) $\oint_\Gamma (y^2 + xy)\,dx + (x^3y)\,dy$ where $\Gamma = \{(2\cos t, \sin t): 0 \le t \le 2\pi\}$.

   (b) $\oint_\Gamma (3x^2 + y)\,dx + (x + 2y)\,dy$ where Γ is the boundary of $T = \{(x, y): |x| + |y| \le 1\}$.

   (c) $\oint_\Gamma \mathbf{f} \cdot \mathbf{t}\,ds$ where $\mathbf{f}(x, y) = (xy, x + 2y)$, $\mathbf{t}$ is the unit tangent vector taken in the positive direction, and Γ is as in (a).

   (d) $\oint_\Gamma \mathbf{f} \cdot \mathbf{n}\,ds$ where $\mathbf{f}$ is as in (c), $\mathbf{n}$ is the unit outward normal to T, and Γ is as in (b).

4. Demonstrate that the identity (6–5.18) holds under each of the following circumstances.

   (a) $\mathbf{f}(x, y, z) = (2x, 3y^2, xz)$ where $S = \{(x, y, x^2 + y^2): x^2 + y^2 \le 1\}$.
   (b) $\mathbf{f}(x, y, z) = (x + 3y, y^2z, 3xy)$ where $S = \{(x, y, z): x^2 + y^2 + z^2 = 1, x \ge 0, y \ge 0, z \ge 0\}$.
   (c) $\mathbf{f}(x, y, z) = (yz, 3xy, xz^2)$ where $S = \{(x, y, z): x = z, x^2 + y^2 \le 1\}$.
   (d) $\mathbf{f}$ as in (a) and $S = \{(x, y, 2 - x^2 - y^2): x^2 + y^2 \le 1\}$.

5. Use Stokes' theorem to change the following line integrals to a surface integral of equal value and evaluate both integrals.

   (a) $\int_\Gamma (yz^2 + 2x)\,dx + (xz^2 + 1)\,dy + (2xyz + x^2)\,dz$ where $\Gamma = \{(\cos t, \sin t, \sin t): 0 \le t \le 2\pi\}$.

   (b) $\int_\Gamma xz\,dx + yz\,dy + xy\,dz$ where Γ is the boundary of the triangle whose vertices are $(1, 0, 0)$, $(0, 1, 0)$, and $(0, 0, 1)$.

   (c) $\int_\Gamma \mathbf{f} \cdot \mathbf{t}\,ds$ where $\mathbf{f}(x, y, z) = (xr^{-3}, yr^{-3}, zr^{-3})$, $r = (x^2 + y^2 + z^2)^{1/2}$, and $\Gamma = \{(x, y, z): z = x^2 + y^2\} \cap \{(x, y, z): 4x^2 + 9y^2 = 36\}$.

   (d) $\int_\Gamma \mathbf{f} \cdot \mathbf{t}\,ds$ where $\mathbf{f}(x, y, z) = (x^2, y + z, x + y)$ and Γ is the same as in (c).

6. Demonstrate the correctness of (6–5.14).

7. Let the orientable finite surface S of Stokes' theorem lie in the (x, y)-plane and let the edge of S, Γ, be positively oriented with respect to the region S of the (x, y)-plane. Show that from this setting we can prove that Green's theorem is a special case of Stokes' theorem.

8. Under the assumptions that have been made in Proposition 6–5.2, give the details to show that (6–5.21) holds.

9. Make a sketch of the curve Γ of Example 6–5.3. Using the function **f** of this same example, sketch as in Figure 6–5.6 the vectors **f**, **t**, and $(\mathbf{f} \cdot \mathbf{t})\mathbf{t}$ at the points of Γ where $\lambda = n\pi/4$, $n = 0, 1, \cdots, 7$. Use a unit for plotting these vectors that is much smaller than the units in plotting Γ so that the sketch will be less cluttered. From the example the circulation along Γ due to **f**, $C(\Gamma; \mathbf{f}) = -\pi$. Calculate the Riemann sum, like (6–5.27), where $n = 8$ and where each of the points P_i is the midpoint of a subarc of Γ of length $\pi/4$. Compare the value of this Riemann sum with $C(\Gamma; \mathbf{f})$.

10. Prove the 2-dimensional analogies to Green's identities. That is, give some sufficient conditions on the region S in the (x, y)-plane and the functions u and v such that

$$\text{I.} \quad \oint_\Gamma u \frac{\partial v}{\partial n}\, ds = \int_S (u\nabla^2 v + \nabla u \cdot \nabla v)\, dA_{xy},$$

$$\text{II.} \quad \oint_\Gamma \left(u \frac{\partial v}{\partial n} - v \frac{\partial u}{\partial n}\right) ds = \int_S (u\nabla^2 v - v\nabla^2 u)\, dA_{xy}$$

where $\partial/\partial n$ means the directional derivative on S in the direction of the outward normal to S. In this case $\nabla = (\partial/\partial x, \partial/\partial y)$ (cf. Problem 10, page 338).

Answers

1.	(a) $\frac{1}{3}$.	(b) -4π.	(c) $-\frac{7}{2}$.	(d) 0.
2.	(a) 0.	(b) 4π.	(c) $\frac{1}{8}$.	(d) 0.
3.	(a) 2π.	(b) 0.	(c) 2π.	(d) 4π.
4.	(a) 0.	(b) $\pm(13\pi - 8)/16$.	(c) 0.	(d) 0.
5.	(a) 0.	(b) 0.	(c) 0.	(d) 0.

6–6. Independence of the Path; Applications

With the completion of the three important integral theorems, Green's theorem, Stokes' theorem and the divergence theorem, we have passed over a threshold into a room that is very rich in physical applications. The many physical circumstances that give rise to a vector field—for example, gravitational, electrostatic, or magnetic attraction and the flow of fluids—frequently require the use of these basic theorems in their analysis.

Our first application of Stokes' theorem is in finding conditions on a vector field **f** such that

$$\int_{\Gamma\ \alpha}^{\beta} \mathbf{f} \cdot \mathbf{dr}$$

is independent of the *path* (piecewise smooth curve) Γ that joins $\boldsymbol{\alpha}$ to $\boldsymbol{\beta}$. Example 5–1.2 demonstrates that this independence is not always the case. The following definition gives explicitly the intended meaning of a line integral being independent of the path.

Definition 6–6.1. *Let* **f** *be continuous in an open region* Ω *in* E^n. *If for any two points* $\boldsymbol{\alpha}$ *and* $\boldsymbol{\beta}$ *in* Ω *the line integral*

$$\int_{\Gamma\ \alpha}^{\beta} \mathbf{f} \cdot \mathbf{dr} \tag{6–6.1}$$

does not depend upon the path (piecewise smooth curve) Γ *that is contained in* Ω *and joins* $\boldsymbol{\alpha}$ *to* $\boldsymbol{\beta}$, *then this line integral is independent of the path in* Ω.

The word path is used sometimes for a more general curve, but in our discussion it is sufficiently general for a path to be a piecewise smooth curve. Notice that sufficient hypothesis is given to insure that the integral of (6–6.1) always exists (cf. Section 5–1).

The following proposition gives a necessary and sufficient condition for a line integral to be independent of its path in a given region.

Proposition 6–6.1. *Let* **f** *be continuous in an open region* Ω *in* E^n. *The line integral*

$$\int_{\Gamma\ \alpha}^{\beta} \mathbf{f} \cdot \mathbf{dr}$$

is independent of its path Γ *in* Ω *if and only if there exists a scalar function* $\psi \in C^1(\Omega)$ *such that* $\nabla\psi = \mathbf{f}$ *in* Ω.

Proof for $n = 2$. Assume first that the line integral is independent of its path in Ω. Let $\boldsymbol{\alpha}$ be a fixed point in Ω and let Γ be a path in Ω joining $\boldsymbol{\alpha}$ to $\boldsymbol{\beta} = (x, y)$ (cf. Figure 6–6.1). Then let ψ be defined by

$$\psi(x, y) = \int_{\Gamma\ \alpha}^{\beta} \mathbf{f} \cdot \mathbf{dr} = \int_{\Gamma\ \alpha}^{(x,y)} P\,dx + Q\,dy$$

where $\mathbf{f} = (P, Q)$. Let $\Gamma_1 = \{(x + t, y): 0 \le t \le \lambda\}$ where λ is chosen as a positive number that is small enough such that $\Gamma_1 \subset \Omega$. This is possible since Ω is an open set. Let $\boldsymbol{\gamma} = (x + \lambda, y)$ and let Γ_2 be the path that joins

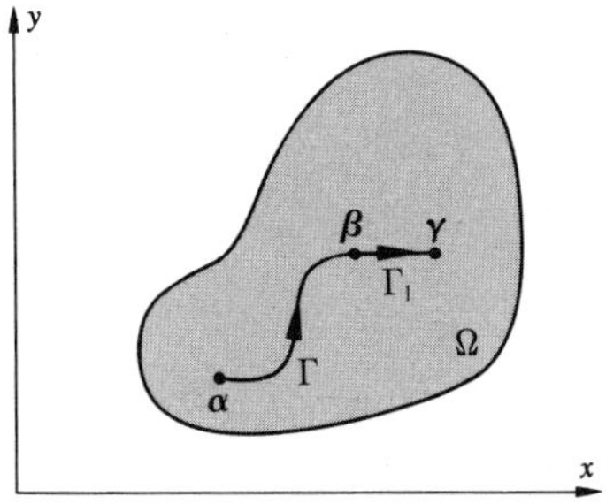

Figure 6–6.1

α to **γ** by first going along Γ to **β** and then along Γ_1 to **γ**. By (5–1.12),

$$\begin{aligned}\psi(x+\lambda, y) - \psi(x, y) &= \int_{\Gamma_2 \boldsymbol{\beta}}^{\boldsymbol{\gamma}} \mathbf{f} \cdot d\mathbf{r} - \int_{\Gamma \boldsymbol{\alpha}}^{\boldsymbol{\beta}} \mathbf{f} \cdot d\mathbf{r} \\ &= \int_{\Gamma_1 \boldsymbol{\beta}}^{\boldsymbol{\gamma}} \mathbf{f} \cdot d\mathbf{r} \\ &= \int_{\Gamma_1 \boldsymbol{\beta}}^{\boldsymbol{\gamma}} P\, dx + Q\, dy. \qquad (6\text{–}6.2)\end{aligned}$$

Using Proposition 5–1.1, the last integral of (6–6.2) becomes

$$\int_0^\lambda P(x+t, y)\, dt,$$

since $\mathbf{r}_2'(t) = 0$ in the parametric representation of Γ_1. Therefore, for $\lambda > 0$,

$$\frac{\psi(x+\lambda, y) - \psi(x, y)}{\lambda} = \frac{1}{\lambda} \int_0^\lambda P(x+t, y)\, dt.$$

Using the mean-value theorem for an integral of a single variable for $0 \leq t \leq \lambda$, we have

$$\begin{aligned}\frac{\psi(x+\lambda, y) - \psi(x, y)}{\lambda} &= \frac{1}{\lambda} P(x+\bar{t}, y) \int_0^\lambda dt \\ &= P(x+\bar{t}, y) \qquad (6\text{–}6.3)\end{aligned}$$

where $0 < \bar{t} < \lambda$. Taking the limit on both sides of (6–6.3) as $\lambda \to 0$ (which implies $\bar{t} \to 0$) gives

$$\psi_x(x, y) = P(x, y),$$

since P is continuous in Ω. The point (x, y) was arbitrarily chosen in Ω and hence $\psi_x = P$ in Ω. Similarly, $\psi_y = Q$ and hence $\nabla\psi = (P, Q) = \mathbf{f}$ in Ω.

Suppose now that there exists a function $\psi \in C^1(\Omega)$ such that $\nabla\psi = \mathbf{f}$ in Ω. Let **α** and **β** be two points in Ω and let Γ be any path that joins **α** to **β**. If

$$\Gamma = \{(x(t), y(t)): a \leq t \leq b\}$$

is a parametric representation of Γ, then by Proposition 5–1.1

$$\int_{\Gamma\,\boldsymbol{\alpha}}^{\boldsymbol{\beta}} \mathbf{f} \cdot d\mathbf{r} = \int_{\Gamma\,\boldsymbol{\alpha}}^{\boldsymbol{\beta}} \nabla\psi \cdot d\mathbf{r} = \int_a^b \left(\frac{\partial\psi}{\partial x}\frac{dx}{dt} + \frac{\partial\psi}{\partial y}\frac{dy}{dt}\right) dt.$$

However, by the chain rule, the integrand can be condensed to ψ', and we have

$$\begin{aligned}\int_{\Gamma\,\boldsymbol{\alpha}}^{\boldsymbol{\beta}} \mathbf{f} \cdot d\mathbf{r} &= \int_a^b \psi'\, dt \\ &= \psi(x(b), y(b)) - \psi(x(a), y(a)) \\ &= \psi(\boldsymbol{\beta}) - \psi(\boldsymbol{\alpha}).\end{aligned}$$

Therefore, the value of the line integral depends only on $\boldsymbol{\alpha}$ and $\boldsymbol{\beta}$. ∎

Even though the proof is restricted to two dimensions, the idea could easily be extended to E^n for $n > 2$. The function ψ is called a *potential function* relative to **f**.

Whereas this proposition does give us a necessary and sufficient condition that the value of a certain line integral is independent of the path, it falls far short of being a desirable type of condition. The problem is that, for a given line integral, we are not likely to know whether or not a function ψ of the desired type exists. The following proposition, which also gives a necessary and sufficient condition for a line integral to be independent of the path, has a similar shortcoming. However, these propositions do have merit in that, when a condition of the desired type is found, they tell us that certain additional things are then true.

Proposition 6–6.2. *Let* **f** *be continuous in an open region* Ω *in* E^n. *The line integral*

$$\int_{\Gamma\,\boldsymbol{\alpha}}^{\boldsymbol{\beta}} \mathbf{f} \cdot d\mathbf{r}$$

is independent of its path Γ *in* Ω *if and only if* $\int_C \mathbf{f} \cdot d\mathbf{r} = 0$ *for every simple closed path in* Ω.

Proof. Let C be an arbitrary simple closed path in Ω and let $\boldsymbol{\alpha}$ and $\boldsymbol{\beta}$ be two points of C (cf. Figure 6–6.2(a)). Let Γ_1 be the path given by C that joins $\boldsymbol{\alpha}$ to $\boldsymbol{\beta}$ and let Γ_2 be the path given by C that joins $\boldsymbol{\beta}$ to $\boldsymbol{\alpha}$. Then by (5–1.12),

$$\int_C \mathbf{f} \cdot d\mathbf{r} = \int_{\Gamma_1\,\boldsymbol{\alpha}}^{\boldsymbol{\beta}} \mathbf{f} \cdot d\mathbf{r} + \int_{\Gamma_2\,\boldsymbol{\beta}}^{\boldsymbol{\alpha}} \mathbf{f} \cdot d\mathbf{r}. \qquad (6\text{–}6.4)$$

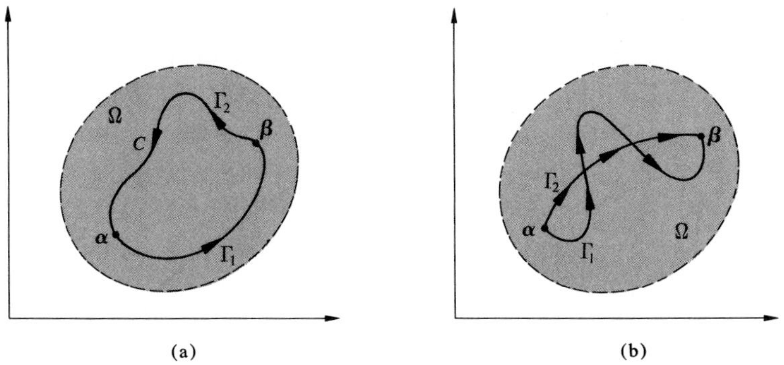

Figure 6–6.2

However, if we assume that the line integral is independent of the path between these points $\boldsymbol{\alpha}$ and $\boldsymbol{\beta}$, we get

$$_{\Gamma_1}\int_{\alpha}^{\beta} \mathbf{f} \cdot d\mathbf{r} = {}_{\Gamma_2}\int_{\alpha}^{\beta} \mathbf{f} \cdot d\mathbf{r}.$$

With this result and the fact that reversing the direction on a line integral changes the sign of the integral, we can substitute into (6–6.4) and get

$$\int_C \mathbf{f} \cdot d\mathbf{r} = 0$$

(cf. (5–1.3)).

For the other part of the theorem let Γ_1 and Γ_2 be any two paths in Ω that join $\boldsymbol{\alpha}$ to $\boldsymbol{\beta}$ (cf. Figure 6–6.2(b)). If we assume that

$$\int_C \mathbf{f} \cdot d\mathbf{r} = 0$$

for every *closed* path (not necessarily simple) in Ω, then it would follow that for C, the curve Γ_1 followed by Γ_2 with its direction reversed,

$$_{\Gamma_1}\int_{\alpha}^{\beta} \mathbf{f} \cdot d\mathbf{r} + {}_{\Gamma_2}\int_{\beta}^{\alpha} \mathbf{f} \cdot d\mathbf{r} = {}_{\Gamma_1}\int_{\alpha}^{\beta} \mathbf{f} \cdot d\mathbf{r} - {}_{\Gamma_2}\int_{\alpha}^{\beta} \mathbf{f} \cdot d\mathbf{r} = 0$$

or

$$_{\Gamma_1}\int_{\alpha}^{\beta} \mathbf{f} \cdot d\mathbf{r} = {}_{\Gamma_2}\int_{\alpha}^{\beta} \mathbf{f} \cdot d\mathbf{r}.$$

Thus

$$_{\Gamma}\int_{\alpha}^{\beta} \mathbf{f} \cdot d\mathbf{r}$$

is independent of its path in Ω.

However, the theorem requires that C be a simple closed path. If Γ_1 and Γ_2 are such that they intersect a finite number of times, as they do in Figure 6–6.2(b), it is possible to consider the "parts" of Γ_1 and Γ_2 between successive intersections. In such a case these "parts" of the two paths form a simple closed path. With the hypothesis of the proposition, we can conclude upon the basis of the earlier argument that the line integrals along these "parts" of Γ_1 and Γ_2 between successive intersection points are equal. It is then easy to establish the equality of the line integrals along Γ_1 and Γ_2 from $\boldsymbol{\alpha}$ to $\boldsymbol{\beta}$ upon the equality of the line integrals along the "parts" of Γ_1 and Γ_2 between successive intersections of Γ_1 and Γ_2.

If Γ_1 and Γ_2 have more than a finite number of points in common, then the proof can be completed by selecting paths C_1 and C_2 "close" to Γ_1 and Γ_2, respectively, that intersect only a finite number of times and show that the values of the line integrals for C_i and Γ_i, $i = 1, 2$, are "close" to one another (cf. Problem 18). ▮

This proposition again does not give us the type of condition that we want, since we are not any more likely to know that the value of the line integral around every simple closed path in Ω is zero than we are to know that the line integral is independent of the path in Ω. Before stating the next theorem, which does give a desirable necessary and sufficient condition for a line integral to be independent of its path, it is essential to define a simply connected region.

Definition 6–6.2. *An open region Ω in E^n is simply connected if every simple closed path in Ω is the edge of an orientable finite surface in Ω.*

Intuitively, a simply connected region in E^2 means that Ω has no "holes." In Figure 6–6.3, (a) illustrates a simply connected region and (b) illustrates an open region that is not simply connected. We can now state the basic theorem about a line integral being independent of its path.

Theorem 6–6.1. *Let $\mathbf{f} \in C^1(\Omega)$ where Ω is an open region in (x, y, z)-space (or (x, y)-space). If the line integral*

$$\int_{\Gamma\,\boldsymbol{\alpha}}^{\boldsymbol{\beta}} \mathbf{f} \cdot d\mathbf{r} \tag{6–6.5}$$

is independent of its path in Ω, then **curl f** $= \mathbf{0}$ *(or $(\partial f_2/\partial x) - (\partial f_1/\partial y) = 0$) in Ω. If Ω is a simply connected region and* **curl f** $= \mathbf{0}$ *(or $(\partial f_2/\partial x) - (\partial f_1/\partial y) = 0$) in Ω, then the value of the line integral of (6–6.5) is independent of its path in Ω.*

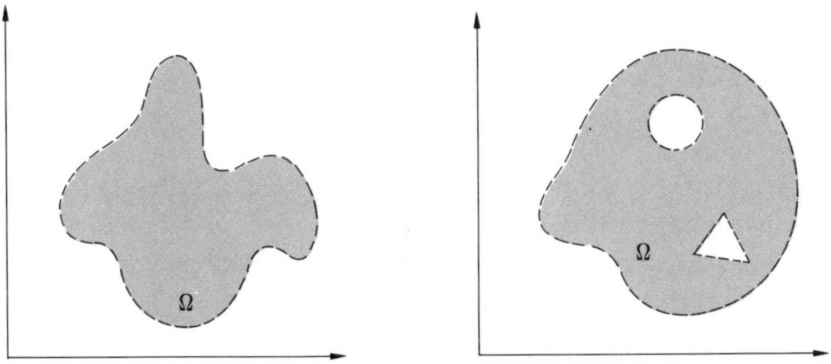

(a) (b)

Figure 6–6.3

Proof. In (x, y, z)-space, if the value of the line integral is independent of its path, then by Proposition 6–6.1 there exists a potential function ψ such that in Ω, $\mathbf{f} = \nabla\psi$. Then $\mathbf{curl\,f} = \nabla \times \nabla\psi = \mathbf{0}$ by (6–2.21).

In (x, y)-space let $\mathbf{g} = (f_1, f_2, 0)$. According to the hypothesis of the theorem, the integral

$$_C\int_\alpha^\beta \mathbf{g} \cdot d\mathbf{r}$$

is independent of its path in $\Omega' = \{(x, y, z) : (x, y) \in \Omega\}$. Consequently,

$$\mathbf{curl\,g} = \left(0, 0, \frac{\partial f_2}{\partial x} - \frac{\partial f_1}{\partial y}\right) = \mathbf{0}$$

in Ω', or

$$\frac{\partial f_2}{\partial x} - \frac{\partial f_1}{\partial y} = 0$$

in Ω.

In (x, y, z)-space let Γ be a simple closed path in Ω. Since Ω is simply connected, there exists an orientable finite surface S in Ω whose edge is Γ. Applying Stokes' theorem to S with edge Γ, we have

$$\int_S (\nabla \times \mathbf{f}) \cdot \mathbf{n}\, dA = \int_\Gamma (\mathbf{f} \cdot \mathbf{t})\, ds.$$

Since $\mathbf{curl\,f} = \mathbf{0}$ in Ω, it follows that

$$0 = \int_\Gamma \mathbf{f} \cdot \mathbf{t}\, ds = \int_\Gamma \mathbf{f} \cdot d\mathbf{r}.$$

Therefore, by Proposition 6–6.2 we can conclude that the value of the line integral of (6–6.5) is independent of its path in Ω. In (x, y)-space the argument is similar except that Green's theorem is used instead of Stokes' theorem. ∎

Notice that this theorem required that $\mathbf{f} \in C^1(\Omega)$ instead of just being continuous in Ω, as was required in Propositions 6–6.1 and 6–6.2. The requirement that Ω be simply connected is an essential part of the theorem. This is demonstrated by Example 6–5.2. For in this example, if

$$\Omega = \{(x, y): 0 < x^2 + y^2 < 2\},$$

then Ω is an open region that is not simply connected, and we have the simple closed path

$$\Gamma = \{(x, y): x^2 + y^2 = 1\}$$

such that

$$\int_\Gamma \mathbf{f} \cdot \mathbf{dr} = 2\pi \neq 0.$$

Yet

$$\frac{\partial f_2}{\partial x} - \frac{\partial f_1}{\partial y} = 0$$

in Ω.

The following example illustrates the use of Theorem 6–6.1.

Example 6–6.1. Consider the line integral

$$\Gamma\int_\alpha^\beta \mathbf{f} \cdot \mathbf{dr}$$

where $\mathbf{f}(x, y) = (y^2 + 2x, 2xy)$, $\boldsymbol{\alpha} = (0, 0)$, $\boldsymbol{\beta} = (2, 1)$, and

$$\Gamma = \left\{(x, y): \frac{1}{\pi}\sin^{-1}\frac{x}{2} + \frac{x}{4}\log\left(\frac{e(x+1)}{3}\right), 0 \leq x \leq 2\right\}.$$

Then since

$$\frac{\partial}{\partial x} f_2(x, y) - \frac{\partial}{\partial y} f_1(x, y) = (2y - 2y) = 0$$

in all of (x, y)-space, it follows that the given line integral is independent of the path. Hence let us choose the path $\Gamma_1 \cup \Gamma_2$ where

$$\Gamma_1 = \{(x, y): y = 0, 0 \leq x \leq 2\} \quad \text{and} \quad \Gamma_2 = \{(x, y): x = 2, 0 \leq y \leq 1\}.$$

Therefore,

$$\Gamma\int_\alpha^\beta \mathbf{f} \cdot \mathbf{dr} = \int_0^2 2x\,dx + \int_0^1 4y\,dy = 6.$$

In this way the integration is simpler.

Relative to the subject of line integrals being independent of their path is the problem of solving total differential equations (or Pfaffian differential equations). The following example illustrates such a problem.

Example 6–6.2. Consider the total differential equation,

$$P(x, y)\,dx + Q(x, y)\,dy = (1 + 2xy)\,dx + (x^2 + 2y)\,dy = 0. \qquad (6\text{–}6.6)$$

Try to find a function $\psi \in C^1(\Omega)$ where Ω is an open region of the (x, y)-plane where the functions P and Q are defined such that

$$d\psi(x, y; dx, dy) = P(x, y)\,dx + Q(x, y)\,dy \qquad (6\text{–}6.7)$$

for all $(x, y) \in \Omega$. Thus in Ω, $\psi_x(x, y) = P(x, y)$ and $\psi_y(x, y) = Q(x, y)$.

In order to try to find ψ, first find an indefinite integral of $P(x, y)$ with respect to x. Thus, let

$$\psi(x, y) = \int P(x, y)\,dx = x + x^2y + g_1(y)$$

where g_1 is an arbitrary differentiable function of a single variable. Then $\psi_y(x, y) = Q(x, y)$ implies

$$x^2 + g_1'(y) = x^2 + 2y.$$

Thus, $g_1'(y) = 2y$ or $g_1(y) = y^2 + c$ where c is a constant. Taking $c = 0$ gives

$$\psi(x, y) = x + x^2y + y^2, \qquad (6\text{–}6.8)$$

a solution of (6–6.6).

The expression of the left side of the total differential equation (6–6.6), $P(x, y)\,dx + Q(x, y)\,dy$, is a two-variable differential form (or Pfaffian form). Notice that the functions P and Q are defined on the open region, (x, y)-space. If $\mathbf{f} \in C(\Omega)$ where $\mathbf{f}\colon \Omega \to E^n$ and Ω is an open region of E^n, then $\mathbf{f} \cdot \mathbf{dx}$ is an n variable *differential form*, and $\mathbf{f} \cdot \mathbf{dx} = 0$ is an n variable *total differential equation*. If $\psi \in C^1(\Omega)$ and

$$d\psi(\mathbf{x}; \mathbf{dx}) = \mathbf{f}(\mathbf{x}) \cdot \mathbf{dx}, \qquad (6\text{–}6.9)$$

then ψ is a *solution* of the total differential equation, and the differential form is said to be *exact* in Ω.

From Proposition 6–6.1 it is clear that $\mathbf{f}(\mathbf{x}) \cdot \mathbf{dx}$ is an exact differential form in Ω if and only if

$$\int_\Gamma \mathbf{f} \cdot \mathbf{dx}$$

is independent of its path in Γ. Similarly, by Proposition 6–6.2, $\mathbf{f}(\mathbf{x}) \cdot \mathbf{dx}$ is an exact differential form in Ω if and only if

$$\int_{\Gamma} \mathbf{f} \cdot \mathbf{dx} = 0$$

around every simple closed path in Ω. Lastly, by Theorem 6–6.1 it is clear that a three-(two-) variable differential form is exact in a simply connected region if and only if

$$\mathbf{curl\ f} = \mathbf{0} \left(\frac{\partial f_2}{\partial x} - \frac{\partial f_1}{\partial y} = 0 \right)$$

in Ω.

Since some line integrals are not independent of their paths in a given open region, then some differential forms are not exact, and the associated total differential equation does not immediately have a solution. However, it is sometimes possible to multiply the given total differential equation by a certain function that will give an exact differential form. The following example illustrates this point.

Example 6–6.3. Consider the following total differential equation:

$$\mathbf{f}(x, y, z) \cdot (dx, dy, dz) = y(y - z)\, dx + z(z - x)\, dy + y(x - y)\, dz = 0. \tag{6–6.10}$$

Since $\mathbf{curl\ f}(x, y, z) = 2(x - y - z, -y, -y) \neq \mathbf{0}$, it follows by Theorem 6–6.1 that the differential form of (6–6.10) is not exact. However, if equation (6–6.10) is multiplied by $\mu(x, y, z) = (y - z)^{-2}$, the new differential form is exact, and the new total differential equation has a solution—namely,

$$\psi(x, y, z) = \frac{y(x - y)}{y - z}$$

for $\Omega = \{(x, y, z) : y > z\}$ (or $\{(x, y, z) : y < z\}$).

A function μ such that $(\mu\mathbf{f}) \cdot \mathbf{dx}$ is an exact differential form in Ω is an *integrating factor* of $\mathbf{f} \cdot \mathbf{dx} = 0$ in Ω, and any total differential equation for which an integrating factor exists in Ω is said to be *integrable* in Ω. A solution ψ of $(\mu\mathbf{f}) \cdot \mathbf{dx}$ is also said to be a solution of $\mathbf{f} \cdot \mathbf{dx}$. In Sneddon's *Elements of Partial Differential Equations* (pp. 18ff.), necessary and sufficient conditions are given for a three-variable total differential equation being integrable. Also several methods of finding solutions are given. However, should the total differential equation be exact a method similar to that of Example 6–6.2 can be used to find a solution.

If $\mathbf{f} \in C(\Omega)$ where Ω is an open region of E^n (n is usually 2 or 3), then the vector field given by $\mathbf{f}$ is *irrotational* if

$$\int_\Gamma \mathbf{f} \cdot \mathbf{dr} = 0$$

for every closed path in Ω. The word irrotational is motivated by the fact that in an irrotational velocity vector field the circulation along every closed path is zero (cf. (6–5.28)). The following theorem summarizes the conditions of an irrotational field.

Theorem 6–6.2. *If* $\mathbf{f} \in C^1(\Omega)$ *and* Ω *is a simply connected region in* E^3, *then the following statements are equivalent:*

(a) $\int_\Gamma \mathbf{f} \cdot \mathbf{dr}$ is independent of its path in Ω.

(b) *There exists a potential function* $\psi \in C^1(\Omega)$ *such that* $\nabla\psi = \mathbf{f}$ *in* Ω.

(c) $\int_\Gamma \mathbf{f} \cdot \mathbf{dr} = 0$ *for every simple closed path in* Ω.

(d) $\mathbf{curl}\, \mathbf{f} = \mathbf{0}$ *in* Ω.

(e) $\mathbf{f} \cdot \mathbf{dx}$ *is an exact differential form in* Ω.

(f) $\mathbf{f}$ *is irrotational in* Ω.

The second application of our integral theorems is to find conditions on a vector field $\mathbf{f} \in C(\Omega)$ where Ω is an open region in (x, y, z)-space such that

$$\int_S \mathbf{f} \cdot \mathbf{n}\, dA = 0 \qquad (6\text{–}6.11)$$

where S is any one of a certain class of closed finite surfaces in Ω. More precisely, S must be the boundary of a region $\Omega_1 \subset \Omega$ such that (6–4.6) holds. The vector $\mathbf{n}$ of (6–6.11) is the outward normal of Ω_1. Notice that if $\mathbf{f}$ is a velocity field, then the integral of (6–6.11) gives the net outward flow from Ω_1 per unit of time due to $\mathbf{f}$ (cf. Figure 6–4.4). A vector field $\mathbf{f} \in C(\Omega)$ such that equation (6–6.11) holds for each finite closed surface S in Ω of the aforementioned type is *solenoidal* in Ω. This word is motivated by the fact that the force field due to a solenoid has this property. Intuitively, we can sense a velocity field as being solenoidal in Ω if the streamlines do not "begin" or "end" in any region $\Omega_1 \subset \Omega$.

The following theorem, which is analogous to Theorem 6–6.2, summarizes the conditions relative to a solenoidal field.

Theorem 6–6.3. *If* $\mathbf{f} \in C^1(\Omega)$ *where* Ω *is an open region in* (x, y, z)*-space, then the following conditions are equivalent:*

(a) $\int_S \mathbf{f} \cdot \mathbf{n}\, dA = 0$ *for every closed finite surface* $S \subset \Omega$ *that is the boundary of a region* $\Omega_1 \subset \Omega$ *such that* (6–4.6) *holds.*

(b) *There exists* $\mathbf{F} \in C^2(\Omega)$ *such that* $\mathbf{curl}\, \mathbf{F} = \mathbf{f}$ *in* Ω.

(c) *div* $\mathbf{f} = 0$ *in* Ω.

(d) $\mathbf{f}$ *is solenoidal in* Ω.

Proof. Clearly, (d) and (a) are equivalent, and (c) implies (a). However, by use of the results of Problem 9 of Section 6–4, it is possible to show that (a) implies (c). Assuming (b), then (c) follows from (6–2.22). The proof is complete if it can be shown that (c) implies (b).

If there exist two scalar functions $\phi \in C^2(\Omega)$ and $\psi \in C^3(\Omega)$ such that $\mathbf{f} = \nabla\phi \times \nabla\psi$ in Ω, then we could take $\mathbf{F} = \phi\, \nabla\psi$. By (6–2.15) it would follow that $\nabla \times \mathbf{F} = \mathbf{f}$.

In order to show that $\mathbf{f}$ can be written as the cross product of two gradient vectors, consider the system

$$\frac{dx}{f_1} = \frac{dy}{f_2} = \frac{dz}{f_3}. \tag{6–6.12}$$

Using the technique of Example 6–1.2, we can obtain two functionally independent first integrals of this system, for example, $u_1(x, y, z) = k_1$ and $u_2(x, y, z) = k_2$. The intersection of any two of these surfaces gives a curve that is a solution curve of this system. Therefore, the direction of the curve of intersection, $\pm\nabla u_1 \times \nabla u_2$ is in the same direction as $\mathbf{f}$. In other words,

$$f = \alpha\, \nabla u_1 \times \nabla u_2 \tag{6–6.13}$$

where α is a scalar function.

If α is a constant we can write $\mathbf{f} = \nabla(\alpha u_1) \times \nabla u_2$, and we have $\mathbf{f}$ in the required form. On the other hand, by (6–2.14), (6–2.16), and (6–2.21)

$$\begin{aligned} 0 &= \nabla \cdot \mathbf{f} \\ &= \nabla \cdot (\alpha\, \nabla u_1 \times \nabla u_2) \\ &= \nabla \alpha \cdot (\nabla u_1 \times \nabla u_2). \end{aligned} \tag{6–6.14}$$

However, if we write (6–6.14) as

$$0 = \frac{\partial(\alpha, u_1, u_2)}{\partial(x, y, z)},$$

then it is clear that we have the condition that the functions α, u_1, and u_2 are functionally dependent (cf. Proposition 4–3.2). Thus there exists a function

g such that $\alpha = g(u_1, u_2)$. Let

$$G(u_1, u_2) = \int g(u_1, u_2)\, du_1. \qquad (6\text{–}6.15)$$

By the chain rule

$$\nabla G = \frac{\partial G}{\partial u_1} \nabla u_1 + \frac{\partial G}{\partial u_2} \nabla u_2$$

or by (6–6.15) we have

$$\nabla G = g \nabla u_1 + \frac{\partial G}{\partial u_2} \nabla u_2. \qquad (6\text{–}6.16)$$

Then by (6–2.13)

$$\nabla G \times \nabla u_2 = g \nabla u_1 \times \nabla u_2,$$

but since $g = \alpha$, it follows from (6–6.13) that

$$\mathbf{f} = \nabla G \times \nabla u_2. \quad \blacksquare$$

Notice that the function $\mathbf{F}$ that satisfies (b) is not unique, for if $\chi \in C^3(\Omega)$, then $\mathbf{F} + \nabla\chi$ is also a solution of the equation of (b) since $\nabla \times \nabla\chi = \mathbf{0}$.

Problems

1. Find a maximal open region in the space of the domain of the integrand functions on which the following line integrals are independent of their path. A set is *maximal* in a class of sets if it is not properly contained in another set of this same class. In case the line integral is not independent of its paths in any open region, the answer is the empty set.

(a) $\int_\Gamma y^3\, dx + 2xy^2\, dy,$

(b) $\int_\Gamma \frac{-y}{(x-y)^2}\, dx + \frac{x}{(x-y)^2}\, dy,$

(c) $\int_\Gamma (2xy + y + 1)\, dx + (x^2 + 1)\, dy,$

(d) $\int_\Gamma z\, dx + (z^2 + z)\, dy + (x + y)\, dz,$

(e) $\int_\Gamma (\log y)(1 - x^2)^{-1}\, dx + (y^{-1} \sin^{-1} x)\, dy,$

(f) $\int_\Gamma x(x^2 + y^2)^{-1}\, dx + y(x^2 + y^2)^{-1}\, dy.$

2. Find a maximal open region in which the following differential forms are exact. Use the technique of Example 6–6.2 to find a solution of the associated total differential equation.
   (a) $(y + 1)\,dx + (x + 2y)\,dy$,
   (b) $(3x^2 + y^2 + 2xz)\,dx + 2xy\,dy + x^2\,dz$,
   (c) $2xe^{-x^2}\log(y + 1)\,dx + e^{-x^2}(y + 1)^{-1}\,dy$,
   (d) $y(y + z)^{-1}\,dx + xz(y + z)^{-2}\,dy - xy(y + z)^{-2}\,dz$.

3. The gravitational force on a unit particle at $\mathbf{r}_0 = (x_0, y_0, z_0)$ due to a wire of linear density λ occupying the position of the curve $\Gamma = \{\mathbf{r}(s): a \le s \le b\}$ is

$$\mathbf{f}(\mathbf{r}_0) = \int_\Gamma \frac{\lambda(\mathbf{r})(\mathbf{r} - \mathbf{r}_0)}{\|\mathbf{r} - \mathbf{r}_0\|^3}\,ds.$$

   Find $\mathbf{f}(\mathbf{r}_0)$ for each of the following cases:
   (a) Let Γ be the line segment from $(1, 0, 0)$ to $(2, 1, 2)$, $\lambda = 1$ and $\mathbf{r}_0 = (0, 0, 0)$.
   (b) Let $\Gamma = \{(\cos t, \sin t): 0 \le t \le \pi/3\}$, $\lambda = 1$ and $\mathbf{r}_0 = (0, 0, 1)$.
   (c) Let $\Gamma = \{(t, t^2, t^3): 0 \le t \le 1\}$, $\lambda = x$ and $\mathbf{r}_0 = (0, 1, 0)$.

4. Define the gravitational force on a unit particle at $\mathbf{r}_0 = (x_0, y_0, z_0)$ due to a lamina of surface density σ occupying the position of a finite surface S. Use this formula to calculate $\mathbf{f}(\mathbf{r})$ for each of the following cases (cf. Problem 3).
   (a) Let $S = \{(x, y, z): x^2 + y^2 \le 1, z = 0\}$, $\sigma = 1$ and $\mathbf{r}_0 = (0, 0, z_0)$.
   (b) Let $S = \{(x, y, z): x = 1, 0 \le y \le 2, 0 \le z \le 1\}$, $\sigma = 1$ and $\mathbf{r}_0 = (0, 0, 0)$.

5. Define the gravitational force on a unit particle at $\mathbf{r}_0 = (x_0, y_0, z_0)$ due to a volume distribution of mass of volume density ρ occupying the position of the finite space region Ω.
   (a) Let $\Omega = \{(x, y, z): x^2 + y^2 + z^2 \le a^2\}$, $\lambda = 1$ and $\mathbf{r}_0 = (x_0, y_0, z_0)$ where $\|\mathbf{r}_0\| > a$.
   (b) Let $\Omega = \{(x, y, z): x^2 + y^2 \le z^2, 0 \le z \le 1\}$, $\lambda = 1$ and $\mathbf{r}_0 = (x_0, 0, 0)$ where $x_0 \ne 0$.

6. The Newtonian potential at $\mathbf{r}_0 = (x_0, y_0, z_0)$ due to a particle of mass m_i at $\mathbf{r}_i$ is

$$\phi(\mathbf{r}_0) = \frac{m_i}{\|\mathbf{r}_i - \mathbf{r}_0\|}.$$

   (a) What is the (Newtonian) potential at $\mathbf{r}_0$ due to n particles if the mass of ith particle is m_i and the ith particle is located at $\mathbf{r}_i$, $1 \le i \le n$?
   (b) Evaluate the following limit

$$\lim_{\lambda\to 0}\left(\frac{m}{\|\mathbf{r}_0 - \lambda\mathbf{u}\|} - \frac{m}{\|\mathbf{r}_0 + \lambda\mathbf{u}\|}\right)$$

where $m\lambda = M$, a positive constant. The result is called the potential at $\mathbf{r}_0$ due to a *doublet* (*dipole*) of moment $\mathbf{M} = M\mathbf{u}$ located at the origin.

7. The Newtonian potential at $\mathbf{r}_0$ due to a wire of linear density λ occupying the position of the curve $\Gamma = \{\mathbf{r}(s): a \leq s \leq b\}$ is

$$\phi(\mathbf{r}_0) = \int_\Gamma \frac{\lambda(\mathbf{r})}{\|\mathbf{r} - \mathbf{r}_0\|}\, ds.$$

For each of the circumstances of Problem 3, find $\phi(\mathbf{r}_0)$.

8. Define the Newtonian potential at $\mathbf{r}_0$ due to a surface distribution of mass as described in Problem 4. For each of the specific circumstances given in Problem 4, find $\phi(\mathbf{r}_0)$.

9. Define the Newtonian potential at $\mathbf{r}_0$ due to a volume distribution of mass of volume density ρ occupying the position of the finite space region Ω. For each of the specific circumstances of Problem 5, find $\phi(\mathbf{r}_0)$.

10. Find the potential at $\mathbf{r}_0 = (x_0, y_0, 0)$, $\mathbf{r}_0 \neq \mathbf{0}$ due to an infinite wire in the position of the z-axis with linear density $\lambda = 1$. This is called the *logarithmic potential* at (x_0, y_0) due to a *logarithmic particle* of mass 2λ at $(0, 0)$.

11. Test each of the following fields as to whether it is irrotational or solenoidal. Indicate the open region Ω upon which you are considering the field.

    (a) $(0, 0, -g)$, is a positive constant,

    (b) $\dfrac{\mathbf{r}}{\|\mathbf{r}\|^3}$, $\mathbf{r} = (x, y, z)$,

    (c) $\dfrac{\mathbf{r}}{\|\mathbf{r}\|^2}$, $\mathbf{r} = (x, y, z)$

    (d) $(2x + y, x, 3z^2)$,
    (e) $(2xy - 1, -y^2, 1 - 2x)$,
    (f) $(y \sin z, x \sin z, xy \cos z)$.

12. The *work* done by a force field $\mathbf{f}$ on a particle as it moves along Γ from $\boldsymbol{\alpha}$ to $\boldsymbol{\beta}$ is

$$W(\mathbf{f}; \Gamma) = \int_{\boldsymbol{\alpha}}^{\boldsymbol{\beta}} \mathbf{f} \cdot d\mathbf{r}.$$

    (a) Justify this definition assuming that work is force times distance.
    (b) A field $\mathbf{f}$ is *conservative* in an open region Ω if $W(\mathbf{f}; \Gamma) = 0$ for every closed path Γ in Ω. Of the fields in Problem 11, find those that are conservative on a nonempty region and indicate the region that is being considered.
    (c) Find $W(\mathbf{f}; \Gamma)$, if $\mathbf{f}$ is as in Problem 11(d) and Γ is the line joining $(1, 1, 2)$ to $(-3, 2, -1)$.

13. If the force field $\mathbf{f}$ is conservative in a simply connected domain in (x, y, z)-space, the *potential* at $\mathbf{r}_0 = (x_0, y_0, z_0)$ due to the field $\mathbf{f}$ is

$$\phi(\mathbf{r}_0) = -\int_{\Gamma\ \mathbf{r}_0}^{\mathbf{r}_1} \mathbf{f} \cdot \mathbf{dr}$$

where Γ is any path in Ω from $\mathbf{r}_0$ to $\mathbf{r}_1$ (cf. Theorem 6–6.2). Clearly, the potential is zero at $\mathbf{r}_1$. For a mass distribution that is bounded, $\mathbf{r}_1$ is "taken at ∞." Thus we take the limit of the above integral as $\|\mathbf{r}_1\| \to \infty$. Find the potential at $\mathbf{r}_0$ due to the field $\mathbf{f}(\mathbf{r}) = -\mathbf{r}/\|\mathbf{r}\|^3$ where $\mathbf{r} = (x, y, z)$ and $\Omega = \{\mathbf{r} : \mathbf{r} \neq \mathbf{0}\}$.

14. Let $v \in C^1(\Omega_1)$ be a velocity field (dynamic) where Ω_1 is an open region in (x, y, z)-space. Let Ω be a subregion of Ω_1 such that (6–4.1) holds.
    (a) Assuming that no mass is created or destroyed in Ω show that the net change of mass in Ω per unit of time is both

$$-\int_S (\rho\mathbf{v}) \cdot \mathbf{n}\, dA \quad \text{and} \quad \frac{\partial}{\partial t}\int_\Omega \rho\, dV$$

where $\rho \in C^1(\Omega \times I)$ is the scalar function that gives the pressure at $(x, y, z) \in \Omega$ and at time $t \in I$.
    (b) Use the result of (a) to establish that the *equation of continuity*,

$$\frac{\partial \rho}{\partial t} + \operatorname{div}(\rho\mathbf{v}) = 0,$$

holds in $\Omega \times I$.
    (c) The fluid flow is *incompressible* if $D\rho/Dt = 0$ in $\Omega \times I$. Use (6–1.11) and (6–2.15) to show that

$$D\rho/Dt = -\rho\nabla \cdot \mathbf{v}$$

and the fluid flow is incompressible if and only if $\operatorname{div} \mathbf{v} = 0$.

15. Prove the following corollaries to Proposition 6–6.1:
    (a) Let $\psi \in C^1(\Omega)$ where Ω is an open region in E^n. Show that if in Ω, $\nabla\psi = 0$, then in Ω, $\psi = c$, a constant.
    (b) Let ψ_1 and $\psi_2 \in C^1(\Omega)$ where Ω is an open region in E^n. If $\nabla\psi_1 = \nabla\psi_2$ in Ω, then $\psi_1 = \psi_2 + c$ where c is a constant.

16. Prove that if μ is an integrating factor of the total differential equation $\mathbf{f} \cdot \mathbf{dx} = 0$, then the composite function $\Phi\mu$ is also an integrating factor if Φ is any differentiable function of a single variable.

17. Proposition 6–6.1 was given for $n = 2$. Give the proof for this proposition for $n = 3$.

18. The proof of Proposition 6–6.2 is incomplete in that the details are not given to show that the line integral has the same value over two paths, Γ_1 and Γ_2 joining $\boldsymbol{\alpha}$ to $\boldsymbol{\beta}$, that intersect infinitely many times. Complete in the following manner the details of the proof of this proposition for the case $n = 2$.

(a) Show that for a given path Γ in an open region Ω and for a given positive number ε there exists a *step-polygon path* Π (a path made up of line segments such that each segment is parallel to a coordinate axis), that joins $\boldsymbol{\alpha}$ to $\boldsymbol{\beta}$ such that each point of Π is within δ of a point of Γ.

(b) Show that any two step-polygons Π_1 and Π_2 that join $\boldsymbol{\alpha}$ to $\boldsymbol{\beta}$ "enclose" a finite number of simply connected regions.

(c) Given $\varepsilon > 0$, show that there exists $\delta > 0$ such that if Π is within δ of Γ as in (a) then

$$\left| {}_{\Gamma}\int_{\boldsymbol{\alpha}}^{\boldsymbol{\beta}} \mathbf{f}\cdot d\mathbf{r} - {}_{\Pi}\int_{\boldsymbol{\alpha}}^{\boldsymbol{\beta}} \mathbf{f}\cdot d\mathbf{r} \right| < \varepsilon.$$

19. In the proof of Theorem 6–6.1 for the case of Ω being a region in (x, y)-space a plan for the proof was suggested, but the details were omitted. Supply the details for the proof of this portion of the theorem.

20. Establish Green's identities I and II for the exterior of a quasi-regular finite space region Ω in (x, y, z)-space (cf. Problem 10 of Section 6–4). Assume that the functions u and v are such that ru, rv, $r^2\|\nabla u\|$, and $r^2\|\nabla v\|$, $r = \|\mathbf{r}\|$, are bounded outside of a sphere of arbitrarily large radius.

Answers

1. (a) (x, y)-space, (b) $\{(x, y): y > x\}$ or $\{(x, y): y < x\}$,
(c) $\emptyset$, (d) (x, y, z)-space, (e) $\{(x, y): |x| < 1, y > 0\}$,
(f) $\{(x, y): (x, y) \neq (x, 0), x \leq 0\}$ is one of many possible answers.

2. (a) $xy + y^2 + x$, (x, y)-space,
(b) $x^3 + xy^2 + x^2z$, (x, y, z)-space,
(c) $e^{-x^2}\log(y + 1)$, $\{(x, y): y > -1\}$,
(d) $xy(y + z)^{-1}$, $\{(x, y, z): y + z > 0\}$ or $\{(x, y, z): y + z < 0\}$.

Bibliography

Apostol, Tom M. *Mathematical Analysis.* Addison-Wesley Publishing Company, Inc., Reading, Mass., 1957.

Buck, R. Creighton. *Advanced Calculus,* 2d ed. McGraw-Hill Book Company, Inc., New York, 1965.

Burington, R. S., and C. C. Torrance. *Higher Mathematics with Applications to Science and Engineering.* McGraw-Hill Book Company, Inc., New York, 1939.

Courant, R. *Differential and Integral Calculus,* Volumes I and II. Interscience Publishers, Inc., New York, 1947.

Duff, G. F. D. *Partial Differential Equations.* University of Toronto Press, Toronto, 1956.

Fulks, Watson. *Advanced Calculus.* John Wiley & Sons, Inc., New York, 1961.

Hohn, Franz E. *Elementary Matrix Algebra,* 2d ed. The Macmillan Company, New York, 1964.

James, Robert C. *Advanced Calculus.* Wadsworth Publishing Company, Inc., Belmont, Calif., 1966.

James, Robert C. *University Mathematics.* Wadsworth Publishing Company, Inc., Belmont, Calif., 1963.

Kaplan, Wilfred. *Advanced Calculus.* Addison-Wesley Publishing Company, Inc., Reading, Mass., 1952.

Kellogg, Oliver Dimon. *Foundations of Potential Theory.* Frederick Ungar Publishing Company, Inc., New York, 1929.

Lass, Harry. *Elements of Pure and Applied Mathematics.* McGraw-Hill Book Company, Inc., New York, 1957.

Leighton, Walter. *Ordinary Differential Equations,* 2d ed. Wadsworth Publishing Company, Inc., Belmont, Calif., 1966.

Milne, W. E. *Numerical Calculus.* Princeton University Press, Princeton, N.J., 1949.

Olmsted, John M. H. *Advanced Calculus.* Appleton-Century-Crofts, Inc., New York, 1961.

Phillips, H. B. *Vector Analysis.* John Wiley & Sons, Inc., New York, 1933.

Sneddon, Ian N. *Elements of Partial Differential Equations.* McGraw-Hill Book Company, Inc., New York, 1957.

Taylor, Angus E. *Advanced Calculus.* Ginn and Company, Boston, 1955.

Widder, David V. *Advanced Calculus,* 2d ed. Prentice-Hall, Inc., Englewood Cliffs, N.J., 1961.

Index

Symbols

† $A \cap B$, the intersection of the sets A and B, consists of those elements that are in A *and* in B.

† $A \times B$, the Cartesian product of the sets A and B, consists of those ordered pairs (a, b) such that $a \in A$ and $b \in B$.

‡ $A \cup B$, the union of the sets A and B, consists of those elements that are in A *or* in B.

Subjects

I

J

K

L

M

Q

R

S